Modern Radar Detection Theory

Modern Radar Detection Theory

Edited by

Antonio De Maio

Dipartimento di Ingegneria Elettrica e delle Tecnologie dell'Informazione, Università degli Studi di Napoli "Federico II"

Maria Sabrina Greco

Dipartimento di Ingegneria dell'Informazione, Università di Pisa

SciTech Publishing
an imprint of the IET

theiet.org

SciTech PUBLISHING
an imprint of the IET

Published by SciTech Publishing, an imprint of the IET
www.scitechpub.com
www.theiet.org

ISBN 978-1-61353-199-0 (hardback)
ISBN 978-1-61353-200-3 (PDF)

Typeset in India by MPS Limited
Printed in the UK by CPI Antony Rowe Ltd

To my wife Luisa and my children Claudia and Michele
Antonio De Maio

To my family
Maria Sabrina Greco

Contents

1 Introduction to Radar Detection 1

Antonio De Maio, Maria S. Greco, and Danilo Orlando

1.1 Historical Background and Terminology 1

1.2 Symbols 5

1.3 Detection Theory 6

1.3.1 Signal and Interference Models 7

1.3.2 Basic Concepts 9

1.3.3 Detector Design Criteria 11

1.3.4 CFAR Property and Invariance in Detection Theory 13

1.4 Organization, Use, and Outline of the Book 14

1.5 References 16

References 17

2 Radar Detection in White Gaussian Noise: A GLRT Framework 21

Ernesto Conte, Antonio De Maio, and Guolong Cui

2.1 Introduction 21

2.2 Problem Formulation 22

2.3 Reduction by Sufficiency 24

2.4 Optimum NP Detector and Existence of the UMP Test 26

2.4.1 Coherent Case 26

2.4.2 Non-coherent Case 27

2.5 GLRT Design 28

2.6 Performance Analysis 32

2.6.1 Coherent Case 32

2.6.2 Non-coherent Case 36

2.7 Conclusions and Further Reading 40

References 41

3 Subspace Detection for Adaptive Radar: Detectors and Performance Analysis **43**

Ram S. Raghavan, Shawn Kraut, and Christ D. Richmond

3.1 Introduction 43

3.2 Introduction to Signal Detection in Interference and Noise 45

3.2.1 Detecting a Known Signal in Colored Gaussian Noise 46

3.2.2 Detecting a Known Signal with Unknown Phase in Zero-Mean Colored Gaussian Noise 47

3.3 Subspace Signal Model and Invariant Hypothesis Tests 48

3.3.1 Subspace Signal Model 49

3.3.2 A Rationale for Subspace Signal Model 49

3.3.3 Hypothesis Test 51

3.3.4 Maximum Invariants for Subspace Signal Detection in Interference and Noise 53

3.4 Analytical Expressions for P_D and P_{FA} 54

3.4.1 P_D and P_{FA} for Subspace GLRT 54

3.4.2 P_D and P_{FA} for Subspace AMF Test 55

3.4.3 P_D and P_{FA} for Subspace ACE Test 56

3.5 Performance Results of Adaptive Subspace Detectors 57

3.6 Summary and Conclusions 70

Appendix 3.A 71

Appendix 3.B 74

Appendix 3.C 75

Appendix 3.D 79

References 80

4 Two-Stage Detectors for Point-Like Targets in Gaussian Interference with Unknown Spectral Properties **85**

Antonio De Maio, Chengpeng Hao, and Danilo Orlando

4.1 Introduction: Principles of Design 85

4.2 Two-Stage Architecture Description, Performance Analysis, and Comparisons 91

4.2.1 The Adaptive Sidelobe Blanker 93

4.2.2 Modifications of the ASB towards Robustness: The Subspace-Based ASB 97

4.2.3 Modifications of the ASB towards Selectivity 104

4.2.4 Modifications of the ASB towards both Selectivity and Robustness 117

4.2.5 Selective Two-Stage Detectors 125

4.3 Conclusions 128

References 129

5 Bayesian Radar Detection in Interference 133

Pu Wang, Hongbin Li, and Braham Himed

5.1 Introduction 133

5.2 General STAP Signal Model 134

5.3 KA-STAP Models 136

5.3.1 Knowledge-Aided Homogeneous Model 136

5.3.2 Bayesian GLRT (B-GLRT) and Bayesian AMF (B-AMF) 138

5.3.3 Selection of Hyperparameter 140

5.3.4 Extensions to Partially Homogeneous and Compound-Gaussian Models 144

5.4 Knowledge-Aided Two-Layered STAP Model 147

5.5 Knowledge-Aided Parametric STAP Model 151

5.6 Summary 159

Appendix 5.A 159

Appendix 5.B 160

References 162

6 Adaptive Radar Detection for Sample-Starved Gaussian Training Conditions 165

Yuri I. Abramovich and Ben A. Johnson

6.1 Introduction 165

6.2 Improving Adaptive Detection Using EL-Selected Loading 168

6.2.1 Single Adaptive Filter Formed with Secondary Data, Followed by Adaptive Thresholding Using Primary Data 168

6.2.2 Different Adaptive Process per Test Cell with Combined Adaptive Filtering and Detection Using Secondary Data 170

6.2.3 Observations 209

6.3 Improving Adaptive Detection Using Covariance Matrix Structure 212

6.3.1 Background: TVAR(m) Approximation of a Hermitian Covariance Matrix, ML Model Identification and Order Estimation 215

6.3.2 Performance Analysis of TVAR(m)-Based Adaptive Filters and Adaptive Detectors for TVAR(m) or AR(m) Interferences 218

6.3.3 Simulation Results of TVAR(m)-Based Adaptive Detectors for TVAR(m) or AR(m) Interferences 223

6.3.4 Observations 237

6.4 Improving Adaptive Detection Using Data Partitioning 239

6.4.1 Analysis Performance of "One-Stage" Adaptive CFAR Detectors versus "Two-Stage" Adaptive Processing 242

6.4.2 Comparative Detection Performance Analysis 247

6.4.3 Observations 255

References 257

7 Compound-Gaussian Models and Target Detection: A Unified View 263

K. James Sangston, Maria S. Greco, and Fulvio Gini

7.1 Introduction 263

7.2 Compound-Exponential Model for Univariate Intensity 264

7.2.1 Intensity Tail Distribution and Completely Monotonic Functions 264

7.2.2 Examples 265

7.3 Role of Number Fluctuations 266

7.3.1 Transfer Theorem and the CLT 267

7.3.2 Models for Number Fluctuations 269

7.4 Complex Compound-Gaussian Random Vector 270

7.5 Optimum Detection of a Signal in Complex Compound-Gaussian Clutter 272

7.5.1 Likelihood Ratio and Data-Dependent Threshold Interpretation 273

7.5.2 Likelihood Ratio and the Estimator-Correlator Interpretation 275

7.6 Suboptimum Detectors in Complex Compound-Gaussian Clutter 275

7.6.1 Suboptimum Approximations to Likelihood Ratio 276

7.6.2 Suboptimum Approximations to the Data-Dependent Threshold 277

7.6.3 Suboptimum Approximations to Estimator-Correlator 278

7.6.4 Performance Evaluation of Optimum and Suboptimum Detectors 280

7.7 New Interpretation of the Optimum Detector 281

7.7.1 Product of Estimators Formulation 281

7.7.2 General Properties of Product of Estimators 283

Appendix 7.A 290

References 292

8 Covariance Matrix Estimation in SIRV and Elliptical Processes and Their Applications in Radar Detection 295

Jean-Philippe Ovarlez, Frédéric Pascal, and Philippe Forster

8.1 Background and Problem Statement 295

8.1.1 Background Parameter Estimation in Gaussian Case 296

8.1.2 Optimal Detection in Gaussian Case 297

8.2 Non-Gaussian Environment Modeling 300
8.2.1 CES Distribution 300
8.2.2 The Subclass of SIRV 301
8.3 Covariance Matrix Estimation in CES Noise 302
8.3.1 M-Estimators 303
8.3.2 Properties of the M-Estimators 304
8.3.3 Asymptotic Distributions of the M-Estimators 305
8.3.4 Link to M-Estimators in the SIRV Framework 307
8.4 Optimal Detection in CES Noise 312
8.5 Persymmetric Structured Covariance Matrix Estimation 313
8.5.1 Detection in Circular Gaussian Noise 315
8.5.2 Detection in Non-Gaussian Noise 315
8.6 Radar Applications 317
8.6.1 Ground-Based Radar Detection 317
8.6.2 Nostradamus Radar Detection 319
8.6.3 STAP Detection 320
8.6.4 Robustness of the FPE 323
8.7 Conclusion 327
References 327

9 Detection of Extended Target in Compound-Gaussian Clutter 333
Augusto Aubry, Javier Carretero-Moya, Antonio De Maio, Antonio Pauciullo, Javier Gismero-Menoyo, and Alberto Asensio-Lopez

9.1 Introduction 333
9.2 Distributed Target Coherent Detection 334
9.2.1 Overview 334
9.2.2 Rank-One Steering 336
9.2.3 Subspace Steering 338
9.2.4 Covariance Estimation 340
9.3 High-Resolution Experimental Data 342
9.3.1 Sea-Clutter Data 343
9.3.2 Maritime Target Data 345
9.4 Experimental CFAR Behavior 350
9.5 Detection Performance 355
9.5.1 Detection Probability: Simulated Target and Real Clutter 355
9.5.2 Detection Maps: Real Target and Clutter Data 358

9.6 Conclusions 359
Appendix 9.A 360
References 367

Index **375**

CHAPTER 1

Introduction to Radar Detection

Antonio De Maio[1], *Maria S. Greco*[2], *and Danilo Orlando*[3]

1.1 Historical Background and Terminology

Radar is the acronym of the words RAdio Detection And Ranging. As the term suggests, the device has been conceived to detect the presence of a target and to measure its range [1]. The basic operating principle of a radar system was largely known since the days of physicists James Clerk Maxwell and Heinrich Hertz. It relies on the fact that electromagnetic waves interact with matter and a portion of the transmitted energy is intercepted by a prospective object and re-radiated in many directions. As a consequence, the energy backscattered towards the irradiating source can be properly used to infer the presence and the distance of the object [2, and references therein].

In his seminal work on electromagnetism, Maxwell glimpsed the above mechanism, and at the end of the nineteenth century, Hertz conducted a set of experiments showing that radio waves were completely reflected by metallic objects. During 1920s, Guglielmo Marconi, the well-known pioneer of wireless radio, clearly understood the possibility to reveal the presence of far objects by means of radio waves; this appeared clear from his papers, where he described the system used in his experiments along with the experimental results. In addition, he pointed out the utility of this system in the context of navigation safety in adverse weather conditions (e.g., foggy weather), and in 1922, he strongly urged its use in a speech delivered at the Institute of Radio Engineers (now the Institute of Electrical and Electronic Engineers). Another important pioneer of radar systems is the German inventor Christian Hülsmeyer. In 1903, he built up the first ship collision avoidance system exploiting radio echoes, which was formed by a transmitting apparatus and a separated coherent receiver. Each of them was equipped with the respective dipole antenna. The system was capable of detecting ships up to 3 km but it did not provide any range information.

Radar became an operational reality in the late 1920s and early 1930s especially due to military pressures, since the wind of a Second World War (WW2) was in the air. In this period,

[1]Dipartimento di Ingegneria Elettrica e delle Tecnologie dell'Informazione, Università degli Studi di Napoli "Federico II", Napoli, Italy
[2]Dipartimento di Ingegneria dell'Informazione, Università di Pisa, Pisa, Italy
[3]Facoltà di Ingegneria, Università degli Studi "Niccolò Cusano", Roma, Italy

the nations, which played an important role in the course of WW2, independently and in great secrecy, developed systems of this type.

In the United Kingdom, the first studies on the use of radio signals were carried out by Sir Robert Alexander Watson-Watt, engaged by the Meteorological Office, where his relevant activity concerned the exploitation of radio signals emitted during a thunderstorm to locate it. Afterwards, he became a superintendent of the Radio Research Station and was involved in scientific research of military interest. As war clouds gathered over Great Britain, the British scientific community focused its efforts on the detection problem of radio signals reflected by possible threatening objects as aircrafts. To this end, the lightning detection concept of Watson-Watt was exploited to conceive pulsed-based detection systems, which were capable of providing distance measurements. Further developments took place as the WW2 was approaching.

In the United States, the development of radar systems had separately involved both the Navy and the Army with a little coordination of their efforts. The first systems aimed at detecting the presence of a far object were based on the interferences observed on the transmitted signals. The Naval Research Laboratory (NRL) devised a simple wave-interference apparatus for detecting ships and aircrafts, but it cannot determine the respective location or velocity. As it happened in the United Kingdom, this information was available with the invention of pulsed radars. Another important technological advance introduced by NRL was the *duplexer* which allowed the transmitter and the receiver to use the same antenna. As to the Army, the Signal Corps Laboratories significantly contributed to the design of radar systems also exploiting the results obtained by NRL.

As already stated at the beginning of this section, in Germany, Hülsmeyer was the inventor of the first radio-based device for remotely indicating the presence of ships (1903). However, this apparatus cannot be defined a radar, since it does not provide any range information. Over the following three decades, the radio-based detection systems proliferated in Germany, but none of them were true radars. Again, with the approaching of WW2, the scientific efforts on the design of radio-based detection systems became more and more intensive. As a consequence, German industries such as GEMA, Telefunken, and Lorenz started a collaboration with the German military aimed at the realization of true radar systems with excellent performances in terms of detection and localization of aircrafts and ships.

The first radio-based detection system developed in Italy is the so-called *radioecometro* due to the work of Marconi around 1933. In May 1935, Marconi demonstrated his system to the Fascist dictator Benito Mussolini and members of the military General Staff; however, the output power was insufficient for military use. While Marconi's demonstration raised considerable interest, little more was done with his apparatus. Subsequently, the radio-based technology developments were assigned by Mussolini to the *Regio Instituto Electrotecnico e delle Comunicazioni* (RIEC) which was established in 1916 on the campus of the Italian Naval Academy in Livorno. Lieutenant Ugo Tiberio, a physics and radio-technology instructor at the Academy, was assigned to head the project on a part-time basis. Tiberio prepared a report on developing an experimental apparatus that he called *telemetro radiofonico del rivelatore* (RDT). Before the end of 1936, Tiberio and Nello Carrara, a civilian physics instructor at RIEC who was added to be responsible for developing the RDT transmitter, had demonstrated the EC-1, the first Italian RDT system. The EC-1 did not provide a range measurement; as previously seen in other countries, to add this capability, the development of a pulsed system was initiated in 1937. Finally, the evolution EC-3/ter (GUFO) released in 1941 can be considered as the first operational Italian Radar System.

On the German side, FREYA radars produced by GEMA were used to form German air defense early warning radar network [3, and references therein]. These radars used antenna arrays operating at a frequency around 120–130 MHz and were in service from the beginning of the WW2. A number of different types of this radar were developed over the course of the war. German ground-based fighter control used radars called WÜRZBURG and later WÜRZBURG-RIESE, operating in the frequency band around 560 MHz. The overall German air defense system was a network of radar stations, each known as a *Stellung* (site) and denoted by codenames. There were three orders of sites. Specifically, the third-order sites reported information using coded radio transmissions to the first-order sites; the second-order sites formed an air picture and transmitted that to the first-order sites. Finally, the first-order sites fused information received from the other sites (including their own information) and transmitted the combined information to a *Himmelbett* Operations Room, where the overall picture was assembled and passed back to the sites.

Espionage information on German air defense system pushed Allied scientists to devise jamming and deception techniques. One of the first jammers, a noise barrage jammer (MANDREL), was developed by British scientists and employed against FREYA and its derivatives. In response, Germans widened the band over which FREYA could operate and, hence, the frequency coverage of MANDREL had to be increased accordingly. Subsequently, German scientists learned about British air defense whose codename was Chain Home. This system was the ring of coastal early warning radar stations built by the British before and during the WW2. The Chain Home stations were designed to operate at 20–50 MHz (although typical operations were at 20–30 MHz). The availability of multiple operating frequencies gave some protection from jamming. The detection range was typically 190 km. The observations of the Chain Home coupled with the realization of the need to devise early warning radar that did not suffer from the various Allied countermeasures led to the concept of completely passive radar receiver. To this end, Telefunken developed the first bistatic radar of the world, whose name was Klein Heidelberg. The advantages of bistatic architectures were manifold. More precisely, it allowed for covert operations against emitter locators, jammers, and antiradiation missiles exploiting ambient transmissions; very long detection ranges; lower cost, size, weight, and power than an equivalent monostatic radar. In the next years, the idea behind the bistatic architecture was further investigated [4] giving rise to the modern multistatic architectures, which involve multiple entities transmitting signals and receiving echoes. Again, receivers can be kept covert if they are spatially separated from the transmitters.

From the inspection of the radar system evolution over the years, it is apparent that, after the WW2, radar capabilities and applications rapidly continued to advance up to nowadays and are expected to continue. The major areas of radar application refer to

- Military field (air defense systems, offensive missiles, and other weapons);
- Environmental remote sensing (weather observation, planetary observation, and mapping of sea ice to route shipping in an efficient manner);
- Air traffic control;
- Highway safety (radar speed meter and control systems to avoid vehicle collisions);
- Ship safety and other.

Despite these numerous and different application areas, the radars perform three general functions, which are *search, track,* and *image* [5]. The most important operation of a search radar is the processing aimed at target detection, while the primary purpose of a tracking radar is to make measurements of the target state in range, azimuth angle, elevation angle, and Doppler frequency offset within preassigned precision and accuracy. However, the above statements do not preclude that a search radar will provide target measurements, which can be used by other sensors, and that a tracking radar will perform the detection process. In many existing tracking systems, the tracking function is performed by a set of analog circuits which control the antenna and the range servo in order to continually point the antenna beam at the target position verifying that the range measurement and the direction of arrival coincide with the target position. However, in modern systems, the tracking function is accomplished by processing a sequence of target measurements provided by the sensor. Tracking algorithms are usually implemented in software and develop an accurate state vector (position, velocity, and acceleration) for the target. This state estimate becomes an integral part of a fire control system, directing a weapon or cueing another sensor to the target state. Once a target is detected in track, the imaging function comes in handy for target classification, discrimination, and/or identification purposes. As a matter of fact, it provides high-resolution data in range, azimuth, elevation, and sometimes Doppler. Generally speaking, imaging radar attempts to form an image of the object by mapping the electromagnetic scattering coefficients onto a two-dimensional plane. Objects with a higher coefficient are assigned a higher optical reflective index creating an optical image. An example of imaging radar is represented by the *Synthetic Aperture Radar* (SAR) [6]. SAR systems combine radar hardware, waveforms, signal processing, and relative motion to create photograph-like renderings of stationary targets and scenes of interest. The principal product of any basic SAR implementation is a fine-resolution two-dimensional intensity image of the illuminated scene. SAR is widely employed by the remote sensing community for mapping and land-use surveying and by the military for detection, location, identification, and assessment of fixed targets.

The focus of this book is on modern processing techniques for radar detection. In this context, the effectiveness of a radar system is measured by its ability to distinguish the target from the *clutter* and/or other kind of interference. In order to improve the detection capabilities in practical scenarios, radar systems often process several pulses while the antenna beam is pointed at a target. In this case, the time taken to transmit (and receive) the N pulses used for detection (see also Chapter 2) is called *dwell time*. If these pulses are coherently integrated (i.e., both the amplitude and the phase of the received signals are used in the processing so that the signal contributions can be made to add in phase with one another), the time duration of the processing is often called *coherent processing interval*.

On the other hand, the term clutter denotes unwanted echoes (due to the environment) that interfere or obscure the target. There exist several types of clutter which differ in their spectral properties. For instance, clutter can be caused by the earth (land and/or sea), other targets that are not of strategical interest, chaff, rain, clouds, birds, insects, and more. An important kind of clutter is the *ground clutter* [7], which is caused by echoes backscattered from the ground. It can be quite detrimental in littoral environment [8]. A non-clutter source of interference to all radars is intentional (jamming) or unintentional interferers by a source of electromagnetic radiation located at some distance from the radar but transmitting signals in the radar's transmit/receive frequency band. This deleterious energy could be produced by hostile sources (as seen in WW2) or from radiating equipments made for other purposes (for instance, telecommunication systems) that happen to have a similar interference effect on the radar. The problems of clutter and jamming (or other source of interference) suppression have dominated the minds of radar engineers since the beginnings of radar.

Space–Time Adaptive Processing (STAP) and related techniques play a fundamental role for clutter and jamming suppression in radar. The first space-time processor, proposed by Reed, Mallett, and Brennan in References 9 and 10 (RMB test), is a matched filter followed by an envelope detector. It requires knowledge of the covariance matrix of the disturbance; in practice, the covariance matrix is unknown and must be estimated from the observables. It is customary to assume that a set of secondary data, representative of the noise in the cell under test, is available; this scenario will be referred to in the following as homogeneous environment. The RMB test can be properly normalized by the *output power* in order to guarantee the Constant False Alarm Rate (CFAR) property. The corresponding detector is called Adaptive Matched Filter [11]. In Reference 12, the Generalized Likelihood Ratio Test (GLRT) is used to conceive an adaptive decision scheme capable of detecting coherent pulse trains in the presence of Gaussian disturbance with unknown spectral properties. The above work by Kelly represents a milestone in radar adaptive detection, since most of the more recent papers follow its lead. As a matter of fact, problems encountered in practice gave rise to manifold derivatives of the model proposed in Reference 12. For instance, the uncertainties on the actual direction of arrival of target returns led to the application of the subspace idea to the signal model (see Chapters 3 and 4) and, subsequently, to the design of the so-called *tunable receivers* (see Chapter 4). Other recent approaches make use of a priori knowledge to properly select data for estimation purposes and, consequently, to improve detection performances of STAP systems (see Chapter 5). Further studies facing with the lack of secondary data can be found in open literature (see Chapter 6). The model of the disturbance occurring in radar applications is another issue addressed by radar community. Indeed, due to the increased resolution capabilities of modern radars, the Gaussian model becomes inadequate. To cope with this drawback, the compound-Gaussian model and, more generally, the Complex Elliptical Symmetrical (CES) distributions are investigated (see Chapters 7 and 8). Finally, it is important to observe that an increased radar resolution yields to the so-called range-spread targets, which require additional assumptions at the design stage of a radar detector (see Chapter 9).

In the sequel, a brief review of detection theory and statistical tools used to design decision architectures is given.

1.2 Symbols

The following symbols are used throughout the book.

$\mathcal{R}$	Set of real numbers
$\mathcal{C}$	Set of complex numbers
$\mathcal{N}$	Set of Natural numbers
$\mathcal{R}^{m\times n}$	Set of real matrices (vectors) of dimension $m \times n$ ($m \times 1$)
$\mathcal{H}$	Denotes a Hilbert space
j	the imaginary unit
$(\cdot)^T$	Matrix transpose
$(\cdot)^\dagger$	Hermitian operation for a matrix
$(\cdot)^*$	Complex conjugation
$\boldsymbol{x}$	Boldface and lower case letter denotes a vector
$\boldsymbol{X}$	Boldface and upper case letter denotes a matrix
x	Letter in normal font denotes a scalar
x_i or $x(i)$	The ith component of vector $\boldsymbol{x}$
$\boldsymbol{X}(n)$ or $\boldsymbol{x}(n)$ or $x_i(n)$ or $x(n)$	Time dependence on the time index n, depending on if it is a matrix, vector, vector component, or scalar
$\nabla f(\boldsymbol{x})$ or $\nabla_{\boldsymbol{x}} f(\boldsymbol{x})$ or $\frac{\partial f(\boldsymbol{x})}{\partial \boldsymbol{x}}$	The gradient of a function
$X(i,j)$ or X_{ij}	The (i,j)th element of matrix $\boldsymbol{X}$

$X_{ij}(n)$	Time dependence on the time index n of the (i, j)th element of a matrix $\boldsymbol{X}$
$\mathbb{E}[\cdot]$	Denotes statistical expectation
$\hat{x}$	Denotes the estimate of a variable x
$\boldsymbol{x} \perp \boldsymbol{y}$	Denotes orthogonal vectors
$\lvert x\rvert$ and $\lvert z\rvert$	Absolute values of a real number x and modulus of a complex number z
$\lVert\boldsymbol{x}\rVert$ or $\lVert\boldsymbol{x}\rVert_2$	Denotes Euclidean norm
$\lVert\boldsymbol{A}\rVert_F$	Denotes Frobenius norm
$\lVert\boldsymbol{x}\rVert_p$	Denotes l_p norm
$\mathrm{Re}(z)$	Denotes the real part of a complex number
$\mathrm{Im}(z)$	Denotes the imaginary part of a complex number
$\boldsymbol{I}_n$ or $\boldsymbol{I}$	Denotes the identity matrix of dimension $n \times n$ or of proper dimensions
$\boldsymbol{0}_{n,m}$ or $\boldsymbol{0}$	Denotes the null matrix of dimension $n \times m$ or of proper dimensions
$\mathrm{vec}(\boldsymbol{A})$	Column vector formed by stacking the columns of $\boldsymbol{A}$
$\mathrm{diag}(\boldsymbol{A})$	Column vector with the diagonal entries of $\boldsymbol{A}$
$\mathrm{rank}(\boldsymbol{A})$	Rank of matrix $\boldsymbol{A}$
$\boldsymbol{A} \otimes \boldsymbol{B}$	The Kronecker product of two matrices
$\det(\cdot)$	The determinant of a matrix
$\mathrm{Tr}(\cdot)$	The trace of a matrix
$\lambda_i, i = 1, \ldots, m$	The eigenvalues of an $m \times m$ matrix
$R(\boldsymbol{A})$	Range space of $\boldsymbol{A}$
$N(\boldsymbol{A})$	Null space of $\boldsymbol{A}$
$P(\cdot)$	Probability of a discrete event
$p(\cdot)$ or $f(\cdot)$	Probability density function of a random variable
σ^2 or σ_x^2	The variance of the random variable x
$\log(a)$	The logarithm of a relative to base 10
$\ln(a)$	The natural logarithm of a
$\exp(a)$	The exponential function
$\boldsymbol{A} \succ \boldsymbol{B}$	Means that $\boldsymbol{A} - \boldsymbol{B}$ is positive definite
$\boldsymbol{A} \succeq \boldsymbol{B}$	Means that $\boldsymbol{A} - \boldsymbol{B}$ is positive semidefinite

Notation of a matrix $\boldsymbol{A}$

$$\boldsymbol{A} = \begin{bmatrix} A(1,1) & A(1,2) & \cdots & A(1,m) \\ A(2,1) & A(2,2) & \cdots & A(2,m) \\ \vdots & \vdots & \vdots & \vdots \\ A(m,1) & A(m,2) & \cdots & A(m,m) \end{bmatrix}$$

Notation of a vector $\boldsymbol{x}$

$$\boldsymbol{x} = [x_1, x_2, \ldots, x_m]^T = \begin{bmatrix} x_1 \\ x_2 \\ \vdots \\ x_m \end{bmatrix}$$

1.3 Detection Theory

Radar systems are faced with the problem of discriminating the useful target echoes from the interference background, which hides the target making its detection difficult. Thus, the task of a radar processor is to decide whether or not the collected returns contain useful signal components. This problem can be formulated in terms of a *binary hypothesis test* [13]

$$\begin{cases} H_0 : \text{data contain interference only,} \\ H_1 : \text{data contain useful target and interference,} \end{cases} \tag{1.1}$$

where[1] H_0 is referred to as the *null hypothesis* and H_1 as the *alternative hypothesis*.

1.3.1 Signal and Interference Models

Let us focus on the target response and notice that several models are possible according to the operating scenarios, the specific application, the range resolution, the transmitted waveforms, and, generally speaking, uncertainty factors. The ratio between the physical size of a target and the range resolution specifies the number of range bins wherein the target is present. More precisely, when the above ratio is less than one, the target response is produced by a *point-like* scatterer belonging to a given range bin; on the other hand, when the ratio is greater than one, a suitable target model can be developed observing that useful returns can originate from several adjacent range bins, namely the target occupies more than one range cell (*range-spread targets*) [14]. From a theoretical point of view, if $\boldsymbol{Z}$ denotes the matrix of the collected returns, the following two cases can be considered

$$\begin{cases} \boldsymbol{Z} \in \mathcal{C}^{N\times 1} & \text{for a point-like target,} \\ \boldsymbol{Z} \in \mathcal{C}^{N\times K_p}, \quad K_p \in \mathcal{N} \setminus \{1\}, & \text{for a range-spread target,} \end{cases} \tag{1.2}$$

where $\setminus$ denotes the set difference, K_p is the number of range cells containing target returns, and N is the dimension of the data vectors. In addition, it is important to highlight that the latter parameter depends on the number of sensors composing the system, N_a say, and the number of transmitted pulses, N_p say, to sense the surveillance area. The adjustment of these two quantities entails the form of the *target steering/direction vector*, $\boldsymbol{p}$ say, whose size is $N = N_a \times N_p$ [7, 8]. According to the values of (N_a, N_p), the steering vector is said *space–time* if $N_a > 1$ and $N_p > 1$, *temporal* if $N_a = 1$ and $N_p > 1$, and *spatial* if $N_a > 1$ and $N_p = 1$.

In realistic scenarios, the steering of the signal backscattered from a target illuminated by the system may be different from the nominal steering vector due to several environmental and/or instrumental factors. As a consequence, receivers designed under the assumption that the steering vector is perfectly known may exhibit significant performance degradation. An approach to overcome this drawback is the *Subspace Detection Paradigm* [13, 15], which consists in modeling the target as a linear combination of known basis vectors. In other words, it is possible to capture the target energy in the presence of steering mismatches assuming that the latter belongs to a known observation subspace. It follows that at the design stage, the target direction vector can be modeled as

$$\boldsymbol{p} = \boldsymbol{H}\boldsymbol{\theta}, \tag{1.3}$$

[1] In radar terminology, H_0 and H_1 are usually called *noise-only hypothesis* and *signal-plus-noise hypothesis*, respectively.

where $\boldsymbol{H} \in \mathcal{C}^{N\times r}$ is a known full-column-rank matrix representing the subspace of the target, and $\boldsymbol{\theta} \in \mathcal{C}^{r\times 1}$ is an unknown vector containing the coefficients of the linear combination which yields the steering vector. Other models are also possible to take into account steering mismatches [16].

Target reflectivity is another important aspect that should be considered at the design stage. To this end, a factor $\alpha \in \mathcal{C}$ is introduced to model the signal returns at a specific range, and it can be analytically described in several ways. More specifically, denoting the amplitude and the phase of α by A and φ, respectively, the following instances could be envisaged

(1) A and φ are known parameters;

(2) A and φ are unknown deterministic parameters;

(3) A is an unknown deterministic parameter, and φ is a random variable uniformly distributed in $(0, 2\pi)$;

(4) A is a random variable ruled by a known probability density function (pdf) (or a pdf with unknown parameters belonging to a known family of pdfs), and φ is a random variable uniformly distributed in $(0, 2\pi)$ and independent[2] of A.

Modeling A and φ as known parameters is not reasonable in practice, but it is of interest from a theoretical point of view in order to [13, 14, 19]

- determine the Most Powerful (MP) test, which represents an upper bound for the achievable performance of any other test.
- ascertain whether or not the Uniformly Most Powerful (UMP) test with respect to A and/or φ exists;

On the other hand, the remaining models are viable means to address several realistic scenarios. However, it is important to highlight that modeling φ as a random variable can make the evaluation of the likelihood function under H_1 an intractable problem.

Before concluding this section, recall that the signals collected by the radar antennas in general contain interference components, which are due to the receiver electronics (thermal noise), ambient reverberation (clutter), and masking and/or deception jamming signals. Observe that deception jamming signals are assumed to be present only in one or a few range cells of the radar while masking barrage jamming signals are common to all of the radar range cells. Another important remark is that the extent in range of clutter is a function of the grazing angle and radar Pulse Repetition Frequency (PRF). For instance, in the case of low-PRF systems, the radar unambiguous range is greater than the radar horizon and, as a consequence, beyond the horizon, the range gates are corrupted only by jamming signals and noise.

As to the clutter response, the amplitude of the complex envelope is commonly assumed to be Rayleigh distributed, whereas the corresponding quadrature components are jointly Gaussian. The above assumption is usually justified when clutter arises from the superposition of returns from a large number of equivalent elementary scatterers, independent of one another, whence the resulting process is Gaussian as a consequence of the central limit theorem. However, experimental data indicate that large deviations from the Rayleigh statistics are

[2]When the radar cross section of the target obeys the (real) central chi-square distribution, several combinations (Swerling models) are possible according to the degrees of freedom of the pdf and the correlation of target returns (pulse-to-pulse or scan-to-scan fluctuations) [17, 18].

possible; for instance, deviations are observed for low grazing angles and/or high-resolution radars [20]. It also turns out that conventional detectors, namely the ones designed under the assumption of Gaussian disturbance, suffer false alarms inflation and/or marked detection performance degradation as the actual amplitude probability density function (apdf) of the clutter significantly deviates from the Rayleigh law. In such cases, the actual apdf of the clutter is better described in terms of families of apdfs ruled by shape parameters in addition to the scale one as, for instance, the Weibull and the K biparametric families. Moreover, the baseband equivalent of the disturbance can be generally described as a compound-Gaussian process, namely the product of a rapidly varying (complex) Gaussian process times an independent, slowly varying, non-negative modulating process (accounting for the reflectivity of the illuminated patch) [21, 22, and references therein]. On a properly short observation time, such a process degenerates into a spherically invariant random process [20].

Finally, as to the coherent interferer, it can be modeled as a target by means of a steering vector and a complex amplitude, which is reasonably unknown at the receiver and, hence, must be estimated from data (for instance, the subspace idea can be applied to find the jammer direction of arrival).

1.3.2 Basic Concepts

In the sequel, a brief review of the basic concepts concerning decision theory is provided. For the sake of simplicity, let $\boldsymbol{\theta} \in \mathcal{C}^{N\times 1}$ the parameter vector for the observed data distribution and consider the following problem

$$\begin{cases} H_0 : \boldsymbol{\theta} = \boldsymbol{\theta}_0, \\ H_1 : \boldsymbol{\theta} = \boldsymbol{\theta}_1. \end{cases} \tag{1.4}$$

It is clear that the above hypotheses completely specify the distribution of data, and for this reason, they are said to be *simple*. Generally speaking, each simple hypothesis maps into a point of the parameter space, that is given by

$$\Theta = \{\Theta_0, \Theta_1\}, \tag{1.5}$$

where $\Theta_i = \{\boldsymbol{\theta}_i\}$, $i = 0, 1$. On the other hand, if Θ_i contains more than one point of the parameter space, H_i is called *composite* and maps into a subset of the parameter space. A typical example is

$$\begin{cases} H_0 : \boldsymbol{\theta} \leq \boldsymbol{\theta}_r, \\ H_1 : \boldsymbol{\theta} > \boldsymbol{\theta}_r, \end{cases} \tag{1.6}$$

where $\boldsymbol{\theta}_r$ is a specific value of the parameter vector. The case of composite hypotheses is useful to face with scenarios of practical interest, where the unknown relevant parameters have to be properly estimated.

In radar detection, due to the presence of the so-called nuisance parameters that are not relevant to the decision problem, both the null and the alternative hypotheses are very often composite. Moreover, the goal of a radar system is to use the received data as efficiently as possible in deciding which hypothesis is in force, according to an optimality criterion.

To this end, it is worth distinguishing two types of errors that can occur in the decision process, i.e.,

(1) *Type I error*. It consists in mistaking the interference for signal when only the interference is present, i.e., when we decide H_1 but H_0 is true. In radar, this type of error is named *false alarm*.

(2) *Type II error*. It occurs when a signal is present, but it is erroneously considered interference, i.e., when we decide H_0 but H_1 is true. It is also called *missed detection*.

A guideline to the design of decision architectures could be to simultaneously reduce both error probabilities, but, in many physical situations, these objectives are in conflict [14]. Thus, a reasonable criterion consists in keeping one error probability constant while minimizing the other one (in the sequel, a formal definition of this criterion is given).

Before proceeding, let us introduce the performance parameters that come in handy for the design of the best test (if any) or at least optimized detection strategies. More precisely, a decision rule $\Phi(\boldsymbol{r})$ is a function of the observed data, $\boldsymbol{r}$ say, such that

$$\Phi(\boldsymbol{r}) = \begin{cases} 0 \equiv H_0, & \text{if } \boldsymbol{r} \in \mathbb{R}_0, \\ 1 \equiv H_1, & \text{if } \boldsymbol{r} \in \mathbb{R}_1, \end{cases} \tag{1.7}$$

where $\{\mathbb{R}_0, \mathbb{R}_1\}$ is a partition of the space of the observables.

When H_0 is simple, namely $\Theta_0 = \{\boldsymbol{\theta}_0\}$, then the *probability of false alarm* (P_{fa}) or size, ϱ say, is defined as

$$\varrho \triangleq P_{\theta_0}[\Phi(\boldsymbol{r}) = 1] = E_{\theta_0}[\Phi(\boldsymbol{r})] = P_{fa}, \tag{1.8}$$

where $P_\theta[\cdot]$ and $E_\theta[\cdot]$ denote the probability measure and the statistical expectation, respectively, when the parameter is $\boldsymbol{\theta}$. On the other hand, if H_0 is composite, then the size of $\Phi(\boldsymbol{r})$ is defined using the supremum, i.e.,

$$\varrho \triangleq \sup_{\boldsymbol{\theta} \in \Theta_0} P_\theta[\Phi(\boldsymbol{r}) = 1] = \sup_{\boldsymbol{\theta} \in \Theta_0} E_\theta[\Phi(\boldsymbol{r})]. \tag{1.9}$$

It is important to observe that in radar, the decision rules compare a statistic (function of data) with a threshold set to ensure a given level of P_{fa}. To this end, the threshold is usually tuned assuming that H_0 is simple, even though it actually is not. As a consequence, the threshold depends on the value of the parameter $\boldsymbol{\theta}_0$ used to evaluate it. The above dependence could be removed exploiting (1.9), which yields a detection threshold (if any) allowing to control the worst-case probability of accepting H_1 given H_0.

The *power* of a test (also referred to as *probability of detection* (P_d)) is

$$\beta(\boldsymbol{\theta}) \equiv P_d(\boldsymbol{\theta}) \triangleq P_\theta[\Phi(\boldsymbol{r}) = 1] = E_\theta[\Phi(\boldsymbol{r})], \quad \boldsymbol{\theta} \in \Theta_1. \tag{1.10}$$

Finally, we say that a test Φ of size ϱ is UMP if

$$P_\theta[\Phi(\boldsymbol{r}) = 1] \geq P_\theta[\Phi'(\boldsymbol{r}) = 1], \quad \forall \boldsymbol{\theta} \in \Theta_1, \tag{1.11}$$

with Φ' any other test of size $\varrho' \leq \varrho$. Note that if the hypotheses are simple, the UMP test is the MP test.

1.3.3 Detector Design Criteria

As stated above, it is desirable to devise powerful decision schemes of small preassigned size. In the case of simple hypotheses, we say that a test Φ is the *best* test of size ϱ if it is the most powerful among all tests of size less than or equal to ϱ. The *Neyman–Pearson* criterion tells us how to find this test.

Theorem 1.1. (Neyman–Pearson lemma): *Let* $\Theta = \{\boldsymbol{\theta}_0, \boldsymbol{\theta}_1\}$ *and denote by* $f(\boldsymbol{r}; \boldsymbol{\theta}_i)$ *the pdf or probability mass function of* $\boldsymbol{r}$ *for* $\boldsymbol{\theta}_i$, $i=0,1$. *A test of the form*

$$\Phi(\boldsymbol{r}) = \begin{cases} 1, & f(\boldsymbol{r}; \boldsymbol{\theta}_1) > \eta f(\boldsymbol{r}; \boldsymbol{\theta}_0), \\ \gamma, & f(\boldsymbol{r}; \boldsymbol{\theta}_1) = \eta f(\boldsymbol{r}; \boldsymbol{\theta}_0), \\ 0, & f(\boldsymbol{r}; \boldsymbol{\theta}_1) < \eta f(\boldsymbol{r}; \boldsymbol{\theta}_0), \end{cases} \tag{1.12}$$

for some $\eta \geq 0$ *and* $\gamma \in [0, 1]$ *is the MP test of size* $\varrho \in (0, 1)$ *for testing* $H_0 : \boldsymbol{\theta} = \boldsymbol{\theta}_0$ *vs* $H_1 : \boldsymbol{\theta} = \boldsymbol{\theta}_1$.

Note that $\Phi(\boldsymbol{r}) = i$, $i = 0, 1$, means that we choose H_i; on the other hand, $\Phi(\boldsymbol{r}) = \gamma$ means that we choose H_1 with probability γ and H_0 with probability $1 - \gamma$. As to the choice of η and γ, observe that the size of the test can be evaluated as

$$\varrho = E_{\theta_0}[\phi(\boldsymbol{r})] = P_{\theta_0}[f(\boldsymbol{r}; \boldsymbol{\theta}_1) > \eta f(\boldsymbol{r}; \boldsymbol{\theta}_0)] + \gamma P_{\theta_0}[f(\boldsymbol{r}; \boldsymbol{\theta}_1) = \eta f(\boldsymbol{r}; \boldsymbol{\theta}_0)] \tag{1.13}$$

$$= 1 - P_{\theta_0}[f(\boldsymbol{r}; \boldsymbol{\theta}_1) \leq \eta f(\boldsymbol{r}; \boldsymbol{\theta}_0)] + \gamma P_{\theta_0}[f(\boldsymbol{r}; \boldsymbol{\theta}_1) = \eta f(\boldsymbol{r}; \boldsymbol{\theta}_0)]. \tag{1.14}$$

Now, if there exists a η_0 such that

$$P_{\theta_0}[f(\boldsymbol{r}; \boldsymbol{\theta}_1) \leq \eta_0 f(\boldsymbol{r}; \boldsymbol{\theta}_0)] = 1 - \varrho, \tag{1.15}$$

we set $\eta = \eta_0$ and $\gamma = 0$. On the other hand, if there exists a η_0 such that

$$P_{\theta_0}[f(\boldsymbol{r}; \boldsymbol{\theta}_1) < \eta_0 f(\boldsymbol{r}; \boldsymbol{\theta}_0)] < 1 - \varrho < P_{\theta_0}[f(\boldsymbol{r}; \boldsymbol{\theta}_1) \leq \eta_0 f(\boldsymbol{r}; \boldsymbol{\theta}_0)], \tag{1.16}$$

$\eta = \eta_0$ and γ is given by

$$\gamma P_{\theta_0}[f(\boldsymbol{r}; \boldsymbol{\theta}_1) = \eta_0 f(\boldsymbol{r}; \boldsymbol{\theta}_0)] = P_{\boldsymbol{\theta}_0}[f(\boldsymbol{r}; \boldsymbol{\theta}_1) \leq \eta_0 f(\boldsymbol{r}; \boldsymbol{\theta}_0)] - (1 - \varrho). \tag{1.17}$$

It is important to observe that the best test according to the Neyman–Pearson criterion is the likelihood ratio test of the form

$$\frac{f(\boldsymbol{r}; \boldsymbol{\theta}_1)}{f(\boldsymbol{r}; \boldsymbol{\theta}_0)} \underset{H_0}{\overset{H_1}{\gtrless}} \eta, \tag{1.18}$$

where η is the threshold to be set in order to guarantee the preassigned P_{fa}.

In realistic decision problems, the distribution of the received data is not completely known and, hence, we have to deal with composite hypothesis testing problems for which the UMP test may not exist. In addition, the presence of parameters that do not enter into the hypothesis testing makes the design of practically useful tests even more difficult. In this case, $\boldsymbol{\theta}_0$ and $\boldsymbol{\theta}_1$ share common components represented by the nuisance parameters. As a consequence, the design of good detectors when the pdfs have unknown parameters becomes of great practical importance.

In the sequel, we review the most common approaches to composite hypothesis testing problem. For simplicity, we assume that nuisance parameters are not present and refer the reader to Reference 23 for further details on the same design criteria in the presence of nuisance parameters. More precisely, we focus on suboptimal tests such as the GLRT, the Rao test, and the Wald test, which are asymptotically equivalent [23]. Among them, the GLRT has been the most commonly employed in statistical signal processing, even if it shares no known optimality properties for a finite number of observations.

Assuming simple H_0 and composite H_1 (the generalizations accounting for nuisance parameters can be found in Reference 23), the GLRT replaces the unknown parameters with the respective Maximum Likelihood (ML) estimates and has the following expression

$$\Lambda(\boldsymbol{r}) = \frac{\max\limits_{\boldsymbol{\theta}\in\Theta_1} f(\boldsymbol{r};\boldsymbol{\theta})}{f(\boldsymbol{r};\boldsymbol{\theta}_0)} = \frac{f(\boldsymbol{r};\widehat{\boldsymbol{\theta}}_1)}{f(\boldsymbol{r};\boldsymbol{\theta}_0)} \underset{H_0}{\overset{H_1}{\gtrless}} \eta, \tag{1.19}$$

where

$$\widehat{\boldsymbol{\theta}}_1 = \arg\max_{\boldsymbol{\theta}\in\Theta_1} f(\boldsymbol{r};\boldsymbol{\theta}) \tag{1.20}$$

is the ML estimate of $\boldsymbol{\theta}$ under the H_1 hypothesis. Although there is no optimality associated with the GLRT, in most cases of practical interest it gives satisfactory results. The Wald and Rao tests are alternative to the GLRT and may be easier to compute in practice; the expression of the Wald test is

$$(\widehat{\boldsymbol{\theta}}_1 - \boldsymbol{\theta}_0)^T \boldsymbol{F}(\widehat{\boldsymbol{\theta}}_1)(\widehat{\boldsymbol{\theta}}_1 - \boldsymbol{\theta}_0) \underset{H_0}{\overset{H_1}{\gtrless}} \eta, \tag{1.21}$$

while the Rao test is given by

$$\left.\frac{\partial \ln f(\boldsymbol{r};\boldsymbol{\theta})}{\partial \boldsymbol{\theta}}\right|_{\boldsymbol{\theta}=\boldsymbol{\theta}_0}^T \boldsymbol{F}^{-1}(\boldsymbol{\theta}_0) \left.\frac{\partial \ln f(\boldsymbol{r};\boldsymbol{\theta})}{\partial \boldsymbol{\theta}}\right|_{\boldsymbol{\theta}=\boldsymbol{\theta}_0} \underset{H_0}{\overset{H_1}{\gtrless}} \eta, \tag{1.22}$$

where $\boldsymbol{F}(\boldsymbol{\theta})$ denotes the Fisher information matrix [24].

It is important to underline that under the assumption of Gaussian interference when α and the interference covariance matrix, $\boldsymbol{M}$ say, are unknown, the GLRT does not exist (with just a data vector under test, the likelihood function becomes unbounded). In order to circumvent this drawback, it is customary to assume that a set of $K \geq N$ *secondary data*, namely returns free of signal components, but sharing the spectral properties of the interference in the data under test (or *primary data*), is available. Secondary data could be obtained processing range gates in spatial proximity with that under test. This scenario is usually referred to as *homogeneous environment*. In addition to the classical homogeneous scenario, it is also reasonable to consider the *partially homogeneous environment* [25], where the power value of primary and secondary data vectors is not the same; more precisely, the two covariance matrices coincide only up to a scaling factor. As a consequence, the detection problem at hand can be re-formulated as

$$\begin{cases} H_0 : \begin{cases} \boldsymbol{r} = \boldsymbol{n}, \\ \boldsymbol{r}_k = \boldsymbol{n}_k, \quad k = 1, \ldots, K, \end{cases} \\ \\ H_1 : \begin{cases} \boldsymbol{r} = \alpha\boldsymbol{p} + \boldsymbol{n}, \\ \boldsymbol{r}_k = \boldsymbol{n}_k, \quad k = 1, \ldots, K, \end{cases} \end{cases} \tag{1.23}$$

where $\boldsymbol{n}, \boldsymbol{n}_k, k = 1, \ldots, K$ are the interference components that are supposed statistically independent with zero mean and sharing the same covariance matrix in homogeneous environment, whereas in partially homogeneous environment

$$E[\boldsymbol{n}\boldsymbol{n}^\dagger] = \sigma^2 \boldsymbol{M} \quad \text{and} \quad E[\boldsymbol{n}_k \boldsymbol{n}_k^\dagger] = \boldsymbol{M}, \quad k = 1, \ldots, K, \tag{1.24}$$

with $\sigma^2 > 0$. Note that the homogeneous environment is a special case of the partially homogeneous one; indeed, if $\sigma^2 = 1$ primary and secondary data share the same statistical characterization of the noise. Finally, it is worth mentioning the *heterogeneous environment*, where

$$E[\boldsymbol{n}\boldsymbol{n}^\dagger] \neq E[\boldsymbol{n}_1 \boldsymbol{n}_1^\dagger] \neq \cdots \neq E[\boldsymbol{n}_K \boldsymbol{n}_K^\dagger]. \tag{1.25}$$

The reader is referred to Chapters 7–9 for further details on the above model, while in Reference 26 the reader can find a list of papers concerning adaptive radar detection of targets against ground and sea clutters, clustered according to several related issues.

1.3.4 CFAR Property and Invariance in Detection Theory

A very challenging problem arising in radar signal processing is to achieve a reliable target detection in the presence of severe clutter backgrounds. When the clutter statistics are unknown or highly variable, the false alarm rate of the conventional square law detector, matched filter, and in general nonadaptive radar receivers cannot be controlled and target decisions become often unreliable. This is due to the lack of robustness of the quoted receivers with respect to possible mismatches between the design and operating conditions as well as variations in the clutter statistics. In order to overcome these drawbacks, adaptive systems have to be adopted, namely wherein the unknown interference parameters are suitably estimated and the detection threshold is adjusted accordingly. In this case, we say that the receiver ensures the CFAR property.

Formally speaking, the aim of CFAR detection theory is to design decision rules whose pdfs, under the null hypothesis H_0, are functionally independent of the unknown nuisance parameters (for example, the disturbance power and covariance matrix). As a consequence, the more remarkable property of a CFAR test is the possibility to set the detection threshold to guarantee a preassigned P_{fa} independent of the actual disturbance parameters. Much work has been directed so far towards the design and the assessment of systems capable of ensuring CFARness subject to the commonly encountered instances of clutter models. Basically, most of the proposed detectors develop estimates of the clutter parameters processing the secondary data that are representative of the interference of the cell under test.

The most common CFAR algorithms deal with scalar data at the output of a linear, square-law, or logarithmic detector. Further details on these techniques can be found in Chapter 16 of [5], where several CFAR architectures are examined starting from the Cell-Averaging CFAR up to the Order Statistic CFAR algorithms. Unlike the standard approaches to CFARness, in this book, processed data are vectors and matrices, which stem from the structure of the sensing system. As a consequence, the generalization of CFAR processors to multiple dimensions is herein presented and analyzed.

Unfortunately, the CFAR property is not granted by a generic detection scheme. Otherwise stated, before claiming the CFARness for a given receiver, it must be proved that its decision statistic does not depend on the interference parameters under the null hypothesis H_0 (see, for instance, References 11, 12, and 27). On the other hand, CFARness can be obtained properly modifying the decision statistic with normalization factors (or other heuristic tricks), which remove the dependence on the unknown interference parameters, or forcing suitable symmetry properties at the design stage, which, in turn, imply the CFAR property. This last point can be accomplished by means of the *Theory of Invariance* in hypothesis testing [19, 28], which allows to restrict the attention to invariant tests, namely decision rules that do not distinguish between scenarios differing in their nuisance parameters, and to define optimality criteria within this family. As a matter of fact, the invariance criterion equalizes performance among diverse scenarios sharing same significant parameters (e.g., the signal-to-noise ratio). It can be shown in certain cases that the UMP invariant test (if it exists) yields the minimax test for the problem in the sense of maximizing the minimum (over parameter values) of P_d for a given P_{fa}. Moreover, it is important to observe that under some mild technical assumptions, the GLRT, the Rao test, and often the Wald test lead to invariant architectures [29].

Several applications of this theory to radar detection can be found in the open literature; just to give some examples see References 30–37. Applying this criterion has the effect of drastically reducing the problem size since it can be shown that all invariant test statistics can be expressed in terms of a statistic (having much fewer dimensions than the data) called the maximal invariant, which organizes the original data into equivalence classes. Therefore, characterizing this statistic greatly facilitates and sometimes completes the search for a good test statistic. Remarkably, the distribution of the maximal invariant is parameterized by another low-dimensional function on the parameter space (called the induced maximal invariant). In this way, most of the nuisance parameters are removed from the problem. Finally, it is important to recall that forcing suitable symmetry properties with respect to the nuisance parameters to a decision statistic implies as a by-product some important features of practical interest such as the CFAR behavior.

1.4 Organization, Use, and Outline of the Book

The book is organized so that it can be used in a variety of ways for a variety of purposes. Sequential reading is not necessary. Extensive cross referencing has been attempted so that related material can easily be found; we hope this will enhance the book's value as a reference. For the same purpose, a detailed table of contents and an extensive index are also provided. This book has been conceived to provide a comprehensive reference on the last developments in adaptive radar detection, and it should be useful to post-graduate students, PhD students, researchers, and, more generally, engineers working on statistical signal processing and its applications to the radar systems. Finally, we assume that the reader has a quite solid background in probability theory and random processes, matrix theory, linear algebra, and mathematical analysis.

The remainder of this book is organized as follows:

- Chapter 2 deals with the theory of radar detection in white Gaussian noise. The problem is formulated in terms of a binary hypothesis test where the null hypothesis of target absence is tested versus the composite alternative accounting for target presence. The interference experienced by the radar is modeled as an additive zero-mean complex circular Gaussian random process with known power spectral density, whereas target returns are supposed known up to a multiplicative factor accounting for the target reflectivity and channel propagation effects. Practically implementable receivers can be devised by means of two different approaches. Specifically, the former is a Bayesian framework and models the target unknown parameters as random quantities, the latter is a robust approach and supposes the target parameters deterministic and unknown. Finally, the performance of the considered receivers is assessed in terms of P_{fa} and P_d either using analytic equations or resorting to numerical simulations.

- Chapter 3 addresses coherent multichannel detection of a signal contained in a known subspace and competing with unknown interference from the perspective of invariant hypothesis testing for radar and sonar applications. The considered framework provides a systematic approach for dealing with the large number of nuisance parameters typically inherent in these problems. Example results are provided to illustrate the application of subspace detectors to the problem of mitigating the detection loss resulting from signal mismatch errors.
- Chapter 4 focuses on the design and the analysis of two-stage detectors. These architectures are formed cascading two structures with opposite behaviors in terms of directivity. The overall system declares the presence of a target in the cell under test only when data survive both detection thresholdings. Their directivity can be adjusted by a proper selection of the two thresholds, so trading good rejection capabilities of sidelobe interferers for acceptable detection loss of matched signals. This chapter is aimed at providing a complete review of various two-stage solutions, which can be conceived by coupling existing detectors. Moreover, closed-form expressions for the probability of false alarm and the probability of detection (for matched and mismatched signals) are derived and evaluated resorting to standard numerical integration techniques. The performance analysis highlights that the two-stage detectors are flexible architectures which provide a wide range of directivity while retaining good matched detection performances.
- Chapter 5 aims to discuss recent advances on Bayesian Knowledge Aided-STAP (KA-STAP) techniques. Such techniques advocate an intelligent utilization of a priori knowledge from various sources, such as previous measurements, digital geographic maps, and real-time radar platform parameters. A natural and systematic way for incorporating such a priori knowledge for detection is to employ a Bayesian inference framework. This is appealing for KA-STAP because the Bayesian method not only allows for the formal and systematic use of prior information (on the interference covariance matrix) but also quantifies uncertainties presented in the prior knowledge through hyperparameters. This chapter unfolds as the classical STAP signal model evolves into a framework of KA-STAP model including a knowledge-aided homogeneous model, a knowledge-aided partially homogeneous model, and a knowledge-aided compound-Gaussian model. Moreover, a hierarchical two-layered STAP model is discussed which provides a new way to describe the non-homogeneity between the test data and training data. Finally, parametric Bayesian detectors are presented, which exploit the structural space–time information for the interference model allowing a fast implementation and further reduction in the amount of training data, along with the consequent Bayesian estimation.
- Chapter 6 is focused on sample-starved scenarios characterized by a number of independent identically distributed training samples that is comparable with the dimension of the data. Then, it presents an alternative adaptive detection framework which heavily relies on more efficient covariance matrix estimates in the sample deficient environment, namely when the number of secondary data is not enough for reliable estimates, using the expected likelihood principle. In the second part of the chapter, a time-varying autoregressive model is exploited to account for a wider range of interference scenarios and array structures.

Finally, the proposed detectors are then combined within a two-stage approach to allow the use of highly effective quasi-CFAR designs, and under these sample-starved conditions, the resultant detectors exhibit superior behavior to the classical detectors designed using asymptotically efficient ML techniques.

- Chapter 7 focuses on the compound-Gaussian model for spiky radar clutter. Due to increased resolution capabilities, the exponential model for the clutter returns does not reveal adequate. In particular, for sea clutter, large deviations from this model can be observed at low grazing angles. Several models have been proposed to obtain more accurate descriptions of the clutter, and these investigations led to the concept that the distribution of radar clutter returns should be modeled by a mixture of exponential intensity distributions. Such models are called compound-Gaussian models.

  The chapter reviews the general compound-Gaussian model and shows that various generalizations of the multivariate Gaussian distribution are specific instances of the compound-Gaussian model. Then, for each of these instances, the general optimal detector is devised. In addition, several examples of sub-optimal detectors are also provided. The final part of this chapter deals with the limitations of the compound-Gaussian model and proposes a research area that may lead to the development of more general random process models to describe the radar clutter returns.
- Chapter 8 deals with the disturbance covariance matrix estimation theory when the disturbance is CES distributed. The CES family is a broad one whose compound-Gaussian distributions introduced in Chapter 7 are particular cases. The robust M-estimators are presented and their statistical properties are summarized, then some radar detectors using these matrix estimators are applied to real ground- and sea-clutter data.
- Chapter 9 designs some detectors for range-spread targets and analyzes their performance on recorded live data. To this end, real target and sea-clutter data collected by a fully coherent Ka-band radar system, featuring sub-meter range resolution, are used. The study is of particular interest for homeland security radar applications where a careful coastal control is necessary to prevent the arrival of non-authorized small boats. The performance of both rank-one and subspace range-spread target detection strategies is analyzed, both in terms of CFAR behavior and in terms of detection capabilities. With reference to the former issue, clutter-only datafiles are used, whereas concerning the latter, data containing both real target and clutter are used. The target returns come from typical small boats (such as inflatable, wooden, and patrol boats) appearing range-distributed at the resolution of the exploited radar system. Range-time detection maps are shown, assessing the capability of the analyzed processors to detect the aforementioned targets of great interest for homeland coastal security. Finally, the performance improvement achieved by over-resolving the target is quantified.

1.5 References

The common denominator of this book is the detection theory (or hypothesis testing) which is a branch of the statistical signal processing. A domain of application for detection theory is radar and sonar systems or, generally speaking, sensing systems. As a matter of fact, the main task of such systems is to decide for the presence of useful signals embedded in interference background. The book by Lehman [19] represents a relevant contribution to detection theory and treats the mathematical foundations of hypothesis testing and parameter estimation. The works by Kay [23, 24] strike a balance between highly theoretical expositions and the more practical treatments, focusing extensively on real-world signal processing applications. The mentioned books are complementary and oriented to practicing electrical engineers, PhD students, as well as researchers, providing an invaluable introduction to parameter estimation and detection theory. The book by Van Trees [14] applies hypothesis testing and parameter estimation to detection, estimation, and modulation of continuous-time waveforms. Influential and scholarly engineering texts are the book by Scharf [13] and Helstrom [38]. The former covers four distinct topical lines: a preliminary review of the mathematical and statistical tools used throughout the book, detection theory, estimation theory, and time series analysis. The latter provides an introduction to signal-detection with emphasis on the design of optimal and near-optimal detectors of weak signals in the presence of random noise for communication, radar, sonar, and optical applications. Finally, textbooks [1,5,39,40] deal with several aspects of modern radar detection systems and are oriented to the radar engineers.

References

[1] M. I. Skolnik, *Introduction to Radar Systems*, 3rd ed., McGraw-Hill, New York, 2001.

[2] *History of radar*, http://en.wikipedia.org/wiki/History_of_radar.

[3] H. Griffith and N. Willis, "Klein Heidelberg – The world's first modern bistatic radar system," *IEEE Transactions on Aerospace and Electronic Systems*, Vol. 46, No. 4, pp. 1571–1588, October 2010.

[4] J. Brown, K. Woodbridge, H. Griffiths, A. Stove, and S. Watts, "Passive bistatic radar experiments from an airborne platform," *IEEE Aerospace and Electronic Systems Magazine*, Vol. 47, No. 11, pp. 50–55, November 2012.

[5] M. A. Richards, J. A. Scheer, and W. A. Holm, *Principles of Modern Radar: Basic Principles*, Scitech Publishing, Raleigh, NC, 2010.

[6] W. L. Melvin and J. A. Scheer, *Principles of Modern Radar: Advanced Techniques*, Scitech Publishing, Edison, NJ, 2013.

[7] R. Klemm (ed.), *Applications of Space–Time Adaptive Processing*, IEE Publishing, London, United Kingdom, 2004.

[8] J. Ward, "Space-time adaptive processing for airborne radar," MIT, Lexington, Tech. Rep. No. 1015, December 13, 1994.

[9] L. E. Brennan and I. S. Reed, "Theory of adaptive radars," *IEEE Transactions on Aerospace and Electronic Systems*, Vol. 9, No. 2, pp. 237–252, March 1973.

[10] I. S. Reed, J. D. Mallett, and L. E. Brennan, "Rapid convergence rate in adaptive arrays" *IEEE Transactions on Aerospace and Electronic Systems*, Vol. 10, No. 4, pp. 853–863, November 1974.

[11] F. C. Robey, D. L. Fuhrman, E. J. Kelly, and R. Nitzberg, "A CFAR adaptive matched filter detector," *IEEE Transactions on Aerospace and Electronic Systems*, Vol. 29, No. 1, pp. 208–216, January 1992.

[12] E. J. Kelly, "An adaptive detection algorithm," *IEEE Transactions on Aerospace and Electronic Systems*, Vol. 22, No. 2, pp. 115–127, March 1986.

[13] L. L. Scharf, *Statistical Signal Processing: Detection, Estimation, and Time Series Analysis*, Addison-Wesley, United States, 1991.

[14] H. L. Van Trees, *Detection, Estimation, and Modulation Theory, Part I*, John Wiley & Sons, United States, 2002.

[15] E. J. Kelly and K. Forsythe, "Adaptive detection and parameter estimation for multidimensional signal models," Lincoln Lab, MIT, Lexington, Tech. Rep. No. 848, April 19, 1989.

[16] F. Bandiera, D. Orlando, and G. Ricci, "*Advanced radar detection schemes under mismatched signal models*," Synthesis Lectures on Signal Processing No. 8, Morgan & Claypool Publishers, San Rafael, United States, 2009.

[17] P. Swerling, "Probability of detection for fluctuating targets," *IRE Transactions on Information Theory*, Vol. 6, No. 2, pp. 269–308, April 1960.

[18] P. Swerling, "More on detection of fluctuating targets," *IEEE Transactions on Information Theory*, Vol. 11, No. 3, pp. 459–460, July 1965.

[19] E. L. Lehmann, *Testing Statistical Hypotheses*, Springer-Verlag, 2nd ed., New York, United States, 1986.

[20] K. D. Ward, C. J. Baker, and S. Watts, "Maritime surveillance radar Part 1 : Radar scattering from the ocean surface," *IEE Proceedings-F, Radar and Signal Processing*, Vol. 137, No. 2, pp. 51–62, April 1990.

[21] E. Conte, M. Longo, and M. Lops, "Modelling and simulation of non-Rayleigh radar clutter," *IEE Proceedings-F, Radar and Signal Processing*, Vol. 138, No. 2, pp. 121–130, April 1991.

[22] E. Conte and G. Ricci, "Sensitivity Study of GLRT Detection in Compound-Gaussian Clutter," *IEEE Transactions on Aerospace and Electronics Systems*, Vol. 34, No. 1, pp. 308–316, January 1998.

[23] S. M. Kay, *Fundamentals of Statistical Signal Processing, Detection Theory*, Prentice-Hall, Englewood Cliffs, NJ, 1998, Vol. II.

[24] S. M. Kay, *Fundamentals of Statistical Signal Processing, Estimation Theory*, Prentice-Hall, Upper Saddle River, NJ, 1993, Vol. I.

[25] S. Kraut and L. L. Scharf, "The CFAR adaptive subspace detector is a scale-invariant GLRT," *IEEE Transactions on Signal Processing*, Vol. 47, No. 9, pp. 2538–2541, September 1999.

[26] F. Gini, A. Farina, and M. Greco, "Selected list of references on radar signal processing," *IEEE Transactions on Aerospace and Electronic Systems*, Vol. 37, No. 1, pp. 329–359, January 2001.

[27] E. Conte, M. Lops, and G. Ricci, "Asymptotically optimum radar detection in compound Gaussian noise," *IEEE Transactions on Aerospace and Electronic Systems*, Vol. 31, No. 2, pp. 617–625, April 1995.

[28] R. J. Muirhead, *Aspects of Multivariate Statistical Theory*, John Wiley & Sons, New Jersey, United States, 1982.

[29] S. M. Kay and J. R. Gabriel, "An invariance property of the generalized likelihood ratio test," *IEEE Signal Processing Letters*, Vol. 10, No. 12, pp. 352–355, December 2003.

[30] S. Bose and A. O. Steinhardt, "A maximal invariant framework for adaptive detection with structured and unstructured covariance matrices," *IEEE Transactions on Signal Processing*, Vol. 43, No. 9, pp. 2164–2175, September 1995.

[31] S. M. Kay and J. R. Gabriel, "Optimal invariant detection of a sinusoid with unknown parameters," *IEEE Transactions on Signal Processing*, Vol. 50, No. 1, pp. 27–40, January 2002.

[32] A. De Maio, S. M. Kay, and A. Farina, "On the invariance, coincidence, and statistical equivalence of the GLRT, Rao test, and Wald test," *IEEE Transactions on Signal Processing*, Vol. 58, No. 4, pp. 1967–1979, April 2010.

[33] A. De Maio and E. Conte, "Adaptive detection in Gaussian interference with unknown covariance after reduction by invariance," *IEEE Transactions on Signal Processing*, Vol. 58, No. 6, pp. 2925–2934, June 2010.

[34] R. S. Raghavan, "Maximal invariants and performance of some invariant hypothesis tests for an adaptive detection problem," *IEEE Transactions on Signal Processing*, Vol. 61, No. 14, pp. 3607–3619, July 2013.

[35] L. L. Scharf and B. Friedlander, "Matched subspace detectors," *IEEE Transactions on Signal Processing*, Vol. 42, No. 8, pp. 2146–2157, August 1994.

[36] O. Besson, L. L. Scharf, and S. Kraut, "Adaptive detection of a signal known only to lie on a line in a known subspace, when primary and secondary data are partially homogeneous," *IEEE Transactions on Signal Processing*, Vol. 54, No. 12, pp. 4698–4705, December 2006.

[37] S. K. Hyung and A. O. Hero, "Comparison of GLR and invariant detectors under structured clutter covariance," *IEEE Transaction on Image Processing*, Vol. 10, No. 10, pp. 1509–1520, October 2001.

[38] C. W. Helstrom, *Elements of Signal Detection and Estimation*, Prentice-Hall PTR, New Jersey, United States, 1994.

[39] D. K. Barton, *Radar System Analysis and Modeling*, Artech House, Norwood, MA, 2005.

[40] A. Farina, "*Optimized radar processors*," IEE Radar, Sonar, Navigation & Avionics Series, IET Publication, London, United Kingdom, 1987.

CHAPTER 2

Radar Detection in White Gaussian Noise: A GLRT Framework

Ernesto Conte[1], Antonio De Maio[1], and Guolong Cui[2]

2.1 Introduction

This chapter is focused on the theory of radar detection in white Gaussian noise. The problem is formulated in terms of a binary hypothesis test where the null hypothesis of target absence is tested versus the composite alternative accounting for target presence. The interference experienced by the radar is modeled as an additive zero-mean complex circular Gaussian random process with known Power Spectral Density (PSD), whereas the useful contribution is supposed known but for a scaling factor accounting for the target reflectivity and channel propagation effects. The approach pursued to achieve target detection is based on the computation of a sufficient statistic which performs data compression while keeping all the information contained in the original observables. Additionally, it allows for a nice interpretation of the detection pre-processing in terms of pulse compression (range processing) and pulse integration. After data compression by sufficiency, the Neyman–Pearson (NP) criterion (explained in Chapter 1) is exploited to devise the optimum radar detector for both coherent and non-coherent pulse trains; unfortunately the resulting test is not Uniformly Most Powerful (UMP). In order to circumvent this drawback and come up with practically implementable receivers, two different approaches are available [1,2]. The former is a Bayesian framework and models the target unknown parameters as random quantities, the latter is a robust approach and supposes the target parameters deterministic and unknown. While the Bayesian technique usually leads to detectors tied up to the chosen a priori distribution for the unknown parameters, the robust approach leads to decision rules independent of any prior assigned to the unknowns. Leveraging on this desirable feature, in this chapter, the robust approach is followed and the Generalized Likelihood Ratio Test (GLRT) criterion (see also Chapter 1) is used to design radar detectors. As a result the classic coherent receiver, the linear and the square-law integrators are

[1]Dipartimento di Ingegneria Elettrica e delle Tecnologie dell'Informazione, Università degli Studi di Napoli "Federico II", Napoli, Italy

[2]School of Electronic Engineering, University of Electronic Science and Technology of China, Chengdu City, China

obtained with reference to coherent (the first) and non-coherent pulse trains (the second and the third). Finally, the performance of the NP detectors and the GLRTs are assessed in terms of false alarm probability (P_{FA}) and detection probability (P_D) either using analytic equations or resorting to numerical simulations.

The chapter is organized as follows. Section 2.2 is devoted to problem formulation while Section 2.3 is focused on reduction by sufficiency. The design of the optimum NP receiver is handled in Section 2.4 whereas the GLRT-based detectors are derived in Section 2.5. The performance analysis is developed in Section 2.6. Finally, concluding remarks and suggestions for further reading are given in Section 2.7.

2.2 Problem Formulation

Let us consider a stationary monostatic radar system that transmits a possibly modulated coherent burst of pulses and collects data from a given azimuth-elevation cell during a time interval usually referred to as dwell time. Denoting by $s_r(t)$ the baseband equivalent of the received signal (obtained for instance through a standard heterodyne receiver), the classic radar detection problem in the specific azimuth-elevation cell can be formalized as the following binary hypothesis test

$$\begin{cases} H_0 : s_r(t) & \text{contains noise only,} \\ H_1 : s_r(t) & \text{contains useful target signal plus noise.} \end{cases} \tag{2.1}$$

Under the target plus noise hypothesis H_1, the baseband equivalent of the received signal can be written as [3]

$$H_1 : s_r(t) = e^{-j2\pi f_0 \tau_0} \sum_{k=0}^{K-1} A_k e^{j\phi_k} p(t - kT - \tau_0) e^{j2\pi f_d (t-\tau_0)} + n(t), \tag{2.2}$$

where

- f_0 is the carrier frequency;
- K denotes the number of pulses;
- T is the pulse repetition time;
- $p(t)$ represents the baseband equivalent of the generic pulse with length T_1. Precisely,

$$p(t) = \sum_{i=0}^{N-1} \boldsymbol{a}(i) u(t - iT_p), \tag{2.3}$$

  with $\boldsymbol{a}(i) \in \mathcal{C}$, $i = 0, \ldots, N-1$, the radar code elements, and $u(t)$ the chip pulse (or sub-pulse) baseband equivalent whose length is $T_p = T_1/N$ and whose energy is assumed, without loss of generality, unitary;
- τ_0 is the round trip delay related to the target range R through the equation $\tau_0 = 2R/c$;
- f_d is the target Doppler frequency that is related to the radial velocity v_r through the equation $f_d = 2v_r/\lambda$ with λ the carrier wavelength;
- the complex amplitudes $A_k e^{j\phi_k}$, $k = 0, \ldots, K-1$, are unknown parameters accounting for the target radar cross section, channel propagation effects, and other terms involved into the

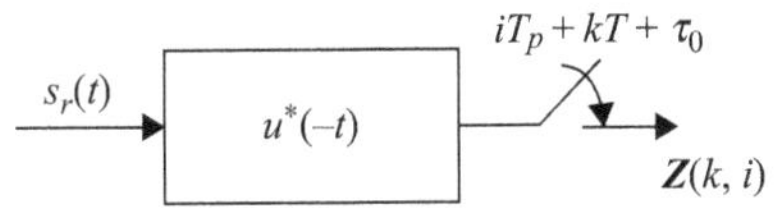

Figure 2.1 Sub-pulse matched filter.

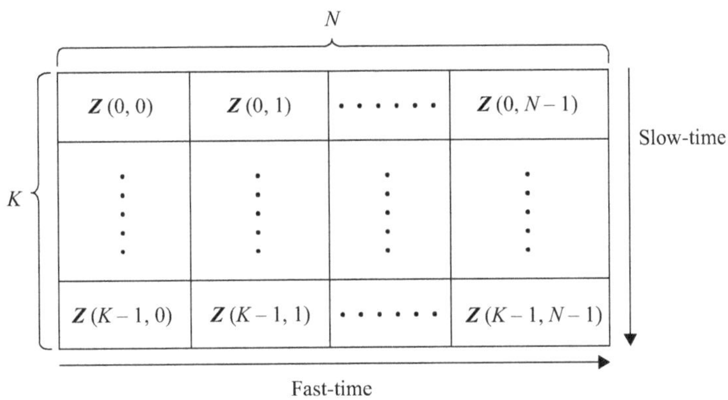

Figure 2.2 Fast-time/slow-time data matrix $\boldsymbol{Z}$ from the range-azimuth-elevation cell under test.

radar range equation. If $A_k e^{j\phi_k} = Ae^{j\phi}$, $k = 0, \ldots, K-1$, then train is coherent, otherwise it is referred to as non-coherent;

- $n(t)$ is the additive noise component modeled as a complex circular zero-mean Gaussian random process with constant and known PSD within the receiver bandwidth.

Under the H_0 hypothesis, $s_r(t)$ contains only noise, namely

$$H_0 : s_r(t) = n(t). \tag{2.4}$$

The received signal is filtered through a sub-pulse matched filter $h(t) = u^*(-t)$ and sampled at the time instants $iT_p + kT + \tau_0$, $i = 0, \ldots, N-1, k = 0, \ldots, K-1$ (see Figure 2.1). The samples $\boldsymbol{Z}(k, i)$, from the range cell under test, are arranged into the fast-time/slow-time matrix $\boldsymbol{Z}$ whose columns contain the slow-time samples whereas its rows include those in fast-time (see Figure 2.2).

Test (2.1) can be rewritten in terms of the data matrix $\boldsymbol{Z}$, of size $K \times N$, as

$$\begin{cases} H_0 : \boldsymbol{Z} = \boldsymbol{N}, \\ H_1 : \boldsymbol{Z} = \boldsymbol{Q}\boldsymbol{p}\boldsymbol{a}^\dagger + \boldsymbol{N}, \end{cases} \tag{2.5}$$

where

- $\boldsymbol{a} = [\boldsymbol{a}(0), \ldots, \boldsymbol{a}(N-1)]^\dagger$ is the code vector assumed, without loss of generality, with unitary norm;
- $\boldsymbol{Q} = \mathrm{diag}(\boldsymbol{q})$ with $\boldsymbol{q} = \bar{\xi}(0, f_d)\left[A_0 e^{j\theta_0}, \ldots, A_{K-1} e^{j\theta_{K-1}}\right]^T$, $\theta_k = \phi_k - 4\pi f_0 R_0 / c$, and

$$\bar{\xi}(\tau, f) = \int_{-\infty}^{\infty} u(\beta) u^*(\beta - \tau) e^{j2\pi f \beta} d\beta.$$

- $\boldsymbol{p} = \left[1, e^{j2\pi f_d T}, \ldots, e^{j2\pi f_d T(K-1)}\right]^T$ is the temporal steering vector;
- the entries of the $K \times N$ matrix $\boldsymbol{N}$, $\boldsymbol{N}(k,i)$ for $k = 0, \ldots, K-1$, $i = 0, \ldots, N-1$, are modeled as independent and identically distributed (i.i.d.) complex circular zero-mean Gaussian random variables with $E[|\boldsymbol{N}(k,i)|^2] = \sigma^2$, i.e., $\boldsymbol{N}(k,i) \sim \mathcal{CN}(0, \sigma^2)$.

In the following, the chip pulse is assumed Doppler tolerant (namely the expected range of Doppler frequencies are such that $\overline{\xi}(0, f_d) \approx \overline{\xi}(0,0) = 1$) and $\exp(j2\pi f_d T_1) \approx 1$. These two conditions are sufficient to ensure a Doppler tolerant pulse $p(t)$. Before concluding, it is worth pointing out that, in radar applications, the parameters $A_k e^{j\theta_k}$ are reasonably unknown and to circumvent this drawback two different approaches are usually exploited. The former is tantamount to modeling $A_k e^{j\theta_k}$ as unknown deterministic parameters whereas the latter is a Bayesian approach that assigns suitable priors to the unknowns. In this last case, the solution to the testing problem is often tied up to the specific priors whereas the deterministic and unknown parameter framework does not require any prior knowledge. Based on this guideline, in this chapter, the robust approach is considered.

2.3 Reduction by Sufficiency

Problem (2.5) is a composite test where the simple hypothesis H_0 is tested versus the composite alternative H_1 with parameter vector $\boldsymbol{q}$ and parameter space $\mathcal{C}^K$. To solve it, we follow the approach of determining first a sufficient statistic that summarizes all the information in the data about the parameters. After data reduction by sufficiency, which realizes a significant data compression, we synthesize the optimum NP detector as the Likelihood Ratio Test (LRT) computed from the sufficient statistic.

To this end, let us write the data probability density function (pdf) under the H_1 hypothesis

$$f_{\boldsymbol{Z}}(\boldsymbol{Z}|H_1) = \frac{1}{\pi^{NK}\sigma^{2NK}} \exp\left\{ -\frac{\mathrm{Tr}\left[\left(\boldsymbol{Z} - \boldsymbol{Q}\boldsymbol{p}\boldsymbol{a}^\dagger\right)\left(\boldsymbol{Z} - \boldsymbol{Q}\boldsymbol{p}\boldsymbol{a}^\dagger\right)^\dagger\right]}{\sigma^2} \right\} \tag{2.6}$$

and recast the argument of the exponential as

$$-\frac{1}{\sigma^2}\mathrm{Tr}\left(\boldsymbol{Z}\boldsymbol{Z}^\dagger\right) - \frac{1}{\sigma^2}\mathrm{Tr}\left(\boldsymbol{Q}\boldsymbol{p}\boldsymbol{a}^\dagger\boldsymbol{a}\boldsymbol{p}^\dagger\boldsymbol{Q}^\dagger\right) + \frac{2}{\sigma^2}\mathrm{Re}\left\{\mathrm{Tr}\left(\boldsymbol{Z}\boldsymbol{a}\boldsymbol{p}^\dagger\boldsymbol{Q}^\dagger\right)\right\}. \tag{2.7}$$

Based on the Fisher–Neyman factorization theorem and minimality arguments concerning regular exponential families [4, Chap. 5], we can claim that

(1) If the train is coherent then $\boldsymbol{Q} = Ae^{j\theta}\boldsymbol{I}_K$; $Ae^{j\theta}$ is the sole unknown parameter; a minimal sufficient statistic is one-dimensional and coincides with

$$L = \mathrm{Tr}\left(\boldsymbol{Z}\boldsymbol{a}\boldsymbol{p}^\dagger\right). \tag{2.8}$$

(2) If the train is non-coherent then $A_k e^{j\theta_k}$ are the unknown parameters; a minimal sufficient statistic is K-dimensional and coincides with

$$\boldsymbol{L} = \left[\boldsymbol{Z}_{0,R}\boldsymbol{a}p^*(0), \ldots, \boldsymbol{Z}_{K-1,R}\boldsymbol{a}p^*(K-1)\right]^T, \tag{2.9}$$

where

$$\mathbf{Z} = \left[\mathbf{Z}_{0,R}^T, \mathbf{Z}_{1,R}^T, \ldots, \mathbf{Z}_{K-1,R}^T\right]^T.$$

Remark: Notice that (2.9) represents a sufficient statistic also for the coherent train detection problem; however, in this last case, it is not minimal. Its computation requires a linear filtering procedure matched to the radar code, namely, in radar jargon, a pulse compression. Besides, with reference to the coherent case, since $\mathrm{Tr}\left[(\mathbf{Z}\mathbf{a})\,\mathbf{p}^{\dagger}\right] = \mathrm{Tr}\left[(\mathbf{p}^{\dagger}\mathbf{Z})\,\mathbf{a}\right]$, from a theoretical point of view it is possible to decide arbitrarily which processing between fast-time (pulse compression) and slow-time (Doppler processing) to implement first. Nevertheless, it is worth pointing out that from a practical point of view it is convenient to implement fast-time processing before the slow-time one in order to avoid latencies. In Figure 2.3(a) and (b), the block schemes for the computation of (2.8) and (2.9), respectively, are reported. The main difference relies on the final coherent integration required in (2.8).

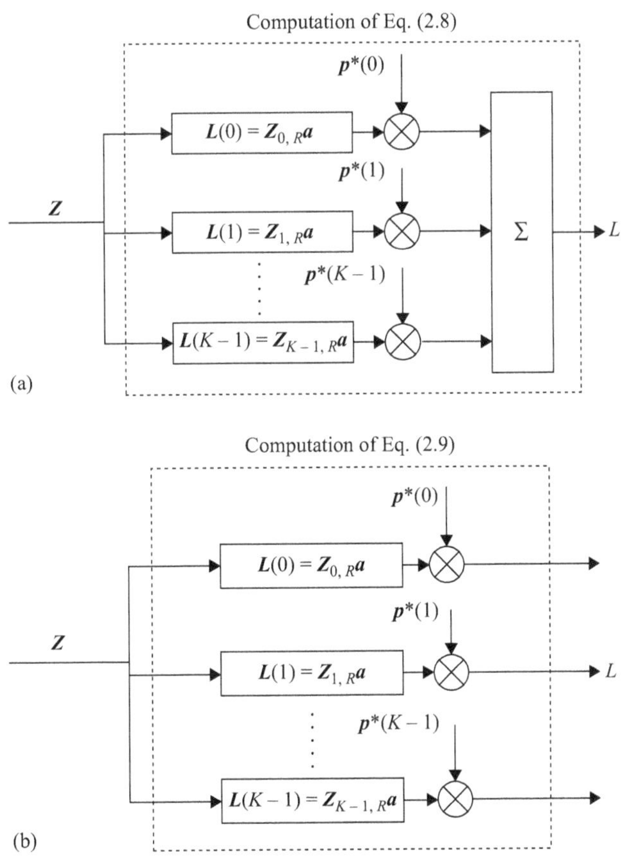

Figure 2.3 Block schemes for the computation of Eq. (2.8) (a) and Eq. (2.9) (b).

2.4 Optimum NP Detector and Existence of the UMP Test

The scope of this section is twofold: first to synthesize the optimum detector, according to the NP criterion for problem (2.5); second, due to the composite nature of the H_1 hypothesis, to ascertain the existence of a UMP test.

The optimum detector, which maximizes the detection performance for a given false alarm probability, is the LRT computed from the sufficient statistic. In order to proceed with the synthesis, let us distinguish between the coherent and non-coherent cases.

2.4.1 Coherent Case

The LRT detector from the sufficient statistic (2.8) can be expressed as

$$\frac{f_L(L|H_1)}{f_L(L|H_0)} \underset{H_0}{\overset{H_1}{\gtrless}} \gamma, \tag{2.10}$$

where $f_L(L|H_0)$ and $f_L(L|H_1)$ are the likelihood functions of the sufficient statistic under the H_0 and the H_1 hypothesis, respectively. Further developments require specifying the aforementioned functions. To this end, we observe that under the H_0 hypothesis, L is distributed as a complex circular zero-mean Gaussian random variable with variance $K\sigma^2$, i.e.,

$$L|H_0 \sim \mathcal{CN}(0, K\sigma^2), \tag{2.11}$$

whereas under the alternative, it is distributed as

$$L|H_1 \sim \mathcal{CN}(KAe^{j\theta}, K\sigma^2). \tag{2.12}$$

It follows that the NP receiver can be written as

$$\frac{f_L(L|H_1)}{f_L(L|H_0)} = \exp\left\{-\frac{|L - KAe^{j\theta}|^2 - |L|^2}{K\sigma^2}\right\} \underset{H_0}{\overset{H_1}{\gtrless}} \gamma. \tag{2.13}$$

After some algebraic manipulations, taking the logarithm and absorbing inessential terms into the detection threshold, (2.13) can be shown equivalent to

$$\text{Re}\{e^{-j\theta}L\} = \text{Re}\left\{e^{-j\theta}\boldsymbol{p}^{\dagger}(\boldsymbol{Za})\right\} \underset{H_0}{\overset{H_1}{\gtrless}} \gamma, \tag{2.14}$$

where γ is a suitable modification of the original detection threshold in (2.10). In Figure 2.4, a block scheme of the NP detector (2.14) for the coherent train is given, evidently it is not implementable since it requires the exact knowledge of the phase θ; otherwise stated the UMP test does not exist. Nevertheless, (2.14) is still important as it provides an upper-bound to the performance of any practically feasible detection schemes.

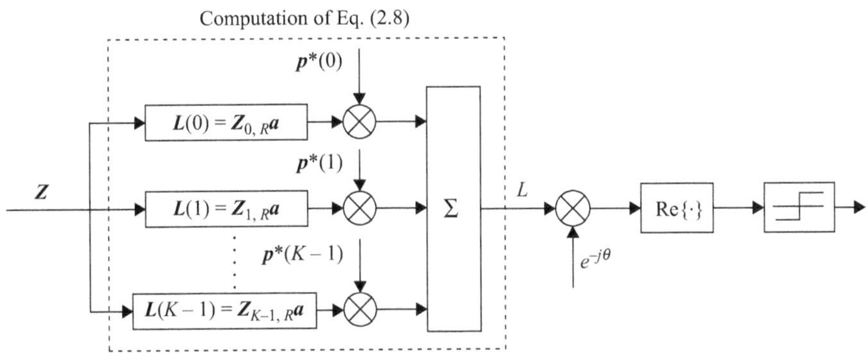

Figure 2.4 Block scheme of the NP receiver (2.14) for the coherent pulse train.

2.4.2 Non-coherent Case

The LRT from the sufficient statistic (2.9) can be expressed as

$$\frac{f_{\mathbf{L}}(\mathbf{L}|H_1)}{f_{\mathbf{L}}(\mathbf{L}|H_0)} \underset{H_0}{\overset{H_1}{\gtrless}} \gamma, \tag{2.15}$$

where the entries of $\mathbf{L}$ in (2.9) are independent complex circular Gaussian random variables. Specifically, under H_0, the k-th entry $\mathbf{Z}_{k,R}\mathbf{a}\mathbf{p}^*(k)$ is distributed as

$$\mathbf{Z}_{k,R}\mathbf{a}\mathbf{p}^*(k) \sim \mathcal{CN}(0, \sigma^2), \tag{2.16}$$

whereas under H_1

$$\mathbf{Z}_{k,R}\mathbf{a}\mathbf{p}^*(k) \sim \mathcal{CN}(A_k e^{j\theta_k}, \sigma^2). \tag{2.17}$$

As a consequence, the NP detector can be recast as

$$\frac{f_{\mathbf{L}}(\mathbf{L}|H_1)}{f_{\mathbf{L}}(\mathbf{L}|H_0)} = \exp\left\{-\frac{\sum_{k=0}^{K-1} |\mathbf{L}(k) - A_k e^{j\theta_k}|^2 - \mathbf{L}^\dagger \mathbf{L}}{\sigma^2}\right\} \underset{H_0}{\overset{H_1}{\gtrless}} \gamma, \tag{2.18}$$

which can be shown equivalent to

$$\sum_{k=0}^{K-1} \mathrm{Re}\left\{A_k e^{-j\theta_k} \mathbf{L}(k)\right\} = \sum_{k=0}^{K-1} \mathrm{Re}\left\{A_k e^{-j\theta_k} \mathbf{p}^*(k) \mathbf{Z}_{k,R}\mathbf{a}\right\} \underset{H_0}{\overset{H_1}{\gtrless}} \gamma, \tag{2.19}$$

where γ is a suitable modification of the original detection threshold in (2.15). In particular, if the amplitudes hold $A_k = A$, for $k = 0, \ldots, K-1$, (2.19) becomes equivalent to

$$\sum_{k=0}^{K-1} \mathrm{Re}\left\{e^{-j\theta_k} \mathbf{L}(k)\right\} \underset{H_0}{\overset{H_1}{\gtrless}} \gamma. \tag{2.20}$$

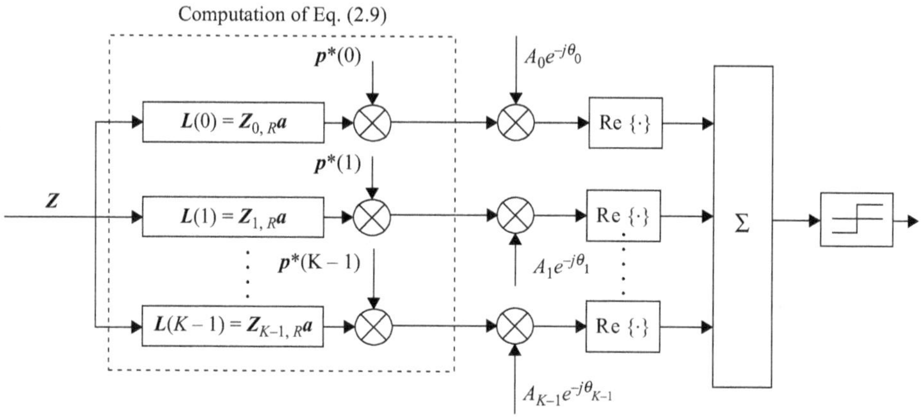

Figure 2.5 Block scheme of the NP receiver (2.19) for the non-coherent pulse train.

Figure 2.5 describes a block scheme of the NP detector (2.20) for the non-coherent train. First of all, fast-time processing is performed. Then the K outputs (corresponding to different pulses) are (1) compensated for the phase terms induced by the lack of coherence; (2) weighted according to the useful signal strength on that specific pulse. Finally, the real parts of the aforementioned quantities are summed and compared with the detection threshold.

Concluding Remark: For both the coherent and the non-coherent cases, the NP detector is not UMP and, hence, it is not practically implementable. In order to circumvent this drawback, we resort to a robust design framework based on the GLRT that is tantamount to substituting the maximum likelihood estimates of the unknown parameters in place of their exact values.

2.5 GLRT Design

This section is devoted to the design of GLRT detectors based on the previously obtained sufficient statistics for both the coherent and the non-coherent pulse train models. We start with the coherent case and write the GLRT as

$$\frac{\max_{A,\theta} f_L(L|H_1)}{f_L(L|H_0)} \underset{H_0}{\overset{H_1}{\gtrless}} \gamma. \tag{2.21}$$

Applying the logarithm to both the sides of (2.21), the GLRT reduces to

$$\max_{A,\theta} \ln f_L(L|H_1) - \ln f_L(L|H_0) \underset{H_0}{\overset{H_1}{\gtrless}} \ln \gamma. \tag{2.22}$$

The left hand side of the above equation can be computed as

$$\max_{A,\theta} \ln f_L(L|H_1) - \ln f_L(L|H_0) = \frac{|L|^2}{K\sigma^2} - \min_{A,\theta} \frac{(L - KAe^{j\theta})^\dagger (L - KAe^{j\theta})}{K\sigma^2} \leq |L|^2 \tag{2.23}$$

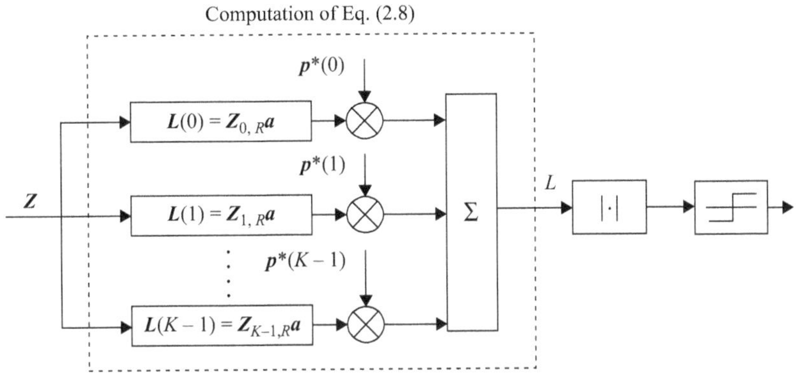

Figure 2.6 Block scheme of GLRT (2.24) for the coherent pulse train.

with equality if and only if $Ae^{j\theta} = L/K$. Hence, the GLRT for the coherent case can be recast as

$$\left| \boldsymbol{p}^{\dagger}(\boldsymbol{Za}) \right| \underset{H_0}{\overset{H_1}{\gtrless}} \gamma, \tag{2.24}$$

with the suitable modified threshold γ. A block scheme of detector (2.24) is shown in Figure 2.6; after the coherent sum performed to compute the sufficient statistic (2.8), the magnitude of the resulting complex number is compared with a detection threshold. Further insights in the required processing can be obtained writing the decision statistic as

$$\left| \sum_{k=0}^{K-1} (\boldsymbol{Z}_{k,R}\boldsymbol{a}) \, e^{-j2\pi k f_d T} \right|,$$

namely as the discrete time Fourier transform of the fast-time samples after pulse compression. This expression explicitly highlights that with the same range processing (namely exploiting the same pulse compression outputs), it is possible to test all the Doppler cells by just changing the value of f_d.

An important remark is now in order: detector (2.24) coincides with the NP test obtained modeling the target phase as a uniformly distributed random variable within the interval $[0, 2\pi[$ [5, Chap. 15, pp. 568].

As to the non-coherent case, we have to distinguish between two different situations:

(1) The amplitudes of the returns are equal, e.g., $A_k = A$, $k = 0, \ldots, K-1$;
(2) The amplitudes A_k represent K unknown parameters without any known a priori relationship.

Case 1, Non-coherent pulse train with one and the same amplitude: The GLRT is

$$\max_{A,\theta_0,\ldots,\theta_{K-1}} \frac{f_{\boldsymbol{L}}(\boldsymbol{L}|H_1)}{f_{\boldsymbol{L}}(\boldsymbol{L}|H_0)} \underset{H_0}{\overset{H_1}{\gtrless}} \gamma, \tag{2.25}$$

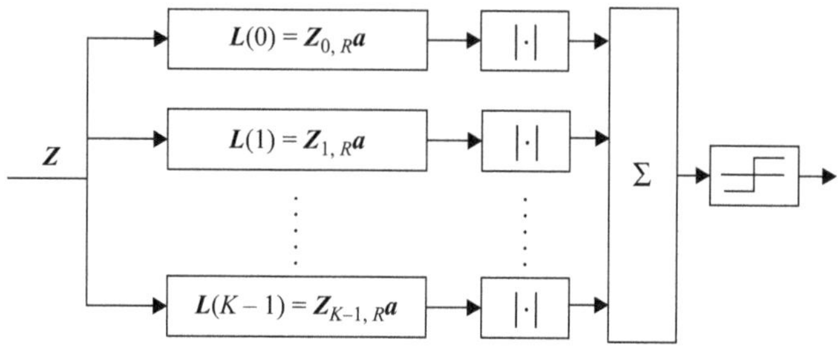

Figure 2.7 Block scheme of the linear integrator (2.29) for the non-coherent pulse train.

or equivalently,

$$-\min_A \sum_{k=0}^{K-1} \left[\min_{\theta_k} \left| L(k) - Ae^{j\theta_k} \right|^2 - |L(k)|^2 \right] \underset{H_0}{\overset{H_1}{\gtrless}} \gamma. \tag{2.26}$$

To perform the optimizations over θ_k, we observe that the reverse triangular inequality [6] implies

$$\left| L(k) - Ae^{j\theta_k} \right|^2 \geqslant ||L(k)| - A|^2, \quad k = 0, \ldots, K-1, \tag{2.27}$$

with equality if $\theta_k = \angle L(k)$. As a consequence, test (2.26) becomes equivalent to

$$-\min_A \sum_{k=0}^{K-1} \left[A^2 - 2A|L(k)| \right] \underset{H_0}{\overset{H_1}{\gtrless}} \gamma. \tag{2.28}$$

Hence, optimizing over A, the GLRT can be finally expressed as

$$\sum_{k=0}^{K-1} |L(k)| \underset{H_0}{\overset{H_1}{\gtrless}} \gamma \tag{2.29}$$

and is usually referred to as linear non-coherent integrator. A block scheme of the linear non-coherent detector is shown in Figure 2.7. Observe that the final multipliers involved in the computation of the sufficient statistic (2.9) are omitted because any phase information is lost due to the presence of the modulus operation, i.e., $|L(k)| = |Z_{k,R}a|$. This also justifies why practically implementable non-coherent processing is unable to provide any information on the actual target Doppler.

Case 2, Non-coherent pulse train with different amplitudes: The GLRT is given by

$$\max_{A_0,\ldots,A_{K-1},\theta_0,\ldots,\theta_{K-1}} \frac{f_L(L|H_1)}{f_L(L|H_0)} \underset{H_0}{\overset{H_1}{\gtrless}} \gamma, \tag{2.30}$$

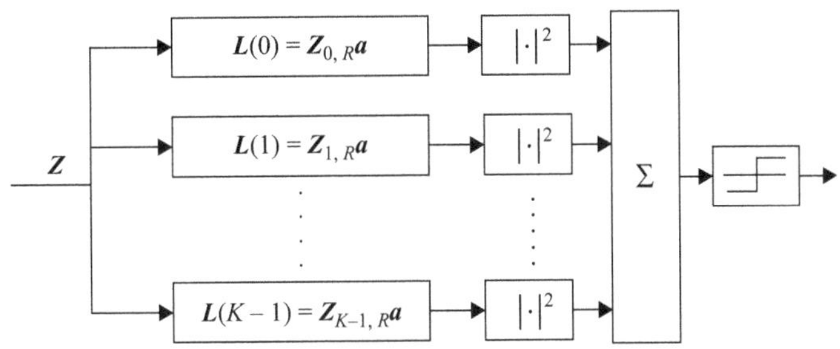

Figure 2.8 Block scheme of the square-law integrator (2.32) for the non-coherent pulse train.

or equivalently,

$$-\sum_{k=0}^{K-1}\left[\min_{A_k,\theta_k}\left|\boldsymbol{L}(k) - A_k e^{j\theta_k}\right|^2 - |\boldsymbol{L}(k)|^2\right] \overset{H_1}{\underset{H_0}{\gtrless}} \gamma. \tag{2.31}$$

Optimizations over A_k and θ_k yield $A_k = |\boldsymbol{L}(k)|$ and $\theta_k = \angle \boldsymbol{L}(k)$ for $k = 0, \ldots, K-1$. The GLRT is thus equivalent to

$$\sum_{k=0}^{K-1} |\boldsymbol{L}(k)|^2 \overset{H_1}{\underset{H_0}{\gtrless}} \gamma, \tag{2.32}$$

which is usually referred to as square-law non-coherent integrator. A block scheme of the linear law detector is shown in Figure 2.8.

Concluding Remarks: The linear detector (2.29) can be also obtained as the high Signal-to-Noise Ratio (SNR) approximation of the NP receiver [5] designed for $A_k = A, k = 0, \ldots, K-1$, and i.i.d. target phases modeled as uniformly distributed within $[0, 2\pi[$. Moreover, under the same assumptions on the amplitudes and phases, (2.32) can be interpreted as the approximation of the NP receiver in the low SNR regime [5]. Finally, detector (2.32) is the optimum NP test designed assuming i.i.d. Rayleigh distributed amplitudes A_k, $k = 0, \ldots, K-1$, and i.i.d. target phases (statistically independent from the amplitudes) uniformly distributed within $[0, 2\pi[$.

Before concluding, a discussion on the robustness of the considered GLRT approach with respect to the dual Bayesian framework is necessary. While a Bayesian receiver is based on the assignment of some priors for the unknown parameters, the GLRT does not require this knowledge. As a result, the former is tied up to the specific parameter fluctuation law whereas the latter is one and the same independent of the considered priors and their distributional parameters. With reference to the specific detection problem considered here, these two considerations are in order.

(1) For the coherent case, assuming a uniformly distributed phase within $[0, 2\pi[$, the resulting Bayesian detector is one and the same independent of the amplitude prior. Moreover, it also coincides with the GLRT.

(2) For the non-coherent case, assuming statistically independent and uniformly distributed phases, the Bayesian approach leads to detectors tied up to the fluctuation laws of the

return amplitudes (specifically on their joint pdf). As a consequence mismatches between the design and actual fluctuation laws may lead to performance degradations. On the contrary, the GLRT is a universal receiver and, as already pointed out, does not require any prior knowledge. Finally, in general, it is not statistically equivalent to the Bayesian detector (an exception is the aforementioned i.i.d. Rayleigh fluctuating amplitudes case).

2.6 Performance Analysis

This section is devoted to the performance analysis of the detectors presented in Sections 2.4 and 2.5. To this end, analytic expressions for the false alarm probability (P_{FA}) and detection probability (P_D) are provided except for the linear non-coherent integrator (2.29) whose performance is obtained via Monte Carlo simulations due to the lack of easily manageable analytic formulas. First of all, the coherent case is considered: the GLRT performance is compared with the optimum benchmark curve ensured by the clairvoyant structure (2.14), and the effect of the coherent integration gain is assessed. Then, the non-coherent situation is studied: a performance comparison among the linear, square-law, and optimum non-coherent integration is conducted. Finally, the effect of non-coherent integration is discussed.

2.6.1 Coherent Case

Let us focus on the NP detector for coherent pulse trains (2.14) and denote by $l_1 = \mathrm{Re}\left\{e^{-j\theta}\boldsymbol{p}^{\dagger}\boldsymbol{Za}\right\}$. Under the H_1 hypothesis, $e^{-j\theta}\boldsymbol{p}^{\dagger}\boldsymbol{Za}$ is a complex circular Gaussian random variable with average KA and variance $K\sigma^2$, i.e., $e^{-j\theta}\boldsymbol{p}^{\dagger}\boldsymbol{Za} \sim \mathcal{CN}(KA, K\sigma^2)$. Hence, l_1 is a real Gaussian random variable with mean $\mu_{l_1} = KA$ and variance $\sigma_{l_1}^2 = K\sigma^2/2$, namely

$$f_{l_1}(t|H_1) = \frac{1}{\sqrt{2\pi}\sigma_{l_1}} \exp\left(-\frac{(t-\mu_{l_1})^2}{2\sigma_{l_1}^2}\right). \tag{2.33}$$

This implies that P_D can be obtained as

$$P_D = \int_{\gamma}^{\infty} f_{l_1}(t|H_1)dx = Q\left(\frac{\gamma - KA}{\sqrt{K\sigma^2/2}}\right), \tag{2.34}$$

where $Q(\cdot)$ is the Q-function [7, p. 6]:

$$Q(x) = \frac{1}{\sqrt{2\pi}} \int_{x}^{\infty} \exp\left(-\frac{t^2}{2}\right) dt. \tag{2.35}$$

As a by-product, P_{FA} can be obtained setting $A = 0$ in (2.34), i.e.,

$$P_{FA} = Q\left(\frac{\gamma}{\sqrt{K\sigma^2/2}}\right). \tag{2.36}$$

Using the inverse Q-function, the Receiver Operating Characteristic (ROC) can be computed as

$$P_D = Q\left(Q^{-1}(P_{FA}) - \sqrt{2K\mathrm{SNR}}\right), \tag{2.37}$$

where SNR per pulse is given by

$$\mathrm{SNR} = \frac{A^2}{\sigma^2}. \tag{2.38}$$

Let us now focus on the GLRT in (2.24) and denote by $g_1 = |\boldsymbol{p}^\dagger \boldsymbol{Z}\boldsymbol{a}|$. As already highlighted, under the H_1 hypothesis, $\boldsymbol{p}^\dagger \boldsymbol{Z}\boldsymbol{a} \sim \mathcal{CN}(KAe^{j\theta}, K\sigma^2)$. Thus, g_1 is Rician distributed

$$f_{g_1}(z|H_1) = \frac{2z}{K\sigma^2} \exp\left(-\frac{z^2 + K^2A^2}{K\sigma^2}\right) I_0\left(\frac{2zA}{\sigma^2}\right) U(z), \tag{2.39}$$

where $U(z)$ is the unit-step function

$$U(z) = \begin{cases} 1, & z \geq 0, \\ 0, & z < 0, \end{cases} \tag{2.40}$$

and $I_v(x)$ is the modified Bessel function of the first kind and order v,

$$I_v(x) = \sum_{m=0}^{\infty} \frac{1}{m!\Gamma(m+v+1)} \left(\frac{x}{2}\right)^{2m+v}. \tag{2.41}$$

It follows that P_D can be obtained as

$$P_D = \int_{\gamma}^{\infty} f_{g_1}(z|H_1)dz = Q_1\left(\sqrt{\frac{2KA^2}{\sigma^2}}, \sqrt{\frac{2\gamma^2}{K\sigma^2}}\right), \tag{2.42}$$

where $Q_m(a, b)$ denotes the generalized Marcum function of order m, which can be expressed in terms of the following integral [8, Eq. (21)]:

$$Q_m(a, b) = \int_b^{\infty} \frac{x^m}{a^{m-1}} \exp\left\{-\frac{x^2 + a^2}{2}\right\} I_{m-1}(ax)dx. \tag{2.43}$$

Under the H_0 hypothesis, $\boldsymbol{p}^\dagger \boldsymbol{Z}\boldsymbol{a} \sim \mathcal{CN}(0, K\sigma^2)$. Hence, g_1 is Rayleigh distributed with parameter $\sigma_{g_1}^2 = K\sigma^2/2$, i.e., its pdf is

$$f_{g_1}(z|H_0) = \frac{2z}{K\sigma^2} \exp\left(-\frac{z^2}{K\sigma^2}\right) U(z). \tag{2.44}$$

Thus, P_{FA} can be computed as

$$P_{FA} = \exp\left(-\frac{\gamma^2}{K\sigma^2}\right). \tag{2.45}$$

Substituting (2.45) into (2.42), the ROC is expressed as

$$P_D = Q_1\left(\sqrt{2K\mathrm{SNR}}, \sqrt{-2\ln P_{FA}}\right), \tag{2.46}$$

where SNR (per pulse) is given in (2.38).

Remark: Equations (2.37) and (2.46) clearly highlight the effect of coherent integration. Precisely, the coherent equivalent SNR ($\mathrm{SNR}_{\mathrm{ceq}}$) after coherent integration can be obtained

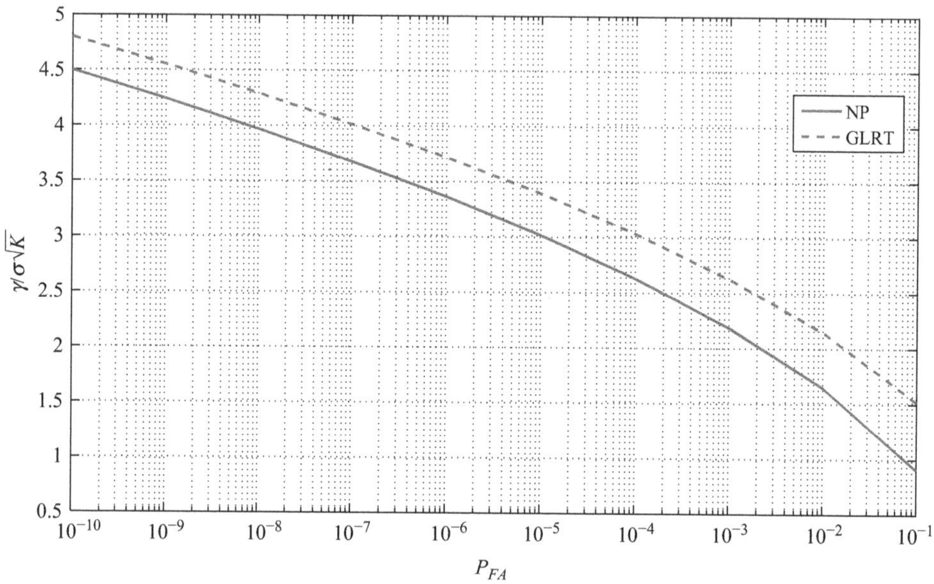

Figure 2.9 Normalized detection threshold $\frac{\gamma}{\sigma\sqrt{K}}$ versus P_{FA} for the NP receiver (2.14) and the GLRT (2.24).

by multiplying the single pulse SNR and the number of transmitted pulses, namely $\mathrm{SNR_{ceq}} = K\mathrm{SNR}$.

In Figure 2.9, the normalized detection thresholds $\frac{\gamma}{\sigma\sqrt{K}}$ are plotted versus P_{FA} using (2.36) for the NP receiver (2.14) and (2.45) for the GLRT (2.24). As expected, being P_{FA} versus the normalized threshold a complementary cumulative distribution function, the larger P_{FA}, the lower the detection threshold. Moreover, the NP receiver (2.14) requires a lower normalized threshold than the GLRT (2.24). Precisely for $P_{FA} \in [10^{-8}, 10^{-4}]$ the threshold percentage change, defined as

$$1 - \frac{Q^{-1}(P_{FA})}{\sqrt{2\log\frac{1}{P_{FA}}}},$$

ranges within the interval [0.08, 0.13]. This induces a penalization in the GLRT detection performance even if the GLRT threshold growth is also due to the structure of the decision statistics in (2.14) and (2.24). In fact, the former is dominated by the latter, namely $\mathrm{Re}\left\{e^{-j\theta}\boldsymbol{p}^{\dagger}\boldsymbol{Za}\right\} \le \left|\boldsymbol{p}^{\dagger}\boldsymbol{Za}\right|$.

In Figure 2.10(a) and (b), the ROC of the NP receiver and GLRT is plotted for SNR $= 3$ dB and $K \in \{1, 4, 16\}$ using two different scales: a linear scale, in Figure 2.10(a), and a logarithmic scale, in Figure 2.10(b), necessary to emphasize the P_D behavior for low P_{FA} values. The curves illustrate that the larger P_{FA} and K, the higher P_D. Specifically, with reference to the NP receiver, P_Ds are approximately 0.04, 0.6, and 1 at $P_{FA} = 10^{-4}$ for $K = 1, 4$, and 16, respectively. As to the GLRT, they are approximately 0.01, 0.4, and 1, respectively.

In Figure 2.11, P_D of the GLRT and the clairvoyant structure is plotted versus KSNR for $P_{FA} = 10^{-6}$. The curves illustrate that the GLRT requires about 0.6 dB higher KSNR than

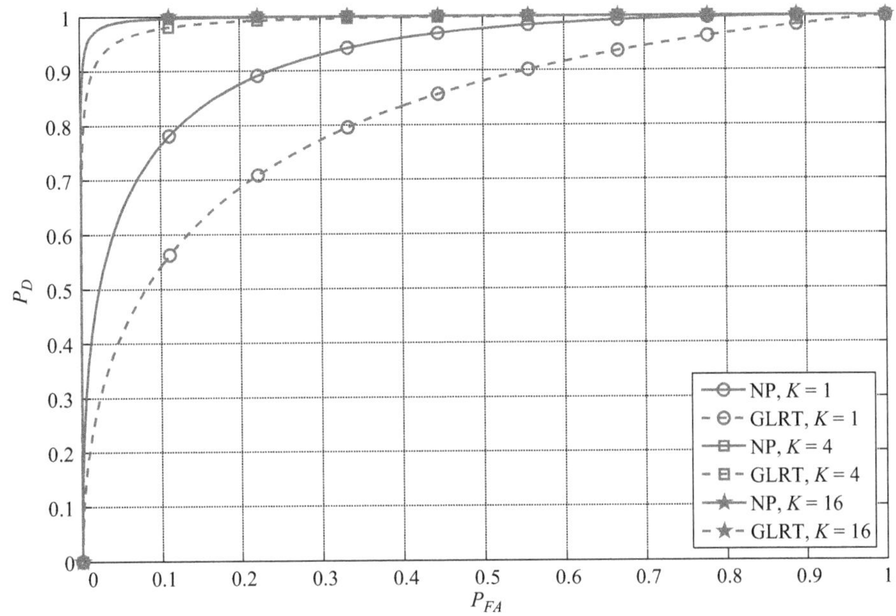

(a)

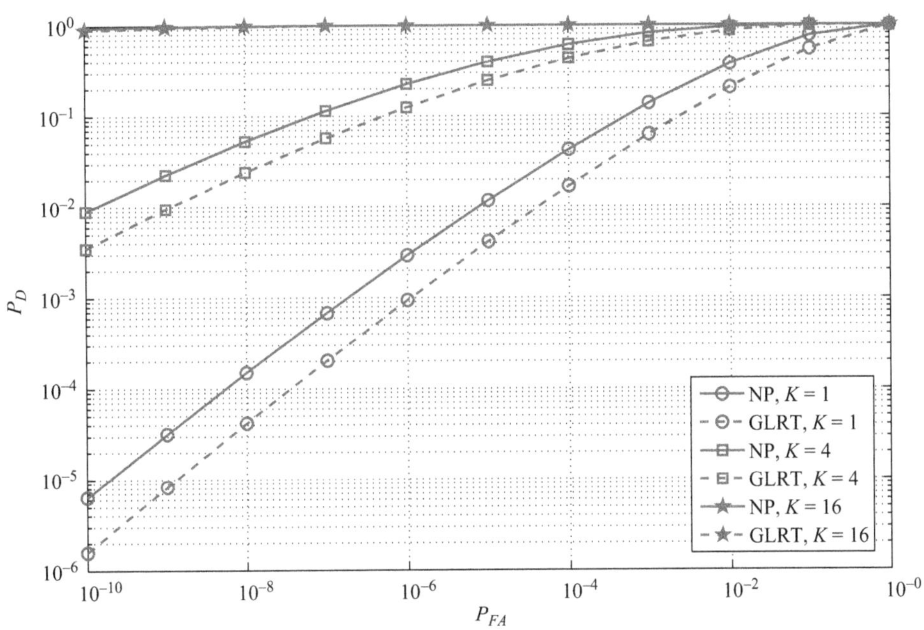

(b)

Figure 2.10 ROC of the NP detector (2.14), obtained using (2.37), and of the GLRT, obtained using (2.46), for SNR $= 3$ dB and $K \in \{1, 4, 16\}$. (a) Linear scale and (b) logarithmic scale.

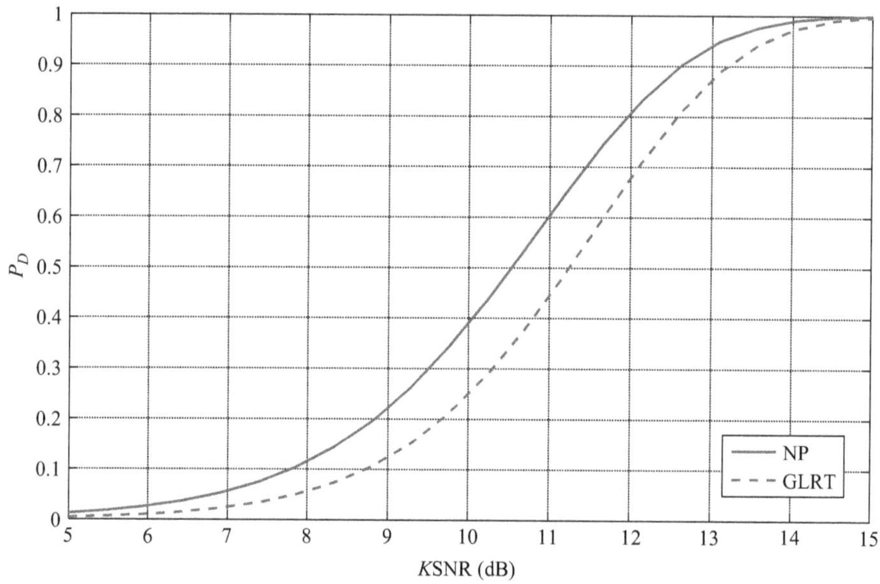

Figure 2.11 P_D versus KSNR for the NP detector (2.14) and the GLRT (2.24), for $P_{FA} = 10^{-6}$.

the NP detector at $P_D = 0.9$ and about 0.7 dB at $P_D = 0.5$. It is finally worth mentioning the universality of the representation in Figure 2.11, in the sense that it completely characterizes the performance of the considered coherent detectors for any values of K and single pulse SNR.

2.6.2 Non-coherent Case

Let us focus on the clairvoyant structure (2.19) and denote by

$$l_2 = \sum_{k=0}^{K-1} \text{Re}\left\{A_k e^{-j\theta_k} \boldsymbol{p}^*(k) \boldsymbol{Z}_{k,R} \boldsymbol{a}\right\},$$

the decision statistic for non-coherent pulse trains. Under the H_1 hypothesis, $\sum_{k=0}^{K-1} \times A_k e^{-j\theta_k} \boldsymbol{p}^*(k) \boldsymbol{Z}_{k,R} \boldsymbol{a}$ is a complex circular Gaussian variable with average $\sum_{k=0}^{K-1} A_k^2$ and variance $\sum_{k=0}^{K-1} A_k^2 \sigma^2$, i.e.,

$$\sum_{k=0}^{K-1} A_k e^{-j\theta_k} \boldsymbol{p}^*(k) \boldsymbol{Z}_{k,R} \boldsymbol{a} \sim \mathcal{CN}\left(\sum_{k=0}^{K-1} A_k^2, \sum_{k=0}^{K-1} A_k^2 \sigma^2\right).$$

Thus, l_2 follows a real Gaussian distribution with average $\mu_{l_2} = \sum_{k=0}^{K-1} A_k^2$ and variance $\sigma_{l_2}^2 = \sum_{k=0}^{K-1} A_k^2 \sigma^2 / 2$. This implies that P_D can be obtained as

$$P_D = Q\left(\frac{\gamma - \sum_{k=0}^{K-1} A_k^2}{\sqrt{\sum_{k=0}^{K-1} A_k^2 \sigma^2 / 2}}\right). \tag{2.47}$$

Under the H_0 hypothesis, $A_k e^{-j\theta_k} \boldsymbol{p}^*(k) \boldsymbol{Z}_{k,R} \boldsymbol{a}$ is a complex circular zero-mean Gaussian variable with variance $A_k^2\sigma^2$, i.e., $A_k e^{-j\theta_k} \boldsymbol{p}^*(k) \boldsymbol{Z}_{k,R} \boldsymbol{a} \sim \mathcal{CN}(0, A_k^2\sigma^2)$. Thus, l_2 follows a real zero-mean Gaussian distribution with variance $\sigma_{l_2}^2 = \sum_{k=0}^{K-1} A_k^2\sigma^2/2$. Hence, P_{FA} can be computed as

$$P_{FA} = Q\left(\frac{\gamma}{\sqrt{\sum_{k=0}^{K-1} A_k^2\sigma^2/2}}\right). \tag{2.48}$$

Note that if $A_k = A$, $k = 0, \ldots, K-1$, the NP detector for the non-coherent case reduces to (2.20), whose P_D and P_{FA} are, respectively, the same as those obtained in the coherent case (2.34) and (2.36).

Using the inverse Q-function and P_{FA} in (2.48), the ROC of (2.47) can be recast as

$$P_D = Q\left(Q^{-1}(P_{FA}) - \sqrt{2K\text{SNR}_{\text{eq}}}\right), \tag{2.49}$$

where SNR_{eq} denotes the equivalent SNR per pulse, defined as

$$\text{SNR}_{\text{eq}} = \frac{1}{K}\sum_{k=0}^{K-1} \frac{A_k^2}{\sigma^2}. \tag{2.50}$$

A careful comparison between (2.37) and (2.49) reveals that they share the same functional expression with the only difference due to the presence of SNR_{eq} (2.50) in place of the single pulse SNR (2.38). In other words, the optimum non-coherent processing still provides a coherent integration gain.

As to the GLRT of case (1) (Eq. (2.29)), there are no known easily manageable analytical expressions P_{FA} and P_D. Hence, in the following, it is only considered the analytic performance evaluation of the square-law detector (2.32), whereas Monte Carlo simulations are used to analyze the behavior of (2.29).[1] To this end, denote by

$$g_2 = \sum_{k=0}^{K-1} |\boldsymbol{L}(k)|^2 \tag{2.51}$$

and define $r = g_2/\sigma^2$ as the normalized decision statistic. Based on the previous assumptions, it is not difficult to show that under H_1, r is a complex non-central chi-square random variable [9] with K Degrees of Freedom (DOFs) and non-centrality parameter $\chi_K = \sum_{k=0}^{K-1} A_k^2/\sigma^2$, i.e.,

$$f_r(r|H_1) = \left(\frac{r}{\chi_K}\right)^{(K-1)/2} e^{-r-\chi_K} I_{K-1}\left(2\sqrt{r\chi_K}\right) U(r). \tag{2.52}$$

[1] For the linear detector, it is set $A_k = A$ for $k = 0, \ldots, K-1$.

As a consequence, P_D can be obtained as [10]

$$\begin{aligned} P_D &= P\{g_2 \geq \gamma | H_1\} = P\{g_2/\sigma^2 \geq \gamma/\sigma^2 | H_1\} \\ &= Q_K\left(\sqrt{2\sum_{k=0}^{K-1} A_k^2/\sigma^2}, \sqrt{2\gamma/\sigma^2}\right) \\ &= Q_K\left(\sqrt{2K\mathrm{SNR}_{\mathrm{eq}}}, \sqrt{2\gamma/\sigma^2}\right). \end{aligned} \tag{2.53}$$

This last expression clearly shows the lack of a coherent integration gain because the functional relationship between P_D and the two arguments in the generalized Marcum function depends on the number of pulses K.

Under the H_0 hypothesis, r is distributed as a complex central chi-square random variable [9] with K DOFs, namely its pdf is given by

$$f_r(r|H_0) = \frac{r^{K-1}}{(K-1)!} e^{-r} U(r). \tag{2.54}$$

Hence, P_{FA} can be computed as

$$\begin{aligned} P_{FA} &= P\{g_2 \geq \gamma | H_0\} = P\{g_2/\sigma^2 \geq \gamma/\sigma^2 | H_0\} \\ &= \frac{1}{\Gamma(K)} \gamma_{\mathrm{inc}}(K, \gamma/\sigma^2), \end{aligned} \tag{2.55}$$

where $\gamma_{\mathrm{inc}}(n, x)$ denotes the incomplete Gamma function [11, 8.350-2, p. 899], i.e.,

$$\gamma_{\mathrm{inc}}(n, x) = \int_x^{\infty} t^{n-1} \exp(-t) dt. \tag{2.56}$$

Moreover, (2.55) can be rewritten as a finite summation, i.e.,

$$P_{FA} = \exp\left(-\frac{\gamma}{\sigma^2}\right) \sum_{k=0}^{K-1} \frac{1}{k!} \left(\frac{\gamma}{\sigma^2}\right)^k. \tag{2.57}$$

Notice that for $K = 1$, (2.57) reduces to (2.45), whereas (2.53) becomes (2.42).

In Figure 2.12, the ROC of the NP detector (using (2.49)), of the linear law detector (exploiting Monte Carlo simulations), and of the square-law detector (using (2.53)) is plotted. Similar behaviors as in Figure 2.10 can be observed. However, the performance gaps between the NP receiver and GLRTs become larger and larger as K increases. This can be explained observing that the clairvoyant structure coherently integrates the available samples whereas the GLRTs, neglecting the phase information on each pulse, only perform a non-coherent integration. Hence, the growth in the performance loss of the GLRTs can be justified observing that the amount of information lost by the non-coherent integration increases as K increases.

In Figure 2.13, P_Ds of the clairvoyant NP receiver, the linear law and the square-law non-coherent detectors are plotted versus $\mathrm{SNR}_{\mathrm{eq}}$ for $P_{FA} = 10^{-6}$ and some values of K. The curves highlight that the larger K, the higher the detection probability. Moreover, the loss of the GLRTs with respect to the NP detector becomes heavier and heavier as K increases, even if the performance gaps between the square-law and the linear detectors are almost insignificant. Otherwise stated, defining the integration gain as the ratio between the single pulse SNR,

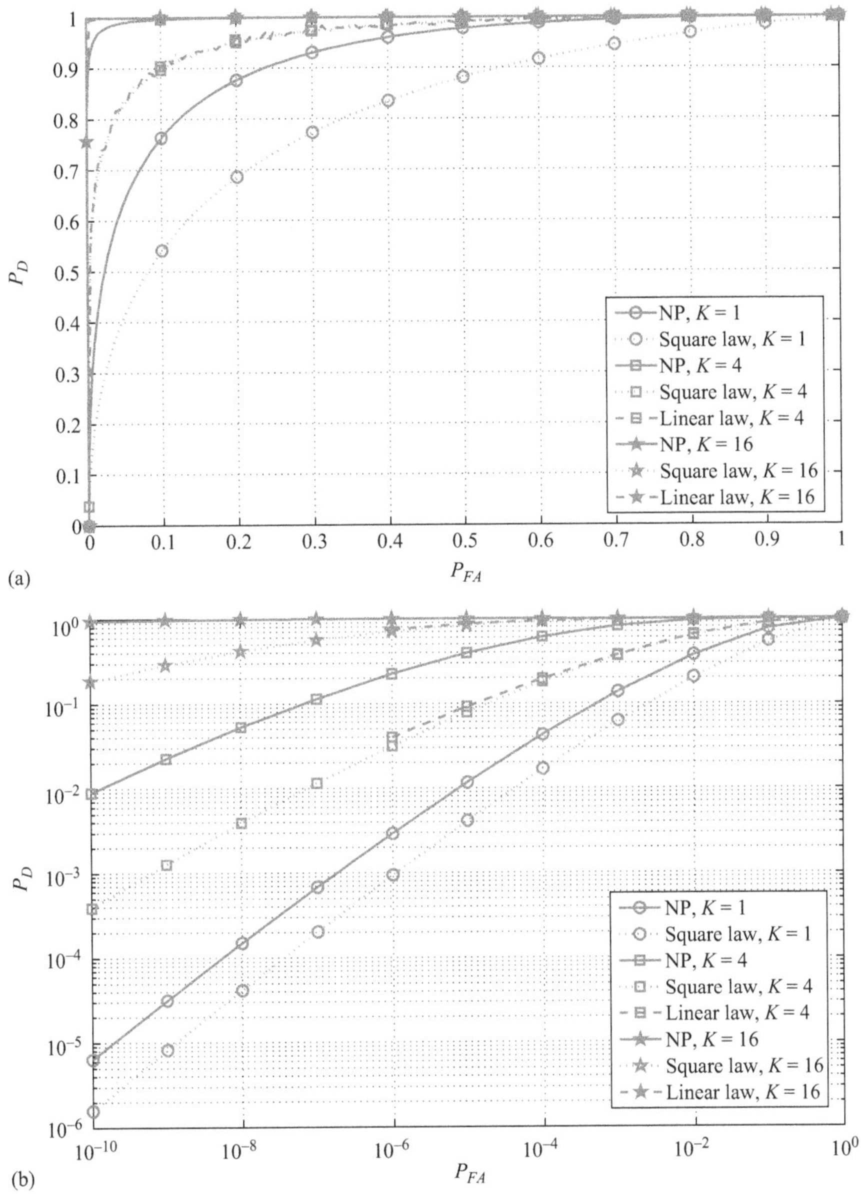

(a)

(b)

Figure 2.12 ROC of the NP detector, obtained using (2.49), of the linear law detector, obtained using Monte Carlo simulations, and of the square-law detector, obtained using (2.53), for $\text{SNR}_{\text{eq}} = 3$ dB and $K \in \{1, 4, 16\}$. (a) Linear scale and (b) logarithmic scale.

necessary to achieve a specified detection performance for a given P_{FA} level, and the pre-integration SNR required such that, after integration of K pulses, the specified detection and false alarm performances are met, it can be observed that the gain experienced by the GLRTs in the non-coherent case is smaller than K which is the value corresponding to coherent detection (namely the value achieved by the clairvoyant receiver).

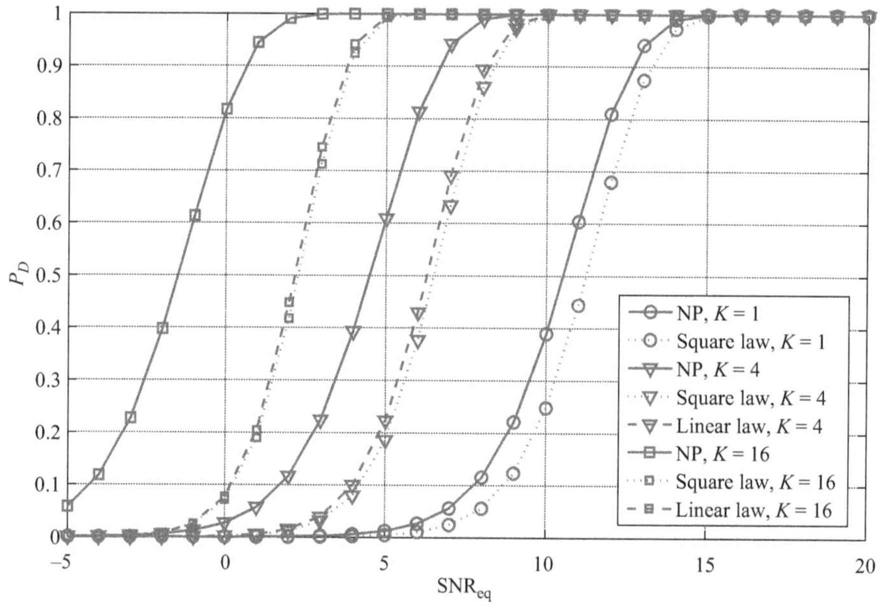

Figure 2.13 P_D versus $\mathrm{SNR}_{\mathrm{eq}}$ for the NP detector (2.19), the linear law detector (2.29), and the square-law detector (2.32), for $P_{FA} = 10^{-6}$, and $K \in \{1, 4, 16\}$.

2.7 Conclusions and Further Reading

This chapter has considered classic radar detection in the presence of white Gaussian noise. The proposed approach first performs a compression of the observation space through reduction by sufficiency. Then detector design is attacked in the reduced dimensional space. The optimum NP detector is unfortunately not implementable for both the coherent and the non-coherent pulse trains. Thus, the robust GLRT-based approach is exploited to design practically implementable receivers that are canonical, namely independent of any prior assigned to the unknown parameters. This construction, depending on the coherency of the train, leads to either the classic coherent radar detector or linear and square-law integrators. Hence, a performance analysis of the obtained detection structures is conducted through analytic equations, where possible, or exploiting numerical simulations.

Before concluding, some additional references to the considered problem are provided. Meyer and Mayer [12] is a classic book on radar detection: therein many detection curves for different radar scenarios can be found. In Reference 13, a concise but effective treatment of the problem is presented discussing also the constant false alarm probability property. Target fluctuation and interesting practical issues associated with the search/detection function are available in References 14 and 15. In Chapter 15 of Reference 5, the dual Bayesian approach is presented modeling the phase of target returns as statistically independent uniformly distributed random variables. The performance analysis of classic radar receivers in the presence of general target fluctuation models is dealt with in References 16–18, and, for possibly correlated target echoes in Reference 8. Finally, a detailed analysis when the interference is no longer white and follows a compound-Gaussian model is presented in Reference 19.

References

[1] H. L. Van Trees, *Detection, Estimation, and Modulation Theory (Part I)*. New York, USA: John Wiley & Sons, Inc., 2001.

[2] S. M. Kay, *Fundamentals of Statistical Signal Processing, Vol. II: Detection Theory*. Upper Saddle River, NJ: Prentice-Hall, 1998.

[3] J. Ward, "Space-time adaptive processing for airborne radar," Lincoln Laboratory, MIT, Technical Report 1015, December 1994.

[4] S. M. Kay, *Fundamentals of Statistical Signal Processing: Estimation Theory*. Upper Saddle River, NJ: Prentice-Hall, 1993.

[5] M. A. Richards, J. A. Scheer, and W. A. Holm, *Principles of Modern Radar: Basic Principles*. Raleigh, NC: Scitech Publishing, Inc., 2010.

[6] M. Beck, G. Marchesi, D. Pixton, and L. Sabalka, *A First Course in Complex Analysis*. Lecture Notes, Preprint, 2002.

[7] M. K. Simon, *Probability Distributions Involving Gaussian Random Variables: A Handbook for Engineers and Scientists*. Boston: Kluwer Academic Publishers, 2002.

[8] G. Cui, A. De Maio, and M. Piezzo, "Performance prediction of the non-coherent radar detector for generalized Swerling-Chi fluctuating targets," *IEEE Transactions on Aerospace and Electronic Systems*, vol. 49, no. 1, pp. 356–368, January 2013.

[9] A. Papoulis and S. U. Pillai, *Probability, Random Variables, and Stochastic Processes*, 4th ed. New York: McGraw-Hill, 2002.

[10] D. A. Shnidman, "The calculation of the probability of detection and the generalized Marcum Q-function," *IEEE Transactions on Information Theory*, vol. 35, no. 2, pp. 389–400, March 1989.

[11] I. S. Gradshteyn and I. M. Ryzhik, *Table of Integrals, Series, and Products*, 7th ed. San Diego: Academic Press, 2007.

[12] D. P. Meyer and H. A. Mayer, *Radar Target Detection: Handbook of Theory and Practice*. Elsevier, New York: Academic Press, 1973.

[13] N. Levanon, *Radar Principles*, 1st ed. New York, USA: Wiley-Interscience, 1988.

[14] J. V. DiFranco and W. L. Rubin, *Radar Detection*. Raleigh, NC: Scitech Publishing, Inc., 2004.

[15] F. E. Nathanson, J. P. Reilly, and M. N. Cohen, *Radar Design Principles: Signal Processing and the Environment*, 2nd ed. Raleigh, NC: SciTech Publishing, 2006.

[16] A. De Maio, A. Farina, and G. Foglia, "Target fluctuation models and their application to radar performance prediction," *IEE Proceedings Radar Sonar and Navigation*, vol. 151, no. 5, pp. 261–269, October 2004.

[17] D. A. Shnidman, "Calculation of probability of detection for log-normal target fluctuations," *IEEE Transactions on Aerospace and Electronic Systems*, vol. 27, no. 1, pp. 172–174, January 1991.

[18] D. A. Shnidman, "Expanded Swerling target models," *IEEE Transactions on Aerospace and Electronic Systems*, vol. 39, no. 3, pp. 1059–1069, July 2003.

[19] E. Conte and G. Ricci, "Performance prediction in compound-Gaussian clutter," *IEEE Transactions on Aerospace and Electronic Systems*, vol. 30, no. 2, pp. 611–616, April 1994.

CHAPTER 3

Subspace Detection for Adaptive Radar: Detectors and Performance Analysis

Ram S. Raghavan[1], *Shawn Kraut*[2], *and Christ D. Richmond*[3]

3.1 Introduction

Coherent processing of various forms of multidimensional signals is commonplace in radar applications. Space–time adaptive processing in radars [1–3] is a well-established example of coherent processing involving the domains of space (multiple receiving antenna elements separated spatially) and time (multiple pulse returns at each antenna element). Examples also include space, time, and transmit/receive polarization domains. More examples of multidimensional processing are being explored for coherently processing signals received from coherent transmission of multiple waveforms in the context of multiple-input multiple-output radars [4, 5]. For multidimensional signals, the usual definition of the Signal of Interest (SOI) as a rank 1 signal needs to be broadened to include signals that belong to a known (or hypothesized) subspace. The subspace signal model includes the usual rank 1 model as a special case and allows for the modeling of received signal in multiple domains. For example, a single transmit signal at high frequencies can be received as sky waves with different polarizations and multiple modes after refraction and reflection from different layers of the Earth's ionosphere as described in References 6–8. The different modes of the received signals span a subspace and can be jointly detected by a subspace detector if the signal subspace is known. The subspace

[1]Air Force Research Laboratory, Wright-Patterson Air Force Base, OH, USA
[2]NorthWest Research Associates, Monterey, CA, USA
[3]MIT Lincoln Laboratory, Lexington, MA, USA
This work was supported by the Air Force Office of Scientific Research under contract 14RY05COR and by the Department of the Air Force under contract FA8721-05-C-0002. Opinions, interpretations, conclusions, and recommendations are those of the authors and are not necessarily endorsed by the United States Government.

signal model in addition provides a means for characterizing uncertainties in representing the received signal in some cases. Uncertainties in modeling the received signal from a hypothesized direction can result from position errors of sensor phase-centers and near-field multipath effects, for example. From the perspective of signal detection in clutter, interference, and noise, the ability to process multidimensional signals coherently needs to take into account the large numbers of nuisance parameters' typically inherent in these problems. These are parameters unknown to the receiver and must be estimated typically for purposes of interference cancelation prior to signal detection. The problem of detecting a subspace signal in a given test data vector can be formulated as a statistical hypothesis testing problem. An approach that has proven effective in dealing with nuisance parameters are invariant hypothesis tests [9, 10]. The general approach is to identify a set of matrices such that the linear transformation of the data by any member of the set leaves the original hypothesis testing problem unchanged, although the original nuisance parameters themselves are changed as a result. These transformations are generally easily determined given the signal model. Any hypothesis test in this class is referred to as an invariant test, and the detection statistic of all invariant tests can be constructed from a maximal invariant statistic. The Generalized Likelihood Ratio Test (GLRT) [11, 12], the Adaptive Matched Filter (AMF) [13], and Adaptive Coherence Estimator (ACE) (see Eq. (22) in Reference 14) [15–18], and references within, are examples of three well-known invariant tests commonly used for the detection of rank 1 signals in clutter, interference, and noise. Also of interest are a number of detectors devised for a variety of constraints [19–26] and references within. If N is the size of the signal vector, maximal invariants for the detection of a subspace signal (of dimension $1 \leq M < N$ in unknown clutter, interference, and noise was shown to be a two-dimensional statistic in Reference 27, where the joint Probability Density Function (PDF) of the two-dimensional statistic was obtained conditioned on both the null hypothesis and the alternative hypothesis. The detection performance of the subspace GLRT was analyzed in Reference 27 for the case of no signal mismatch errors. In this chapter, we extend the three signal detectors above to a subspace signal model. Analytical expressions derived include results of signal mismatch errors. The analysis is applied to an example to illustrate the use of subspace detectors to mitigate detection loss resulting from signal mismatch errors.

The rest of the chapter is organized in the following manner: Section 3.2 provides a brief introduction to the problem of signal detection in zero-mean complex Gaussian noise when the covariance matrix of the noise is known. We introduce the Neyman–Pearson detector that implements the likelihood ratio test and allows the receiver to select a preset probability of false alarm. Two cases are considered: (i) detection of a known signal vector in zero-mean multivariate complex Gaussian noise and (ii) detection of a known signal with unknown overall phase in zero-mean complex Gaussian noise. The hypothesis tests in each case reduce to comparing a statistic referred to as a sufficient statistic to a preset threshold. Expressions for the probability of detection and probability of false alarm for these cases are summarized. The problem of detecting a subspace signal in zero-mean complex Gaussian interference-plus-noise, when the covariance matrix of the interference-plus-noise is unknown, is introduced in Section 3.3. The section describes a subspace signal model and transformations to the data which leave the original hypothesis testing problem unchanged. A two-dimensional statistic referred to as a maximal invariant statistic for the subspace signal detection problem is defined. Hypothesis tests for the subspace signal detection problem reduce to comparing a statistic derived from a maximal invariant statistic to a preset threshold. The key is that the invariant tests enable the test threshold to be set based on a preset probability of false alarm even though the covariance matrix of the clutter-plus-interference-plus-noise is unknown – this feature is referred to as the Constant False Alarm Rate (CFAR) property. Maximal invariant statistic is thus the minimum information (as a statistic) required to implement any invariant hypothesis test. For the subspace signal model, a maximal invariant statistic is shown to be a two-dimensional (i.e. a vector) statistic. A maximal invariant statistic is not a unique vector statistic however, since any reversible transform applied to a maximal invariant statistic is also a maximal invariant statistic. In the case when the signal model is not mismatched with the actual signal model, the two scalar quantities that comprise a maximal invariant statistic introduced in Section 3.3 are: an estimate of signal-to-interference-plus-noise ratio (SINR) and an estimate of the loss in SINR due to finite sample effects. Thus, it is possible to derive an infinite number of invariant hypothesis tests from these two scalar statistics (see Reference 28 for example). We only consider three specific tests from the class of the invariant hypothesis tests in this chapter: (i) the subspace GLRT, (ii) the subspace AMF test, and (iii) the subspace ACE test. The three hypothesis tests and analytical expressions for the probability of detection P_D and probability of false alarm P_{FA} of the tests are given in Section 3.4. Performance results along with an application to illustrate the mitigation of detection performance loss due to signal mismatch errors are given in Section 3.5. Summary and conclusions are provided in Section 3.6. Materials of a supplementary nature are given in the Appendices. The subspace versions of the GLRT, AMF test, and ACE test are derived using the standard approach for the rank 1 signal case in Appendix 3.A for completeness. The proofs for the two-dimensional maximal invariant statistic for the subspace detectors are given in Appendix 3.B, and the joint PDF of maximal invariant statistic conditioned on the null hypothesis and alternative hypothesis including the case of mismatch between the hypothesized signal model and the actual signal model is derived in Appendix 3.C. A number of helpful results and distributions relevant to this work are summarized in Appendix 3.D.

3.2 Introduction to Signal Detection in Interference and Noise

Chapter 2 of this book provided an introduction to radar detection in white Gaussian noise and the GLRT framework. This section provides a very brief review of the detection problem specifically for signals modeled as vectors. Later sections of this chapter extend these ideas to the subspace-based adaptive detection problem. The reader is referred to the following for a thorough introduction to the general problem of signal detection in noise [29–35] for a variety of applications that include statistical communications, radar, and sonar. Classical references from mathematical statistics and multivariate statistical analysis include [36–40]. We begin with the basic problem of signal detection in noise when all parameters are known. The section provides an introduction that leads into the main topic of this chapter, where the problem of detecting a signal contained in a known subspace in complex (circularly symmetric), zero-mean colored Gaussian noise with unknown covariance matrix is considered.

Given a data vector $\boldsymbol{y} \in \mathcal{C}^{N\times 1}$, the basic problem is to determine if (i) the vector comprises only clutter-plus-interference-plus-noise – referred to as the null hypothesis H_0 or (ii) the data vector contains a signal $\boldsymbol{p} \in \mathcal{C}^{N\times 1}$ in addition to the clutter, interference, and noise – referred to as the alternative hypothesis H_1. The detection problem is formulated as a binary hypothesis testing problem, and from the perspective of applications to radar, it is the Neyman–Pearson criterion that is typically used since the approach provides control over the probability of false alarm, which is the conditional probability that the hypothesis test declares the signal to be present in the data vector when in fact no signal is present (i.e. probability of detection conditioned on hypothesis H_0). For applications that typically detect signals of interest continuously over a large surveillance volumes comprising multiple ranges, elevation angles, azimuth angles, and multiple Doppler frequencies, controlling the probability of false alarm is important since it translates to controlling the false alarm rate for a given data throughput. For systems with high input data rates, control over the false alarm rate ensures that the processing is not overwhelmed by detections which may in fact be false alarms.

Let $f(y|H_0)$ and $f(y|H_1)$ denote the PDF of a test vector $\boldsymbol{y}$ conditioned on the null hypothesis H_0 and the alternative hypothesis H_1 respectively. For a given vector $\boldsymbol{y} \in \mathcal{C}^{N\times 1}$, the quantity $f(\boldsymbol{y}|H_k)$; $k = 0, 1$ when viewed as a function of H_k is referred to as a likelihood function. In the Neyman–Pearson detector, the presence of an additive signal $\boldsymbol{p} \in \mathcal{C}^{N\times 1}$ in a given data vector $\boldsymbol{y} \in \mathcal{C}^{N\times 1}$ is formulated as a likelihood ratio test expressed as follows:

$$\mathcal{L}(H|\boldsymbol{y}) = \frac{f(\boldsymbol{y}|H_1)}{f(\boldsymbol{y}|H_0)} \underset{H_0}{\overset{H_1}{\gtrless}} \eta. \tag{3.1}$$

Given a preset threshold η the above detector selects the alternative hypothesis (i.e. sets $H = H_1$) when the likelihood ratio exceeds the threshold η and selects the null hypothesis (i.e. sets $H = H_0$) otherwise. The above likelihood ratio test generally partitions the vector space $\boldsymbol{y} \in \mathcal{C}^{N\times 1}$ into two disjoint regions $\mathcal{D}_0$ and $\mathcal{D}_1$, such that $\mathcal{D}_0 \cup \mathcal{D}_1 = \mathcal{C}^{N\times 1}$ and $\mathcal{D}_0 \cap \mathcal{D}_1 = \phi$ – in most cases of interest the condition where the likelihood ratio exactly equals the threshold has zero probability. The probability of detection P_D, which is the conditional probability that hypothesis H_1 is selected when the data vector has the additive signal and the probability of false alarm P_{FA}, which is the conditional probability that hypothesis H_1 is selected when the data vector comprises only clutter-plus-interference-plus-noise and no additive signal are given by the following:

$$\begin{aligned} P_D &= \int_{\boldsymbol{y}\in\mathcal{D}_1} f(\boldsymbol{y}|H_1)d\boldsymbol{y}, \\ P_{FA} &= \int_{\boldsymbol{y}\in\mathcal{D}_1} f(\boldsymbol{y}|H_0)d\boldsymbol{y}. \end{aligned} \tag{3.2}$$

For a specified probability of false alarm, the threshold level η required in (3.1) is determined from the second expression in (3.2).

In this chapter, clutter, interference, and noise are modeled as complex, circularly symmetric multivariate Gaussian random vectors as summarized in Appendix 3.A. We begin by summarizing the form of the Neyman–Pearson detector for the simple case of signal detection in multivariate zero-mean complex Gaussian noise.

3.2.1 Detecting a Known Signal in Colored Gaussian Noise

For a known signal vector $\boldsymbol{p} \in \mathcal{C}^{N\times 1}$, the hypothesis test is: $H_1 : \boldsymbol{y} \sim \mathcal{CN}(\boldsymbol{p}, \boldsymbol{R})$ vs. $H_0 : \boldsymbol{y} \sim \mathcal{CN}(\boldsymbol{0}, \boldsymbol{R})$, where symbol $\mathcal{CN}(\boldsymbol{m}, \boldsymbol{R})$ denotes the circularly symmetric complex Gaussian distribution with mean $\boldsymbol{m}$ and covariance matrix $\boldsymbol{R}$. The covariance matrix $\boldsymbol{R}$ of the noise is assumed to be a known Hermitian positive definite matrix. Pre-multiplication of the data vector $\boldsymbol{y}$ by $\boldsymbol{R}^{-1/2}$ whitens the noise vector, and the circularly symmetric random variable $\boldsymbol{p}^{\dagger}\boldsymbol{R}^{-1}\boldsymbol{y} = \mathrm{Re}(\boldsymbol{p}^{\dagger}\boldsymbol{R}^{-1}\boldsymbol{y}) + j\,\mathrm{Im}(\boldsymbol{p}^{\dagger}\boldsymbol{R}^{-1}\boldsymbol{y})$ is distributed as: $\boldsymbol{p}^{\dagger}\boldsymbol{R}^{-1}\boldsymbol{y} \sim \mathcal{CN}\left(\boldsymbol{p}^{\dagger}\boldsymbol{R}^{-1}\boldsymbol{p}, \boldsymbol{p}^{\dagger}\boldsymbol{R}^{-1}\boldsymbol{p}\right)$ for H_1 and as $\sim \mathcal{CN}\left(\boldsymbol{0}, \boldsymbol{p}^{\dagger}\boldsymbol{R}^{-1}\boldsymbol{p}\right)$ for H_0. The random variables $\text{Re}(\boldsymbol{p}^{\dagger}\boldsymbol{R}^{-1}\boldsymbol{y})$ and $\text{Im}(\boldsymbol{p}^{\dagger}\boldsymbol{R}^{-1}\boldsymbol{y})$ are statistically independent, and $\text{Im}(\boldsymbol{p}^{\dagger}\boldsymbol{R}^{-1}\boldsymbol{y})$ has the same density function for both hypotheses. The mean of the real-valued statistic $\text{Re}(\boldsymbol{p}^{\dagger}\boldsymbol{R}^{-1}\boldsymbol{y})$ is 0 under hypothesis H_0 and is $\boldsymbol{p}^{\dagger}\boldsymbol{R}^{-1}\boldsymbol{p}$ under hypothesis H_1. The variance of $\text{Re}(\boldsymbol{p}^{\dagger}\boldsymbol{R}^{-1}\boldsymbol{y})$ is $\sigma^2 = \boldsymbol{p}^{\dagger}\boldsymbol{R}^{-1}\boldsymbol{p}/2$ under both the null and the alternative hypotheses. The likelihood ratio test in (3.1) is equivalently expressed as follows – obtained by taking the logarithm of the likelihood ratio:

$$\text{Re}(\boldsymbol{p}^{\dagger}\boldsymbol{R}^{-1}\boldsymbol{y}) \underset{H_0}{\overset{H_1}{\gtrless}} \eta_0. \tag{3.3}$$

The probability of detection and probability of false alarm are given by:

$$\begin{aligned} P_D &= \frac{1}{\sqrt{2\pi}\sigma}\int_{\eta_0}^{\infty} e^{-(z-\boldsymbol{p}^{\dagger}\boldsymbol{R}^{-1}\boldsymbol{p})^2/2\sigma^2}dz = \frac{1}{2}\left[1-\text{erf}\left(\frac{(\eta_0-\boldsymbol{p}^{\dagger}\boldsymbol{R}^{-1}\boldsymbol{p})}{\sqrt{\boldsymbol{p}^{\dagger}\boldsymbol{R}^{-1}\boldsymbol{p}}}\right)\right] = \frac{1}{2}\text{erfc}\left(\frac{(\eta_0-\boldsymbol{p}^{\dagger}\boldsymbol{R}^{-1}\boldsymbol{p})}{\sqrt{\boldsymbol{p}^{\dagger}\boldsymbol{R}^{-1}\boldsymbol{p}}}\right), \\ P_{FA} &= \frac{1}{\sqrt{2\pi}\sigma}\int_{\eta_0}^{\infty} e^{-z^2/2\sigma^2}dz = \frac{1}{2}\left[1-\text{erf}\left(\frac{\eta_0}{\sqrt{\boldsymbol{p}^{\dagger}\boldsymbol{R}^{-1}\boldsymbol{p}}}\right)\right] = \frac{1}{2}\text{erfc}\left(\frac{\eta_0}{\sqrt{\boldsymbol{p}^{\dagger}\boldsymbol{R}^{-1}\boldsymbol{p}}}\right). \end{aligned} \tag{3.4}$$

In the above, $\text{erf}(x)$ and $\text{erfc}(x)$ are the error function and the complementary error function respectively as defined in [41]:

$$\begin{aligned} \text{erf}(x) &= \frac{2}{\sqrt{\pi}}\int_0^x \exp(-y^2)dy, \\ \text{erfc}(x) &= \frac{2}{\sqrt{\pi}}\int_x^{\infty} \exp(-y^2)dy = 1-\text{erf}(x). \end{aligned} \tag{3.5}$$

For a given probability of false alarm (P_{FA}), the threshold η_0 in (3.3) is determined from:

$$\eta_0 = \sqrt{\boldsymbol{p}^{\dagger}\boldsymbol{R}^{-1}\boldsymbol{p}}\quad \text{erfc}^{-1}(2P_{FA}). \tag{3.6}$$

The function $x = \text{erfc}^{-1}(p)$; $0 \le p \le 2$ is the inverse complementary error function and is defined such that $\text{erfc}(x) = p$. As P_{FA} monotonically decreases in the interval $0.5 \ge P_{FA} > 0$ the threshold η_0 monotonically increases in the interval $0 \le \eta_0 < \infty$. Note that the threshold η_0 is negative for $1 \ge P_{FA} > 0.5$.

3.2.2 Detecting a Known Signal with Unknown Phase in Zero-Mean Colored Gaussian Noise

For a known vector $\boldsymbol{p} \in \mathcal{C}^{N\times 1}$, the two hypotheses are: $H_1 : \boldsymbol{y} \sim \mathcal{CN}\left(\boldsymbol{p}\,e^{j\alpha}, \boldsymbol{R}\right)$ vs. $H_0 : \boldsymbol{y} \sim \mathcal{CN}\left(\boldsymbol{0}, \boldsymbol{R}\right)$. The phase α of the signal is unknown. Since the phase of the circularly symmetric noise is also unknown, integrating over the PDF of the phase, the PDF of the random variable $z = |\boldsymbol{p}^{\dagger}\boldsymbol{R}^{-1}\boldsymbol{y}|$ is given by:

$$\begin{aligned} f(z|H_1) &= \frac{z}{\sigma^2}e^{-(a^2+z^2)/2\sigma^2} I_0\left(\frac{za}{2\sigma^2}\right); \quad z \ge 0, \\ f(z|H_0) &= \frac{z}{\sigma^2}e^{-z^2/2\sigma^2}; \quad z \ge 0. \end{aligned} \tag{3.7}$$

In the above, $a = \boldsymbol{p}^{\dagger}\boldsymbol{R}^{-1}\boldsymbol{p}$ and $2\sigma^2 = \boldsymbol{p}^{\dagger}\boldsymbol{R}^{-1}\boldsymbol{p}$.
And the likelihood ratio test in (3.1) is:

$$\mathcal{L}(H|z) = e^{-(\boldsymbol{p}^{\dagger}\boldsymbol{R}^{-1}\boldsymbol{p})} I_0(z) \underset{H_0}{\overset{H_1}{\gtrless}} \eta. \tag{3.8}$$

$I_0(z)$ is the modified Bessel function of order 0 and is a monotonically increasing function of the argument. The detection statistic in the likelihood ratio test can equivalently be replaced by any monotonically increasing function of the argument $z = |\boldsymbol{p}^{\dagger}\boldsymbol{R}^{-1}\boldsymbol{y}|$. Thus the test in (3.8) is equivalent to the following with the normalized statistic:

$$\frac{|\boldsymbol{p}^{\dagger}\boldsymbol{R}^{-1}\boldsymbol{y}|^2}{(\boldsymbol{p}^{\dagger}\boldsymbol{R}^{-1}\boldsymbol{p})} \underset{H_0}{\overset{H_1}{\gtrless}} \eta_0. \tag{3.9}$$

For purposes of detecting the signal $\boldsymbol{p}$, the data vector $\boldsymbol{y}$ can be reduced to a single statistic $|\boldsymbol{q}|^2 = |\boldsymbol{p}^{\dagger}\boldsymbol{R}^{-1}\boldsymbol{y}|^2/(\boldsymbol{p}^{\dagger}\boldsymbol{R}^{-1}\boldsymbol{p})$, which is referred to as a sufficient statistic. Using Proposition 3.1, the statistic $|\boldsymbol{q}|^2 = |\boldsymbol{p}^{\dagger}\boldsymbol{R}^{-1}\boldsymbol{y}|^2/(\boldsymbol{p}^{\dagger}\boldsymbol{R}^{-1}\boldsymbol{p})$ has a non-central complex Chi-squared distribution with 1 complex degree of freedom and non-centrality parameter: $c = (\boldsymbol{p}^{\dagger}\boldsymbol{R}^{-1}\boldsymbol{p})$ for hypothesis H_1 (i.e. $|\boldsymbol{q}|^2 \sim \chi_1^2(c)$), which is also referred to as the Ricean distribution [33]. For hypothesis H_0, the sufficient statistic has a central complex Chi-squared distribution with 1 complex degree of freedom $|\boldsymbol{q}|^2 \sim \chi_1^2$, which is also the exponential density with the expected value of the random variable of 1. The probability of false alarm and probability of detection are given by:

$$\begin{aligned} P_{FA} &= \int_{\eta_0}^{\infty} e^{-y} dy = e^{-\eta_0}, \\ P_D &= \int_{\eta_0}^{\infty} I_0(2\sqrt{yc})\, e^{-(y+c)} dy = Q_1(\sqrt{2c}, \sqrt{2\eta_0}). \end{aligned} \tag{3.10}$$

The second integral above is expressed in terms of the generalized Marcum Q-function $Q_M(a, b)$ defined as [30, 42–45]:

$$Q_M(a, b) = \int_b^{\infty} x \left(\frac{x}{a}\right)^{M-1} e^{-(x^2+a^2)/2} I_{M-1}(ax) dx; \quad M \geq 1. \tag{3.11}$$

3.3 Subspace Signal Model and Invariant Hypothesis Tests

The previous section provided a brief introduction to the problem of detecting an additive signal in zero-mean colored Gaussian noise with known covariance matrix. Radars must detect targets of interest when the interference-plus-noise covariance matrix is unknown. The parameters of the interference-plus-noise covariance matrix are referred to as nuisance parameters, since the unknown parameters must be estimated from available data in order to cancel interference. Since the actual SINR is generally unknown, an estimate of the post-cancelation SINR is maximized (as shown in Appendix 3.A), and a detection statistic is chosen that also takes into account the signal-to-interference-plus-noise loss factor resulting from finite sample size effects. While the procedure cannot guarantee that the SOI is detected, it may not be possible to achieve acceptable detection probability without such an approach and also have a preset probability of false alarm.

A given surveillance area/volume where targets are to be detected is generally divided into resolution cells, with each cell identified by its range, elevation, and azimuth angles in reference to a coordinate system. Targets moving at different radial velocities produce different Doppler shifts to the incident signals, and therefore the presence of a target at different Doppler frequencies must be tested in a given test cell. A set of data vectors from resolution cells in the "vicinity" of a test cell, referred to as secondary data vectors or training data vectors is used for purposes of estimating the nuisance parameters. Adaptive detectors are derived by assuming that the set of secondary data vectors are target-free, statistically independent and identically distributed as the interference and noise in the test vector. Generally, the signal model used assumes that the dimensionality of the signal space is one. A more general signal space model is introduced next for the purpose of deriving adaptive subspace detectors in unknown clutter-plus-interference.

3.3.1 Subspace Signal Model

Let $z \in \mathcal{C}^{N\times 1}$ represent the test vector comprising a possible subspace signal embedded in interference and noise. The term interference is used here more generally and refers to clutter-plus-interference. The interference-plus-noise in the test vector is denoted by $\boldsymbol{x} \in \mathcal{C}^{N\times 1}$. N is the number of samples (multidimensional) that are coherently processed. For spatial processing, N is the number of receiver elements for example; for space–time processing, N is the product of the number of receiver elements and the number of time samples received at each element. Let $\boldsymbol{H} \in \mathcal{C}^{N\times M}$; $M < N$, be a known matrix with linearly independent columns (i.e. rank of $\boldsymbol{H}$ is M). The column space of $\boldsymbol{H}$ therefore defines an M-dimensional subspace in $\mathcal{C}^{N\times 1}$.

The signal vector $\boldsymbol{p}$, to be detected in a given resolution cell is hypothesized to be contained in the column space of a known matrix $\boldsymbol{H} \in \mathcal{C}^{N\times M}$. To allow for possible mismatches in the assumed signal model, the actual signal is modeled to be contained in the column space of a non-singular matrix $\boldsymbol{D} \in \mathcal{C}^{N\times N}$; $N \geq M \geq 1$. The columns of matrix $\boldsymbol{D}$ are partitioned as $\boldsymbol{D} = [\boldsymbol{H}\ \boldsymbol{H}_{\perp}]$, where the column spaces of $\boldsymbol{H}$ and $\boldsymbol{H}_{\perp}$ are assumed to be orthogonal and complimentary subspaces in $\mathcal{C}^{N\times 1}$. The ranks of matrices $\boldsymbol{H}$ and $\boldsymbol{D}$ are M and N respectively. Thus, the hypothesized signal $\boldsymbol{p}$ and the actual signal $\boldsymbol{v}$ are expressed as linear sums of the columns of $\boldsymbol{H}$ and $\boldsymbol{D}$ respectively:

$$\begin{aligned} \boldsymbol{p} &= \boldsymbol{H}\,\boldsymbol{\alpha}, \\ \boldsymbol{v} &= \boldsymbol{D}\,\boldsymbol{\gamma} = [\boldsymbol{H}\ \boldsymbol{H}_{\perp}]\,\boldsymbol{\gamma}. \end{aligned} \tag{3.12}$$

In the above $\boldsymbol{\alpha} \in \mathcal{C}^{M\times 1}$ and $\boldsymbol{\gamma} \in \mathcal{C}^{N\times 1}$ are coefficient vectors. A fixed set of coefficients defines a rank 1 signal and allowing the coefficients to vary defines the subspace signal model.

3.3.2 A Rationale for Subspace Signal Model

Among some of the several assumptions made in assuming a signal space of dimension 1 are: a narrowband transmit signal, a point scatter target at known azimuth/elevation angles with a constant and known radial velocity during the coherent integration interval, a known array manifold, which implies that the array response to a point target at a given azimuth/elevation and Doppler is known. Near-field scatter of signals from objects such as aircraft wings and ship masts is generally neglected. As stated in the previous section, some scenarios inherently require modeling of the signal subspace by a dimension greater than one due to multiple modes.

In cases where the narrowband assumption is not strictly valid, the effective dimension of the signal subspace must be assumed to be greater than 1 [46]. Some targets inherently require subspace signal models [47]. Positioning and/or timing errors at individual sensor elements can result in signal mismatch that results in detection loss. Synchronization errors at different sensors can similarly result in signal mismatch for coherent processing – which in the spatial domain refers to beamforming. These effects were described generally for antenna arrays in Reference 48. Lastly, hypothesis testing of vectors from a grid of resolution cells can inherently imply a mismatch between the actual signal and the steering vector assigned to a cell. While a subspace detector cannot solve all problems, Section 3.5 presents results and conditions where a subspace detector can mitigate the effects of some mismatch errors for a rank 1 signal. Below we develop a simplified model to characterize the signal mismatch errors. For purposes of convenience, the model below considers spatial or temporal processing only, however similar results can be obtained for coherent processing over multiple domains. The hypothesized steering vector $\boldsymbol{p}$ and the actual steering vector $\boldsymbol{v}$ are expressed as below in terms of the phase error vector $\boldsymbol{\zeta}$:

$$\begin{aligned} \boldsymbol{v} &= \boldsymbol{p} \odot \boldsymbol{\zeta}, \\ \boldsymbol{\zeta} &= \left[e^{j2\pi\xi_1/\lambda} \; e^{j2\pi\xi_2/\lambda} \; \ldots \; e^{j2\pi\xi_N/\lambda}\right]^{\dagger}. \end{aligned} \tag{3.13}$$

The vector $\boldsymbol{p}$ is the hypothesized signal vector, the vector $\boldsymbol{\zeta} \in \mathcal{C}^{N\times 1}$ denotes the mismatch due to phase error at the elements. The error quantities $\{\xi_1, \xi_2, \ldots \xi_N\}$ have units of the carrier wavelength λ and denote the projection of the difference between the position vector of a receiving sensor and the equivalent phase-center of the receiving sensor on to the unit vector in the direction of propagation of the signal. The notation $\odot$ refers to the Hadamard product, that is, element wise product in the equation above and referring to (3.12), the vector $\boldsymbol{v}$ is the actual signal vector. Assuming the vector $\boldsymbol{p}$ to be deterministic, the covariance matrix of the actual signal obtained by averaging over the ensemble of phase uncertainties is given by:

$$\mathbb{E}\left[\boldsymbol{v}\boldsymbol{v}^{\dagger}\right] = \boldsymbol{p}\boldsymbol{p}^{\dagger} \odot \mathbb{E}\left[\boldsymbol{\zeta}\boldsymbol{\zeta}^{\dagger}\right] = \boldsymbol{p}\boldsymbol{p}^{\dagger} \odot \boldsymbol{T}. \tag{3.14}$$

The form of the $N \times N$ taper matrix $\boldsymbol{T}$ above depends on the error model: The elements of matrix $\boldsymbol{T}$ can be expressed in terms of the characteristic function [49] of the difference random variable $(\xi_n - \xi_m)$. As an example, for a zero-mean multivariate Gaussian model for the vector $\{\xi_1, \xi_2, \ldots \xi_N\}$, the (n, m)th element of $\boldsymbol{T}$ are:

$$T_{(n,m)} = \mathbb{E}\left[e^{j2\pi(\xi_n - \xi_m)/\lambda}\right] = \begin{cases} 1 & \text{if } n = m, \\ e^{-(2\pi\sigma_{n,m}/\lambda)^2/2} & \text{if } n \neq m. \end{cases} \tag{3.15}$$

In the above $\sigma_{n,m}^2 = \mathbb{E}\left[(\xi_n - \xi_m)^2\right] = 2\sigma_0^2\left[1 - \rho_{n,m}\right]$, where $\sigma_0^2 = \mathbb{E}\left[\xi_n^2\right] = \mathbb{E}\left[\xi_m^2\right]$ and the correlation coefficient $\rho_{n,m} = \sigma_0^2\mathbb{E}[\xi_n\xi_m]$. The eigenvectors corresponding to the dominant eigenvalues of signal covariance matrix define the signal subspace, and the taper matrix increases the dimension of the signal subspace. This is generally the case where the errors lead to a structured form of the taper matrix $\boldsymbol{T}$. The condition $2\pi \max(\sigma_{n,m}) << \lambda$ implies that all elements of the matrix $\boldsymbol{T}$ are approximately 1, and the signal subspace is the same as that of vector $\boldsymbol{p}$. At the other extreme, $2\pi \min(\sigma_{n,m}) >> \lambda$ leads to $\boldsymbol{T} = \boldsymbol{I}_N$ which implies that no coherent processing is possible. It is of interest to note that the signal space model developed here is similar to various forms of covariance matrix taper (CMT) applied to the estimates of the interference-plus-noise covariance matrix [50–52]. For interference cancelation, the application of a CMT produces adapted beam patterns with nulls that are wider that without the application of a CMT and therefore the former approach is more effective in canceling out directional interference.

A second example is a model for phase uncertainty in spatial and/or temporal processing and results from sampling of test vectors over a grid of azimuth angles and/or Doppler frequencies. Sampling over a grid of azimuth angles is useful in some search application where the beam pattern of the transmit array may illuminate a large extent of azimuth angles. Let $\boldsymbol{p}$ be the spatial and/or temporal vector for a given azimuth and/or Doppler. The vector $\boldsymbol{p}$ is used for testing signals from all points in the vicinity of each grid point. The phase difference at the receiver between steering vectors for adjacent points on the grid is $2\pi\Delta$, and over the ensemble of signal azimuth and/or Doppler, the phase error $2\pi\Delta$ is modeled as being uniform in the interval $[-a\pi, a\pi]$. The model parameter a is in the interval $0 \le a \le 1$ and is chosen depending on the fineness ($a \approx 0$) or coarseness ($a \approx 1$) of the azimuth and/or Doppler grid points. The covariance matrix of the signal evaluated on the ensemble is given by (3.14), where

$$\begin{aligned} T_{(n,m)} &= \frac{1}{a}\int_{-a/2}^{+a/2} e^{j2\pi(n-m)\Delta} d\Delta \\ &= \frac{\sin(\pi a(n-m))}{\pi a(n-m)}; \quad n, m = 1, 2, \ldots, N. \end{aligned} \tag{3.16}$$

The eigenvectors corresponding to the dominant eigenvalues of signal covariance matrix define the signal subspace, and the taper matrix increases the dimension of the signal subspace as we illustrate with an example in Section 3.5.

3.3.3 Hypothesis Test

Let H_0 denote the null-hypothesis that z is interference and noise only and H_1 the alternative hypothesis that z is signal-plus-interference-plus-noise. The binary hypothesis test for detecting the signal $\boldsymbol{p}$ in interference and noise is the following:

$$\mathbf{z} = \begin{cases} \boldsymbol{x} & \text{if } H_0, \\ \boldsymbol{x} + \boldsymbol{H}\,\boldsymbol{\alpha}; \quad ||\boldsymbol{\alpha}||_2 > 0 & \text{if } H_1. \end{cases} \tag{3.17}$$

The interference $\boldsymbol{x}$ is a complex, circularly symmetric zero-mean vector with covariance matrix $\boldsymbol{R} \in \mathbb{H}(N)$, where $\mathbb{H}(N)$ denotes the set of Hermitian positive definite matrices of size N. The distribution is denoted by the notation $\boldsymbol{x} \sim \mathcal{CN}(\boldsymbol{0}_{N\times 1}, \boldsymbol{R})$. Note that the circularly symmetric and zero-mean properties imply that the real and imaginary parts of each component of $\boldsymbol{x}$ are such that $\mathbb{E}[\boldsymbol{x}\boldsymbol{x}^T] = \boldsymbol{0}_{N\times N}$. The covariance matrix $\boldsymbol{R}$ is unknown to the receiver, and as stated in the previous section, the elements of the Hermitian positive definite matrix $\boldsymbol{R} \in \mathbb{H}(N)$ are the nuisance parameters in the hypothesis test and are estimated from a given set of $K(\ge N)$ statistically independent and identically distributed secondary data vectors represented by the matrix $\boldsymbol{Y} \in \mathcal{C}^{N\times K}$, with $\boldsymbol{Y} \sim \mathcal{CN}(\boldsymbol{0}_{N\times K}, \boldsymbol{I}_K \otimes \boldsymbol{R})$ for both hypotheses H_0 and H_1. The notation, $\boldsymbol{D} = \boldsymbol{A} \otimes \boldsymbol{B}$ is the Kronecker product of $\boldsymbol{A}$ and $\boldsymbol{B}$. For $\boldsymbol{A} \in \mathcal{C}^{N\times M}$ and $\boldsymbol{B} \in \mathcal{C}^{K\times L}$, the matrix $\boldsymbol{D} \in \mathcal{C}^{NK\times ML}$ comprises blocks of size $K \times L$ and the (n, m)th block is $a_{n,m}\boldsymbol{B}$; $n = 1, 2, \ldots, N$; $m = 1, 2, \ldots, M$, where $a_{n,m}$ is the element in the nth row and mth column of matrix $\boldsymbol{A}$.

The hypothesis test in (3.17) is equivalent to the following:

$$\mathbf{z} : \begin{cases} ||\boldsymbol{\alpha}||_2 = 0 & \text{if } H_0, \\ ||\boldsymbol{\alpha}||_2 > 0 & \text{if } H_1. \end{cases} \tag{3.18}$$

Given a test vector z and secondary data vectors $\boldsymbol{Y}$, we seek to find linear transformations of the data that leave the hypothesis testing problem in (3.18) unchanged. The hypothesis test in (3.18) is equivalent to the test: $||\boldsymbol{p}||_2^2 = \boldsymbol{\alpha}^\dagger \boldsymbol{H}^\dagger \boldsymbol{H}\boldsymbol{\alpha} = 0$ vs. $||\boldsymbol{p}||_2^2 = \boldsymbol{\alpha}^\dagger \boldsymbol{H}^\dagger \boldsymbol{H}\boldsymbol{\alpha} > 0$. Note that the $M \times M$ matrix $\boldsymbol{H}^\dagger \boldsymbol{H}$ is Hermitian positive definite, and therefore the test $||\boldsymbol{p}||_2^2 = 0$ vs. $||\boldsymbol{p}||_2^2 > 0$ is equivalent to the test $||\boldsymbol{\alpha}||_2 = 0$ vs. $||\boldsymbol{\alpha}||_2 > 0$.

Let the Singular Value Decomposition (SVD) of the matrix $\boldsymbol{H}$ be:

$$\boldsymbol{H} = \boldsymbol{U}\boldsymbol{B}\boldsymbol{W}^\dagger. \tag{3.19}$$

$\boldsymbol{U}$ and $\boldsymbol{W}$ are $N \times N$ and $M \times M$ unitary matrices, and $\boldsymbol{B}$ is a $N \times M$ matrix of singular values $b_1 \geq b_2 \geq \cdots \geq b_M > 0$ and is given by: $\boldsymbol{B} = [\mathbf{diag}(b_1, b_2, \ldots, b_M)\ \boldsymbol{0}_{M\times(N-M)}]^\dagger$. Because of the ordering of the singular values, the first M columns of the unitary matrix $\boldsymbol{U}$ are the orthonormal basis for the hypothesized signal subspace. It can be verified that the following reversible linear transformation of the data does not change the hypothesized signal subspace as well as the original hypothesis test. Consider, $\boldsymbol{Q} = \boldsymbol{C}\boldsymbol{U}^\dagger$, where $\boldsymbol{C}$ is a non-singular $N \times N$ matrix given by:

$$\boldsymbol{C} = \begin{bmatrix} \boldsymbol{C}_{11} & \boldsymbol{C}_{12} \\ \boldsymbol{0}_{(N-M)\times M} & \boldsymbol{C}_{22} \end{bmatrix}, \tag{3.20}$$

where the matrices $\boldsymbol{C}_{11} \in \mathcal{C}^{M\times M}$ and $\boldsymbol{C}_{22} \in \mathcal{C}^{(N-M)\times(N-M)}$ are both non-singular and the matrix $\boldsymbol{C}_{12} \in \mathcal{C}^{M\times(N-M)}$. The effect of pre-multiplying (i.e. left multiplication) the test vector and secondary data by any matrix $\boldsymbol{Q} = \boldsymbol{C}\boldsymbol{U}^\dagger$ can be understood as follows:

The first pre-multiplication by $\boldsymbol{U}^\dagger$ results in a coordinate rotation such that the new set of coordinate axes are the orthonormal column vectors of $\boldsymbol{U}$. Thus, the first M axes are the M orthonormal basis vectors of the signal subspace, and the last $N - M$ axes are the basis vectors of the assumed interference-plus-noise-only subspace. The transformed data and signal vectors are:

$$\begin{aligned} z \to \boldsymbol{U}^\dagger z &= \begin{bmatrix} z_1 \\ z_2 \end{bmatrix}; \quad \boldsymbol{Y} \to \boldsymbol{U}^\dagger \boldsymbol{Y} = \begin{bmatrix} \boldsymbol{Y}_1 \\ \boldsymbol{Y}_2 \end{bmatrix}; \\ \boldsymbol{x} \to \boldsymbol{U}^\dagger \boldsymbol{x} &= \begin{bmatrix} \boldsymbol{x}_1 \\ \boldsymbol{x}_2 \end{bmatrix}; \quad \boldsymbol{p} \to \boldsymbol{U}^\dagger \boldsymbol{p} = \begin{bmatrix} \boldsymbol{p}_1 \\ \boldsymbol{0}_{(N-M)\times 1} \end{bmatrix}. \end{aligned} \tag{3.21}$$

In the above, $z_1 \in \mathcal{C}^{M\times 1}$, $z_2 \in \mathcal{C}^{(N-M)\times 1}$, and so on.

The transformed primary and secondary vectors after the coordinate rotation are as shown above. The hypothesis tests developed are invariant to the following transformations:

$$\begin{bmatrix} z_1 & \boldsymbol{Y}_1 \\ z_2 & \boldsymbol{Y}_2 \end{bmatrix} \to \begin{bmatrix} \boldsymbol{C}_{11} & \boldsymbol{C}_{12} \\ \boldsymbol{0}_{(N-M)\times M} & \boldsymbol{C}_{22} \end{bmatrix} \begin{bmatrix} z_1 & \boldsymbol{Y}_1 \\ z_2 & \boldsymbol{Y}_2 \end{bmatrix} \begin{bmatrix} e^{j\alpha} & \boldsymbol{0}_{1\times K} \\ \boldsymbol{0}_{K\times M} & \boldsymbol{V}_{22} \end{bmatrix}. \tag{3.22}$$

In the above, α is an arbitrary phase, $\boldsymbol{V}_{22}$ is a $K \times K$ unitary matrix, and post-multiplication of the secondary data matrices $\boldsymbol{Y}_1$ and $\boldsymbol{Y}_2$ by $\boldsymbol{V}_{22}$ just rotates the K samples in each row of the secondary data matrices. Since the K secondary vectors are assumed to be independent and identically distributed, the row rotations by a unitary matrix do not change the assumed distribution of the secondary data. The matrices: $\boldsymbol{C}_{11} \in \mathcal{C}^{M\times M}$ and $\boldsymbol{C}_{22} \in \mathcal{C}^{(N-M)\times(N-M)}$ are non-singular matrices and $\boldsymbol{C}_{12} \in \mathcal{C}^{M\times(N-M)}$. The above left and right multiplications transform the various components as follows: $z_1 \to (\boldsymbol{C}_{11} z_1 + \boldsymbol{C}_{12} z_2)e^{j\alpha}$; $z_2 \to \boldsymbol{C}_{22} z_2 e^{j\alpha}$; $\boldsymbol{p}_1 \to \boldsymbol{C}_{11}\boldsymbol{p}_1 e^{j\alpha}$ and the transformation $z_1 \to (\boldsymbol{C}_{11} z_1 + \boldsymbol{C}_{12} z_2)e^{j\alpha}$ does not change the original hypothesis test.

Therefore, for any non-singular matrix $C_{11} \in \mathcal{C}^{M\times M}$, the test $||p||_2 = 0$ hypothesis H_0 vs. $||p||_2 > 0$ hypothesis H_1 is the same as $p_1^\dagger C_{11}^\dagger C_{11} p_1 = 0$ vs. $p_1^\dagger C_{11}^\dagger C_{11} p_1 > 0$. Additionally, since $C_{21} = 0_{(N-M)\times M}$, we have $z_2 \to C_{21}z_1 + C_{22}z_2 e^{j\alpha} = C_{22}z_2 e^{j\alpha}$ which does not contain a signal for the hypothesized model. We consider hypothesis tests that are invariant to such transformation.

3.3.4 Maximum Invariants for Subspace Signal Detection in Interference and Noise

We show in Appendix 3.B that any two of the three statistics: $||y||_2^2, y^\dagger P_G y$ and $y^\dagger P_G^\perp y$ comprise a maximal invariant statistic for the detection problem. For a hypothesized signal matrix H and estimate $S = YY^\dagger$, which is K multiplied by the sample covariance matrix estimate, the various quantities in the statistic are defined as:

$$\begin{aligned} y &= S^{-1/2} z, \\ G &= S^{-1/2} H, \\ P_G &= G(G^\dagger G)^{-1} G^\dagger, \\ P_G^\perp &= I_N - P_G. \end{aligned} \tag{3.23}$$

For purposes of evaluating the performance of detection algorithms under conditions of signal model mismatch, it is more convenient to define the following two-dimensional statistic ρ, r as maximal invariants:

$$\begin{aligned} \rho &= \frac{1}{1 + y^\dagger P_G^\perp y}, \\ r &= \frac{1}{1 + \rho(y^\dagger P_G y)}. \end{aligned} \tag{3.24}$$

Note that all three quantities: $||y||_2^2, y^\dagger P_G y, y^\dagger P_G^\perp y$ can be obtained in terms of ρ, r. In the case when the hypothesized signal subspace is matched to the subspace of the actual signal, the statistic $y^\dagger P_G y$ is an estimate of the SINR and the statistic ρ is an estimate of the loss in SINR due to finite sample size effects.

The joint PDF of the statistic ρ, r is derived in Appendix 3.C including the case where the hypothesized signal model is mismatched to the actual signal model. With the data from a given test vector and secondary vectors reduced to the statistic ρ, r, the likelihood ratio test in (3.1) can be simplified to the following, using (3.79), for given SINR δ and zero signal mismatch error $\theta = 0$:

$$\begin{aligned} \mathcal{L}(H|\rho, r) &= \frac{f(\rho, r|\delta, \theta = 0, H_1)}{f(\rho, r|H_0)} \underset{H_0}{\overset{H_1}{\gtrless}} \eta \\ &= e^{-\delta\rho r} \sum_{k=0}^{K-N+1} C_k \left[\delta\rho(1-r)\right]^k \underset{H_0}{\overset{H_1}{\gtrless}} \eta. \end{aligned} \tag{3.25}$$

The quantities C_k in the above expression are given by:

$$C_k = \binom{K-N+1}{k} \frac{\Gamma(M)}{\Gamma(M+k)}; \quad k = 0, 1, \ldots, K-N+1. \tag{3.26}$$

Without knowledge of the SINR δ, there is no function of ρ, r that is monotonically related to the likelihood ratio statistic on the left-hand side of (3.25). As such, there is no uniformly most powerful invariant test for $(1 \leq M < N)$. For the case of $M = N$, however, the maximal invariant statistic is the single statistic: $||\mathbf{y}||_2^2$ (this is equivalent to setting $\boldsymbol{P_G} = \boldsymbol{I}_N$ and $\rho = 1$). For $M = N$, the signal subspace is the space $\mathcal{C}^{N\times 1}$; the GLRT and AMF test are identical in this case and given by the Hotelling's T^2 statistic [37], which is known to be a uniformly most powerful invariant test.

3.4 Analytical Expressions for P_D and P_{FA}

Analytical expressions for the probability of false alarm in terms of the various parameters are useful for the purposes of selecting the detection threshold for a given test. Analytical expressions for P_D and P_{FA} also provide insight into factors that have a major influence on the performance of a detector. Expressions for the probability of detection P_D and the probability of false alarm P_{FA} are derived in this section using the results from Appendix 3.C. Also note that for given signal matrices: $\boldsymbol{H}$ and $\boldsymbol{D}$, interference-plus-noise covariance matrix $\boldsymbol{R}$ and for an assumed $\boldsymbol{\gamma} \in \mathcal{C}^{N\times 1}$, the mismatch angle θ required for computing the probability of detection for the various cases is given in (3.66) of Appendix 3.C. The SINR, δ, can be evaluated from (3.67) as: $\delta = \boldsymbol{\gamma}^\dagger \bar{\boldsymbol{D}}^\dagger \bar{\boldsymbol{D}} \boldsymbol{\gamma} = \boldsymbol{\gamma}^\dagger \boldsymbol{D}^\dagger \boldsymbol{R}^{-1} \boldsymbol{D} \boldsymbol{\gamma}$.

3.4.1 P_D and P_{FA} for Subspace GLRT

The hypothesis test for the signal subspace version of the GLRT is given from (3.47) and as expressed in terms of the maximal invariant statistic (ρ, r) as defined in (3.24) is given by:

$$\frac{\mathbf{y}^\dagger \boldsymbol{P_G} \mathbf{y}}{\left(1 + \mathbf{y}^\dagger \boldsymbol{P}_{\boldsymbol{G}}^{\perp} \mathbf{y}\right)} \underset{H_0}{\overset{H_1}{\gtrless}} \eta_0 - 1. \tag{3.27}$$

Using (3.24) the above is equivalent to the test:

$$r^{-1} \underset{H_0}{\overset{H_1}{\gtrless}} \eta_0. \tag{3.28}$$

Given the joint PDFs of the maximal invariant statistic (ρ, r) conditioned on hypotheses H_1 and H_0 in (3.75) and (3.76) respectively, the probability of detection and probability of false alarm are given by:

$$\begin{aligned} P_D &= \int_0^1 d\rho \int_0^{\eta_0^{-1}} f(r, \rho | H_1) dr, \\ P_{FA} &= \int_0^1 d\rho \int_0^{\eta_0^{-1}} f(r, \rho | H_0) dr. \end{aligned} \tag{3.29}$$

Using Proposition 3.3:

$$P_D = P[r^{-1} > \eta_0 | H_1] = 1 - \frac{(\eta_0 - 1)^M}{\eta_0^{K-N+M}} \sum_{k=0}^{K-N} \binom{K-N+M}{k+M} (\eta_0 - 1)^k S_k,$$

$$S_k = \int_0^1 \left[G_{k+1}\left(\frac{\rho\delta\cos^2\theta}{\eta_0}\right) f_\beta(\rho; K-(N-M)+1, N-M|\delta\sin^2\theta) \right] d\rho. \tag{3.30}$$

The probability of false alarm is obtained by setting $\delta = 0$ in the above and is:

$$P_{FA} = P[r^{-1} > \eta_0 | H_0] = 1 - \frac{(\eta_0 - 1)^M}{\eta_0^{K-N+M}} \sum_{k=0}^{K-N} \binom{K-N+M}{k+M} (\eta_0 - 1)^k$$

$$= \frac{1}{\eta_0^{K-N+M}} \sum_{n=0}^{M-1} \binom{K-N+M}{n} (\eta_0 - 1)^n. \tag{3.31}$$

3.4.2 P_D and P_{FA} for Subspace AMF Test

The subspace AMF test given in (3.49) is: $\boldsymbol{y}^\dagger \boldsymbol{P_G y} \underset{H_0}{\overset{H_1}{\gtrless}} \eta$ and in terms of (ρ, r), the test is written as:

$$r^{-1} - 1 \underset{H_0}{\overset{H_1}{\gtrless}} \rho\eta. \tag{3.32}$$

The probability of detection and probability of false alarm are given by:

$$P_D = \int_0^1 d\rho \int_0^{(1+\rho\eta)^{-1}} f(r, \rho | H_1) dr,$$

$$P_{FA} = \int_0^1 d\rho \int_0^{(1+\rho\eta)^{-1}} f(r, \rho | H_0) dr. \tag{3.33}$$

From (3.75), the statistic r for given ρ is non-central beta: $\beta_{K-N+1,M}(\rho\delta\cos^2\theta)$. The P_D and P_{FA} for the test in (3.32) is obtained in a manner similar to that in (3.28) and is:

$$P_D = P[r < (1+\rho\eta)^{-1} | H_1] = 1 - \frac{1}{L+M} \sum_{k=0}^{L-1} T_k,$$

$$T_k = \int_0^1 f_\beta\left(\frac{1}{1+\rho\eta}; L-k, M+k+1\right) G_{k+1}\left(\frac{\rho\delta\cos^2\theta}{1+\rho\eta}\right) f_\beta(\rho; L+M, N-M|\delta\ \sin^2\theta) d\rho. \tag{3.34}$$

For notational convenience, $L = K - N + 1$ in the above. The probability of false alarm is obtained by setting $\delta = 0$ and is:

$$
\begin{aligned}
P_{FA} &= P[r < (1+\rho\eta)^{-1}|H_0] = 1 - \frac{1}{L+M}\sum_{k=0}^{L-1}\tilde{T}_k, \\
\tilde{T}_k &= \int_0^1 f_\beta\left(\frac{1}{1+\rho\eta}; L-k, M+k+1\right) f_\beta(\rho; L+M, N-M)\,d\rho.
\end{aligned}
\tag{3.35}
$$

3.4.3 P_D and P_{FA} for Subspace ACE Test

The ACE test in (3.51) is equivalent to the following tests:

$$
\begin{aligned}
\frac{y^\dagger P_G y}{y^\dagger P_G^\perp y + y^\dagger P_G y} &\underset{H_0}{\overset{H_1}{\gtrless}} \eta, \\
r^{-1} &\underset{H_0}{\overset{H_1}{\gtrless}} (1-\rho)\eta_0 + 1.
\end{aligned}
\tag{3.36}
$$

The ACE threshold η_0 in the second expression above is given in terms of the quantity η in the first expression above by: $\eta_0 = \eta/(1-\eta)$.

The probability of detection and probability of false alarm are given by:

$$
\begin{aligned}
P_D &= \int_0^1 d\rho \int_0^{(1+(1-\rho)\eta_0)^{-1}} f(r,\rho|H_1)\,dr, \\
P_{FA} &= \int_0^1 d\rho \int_0^{(1+(1-\rho)\eta_0)^{-1}} f(r,\rho|H_0)\,dr.
\end{aligned}
\tag{3.37}
$$

Since the statistic on the left-hand side of the first expression in (3.36) is in the interval (0, 1), the threshold η lies in the interval $0 < \eta < 1$. Expressions for P_D and P_{FA} for the test in (3.36) are:

$$
\begin{aligned}
P_D &= P[r < (\eta_0(1-\rho)+1)^{-1}|H_1] = 1 - \frac{1}{L+M}\sum_{k=0}^{L-1} V_k, \\
V_k &= \int_0^1 f_\beta\left(\frac{1}{1+(1-\rho)\eta_0}; L-k, M+k+1\right) G_{k+1}\left(\frac{\rho\delta\cos^2\theta}{1+(1-\rho)\eta_0}\right) \\
&\quad \times f_\beta(\rho; L+M, N-M|\delta\ \sin^2\theta)\,d\rho.
\end{aligned}
\tag{3.38}
$$

The P_{FA} is given by setting $\delta = 0$ in the above equation and is:

$$
\begin{aligned}
P_{FA} &= P[r < (\eta_0(1-\rho)+1)^{-1}|H_0] = 1 - \frac{1}{L+M}\sum_{k=0}^{L-1}\tilde{V}_k, \\
\tilde{V}_k &= \int_0^1 f_\beta\left(\frac{1}{1+(1-\rho)\eta_0}; L-k, M+k+1\right) f_\beta(\rho; L+M, N-M)\,d\rho.
\end{aligned}
\tag{3.39}
$$

3.5 Performance Results of Adaptive Subspace Detectors

Results illustrating the detection performance of the three adaptive subspace detectors are presented in this section. We consider the case where the coefficients $\boldsymbol{\alpha}$ and $\boldsymbol{\gamma}$ in the signal model (3.12) are deterministic and unknown. As shown in Appendix 3.A, given a test vector $z \in \mathcal{C}^{N\times 1}$ and secondary data vectors $\boldsymbol{Y} \in \mathcal{C}^{N\times K}$; $K \geq N$, the coefficient vector $\boldsymbol{\alpha} \in \mathcal{C}^{M\times 1}$, is estimated as: $\hat{\boldsymbol{\alpha}} = (\boldsymbol{H}^{\dagger}\boldsymbol{S}^{-1}\boldsymbol{H})^{-1}\boldsymbol{H}^{\dagger}\boldsymbol{S}^{-1}z$. No prior model is assumed for $\boldsymbol{\alpha}$ in obtaining the estimate $\hat{\boldsymbol{\alpha}}$. Given the signal model, the estimate of the signal to be detected (to within a non-zero multiplicative constant) is $\hat{\boldsymbol{p}} = [\boldsymbol{H}\hat{\boldsymbol{\alpha}}] \in \mathcal{C}^{N\times 1}$, and the three subspace detectors (i.e. GLRT, AMF, and ACE) each implements the respective test for detecting the rank 1 deterministic signal $\hat{\boldsymbol{p}}$ as shown in Appendix 3.A. For vectors $\boldsymbol{p}$ and $\boldsymbol{v}$ that are inherently random, sample space realizations of these vectors for different Coherent Processing Intervals (CPIs) correspond to randomly scaled sum of the column vectors of $\boldsymbol{H}$ and $\boldsymbol{D}$ respectively. Therefore, it is not possible to use a rank 1 detector matched to any fixed signal to detect a subspace signal for the sample space implied by the model.

In summary, having obtained the estimate $\hat{\boldsymbol{p}} \in \mathcal{C}^{N\times 1}$ the subspace detectors are exactly the corresponding detectors for a rank 1 signal $\hat{\boldsymbol{p}}$, in unknown interference-plus-noise with the exception that the thresholds are chosen from the corresponding expressions for the probability of false alarm for subspace detectors of dimension M. There are two main issues that need to be considered if the hypothesized signal subspace dimension is increased by 1 ($1 \leq M < N$), with the first M dimensions being the same as before. First, if the hypothesized signal $\boldsymbol{p}$ and actual signal $\boldsymbol{v}$ are mismatched, increasing the subspace dimension cannot increase the generalized mismatch angle. If the $(M+1)$th subspace dimension represents the mismatch signal component exactly, the newly hypothesized signal model is matched with the actual signal model, and hence there is no loss in performance due to mismatch. The second issue to consider is the loss in performance due to the increased residual interference-plus-noise level as the hypothesized signal subspace dimension is increased. Figures presented in this section illustrate the result of the two competing effects in determining the detector performance.

Figures 3.1–3.3 illustrate the increase in residual interference-plus-noise power with increasing signal subspace dimension M. The quantity $\log_{10}(P_{FA})$ is shown as a function of the detection threshold, with the subspace dimension M shown as a parameter. Figures 3.1–3.3 are for the GLRT detector in (3.47), AMF detector in (3.49), and ACE detector in (3.51) respectively. In order to hold the probability of false alarm constant, the figures show that the detection threshold needs to be increased as the subspace dimension, M, is increased. Note that for purposes of implementing the computations, the probability of false alarm for a given threshold was evaluated by performing a numerical double integral given in (3.33) for the AMF detector and in (3.37) for the ACE detector. Since the double integral of the joint density function over $0 < r < 1$ and $0 < \rho < 1$ is 1, the evaluation of probability of false alarms (especially for $P_{FA} = 10^{-4}$ or smaller) requires control over numerical errors which can be specified as a parameter in most numerical integration algorithms. The expressions in (3.34) and (3.38) were used to evaluate the probability of detection since the numerical errors due to addition and subtraction of large quantities do not have to be as accurate as that required for the evaluation of the probability of false alarm. For the subspace GLRT the expressions in (3.30) and (3.31) were used for computing the probability of detection and probability of false alarm respectively.

When applied to detecting a rank 1 signal in interference and noise, the decrease in detection performance of subspace detectors as the hypothesized signal subspace dimension is increased

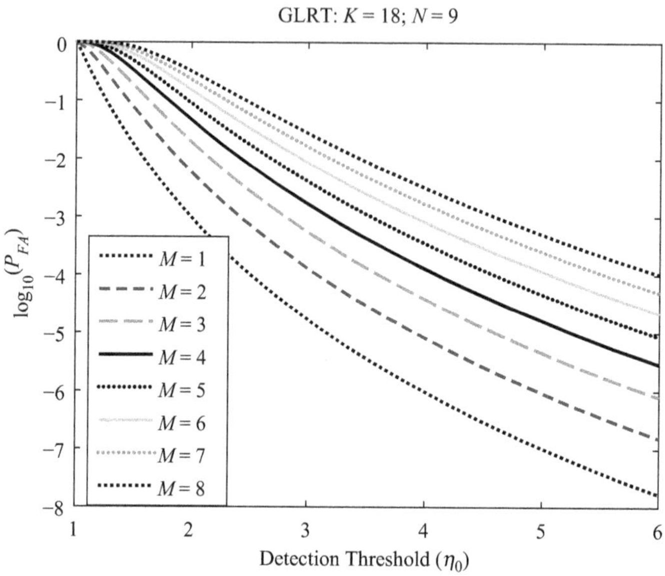

Figure 3.1 $\log_{10}(P_{FA})$ vs. the subspace GLRT threshold η_0 in (3.27) with signal subspace dimension M as a parameter. The figure shows that in order to fix the probability of false alarm at a required level, higher thresholds are required as the signal subspace dimension M is increased.

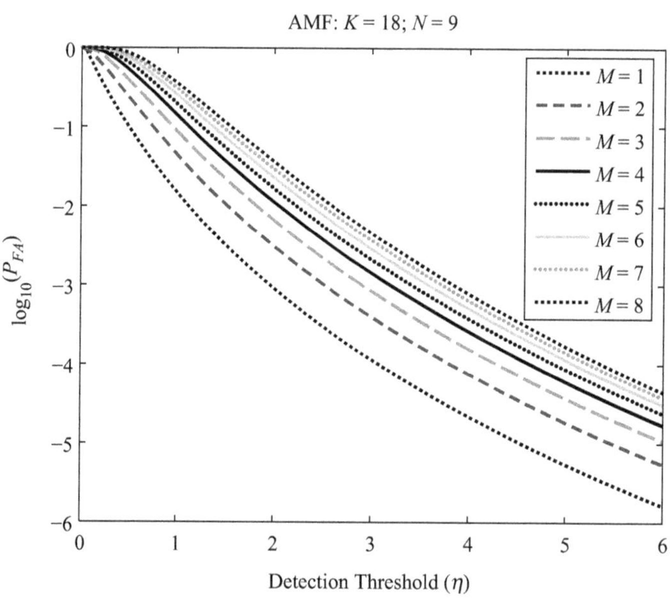

Figure 3.2 $\log_{10}(P_{FA})$ vs. the subspace AMF threshold η in (3.32) with signal subspace dimension M as a parameter. The figure shows that in order to fix the probability of false alarm at a required level, higher thresholds are required as the signal subspace dimension M is increased.

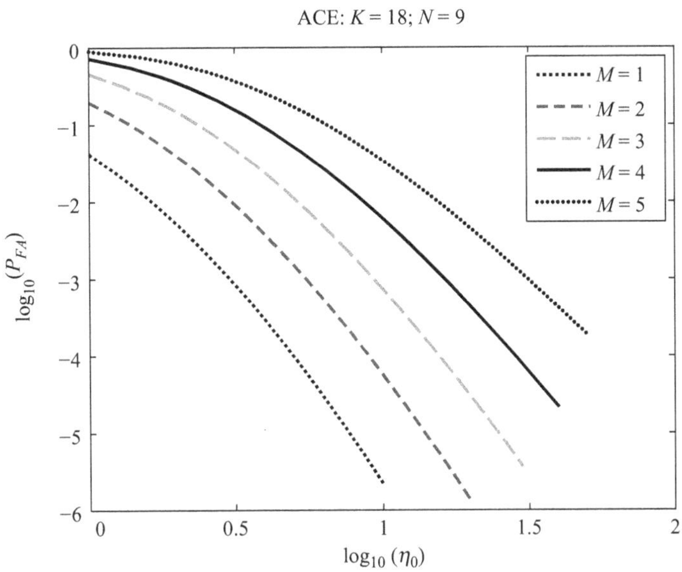

Figure 3.3 $\log_{10}(P_{FA})$ vs. $\log_{10}(\eta_0)$ with signal subspace dimension M as a parameter. The ACE threshold η_0 is given in (3.36). The figure shows that in order to fix the probability of false alarm at a required level, higher thresholds are required as the signal subspace dimension M is increased.

from $M = 1$ to $M = 5$ as illustrated in Figures 3.4–3.6 for the GLRT, AMF, and the ACE detectors. For $M = 1$, the hypothesized signal and the actual signal to be detected are matched and therefore the mismatch angle, $\theta = 0$. The first column of the signal matrix $\boldsymbol{H}$ is matched to the actual signal, the mismatch angle is 0° for all other hypothesized signal subspace dimensions (i.e. M) in these examples. These figures show plots of the probability of detection as a function of SINR (in dB) for the subspace GLRT, AMF, and ACE detectors respectively. The detection thresholds in each case were set such that the probability of false alarm is 10^{-4}. The number of training vectors $K = 18$ and each training vector is of size $N = 9$. The case $K = 2N$ was chosen for purposes of illustration. For each of the three detectors and for a given subspace dimension M, the probability of detection increases monotonically as a function of SINR from P_{FA} to 1. For mismatch angle $\theta = 0°$, these plots show the probability of detection decreasing with increasing subspace dimension M in all cases. Figure 3.7 is a plot of the SINR loss for the three subspace detectors with increasing subspace dimension M for probability of detection of 0.9. Results shown in this figure assume that the generalized mismatch angle $\theta = 0°$ for all three detectors and for all subspace dimensions M. The threshold is chosen in all cases such that the probability of false alarm is 10^{-4}. The SINR loss at a specific probability of detection of a given subspace detector is defined as the additional SINR (in dB) required in comparison to the corresponding detector for $M = 1$ such that the two detectors have the same probability of detection (0.9 in this case), with all other quantities remaining the same (i.e. quantities such as K, N, P_{FA}). By definition therefore, the SINR loss of all three detectors is 0 dB for $M = 1$. The plot of the SINR loss for the subspace GLRT and AMF detector approximately coincides. The SINR loss increases more drastically with increasing M for subspace ACE detector as shown in Figure 3.7.

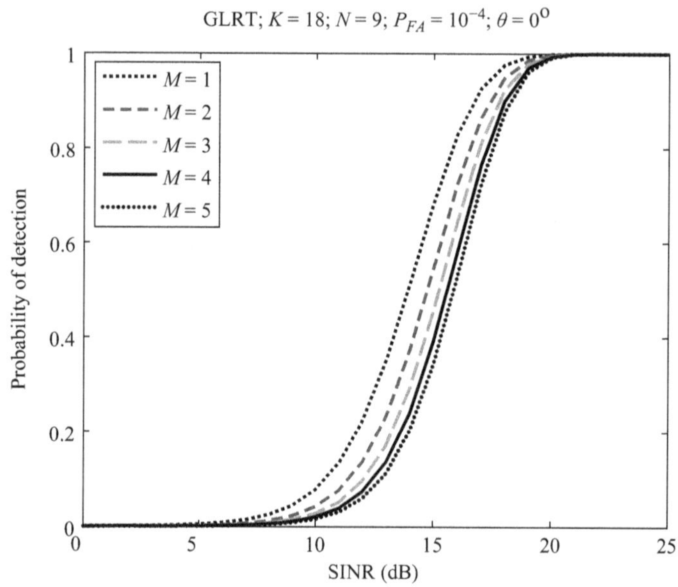

Figure 3.4 P_D vs. SINR (in dB) of the subspace GLRT for different signal subspace dimensions M. The thresholds are chosen such that $P_{FA} = 10^{-4}$ in all cases.

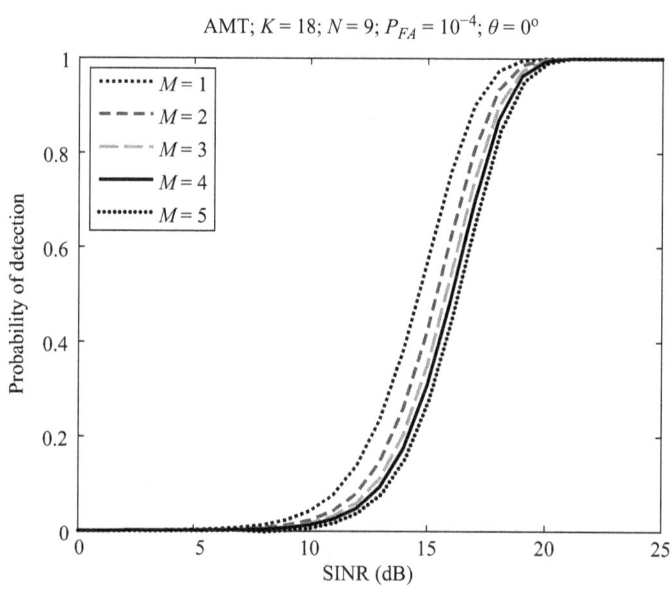

Figure 3.5 P_D vs. SINR (in dB) of the subspace AMF detector for different signal subspace dimensions M. The thresholds are chosen such that $P_{FA} = 10^{-4}$ in all cases.

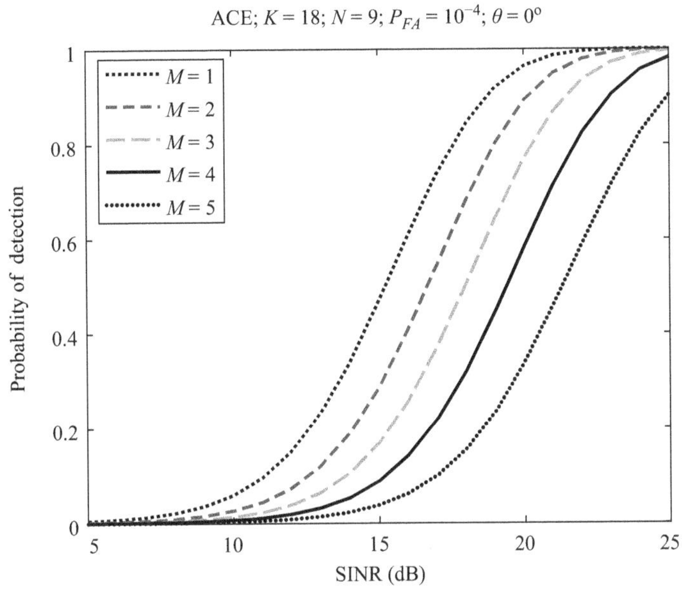

Figure 3.6 P_D vs. SINR (in dB) of the subspace ACE detector for different signal subspace dimensions M. The thresholds are chosen such that $P_{FA} = 10^{-4}$ in all cases.

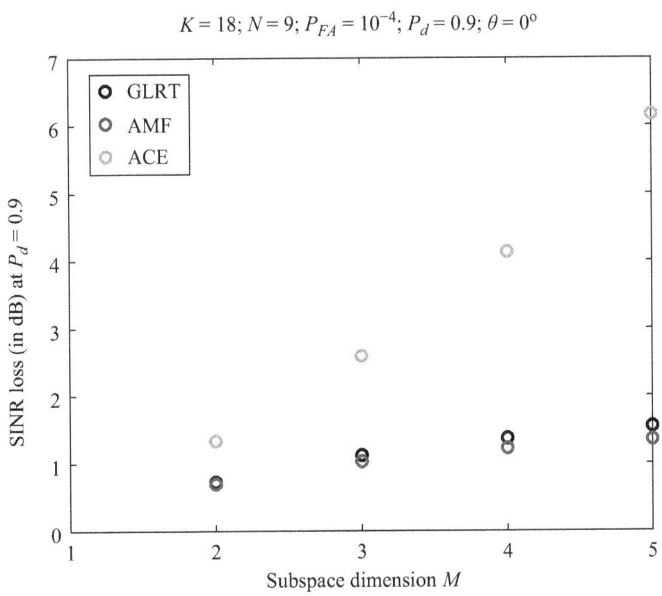

Figure 3.7 Additional SINR required relative to the corresponding detector for $M = 1$ such that $P_d = 0.9$ shown as a function of signal subspace dimension M for subspace GLRT, AMF, and ACE detectors. The mismatch angle $\theta = 0°$ in all cases.

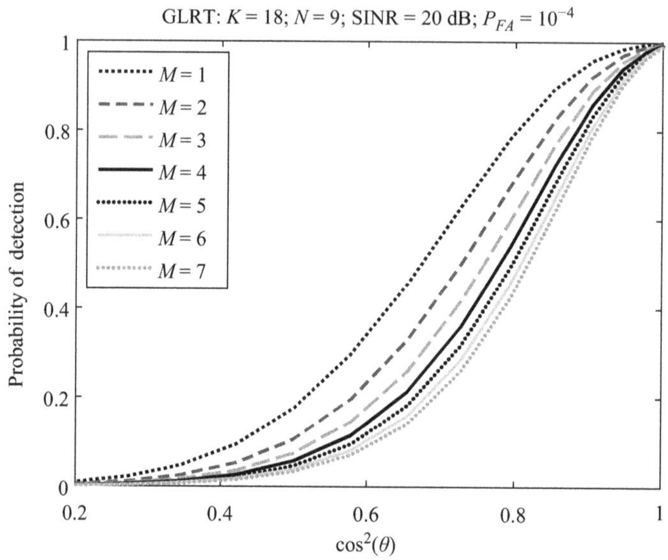

Figure 3.8 Probability of detection for the subspace GLRT as a function of $\cos^2(\theta)$ for different signal subspace dimensions M. The probability of false alarm is 10^{-4} and the SINR = 20 dB in all cases. The figure shows the mismatch signal rejection characteristics of the subspace GLRT.

For a given SINR, Figures 3.8–3.10 are plots of the probability of detection P_D as a function of $\cos^2(\theta)$, the squared-cosine of the generalized mismatch angle for the subspace GLRT, AMF, and ACE detectors respectively. These figures show P_D vs. $\cos^2(\theta)$ for different hypothesized signal subspace dimensions M. A key assumption to keep in mind is that the quantity $\cos^2(\theta)$ is chosen as an independent variable along the abscissa for the various curves in each figure with SINR held fixed. For any specific deterministic signal, the mismatch angle $|\theta|$ would either be constant or decrease as the subspace dimension M is increased. The results are shown for a SINR of 20 dB. Other quantities assumed are: $K = 18$, $N = 9$, and $P_{FA} = 10^{-4}$. With increasing subspace dimension M, the mismatched signal rejection characteristics of the three subspace detectors can be understood in terms of the detection statistic of the three detectors. As shown in Appendix 3.A, the detection statistics for the subspace GLRT, AMF, and ACE detectors are (an equivalent ACE statistic is shown): $(\boldsymbol{y}^\dagger \boldsymbol{P_G y})/(1 + \boldsymbol{y}^\dagger \boldsymbol{P}_G^\perp \boldsymbol{y})$, $(\boldsymbol{y}^\dagger \boldsymbol{P_G y})$, and $(\boldsymbol{y}^\dagger \boldsymbol{P_G y})/(\boldsymbol{y}^\dagger \boldsymbol{P}_G^\perp \boldsymbol{y})$ respectively. $\boldsymbol{P_G}$ and $\boldsymbol{P}_G^\perp$ are the orthogonal projection matrix for the M-dimensional column space of: $\boldsymbol{S}^{-1/2}\boldsymbol{H}$ (denoted by $< \boldsymbol{S}^{-1/2}\boldsymbol{H} >$) and its orthogonal complement subspace in $\mathcal{C}^{N\times 1}$ (dimension $N - M$) respectively. The vector $\boldsymbol{y}$ is given in terms of the test vector $\boldsymbol{z}$ by: $\boldsymbol{y} = \boldsymbol{S}^{-1/2}\boldsymbol{z}$. Mismatched signals are rejected more effectively by the GLRT and ACE detectors in comparison to the AMF detector because of the term $\boldsymbol{y}^\dagger \boldsymbol{P}_G^\perp \boldsymbol{y}$ in the GLRT and ACE statistics. The presence of a strong mismatched signal component contributes significantly to this term, which reduces the detection static of both the GLRT and the ACE detectors. The above term has no effect on the AMF test statistic however. For a high SINR and a mismatch angle chosen independently to be $\theta = 90°$ the AMF detection statistic increases with increasing subspace dimension M as shown in Figure 3.9.

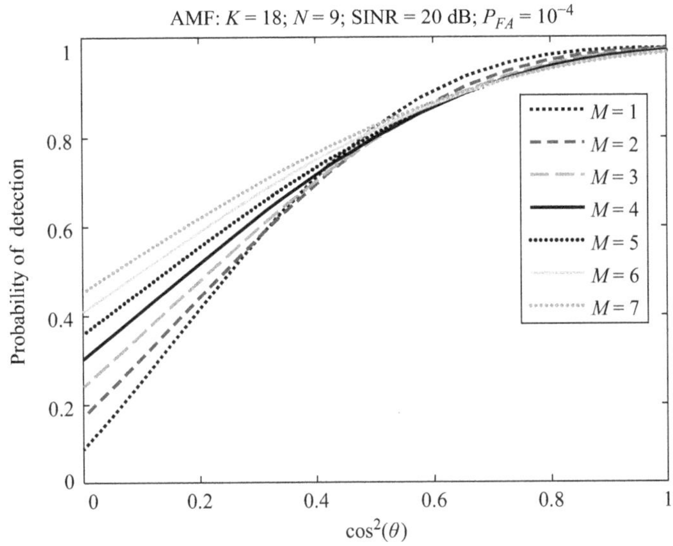

Figure 3.9 Probability of detection for the subspace AMF detector as a function of $\cos^2(\theta)$ for different signal subspace dimensions M. The probability of false alarm is 10^{-4} and the SINR = 20 dB in all cases. The figure shows the mismatch signal rejection characteristics of the subspace AMF detector.

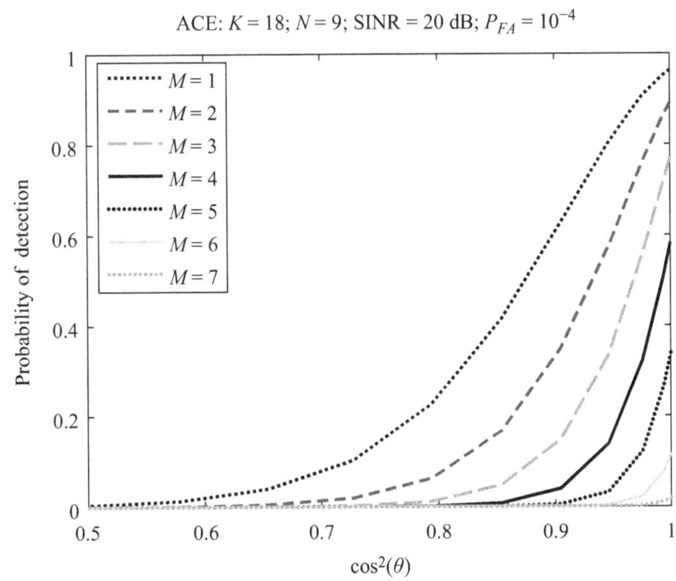

Figure 3.10 Probability of detection for the subspace ACE detector as a function of $\cos^2(\theta)$ for different signal subspace dimensions M. The probability of false alarm is 10^{-4} and the SINR = 20 dB in all cases. The figure shows the mismatch signal rejection characteristics of the subspace ACE detector.

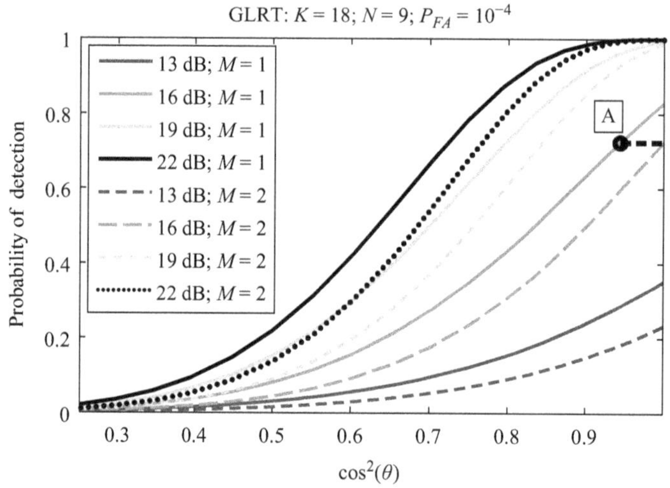

Figure 3.11 P_D for the subspace GLRT $M = 1$ (solid lines) and $M = 2$ (dotted lines) as a function of $\cos^2(\theta)$ for different SINR (in dB). The probability of false alarm is 10^{-4} in all cases.

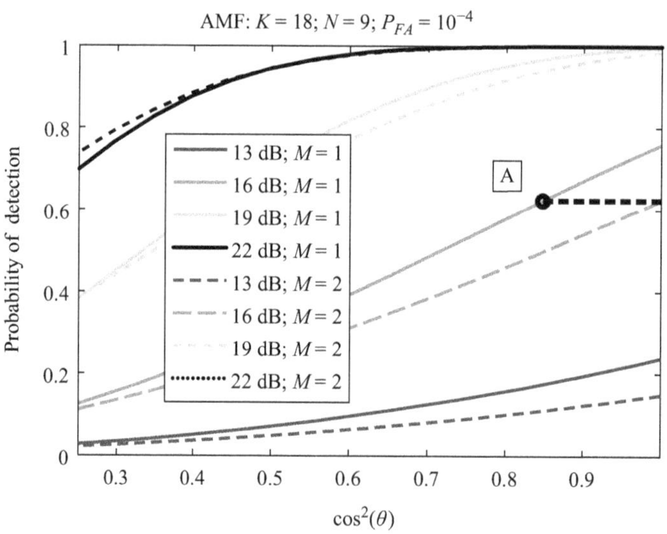

Figure 3.12 P_D for the subspace AMF detector $M = 1$ (solid lines) and $M = 2$ (dotted lines) as a function of $\cos^2(\theta)$ for different SINR (in dB). The probability of false alarm is 10^{-4} in all cases.

For a given SINR and mismatch angle θ (specified as an independent quantity), the probability of detection for the GLRT and ACE detectors decreases as the subspace dimension M is increased as shown in Figures 3.8 and 3.10 respectively. This implies that the smallest subspace dimension is generally preferred for these detectors. By comparison, increasing the subspace dimension M has a different effect on the AMF detector as illustrated in Figure 3.9.

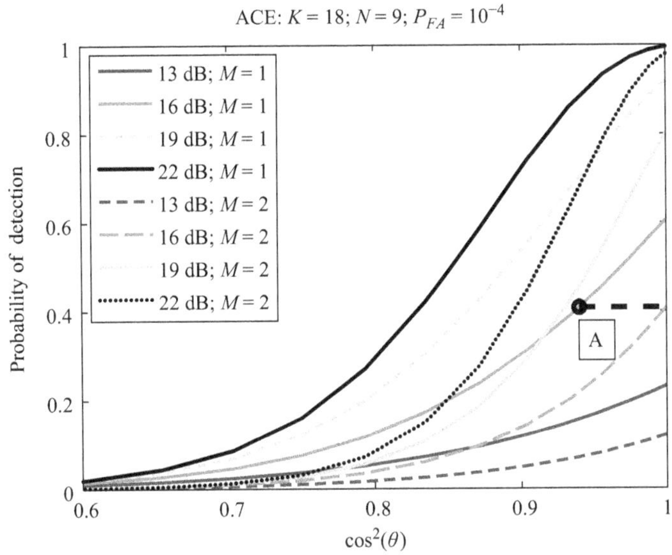

Figure 3.13 P_D for the subspace ACE detector $M=1$ (solid lines) and $M=2$ (dotted lines) as a function of $\cos^2(\theta)$ for different SINR (in dB). The probability of false alarm is 10^{-4} in all cases.

The effect of mismatch between the hypothesized signal and the actual signal is illustrated in Figures 3.11–3.13 for the GLRT, AMF, and ACE detectors respectively. The three figures show the probability of detection as a function of the squared generalized cosine of the mismatch angle $\cos^2(\theta)$, as defined in (3.66). The plots are parameterized in terms of the SINR (in dB) and are shown for the subspace detectors for the hypothesized signal subspace dimensions $M=1$ and $M=2$. For all three detectors, $K=18$, $N=9$, and $P_{FA}=10^{-4}$. For illustrative purposes, consider the point marked by the symbol A on these plots. For all three detectors, the point A appears on the $\cos^2(\theta)$ vs. P_D plot, parameterized by SINR $=16$ dB and $M=1$. The coordinates $(\cos^2(\theta), P_D)$ of the point A are $(0.944, 0.724)$, $(0.85, 0.624)$, and $(0.94, 0.41)$ for the GLRT, AMF, and ACE detectors respectively. The squared generalized cosine of the mismatch angle for point A is chosen such that, the probability of detection for SINR $=16$ dB and $M=1$ is exactly equal to the probability of detection for a corresponding detector for subspace dimension $M=2$, SINR $=16$ dB and mismatch angle $\theta=0°$, with all other quantities held constant.

Thus, the corresponding coordinates of a subspace detector for $M=2$ which incorporates the mismatched signal component as the second column of the signal matrix $\boldsymbol{H}$ are: $(\cos^2(\theta), P_D) = (1, 0.724)$, $(1, 0.624)$, and $(1, 0.41)$ for the GLRT, AMF, and ACE detectors respectively. Note that the probability of detection for the $M=1$ detector with signal mismatch is same as the probability of detection for the corresponding detector with $M=2$ and with no mismatch. These coordinates are also the coordinates of the single point of intersection in $(\cos^2(\theta), P_D)$ space of the horizontal dotted line with the vertical line $\cos^2(\theta)=1$ and the $(\cos^2(\theta)$ vs. $P_D)$ curve for SINR $=16$ dB for $M=2$. It is evident from these figures that a subspace detector for $M=1$ (mismatched case) has a higher probability of detection compared to a subspace detector with $M=2$ (matched case) for points on the SINR $=16$ dB curve that are to the right of A. Similarly, a subspace detector for $M=1$ (mismatched case) has a lower probability

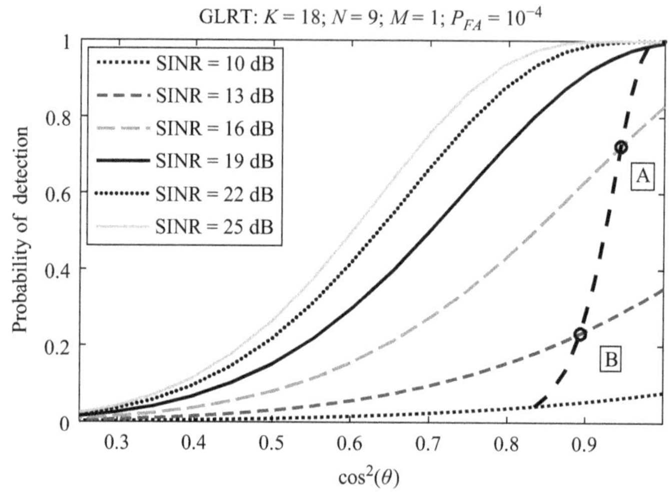

Figure 3.14 P_D for the GLRT ($M=1$) as a function of $\cos^2(\theta)$ for different SINR (in dB). The signal model for $M=2$ is perfectly matched with the hypothesized signal and so the P_D corresponding to the point of intersection of the solid curve (for a given SINR), and the dotted line (e.g., point A for SINR = 16 dB) is the detection probability for the subspace detector corresponding to $M=2$. Thus, for all mismatch angles to the left of the dotted line, increasing the subspace dimension to $M=2$ increases the probability of detection. For mismatch angles to the right of the dotted line, increasing the signal subspace dimension decreases the probability of detection. The probability of false alarm is 10^{-4} in all cases.

of detection compared to a subspace detector with $M=2$ (matched case) for points on the SINR = 16 dB curve that are to the left of A. The point A is where the effects due to the two opposing factors (i.e. the increase of residual interference-plus-noise power as signal subspace dimension M increases vs. reduced signal mismatch errors as signal subspace dimension M, increases balance each other). Note that the point A was selected on the SINR = 16 dB curve for purposes of illustration and the above arguments hold for any other SINR.

As the SINR is varied, the locus of points on the $\cos^2(\theta)$ vs. P_D plots where the P_D for the $M=2$, subspace detector with $\theta=0$ is exactly equal to the P_D for the corresponding mismatched subspace detector for $M=1$ and the same SINR is shown by the dark dotted line in Figures 3.14–3.16 for the subspace GLRT, AMF, and ACE detectors respectively. All other parameters such as K, N, P_{FA} are held constant. The $\cos^2(\theta)$ vs. P_D shown in these figures are for $M=1$ and do not show the results for $M=2$, as in Figures 3.11–3.13 for purposes of clarity. The dark dotted line partitions the P_D vs. $\cos^2(\theta)$ space into two distinct regions. For a given SINR, a subspace detector for $M=1$ with signal mismatch has a higher probability of detection in comparison to that of the corresponding subspace detector for $M=2$ and zero mismatch angle for all points to the right of the line of demarcation. Similarly, for a given SINR, a subspace detector for $M=1$ with signal mismatch has a lower probability of detection in comparison to that of the corresponding subspace detector for $M=2$ and zero mismatch angle for all points to the left of the line of demarcation. The difference in detection probabilities can be significant in the latter case

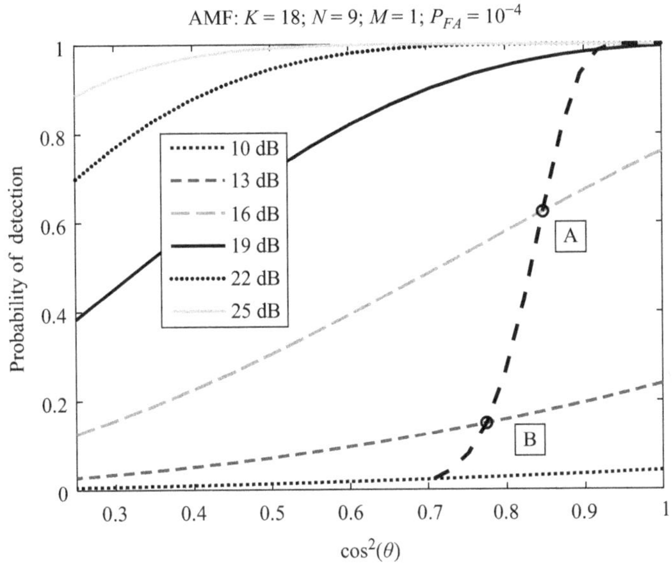

Figure 3.15 Results corresponding to Figure 3.14 for the AMF detector.

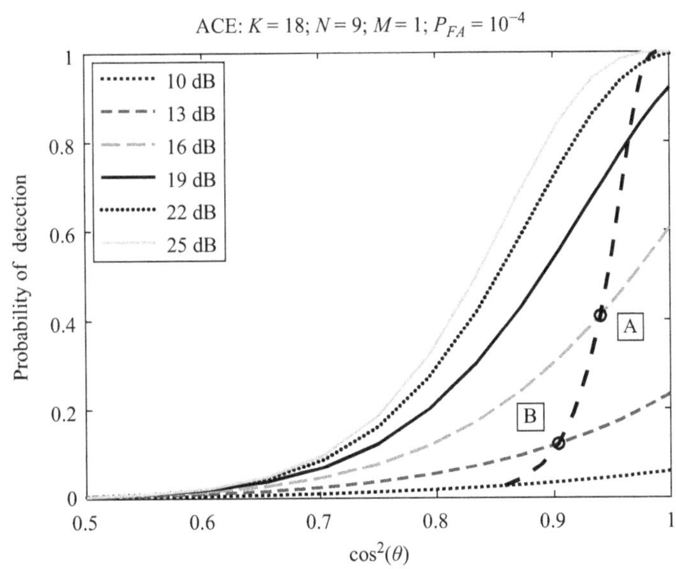

Figure 3.16 Results corresponding to Figure 3.14 for the ACE detector.

as shown by points marked by the symbols A and B on these plots for example which correspond to SINR = 16 dB and SINR = 13 dB respectively. The ordinates of points A and B are the probability of detection for the subspace detector with $M = 2$ (matched case), and the ordinates of points to the left of A and B for the corresponding SINR are the detection

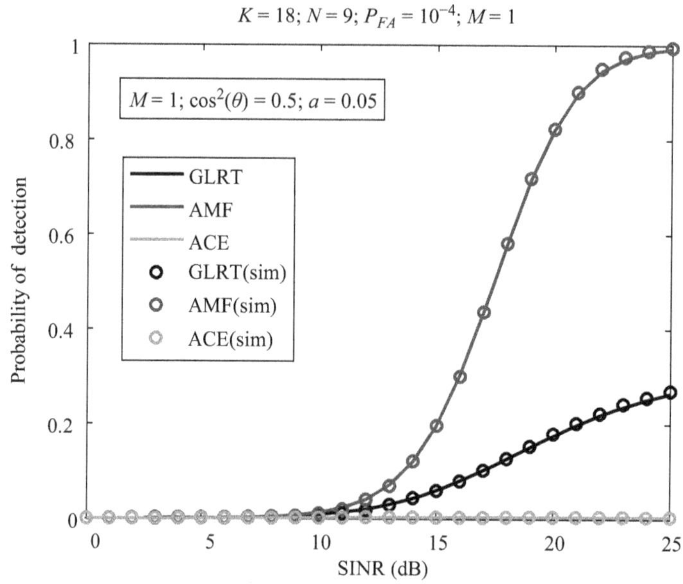

Figure 3.17 P_D vs. SINR (in dB) for the three detectors. The columns of the matrix $\boldsymbol{D}$ are the orthonormal eigenvectors of the signal covariance matrix with the CMT as specified in (3.14) and (3.16) for $a = 0.05$. The results are for the three subspace detectors with the signal space dimension set to $M = 1$. The matrix $\boldsymbol{H}$ is the first column of $\boldsymbol{D}$. The actual signal is mismatched to the hypothesized signal in this example and $\cos^2(\theta) = 0.5$. The probability of false alarm is 10^{-4} in all cases. Solid lines are results from analytical expressions, and symbols are results from computer simulations.

probabilities of a subspace detector for $M = 1$. Thus a subspace detector with a higher signal dimension ($M = 2$ and matched, in this example) can have a significantly higher probability of detection compared to the corresponding subspace detector for $M = 1$, depending on the mismatch angle. Note that signal model assumes that the parameter vector $\boldsymbol{\alpha}$ is deterministic and therefore given a model for mismatch errors, such as those in (3.14) and (3.16) for example, it is possible to increase the subspace dimension to increase detection performance. This is illustrated in Figures 3.17 and 3.18, which are plots of the probability of detection as a function of SINR (in dB), for the three subspace detectors with $M = 1$ and $M = 2$ respectively.

Details of the example are as follows: consider a search application where a signal return from a point situated within a given sector defined with respect to the boresite direction of a uniform linear array of N elements and at a hypothesized range is to be detected in interference and noise. As an example, let the assumed signal return be from the boresite direction and we generate the tapered covariance matrix as specified in (3.14) and (3.16). In this example, the parameter $a = 0.05$. The orthonormal eigenvectors corresponding to the dominant eigenvalues of the signal covariance matrix are the basis vectors of the signal subspace and model the signal vectors from all angles within the search sector more accurately than any single steering vector. The columns of the matrix $\boldsymbol{D}$ are the orthonormal eigenvectors

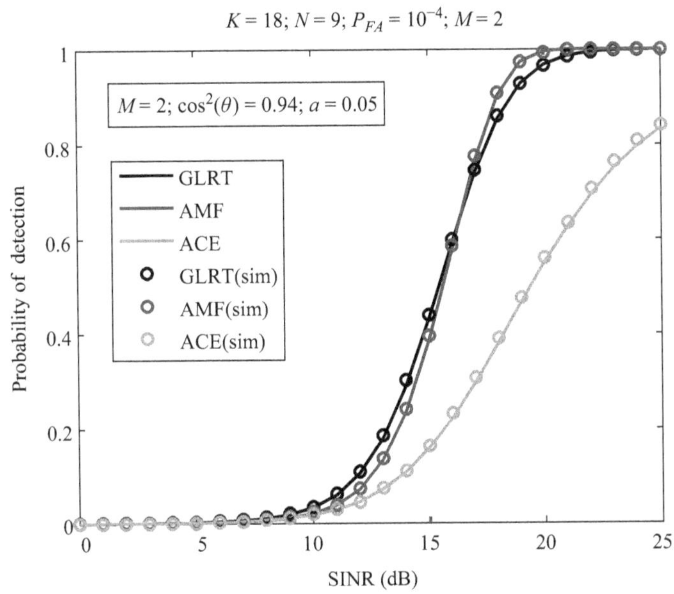

Figure 3.18 P_D vs. SINR (in dB) for the three detectors. The columns of the matrix $\boldsymbol{D}$ are the orthonormal eigenvectors of the signal covariance matrix with the CMT as specified in (3.14) and (3.16) for $a = 0.05$. The results are for the three subspace detectors with the signal space dimension set to $M = 2$. The matrix $\boldsymbol{H}$ is formed by the first two columns of the matrix $\boldsymbol{D}$. The actual signal is the same as that used in Figure 3.17 and increasing the signal space dimension to $M = 2$ reduces the mismatch angle such that $\cos^2(\theta) = 0.94$. The probability of false alarm is 10^{-4} in all cases. Solid lines are results from analytical expressions, and symbols are results from computer simulations.

of the signal covariance matrix with the CMT as specified in (3.14) and (3.16) for $a = 0.05$. The eigenvectors are arranged from left to right in descending order of magnitude of the eigenvalues. Thus, for the hypothesized signal subspace dimension $M = 1$, the hypothesized signal is the eigenvector with the largest eigenvalue of the matrix in (3.14) with the taper in (3.16). As an example, consider the actual signal $\boldsymbol{v} = [1 \ e^{j2\pi a} \ e^{j4\pi a} \ldots e^{j2(N-1)\pi a}]^{\dagger}$ for $a = 0.05$, which corresponds to a steering vector pointed to the azimuth angle $\sin(\phi) = 0.1$. Since the probability of false alarm of the detectors can be set independent of the interference-plus-noise covariance matrix (CFAR property), this matrix is set to the identity matrix in this example. For $M = 1$, the mismatch angle is $\cos^2(\theta) = 0.5$, and Figure 3.17 is a plot of the probability of detection vs. SINR (in dB) of the three subspace detectors. Increasing the hypothesized signal subspace dimension to $M = 2$ in this case reduces the mismatch angle: $\cos^2(\theta) = 0.94$, and Figure 3.18 shows a plot of the probability of detection vs. SINR (in dB) of the three subspace detectors. Note that the SINR is not changed by changing the signal subspace dimension M and it is only the signal mismatch angle θ that is changed. Thus, increasing the subspace dimension by 1 results in significantly improving the detection performance in Figure 3.18 compared to the detection performance of the corresponding detector in Figure 3.17.

3.6 Summary and Conclusions

The chapter begins with a brief review of binary hypothesis testing for detecting a known signal in zero-mean Gaussian noise with known covariance matrix. The chapter introduces the reader to the problem of detecting a signal assumed to belong to a subspace of dimension M; $1 \leq M \leq N$ in unknown interference from an invariance perspective. Excluding the case $M = 1$, subspace signals provide an approach to model uncertainties due to various errors described in the text, which generally do not enable a signal to be known to within a non-zero multiplicative constant. We have shown that the invariance framework in hypothesis testing provides a systematic approach for dealing with the large number of nuisance parameters typically inherent in these problems and that the resulting subspace detectors have the CFAR feature. With the CFAR feature, the receiver can select the detection threshold for any chosen invariant hypothesis test and a hypothesized signal subspace dimension M, without knowledge of the interference-plus-noise covariance matrix so that the probability of false alarm is fixed at a required value. The test statistics of all invariant tests are constructed from maximal invariant statistic and for signal subspace dimension $1 \leq M < N$, maximal invariant statistic was shown to be a two-dimensional statistic. For $M = N$, the maximal invariant statistic reduces to a single scalar statistic. The joint PDFs of maximal invariant statistic conditioned on (i) the null hypothesis H_0 and (ii) the alternative hypothesis H_1 were derived. The analysis of the joint PDF included model mismatch errors where the assumed subspace model for the input signal: $\boldsymbol{p} = \boldsymbol{H}\ \boldsymbol{\alpha}$ is mismatched with the actual input signal which is of the form: $\boldsymbol{v} = [\boldsymbol{H}\ \ \boldsymbol{H}_{\perp}]\ \boldsymbol{\gamma}$ as described in (3.12). The $N \times M$ matrix $\boldsymbol{H}$ is known and comprises linearly independent columns. The column space of $\boldsymbol{H}$ denoted by $<\boldsymbol{H}>$ is the subspace of the hypothesized signal $\boldsymbol{p}$. Similarly, the columns of the matrix $N \times (N - M)$ matrix $\boldsymbol{H}_{\perp}$ denotes a subspace orthogonal to the hypothesized signal subspace. The subspaces $<\boldsymbol{H}>$ and $<\boldsymbol{H}_{\perp}>$ are orthogonal and complementary subspaces of $\mathcal{C}^{N\times 1}$. The vectors $\boldsymbol{\alpha} \in \mathcal{C}^{M\times 1}$ and $\boldsymbol{\gamma} \in \mathcal{C}^{N\times 1}$ are deterministic but unknown and in our model, these vectors introduce the uncertainty in not knowing the exact input signal.

Although an infinite number of invariant tests can be constructed from the two-dimensional maximal invariant statistic, we restricted our analysis of performance to subspace versions of the (i) GLRT, (ii) AMF, and (iii) ACE. Analytical expressions for the probability of false alarm and probability of detection of the three subspace detectors were derived including the case of signal model mismatch errors. We have shown that each of the three subspace detectors is identical in form to the corresponding detector for $M = 1$, where the steering vector $\boldsymbol{p}$ is replaced by the signal vector estimate: $\hat{\boldsymbol{p}} = \boldsymbol{H}(\boldsymbol{H}^{\dagger}\boldsymbol{S}^{-1}\boldsymbol{H})^{-1}\boldsymbol{H}^{\dagger}\boldsymbol{S}^{-1}\boldsymbol{z}$. The threshold levels for each of the subspace detectors needs to be increased in comparison to the corresponding detector for $M = 1$ in order to hold the probability of false alarm constant. The estimated signal vector $\hat{\boldsymbol{p}}$ is random because both the estimated interference-plus-noise covariance matrix $\boldsymbol{S}$ and the interference-plus-noise in the test vector $\boldsymbol{z}$ are random quantities independent of the signal. It is possible to consider random effects inherent in the signal model itself by considering $\boldsymbol{\alpha}$ and $\boldsymbol{\gamma}$ as random vectors. The analysis of random effects in the signal model itself is not carried out and is beyond the scope of the material in this chapter.

For a deterministic signal, we have illustrated that it may be possible to mitigate the loss in detection performance due to signal mismatch errors for $M = 1$ by increasing the subspace dimension M. The improvement in detection performance is possible only if the generalized mismatch angle $(0 \leq \theta \leq \pi/2)$ for $M = 1$ is above a certain mismatch angle threshold. For a given SINR, the threshold angle where the probability of detection of a detector with $M = 1$ equals the detection probability of the corresponding detector with $M = 2$ but with no signal mismatch errors defines the mismatch angle threshold. All other quantities for the $M = 1$ and $M = 2$ except for the mismatch angles are the same. The mismatch angle threshold is a function of SINR and the threshold for the generalized cosine square of mismatch angle generally moves towards $\cos^2(\theta) = 1$ as the SINR increases. Thus, for generalized cosine square of mismatch angle errors that are larger that the threshold, it is not possible to reduce the loss in detection performance due to signal model mismatch errors by increasing the signal subspace dimension alone. A change in the assumed model $\boldsymbol{H}$ is require in these cases. For generalized cosine square of mismatch errors that are smaller than the threshold we have illustrated with an example to mitigate the loss in detection performance by increasing the signal subspace dimension. A CMT is applied to the outer product of a rank 1 signal and eigenvectors of the resulting matrix defines the columns of matrix $\boldsymbol{D}$.

Appendix 3.A

This appendix contains derivations of the subspace signal detectors based on: (i) the GLRT, (ii) the AMF test, and (iii) the ACE test.

We begin by writing the multivariate complex Gaussian PDF:

$$f(z, \boldsymbol{Y}|\boldsymbol{R}, H_0) = \frac{e^{-\mathrm{Tr}\left[\boldsymbol{R}^{-1}[zz^\dagger + \boldsymbol{Y}\boldsymbol{Y}^\dagger]\right]}}{\pi^{(K+1)N}\,[\det(\boldsymbol{R})]^{(K+1)}}, \tag{3.40}$$

where $\boldsymbol{R}$ is the unknown covariance matrix of interference-plus-noise, `Tr[ ]` is the trace operator, and $\dagger$ denotes the Hermitian transpose of a matrix. The interference model used here assumes that the primary vector (i.e. the test vector) and secondary vectors (i.e. training vectors) are statistically independent and have the same covariance matrix. The primary and secondary data under the alternative hypothesis H_1 is given by:

$$f(z, \boldsymbol{Y}|\boldsymbol{R}, \boldsymbol{p}, H_1) = \frac{e^{-\mathrm{Tr}\left[\boldsymbol{R}^{-1}[(z-\boldsymbol{p})(z-\boldsymbol{p})^\dagger + \boldsymbol{Y}\boldsymbol{Y}^\dagger]\right]}}{\pi^{(K+1)N}\,[\det(\boldsymbol{R})]^{(K+1)}}. \tag{3.41}$$

The model for the signal vector $\boldsymbol{p}$ is described in (3.12) of Section 3.3.

3.A.1 Subspace Version of the GLRT

Under hypothesis H_0, the estimate $\hat{\boldsymbol{R}}_0 = (zz^\dagger + \boldsymbol{Y}\boldsymbol{Y}^\dagger)/(K+1)$ maximizes the likelihood function in (3.40). Defining the matrix $\boldsymbol{S} = \boldsymbol{Y}\boldsymbol{Y}^\dagger$, we have

$$\begin{aligned}\max_{\boldsymbol{R}} : f(z, \boldsymbol{Y}|\boldsymbol{R}, H_0) &= f(z, \boldsymbol{Y}|\boldsymbol{R} = \hat{\boldsymbol{R}}_0, H_0)\\ &= c[\det(\boldsymbol{S} + zz^\dagger)]^{-(K+1)}.\end{aligned} \tag{3.42}$$

In the second expression above, c is a constant independent of the data and is therefore unimportant. Similarly, the estimate $\hat{\boldsymbol{R}}_1 = ([z - \boldsymbol{H}\boldsymbol{\alpha}][z - \boldsymbol{H}\boldsymbol{\alpha}]^\dagger + \boldsymbol{Y}\boldsymbol{Y}^\dagger)/(K+1)$ maximizes the likelihood function in (3.41) for a given $\boldsymbol{p} = \boldsymbol{H}\boldsymbol{\alpha}$. Thus,

$$\begin{aligned}\max_{\boldsymbol{R}} : f(z, \boldsymbol{Y}|\boldsymbol{R}, \boldsymbol{\alpha}, H_1) &= f(z, \boldsymbol{Y}|\boldsymbol{R} = \hat{\boldsymbol{R}}_1, \boldsymbol{\alpha}, H_1)\\ &= c[\det(\boldsymbol{S} + [z - \boldsymbol{H}\boldsymbol{\alpha}][z - \boldsymbol{H}\boldsymbol{\alpha}]^\dagger)]^{-(K+1)}.\end{aligned} \tag{3.43}$$

For $K \geq N$, the matrix $\boldsymbol{S}$ is positive definite with probability 1, and therefore the determinant above can be written as:

$$\det(\boldsymbol{S} + [z - \boldsymbol{H}\boldsymbol{\alpha}][z - \boldsymbol{H}\boldsymbol{\alpha}]^{\dagger}) = \det(\boldsymbol{S})\,(1 + [z - \boldsymbol{H}\boldsymbol{\alpha}]^{\dagger}\boldsymbol{S}^{-1}[z - \boldsymbol{H}\boldsymbol{\alpha}]). \tag{3.44}$$

The expression (3.43) is maximized over $\boldsymbol{\alpha}$ by setting $\boldsymbol{H}^{\dagger}\boldsymbol{S}^{-1}[z - \boldsymbol{H}\boldsymbol{\alpha}] = 0$. Substituting the estimate: $\hat{\boldsymbol{\alpha}} = (\boldsymbol{H}^{\dagger}\boldsymbol{S}^{-1}\boldsymbol{H})^{-1}\boldsymbol{H}^{\dagger}\boldsymbol{S}^{-1}z$ in (3.43) and rearranging terms, we have:

$$\begin{aligned}\max_{\boldsymbol{R},\boldsymbol{\alpha}} : f(z, \boldsymbol{Y}|\boldsymbol{R}, \boldsymbol{\alpha}, H_1) &= f(z, \boldsymbol{Y}|\boldsymbol{R} = \hat{\boldsymbol{R}}_1, \boldsymbol{\alpha} = \hat{\boldsymbol{\alpha}}, H_1)\\ &= c(1 + z^{\dagger}\boldsymbol{S}^{-1}z - z^{\dagger}\boldsymbol{S}^{-1}\boldsymbol{H}(\boldsymbol{H}^{\dagger}\boldsymbol{S}^{-1}\boldsymbol{H})^{-1}\boldsymbol{H}^{\dagger}\boldsymbol{S}^{-1}z)^{-(K+1)}.\end{aligned} \tag{3.45}$$

The detection statistic is formed by taking the ratio of the maximized likelihood quantities in (3.45) and (3.42). Taking the $(K+1)$ th root of the ratio, the statistic can be rearranged to obtain:

$$\frac{z^{\dagger}\boldsymbol{S}^{-1}\boldsymbol{H}(\boldsymbol{H}^{\dagger}\boldsymbol{S}^{-1}\boldsymbol{H})^{-1}\boldsymbol{H}^{\dagger}\boldsymbol{S}^{-1}z}{(1 + z^{\dagger}\boldsymbol{S}^{-1}z)} \underset{H_0}{\overset{H_1}{\gtrless}} \eta. \tag{3.46}$$

Define $\boldsymbol{y} = \boldsymbol{S}^{-1/2}z$, $\boldsymbol{G} = \boldsymbol{S}^{-1/2}\boldsymbol{H}$ and the orthogonal projection matrices $\boldsymbol{P_G} = \boldsymbol{G}(\boldsymbol{G}^{\dagger}\boldsymbol{G})^{-1}\boldsymbol{G}^{\dagger}$ and $\boldsymbol{P}_{\boldsymbol{G}}^{\perp} = \boldsymbol{I}_N - \boldsymbol{P_G}$. Note that pre-multiplication of z by $\boldsymbol{S}^{-1/2}$ is for the purpose of "whitening" the interference in the vector. The pre-multiplication of $\boldsymbol{H}$ by the same matrix transforms the column space of $\boldsymbol{H}$, and therefore, $\boldsymbol{P_G}$ is the orthogonal projection matrix to the transformed signal subspace. The subspace GLRT in (3.27) is equivalent to:

$$\frac{\boldsymbol{y}^{\dagger}\boldsymbol{P_G}\boldsymbol{y}}{(1 + \boldsymbol{y}^{\dagger}\boldsymbol{P}_{\boldsymbol{G}}^{\perp}\boldsymbol{y})} \underset{H_0}{\overset{H_1}{\gtrless}} \frac{\eta}{1 - \eta}. \tag{3.47}$$

The above is the subspace signal version of the GLRT. It is of interest to note that a version of the GLRT applicable to multiple CPIs appears in References 53, 54.

3.A.2 Subspace Version of the AMF Test

The subspace version of the AMF test is obtained by deleting the term $(1 + z^{\dagger}\boldsymbol{S}^{-1}z)$ in (3.46):

$$z^{\dagger}\boldsymbol{S}^{-1}\boldsymbol{H}(\boldsymbol{H}^{\dagger}\boldsymbol{S}^{-1}\boldsymbol{H})^{-1}\boldsymbol{H}^{\dagger}\boldsymbol{S}^{-1}z \underset{H_0}{\overset{H_1}{\gtrless}} \eta. \tag{3.48}$$

The above test is also obtained if one were to start from a likelihood ratio formulation and substitute $\hat{\boldsymbol{R}} = \boldsymbol{S}$ for both hypotheses and then maximize the likelihood ratio over the unknown vector $\boldsymbol{\alpha}$.

Using the earlier definitions, the AMF test is also given by:

$$\boldsymbol{y}^{\dagger}\boldsymbol{P_G}\boldsymbol{y} \underset{H_0}{\overset{H_1}{\gtrless}} \eta. \tag{3.49}$$

3.A.3 Subspace Version of the ACE Test

The Matched Subspace Detector (MSD) and the adaptive version of the MSD that is the subspace version of the ACE test are discussed in [55] and [56] respectively. As in the case of

$M = 1$, the subspace version of ACE test can be derived if one assumes that the interference covariance matrix of the primary data is the interference covariance matrix of the secondary data multiplied by an unknown positive constant. Let the covariance matrix of the primary data interference be $q\boldsymbol{R}$, where $q > 0$ and is unknown. The covariance matrix of the interference in the secondary data set is $\boldsymbol{R}$, which is also unknown. To derive the ACE test, we substitute $\hat{\boldsymbol{R}} = \boldsymbol{Y}\boldsymbol{Y}^\dagger = \boldsymbol{S}$ in the likelihood function for H_0 and obtain the Maximum Likelihood Estimate (MLE) of the parameter q – the scale factor K^{-1} in the sample covariance matrix estimate need not be included since it can be accounted for in q. The MLE is $\hat{q}_0 = z^\dagger \boldsymbol{S}^{-1} z$ and next, substitute $\hat{\boldsymbol{R}} = \boldsymbol{Y}\boldsymbol{Y}^\dagger = \boldsymbol{S}$ in the likelihood function for H_1 and obtain the MLEs of both α and q. The estimates are: $\hat{\boldsymbol{\alpha}} = (\boldsymbol{H}^\dagger \boldsymbol{S}^{-1}\boldsymbol{H})^{-1}\boldsymbol{H}^\dagger \boldsymbol{S}^{-1} z$ and $\hat{q}_1 = z^\dagger \boldsymbol{S}^{-1} z - z^\dagger \boldsymbol{S}^{-1}\boldsymbol{H}(\boldsymbol{H}^\dagger \boldsymbol{S}^{-1}\boldsymbol{H})^{-1}\boldsymbol{H}^\dagger \boldsymbol{S}^{-1} z$. Substituting the estimates $\hat{\boldsymbol{R}}, \hat{\boldsymbol{\alpha}}, \hat{q}_0$, and $\hat{q}_1$ into the likelihood ratio, the subspace version of the ACE test is given by:

$$\frac{z^\dagger \boldsymbol{S}^{-1}\boldsymbol{H}(\boldsymbol{H}^\dagger \boldsymbol{S}^{-1}\boldsymbol{H})^{-1}\boldsymbol{H}^\dagger \boldsymbol{S}^{-1} z}{\left(z^\dagger \boldsymbol{S}^{-1} z\right)} \underset{H_0}{\overset{H_1}{\gtrless}} \eta. \tag{3.50}$$

In terms of the orthogonal projection matrix, the above test is:

$$\frac{\boldsymbol{y}^\dagger \boldsymbol{P}_{G}\boldsymbol{y}}{\boldsymbol{y}^\dagger \boldsymbol{y}} \underset{H_0}{\overset{H_1}{\gtrless}} \eta. \tag{3.51}$$

As an example for $M = 1$, set $\boldsymbol{p} = \boldsymbol{H} \in \mathcal{C}^{N\times 1}$ and the subspace signal versions of the GLRT, AMF, and ACE tests in (3.46), (3.48), and (3.50) result in tests in the following familiar form:

$$\begin{aligned}
&\frac{|\boldsymbol{p}^\dagger \boldsymbol{S}^{-1} z|^2}{(\boldsymbol{p}^\dagger \boldsymbol{S}^{-1}\boldsymbol{p})\left(1 + z^\dagger \boldsymbol{S}^{-1} z\right)} \underset{H_0}{\overset{H_1}{\gtrless}} \eta,\\
&\frac{|\boldsymbol{p}^\dagger \boldsymbol{S}^{-1} z|^2}{\boldsymbol{p}^\dagger \boldsymbol{S}^{-1}\boldsymbol{p}} \underset{H_0}{\overset{H_1}{\gtrless}} \eta,\\
&\frac{|\boldsymbol{p}^\dagger \boldsymbol{S}^{-1} z|^2}{(\boldsymbol{p}^\dagger \boldsymbol{S}^{-1}\boldsymbol{p})\left(z^\dagger \boldsymbol{S}^{-1} z\right)} \underset{H_0}{\overset{H_1}{\gtrless}} \eta.
\end{aligned} \tag{3.52}$$

Note that for a given probability of false alarm, the threshold η in the three tests above is different. It is also of interest to note that for the subspace detectors, the unknown parameter vector $\boldsymbol{\alpha} \in \mathcal{C}^{M\times 1}$ is estimated as: $\hat{\boldsymbol{\alpha}} = (\boldsymbol{H}^\dagger \boldsymbol{S}^{-1}\boldsymbol{H})^{-1}\boldsymbol{H}^\dagger \boldsymbol{S}^{-1} z$, the corresponding estimate of the signal vector $\boldsymbol{p} \in \mathcal{C}^{N\times 1}$ is $\hat{\boldsymbol{p}} = \boldsymbol{H}\hat{\boldsymbol{\alpha}} = \boldsymbol{H}(\boldsymbol{H}^\dagger \boldsymbol{S}^{-1}\boldsymbol{H})^{-1}\boldsymbol{H}^\dagger \boldsymbol{S}^{-1} z$, and the three subspace detectors correspond respectively to the following (familiar) rank 1 signal detectors for the signal $\hat{\boldsymbol{p}}$:

$$\begin{aligned}
&\frac{|\hat{\boldsymbol{p}}^\dagger \boldsymbol{S}^{-1} z|^2}{(\hat{\boldsymbol{p}}^\dagger \boldsymbol{S}^{-1}\hat{\boldsymbol{p}})\left(1 + z^\dagger \boldsymbol{S}^{-1} z\right)} \underset{H_0}{\overset{H_1}{\gtrless}} \eta,\\
&\frac{|\hat{\boldsymbol{p}}^\dagger \boldsymbol{S}^{-1} z|^2}{\hat{\boldsymbol{p}}^\dagger \boldsymbol{S}^{-1}\hat{\boldsymbol{p}}} \underset{H_0}{\overset{H_1}{\gtrless}} \eta,\\
&\frac{|\hat{\boldsymbol{p}}^\dagger \boldsymbol{S}^{-1} z|^2}{(\hat{\boldsymbol{p}}^\dagger \boldsymbol{S}^{-1}\hat{\boldsymbol{p}})\left(z^\dagger \boldsymbol{S}^{-1} z\right)} \underset{H_0}{\overset{H_1}{\gtrless}} \eta.
\end{aligned} \tag{3.53}$$

Although the forms of the detectors above are identical to the respective rank 1 detectors, note that the signal $\hat{\boldsymbol{p}} \in \mathcal{C}^{N\times 1}$ is a random vector and the threshold η above is set using the respective equations for the subspace detectors. The signal vector $\boldsymbol{p}$ for $M=1$ in (3.52) on the other hand is a deterministic vector. Both vectors $\boldsymbol{p}$ and $\boldsymbol{v}$ are deterministic vectors, and the randomness of the estimated signal vector $\hat{\boldsymbol{p}} = \boldsymbol{H}(\boldsymbol{H}^{\dagger}\boldsymbol{S}^{-1}\boldsymbol{H})^{-1}\boldsymbol{H}^{\dagger}\boldsymbol{S}^{-1}\boldsymbol{z}$ is only due to the estimated interference-plus-noise covariance matrix $\boldsymbol{S}$ and the interference-plus-noise in the test vector $\boldsymbol{z}$ which are both random quantities independent of the signal.

Appendix 3.B

Given that the matrices $\boldsymbol{P_G}$ and $\boldsymbol{P_G^{\perp}}$ are orthogonal projection matrices of two orthogonal complementary subspaces in $\mathcal{C}^{N\times 1}$, we have $\boldsymbol{P_G} + \boldsymbol{P_G^{\perp}} = \boldsymbol{I}_N$. Therefore for $1 \le M < N$, $||\boldsymbol{y}||^2 = \boldsymbol{y}^{\dagger}\boldsymbol{P_G}\boldsymbol{y} + \boldsymbol{y}^{\dagger}\boldsymbol{P_G^{\perp}}\boldsymbol{y}$ and the detection statistic in (3.47), (3.49), and (3.51) can be written in terms of any two of the three quantities: $||\boldsymbol{y}||_2^2$, $\boldsymbol{y}^{\dagger}\boldsymbol{P_G}\boldsymbol{y}$, $\boldsymbol{y}^{\dagger}\boldsymbol{P_G^{\perp}}\boldsymbol{y}$. Note that for $M=N$, $\boldsymbol{P_G} = \boldsymbol{I}_N$, and therefore the two-dimensional statistic reduces to a scalar statistic $||\boldsymbol{y}||_2^2$ as shown in Reference 57. The column space of a matrix $\boldsymbol{H}$ is denoted by the notation $<\boldsymbol{H}>$.

Let $\boldsymbol{U}$ be a unitary matrix, whose first M columns are an orthonormal basis for $<\boldsymbol{H}>$ and $\boldsymbol{C}$ any non-singular matrix as defined in (3.20). Note that the set of matrices $\boldsymbol{C}$ as described in (3.20) is a group $\mathcal{G}$. This property is useful for proving the results in this appendix. For matrices $\boldsymbol{A} \in \mathcal{G}$, $\boldsymbol{B} \in \mathcal{G}$, and $\boldsymbol{D} \in \mathcal{G}$, we have $\boldsymbol{AB} \in \mathcal{G}$, $(\boldsymbol{AB})\boldsymbol{D} = \boldsymbol{A}(\boldsymbol{BD})$. The identity matrix $\boldsymbol{I}_N \in \mathcal{G}$. For every $\boldsymbol{A} \in \mathcal{G}$, $\boldsymbol{AI}_N = \boldsymbol{I}_N\boldsymbol{A} = \boldsymbol{A}$, $\boldsymbol{A}^{-1} \in \mathcal{G}$ with $\boldsymbol{AA}^{-1} = \boldsymbol{A}^{-1}\boldsymbol{A} = \boldsymbol{I}_N$.

In this appendix, we outline the proof that a two-dimensional statistic $[||\boldsymbol{y}||_2^2\ \boldsymbol{y}^{\dagger}\boldsymbol{P_G^{\perp}}\boldsymbol{y}]$ is a maximal invariant for the subspace signal detection problem. The proof consists of two separate parts as described in Appendix 3.C of [27] and Appendix 3.A of [17]. The first part needs to show that transformation of the data by a matrix $\boldsymbol{CU}^{\dagger}, \boldsymbol{C} \in \mathcal{G}$ leaves the two-dimensional statistic unchanged. In the second part, we need to show that given $||\tilde{\boldsymbol{y}}||_2^2 = ||\boldsymbol{y}||_2^2$ and $\tilde{\boldsymbol{y}}^{\dagger}\boldsymbol{P}_{\tilde{\boldsymbol{G}}}\tilde{\boldsymbol{y}} = \boldsymbol{y}^{\dagger}\boldsymbol{P_G}\boldsymbol{y}$, we can find a matrix $\boldsymbol{Q} \in \mathcal{G}$ such that $\tilde{\boldsymbol{z}} = \boldsymbol{Qz}$ and $\tilde{\boldsymbol{Y}} = \boldsymbol{QYV}$, where $\boldsymbol{V}$ is a unitary matrix of size K.

For the first part, let $\tilde{\boldsymbol{z}} = \boldsymbol{CU}^{\dagger}\boldsymbol{z}$, $\tilde{\boldsymbol{Y}} = \boldsymbol{CU}^{\dagger}\boldsymbol{YV}$, $\tilde{\boldsymbol{H}} = \boldsymbol{CU}^{\dagger}\boldsymbol{H}$, where $\boldsymbol{C} \in \mathcal{G}$. By simple substitution in (3.23), it is easy to verify that: $||\tilde{\boldsymbol{y}}||_2^2 = ||\boldsymbol{y}||_2^2$ and $\tilde{\boldsymbol{y}}^{\dagger}\boldsymbol{P}_{\tilde{\boldsymbol{G}}}\tilde{\boldsymbol{y}} = \boldsymbol{y}^{\dagger}\boldsymbol{P_G}\boldsymbol{y}$ (i.e. transforming the data by any $\boldsymbol{C} \in \mathcal{G}$ does not change the two-dimensional statistic).

In order to keep the proof for the second part brief, we cite the work in Appendix 3.C of [27], where details are given for a similar problem and is directly applicable here. With the orthogonal projection matrix: $\boldsymbol{P_G} = \boldsymbol{S}^{-1/2}\boldsymbol{H}(\boldsymbol{H}^{\dagger}\boldsymbol{S}^{-1}\boldsymbol{H})^{-1}\boldsymbol{H}^{\dagger}\boldsymbol{S}^{-1/2}$ and $\boldsymbol{y} = \boldsymbol{S}^{-1/2}\boldsymbol{z}$ it can be verified that:

$$||\boldsymbol{y}||_2^2 = \boldsymbol{z}^{\dagger}\boldsymbol{S}^{-1}\boldsymbol{z}; \quad \boldsymbol{y}^{\dagger}\boldsymbol{P_G}\boldsymbol{y} = \boldsymbol{z}_{1.2}^{\dagger}\boldsymbol{S}_{1.2}^{-1}\boldsymbol{z}_{1.2}; \quad \boldsymbol{y}^{\dagger}\boldsymbol{P_G^{\perp}}\boldsymbol{y} = \boldsymbol{z}_2^{\dagger}\boldsymbol{S}_{22}^{-1}\boldsymbol{z}_2, \tag{3.54}$$

$$||\tilde{\boldsymbol{y}}||_2^2 = \tilde{\boldsymbol{z}}^{\dagger}\tilde{\boldsymbol{S}}^{-1}\tilde{\boldsymbol{z}}; \quad \tilde{\boldsymbol{y}}^{\dagger}\boldsymbol{P}_{\tilde{\boldsymbol{G}}}\tilde{\boldsymbol{y}} = \tilde{\boldsymbol{z}}_{1.2}^{\dagger}\tilde{\boldsymbol{S}}_{1.2}^{-1}\tilde{\boldsymbol{z}}_{1.2}; \quad \tilde{\boldsymbol{y}}^{\dagger}\boldsymbol{P}_{\tilde{\boldsymbol{G}}}^{\perp}\tilde{\boldsymbol{y}} = \tilde{\boldsymbol{z}}_2^{\dagger}\tilde{\boldsymbol{S}}_{22}^{-1}\tilde{\boldsymbol{z}}_2. \tag{3.55}$$

The partitioned vector $\boldsymbol{z}$ and partitioned matrix $\boldsymbol{S}$ are as shown in (3.68) and (3.69) respectively. The form of the partitioned vector $\tilde{\boldsymbol{z}}$ and partitioned matrix $\tilde{\boldsymbol{S}}$ is similar. Defining $N \times N$ prediction matrices $\boldsymbol{P}, \tilde{\boldsymbol{P}}$, and whitening matrices $\boldsymbol{W}, \tilde{\boldsymbol{W}}$:

$$\boldsymbol{P} = \begin{bmatrix} \boldsymbol{I}_M & -\boldsymbol{S}_{12}\boldsymbol{S}_{22}^{-1} \\ \boldsymbol{0} & \boldsymbol{I}_{(N-M)} \end{bmatrix}; \quad \boldsymbol{W} = \begin{bmatrix} \boldsymbol{S}_{1.2}^{-1/2} & \boldsymbol{0} \\ \boldsymbol{0} & \boldsymbol{S}_{22}^{-1/2} \end{bmatrix}, \tag{3.56}$$

$$\tilde{P} = \begin{bmatrix} I_M & -\tilde{S}_{12}\tilde{S}_{22}^{-1} \\ 0 & I_{(N-M)} \end{bmatrix}; \quad \tilde{W} = \begin{bmatrix} \tilde{S}_{1.2}^{-1/2} & 0 \\ 0 & \tilde{S}_{22}^{-1/2} \end{bmatrix}. \tag{3.57}$$

It can be verified from (3.55) to (3.57) that the condition $||y||_2^2 = ||\tilde{y}||_2^2$ and $\tilde{y}^\dagger P_{\tilde{G}}\tilde{y} = y^\dagger P_G y$ implies:

$$||WPz||_2^2 = ||\tilde{W}\tilde{P}\tilde{z}||_2^2. \tag{3.58}$$

It follows that for a unitary matrix U of the form:

$$U = \begin{bmatrix} U_{1.2} & 0 \\ 0 & U_{22} \end{bmatrix}, \tag{3.59}$$

we have:

$$\tilde{z} = \tilde{P}^{-1}\tilde{W}^{-1}UWPz. \tag{3.60}$$

Defining matrices C and Q as follows:

$$\begin{aligned} C &= \tilde{P}^{-1}\tilde{W}^{-1}UWP, \\ Q &= C, \end{aligned} \tag{3.61}$$

we have:

$$\tilde{z} = Qz. \tag{3.62}$$

Since the matrices: $\{\tilde{P}^{-1}, \tilde{W}^{-1}, U, W, P\} \in \mathcal{G}$, both matrices C and Q above are members of the group $\mathcal{G}$, which specifically identifies the matrix $Q \in \mathcal{G}$ such that: $\tilde{z} = Qz$. Substituting equation (3.62) in $||y||_2^2 = ||\tilde{y}||_2^2$, with $||y||_2^2 = z^\dagger S^{-1}z$, $||\tilde{y}||_2^2 = \tilde{z}^\dagger\tilde{S}^{-1}\tilde{z}$, $\tilde{S} = \tilde{Y}\tilde{Y}^\dagger$ and $S = YY^\dagger$, it can be shown that: $\tilde{Y} = QYV$, where V is a unitary matrix of size K.

Appendix 3.C

The joint PDFs of maximal invariant statistic conditioned on the null hypothesis H_0 and the alternative hypothesis H_1 are derived for the subspace signal model. The case of mismatch between the hypothesized signal subspace and the actual signal subspace is included in the analysis. The joint PDF of maximal invariant statistics can be derived more easily by assuming that a sequence of linear transformations are performed on the given data set. Since the covariance matrix of the interference R is not known, it is important to keep in mind throughout this appendix that these transformations are only for purposes of facilitating the analysis and are not actually performed on the given data. With the exception of the signal matrices H and D, we retain the same notation for the transformed random vectors in this appendix because this does not cause any confusion given the context and also to prevent introducing vectors with new notations. As a notation, the column space of a matrix H is denoted by $<H>$.

As shown in Appendix 3.B, the two-dimensional statistic $[||y||_2^2 \; y^\dagger P_G y]$ is a maximal invariant for the subspace signal detection problem for $1 \leq M < N$. Note that for $M = N$,

$P_G = I_N$ and therefore the maximal invariant is the single scalar statistic $||y||_2^2$ as shown in [57]. It can be verified that the transformation $z \to R^{-1/2}z$, $Y \to R^{-1/2}Y$ and $H \to R^{-1/2}H$, leaves both quantities $||y||_2^2 = z^\dagger S^{-1} z$ and $y^\dagger P_G y = z^\dagger S^{-1} H(H^\dagger S^{-1} H)^{-1} H^\dagger S^{-1} z$ unchanged – the pre-multiplication is by a non-singular matrix and as such does not change the original hypothesis testing problem. For the purposes of analysis therefore, it is convenient to assume that the test vector and secondary data vectors have been pre-whitened by the above transform. The various matrices after the pre-whitening operation are:

$$
\begin{aligned}
\bar{H} &\triangleq R^{-1/2}H, \\
\bar{D}_2 &\triangleq R^{-1/2}H_\perp, \\
\bar{D} &\triangleq R^{-1/2}D = R^{-1/2}[H\ H_\perp] \triangleq [\bar{H}\ \bar{D}_2], \\
P_{\bar{H}} &= \bar{H}(\bar{H}^\dagger \bar{H})^{-1}\bar{H}^\dagger = R^{-1/2}H\left(H^\dagger R^{-1} H\right)^{-1} H^\dagger R^{-1/2}.
\end{aligned}
\tag{3.63}
$$

While the columns of the matrices H and $H_\perp$ span orthogonal and complementary subspaces in $\mathcal{C}^{N\times 1}$, the two subspaces: $<R^{-1/2}H>$ and $<R^{-1/2}H_\perp>$ are not orthogonal for general R. Therefore, we denote the transformed matrix $R^{-1/2}H_\perp$ by $\bar{D}_2$ (and not by $\bar{H}_\perp$) in the above equation. The last equation above defines the orthogonal projection matrix for the subspace $<\bar{H}>$. The distribution of the transformed test vector is: $z \sim \mathcal{CN}(0_{N\times 1}, I_N)$ for hypothesis H_0 and $z \sim \mathcal{CN}(\bar{D}\gamma, I_N)$ for hypothesis H_1. Similarly, the transformed secondary data vectors are distributed as: $Y \sim \mathcal{CN}(0_{N\times K}, I_K \otimes I_N)$ for both hypotheses.

Next, a coordinate rotation is performed such that any vector in $<\bar{H}>$ (i.e. the column space of $\bar{H}$, which is a subspace of dimension M in $\mathcal{C}^{N\times 1}$) is represented by the first M components of the transformed vector. In principle, the SVD of the transformed signal matrix results in: $\bar{H} = \bar{U}\bar{B}\bar{V}^\dagger$, and pre-multiplication by the unitary matrix $\bar{U}^\dagger$ performs the required rotation of coordinate axes. The operations of pre-whitening followed by the rotation of coordinate axes partition all vectors into two components (original notation retained): $z = [z_1^\dagger\ z_2^\dagger]^\dagger$, $z_1 \in \mathcal{C}^{M\times 1}$ and $z_2 \in \mathcal{C}^{(N-M)\times 1}$. Similarly, the components of the pre-whitened and rotated secondary data are $Y_1 \in \mathcal{C}^{M\times K}$ and $Y_2 \in \mathcal{C}^{(N-M)\times K}$. Since the original data are modeled as multivariate Gaussian, the distribution of the transformed data is also multivariate Gaussian and therefore: $H_1 : z \sim \mathcal{CN}\left(\bar{U}^\dagger \bar{D}\,\gamma, I_N\right)$, where $\bar{U}^\dagger \bar{D}\,\gamma = [v_1^\dagger\ v_2^\dagger]^\dagger$, with $v_1 \in \mathcal{C}^{M\times 1}$ and $v_2 \in \mathcal{C}^{(N-M)\times 1}$. The coordinate rotation conserves the squared magnitude of signals and so:

$$
\begin{aligned}
||v_1||_2^2 + ||v_2||_2^2 &= \gamma^\dagger \bar{D}^\dagger \bar{U}\bar{U}^\dagger \bar{D}\,\gamma \\
&= \gamma^\dagger D^\dagger R^{-1/2}\bar{U}\bar{U}^\dagger R^{-1/2} D\gamma \\
&= v^\dagger R^{-1} v = \gamma^\dagger D^\dagger R^{-1} D\gamma.
\end{aligned}
\tag{3.64}
$$

In the above $v = D\gamma$ from (3.12). The quantity $||v_1||_2^2$ above is the squared magnitude of the signal $\bar{U}^\dagger \bar{D}\gamma$ in the subspace $<\bar{U}^\dagger R^{-1/2}H>$ and is the same as the squared magnitude of the signal $\bar{D}\gamma$ in the subspace $<R^{-1/2}H>$. Thus,

$$
\begin{aligned}
||v_1||_2^2 &= ||(\bar{U}^\dagger P_{\bar{H}}\bar{U})\bar{U}^\dagger \bar{D}\gamma||_2^2 = ||P_{\bar{H}}\bar{D}\gamma||_2^2 \\
&= \gamma^\dagger D^\dagger R^{-1} H(H^\dagger R^{-1} H)^{-1} H^\dagger R^{-1} D\gamma.
\end{aligned}
\tag{3.65}
$$

It is useful to define the generalized mismatch angle θ and the SINR δ as follows:

$$\cos^2(\theta) \triangleq \frac{||\boldsymbol{v}_1||_2^2}{||\boldsymbol{v}||_2^2} = \frac{\boldsymbol{\gamma}^\dagger \boldsymbol{D}^\dagger \boldsymbol{R}^{-1} \boldsymbol{H} \left(\boldsymbol{H}^\dagger \boldsymbol{R}^{-1} \boldsymbol{H}\right)^{-1} \boldsymbol{H}^\dagger \boldsymbol{R}^{-1} \boldsymbol{D} \boldsymbol{\gamma}}{\boldsymbol{\gamma}^\dagger \boldsymbol{D}^\dagger \boldsymbol{R}^{-1} \boldsymbol{D} \boldsymbol{\gamma}}. \tag{3.66}$$

$$\delta = \boldsymbol{\gamma}^\dagger \boldsymbol{D}^\dagger \boldsymbol{R}^{-1} \boldsymbol{D} \boldsymbol{\gamma}. \tag{3.67}$$

The pre-whitened and coordinated rotated secondary data matrix is multivariate Gaussian and $\boldsymbol{Y} \sim \mathcal{CN}(\boldsymbol{0}_{N\times K}, \boldsymbol{I}_K \otimes \boldsymbol{I}_N)$ for both hypotheses H_0 and H_1. The transformed matrices and vectors are represented as follows:

$$z = \begin{bmatrix} z_1 \\ z_2 \end{bmatrix}; \quad \boldsymbol{Y} = \begin{bmatrix} \boldsymbol{Y}_1 \\ \boldsymbol{Y}_2 \end{bmatrix}, \tag{3.68}$$

$$\boldsymbol{S} = \begin{bmatrix} \boldsymbol{S}_{11} & \boldsymbol{S}_{12} \\ \boldsymbol{S}_{12}^\dagger & \boldsymbol{S}_{22} \end{bmatrix} = \begin{bmatrix} \boldsymbol{Y}_1\boldsymbol{Y}_1^\dagger & \boldsymbol{Y}_1\boldsymbol{Y}_2^\dagger \\ \boldsymbol{Y}_2\boldsymbol{Y}_1^\dagger & \boldsymbol{Y}_2\boldsymbol{Y}_2^\dagger \end{bmatrix}. \tag{3.69}$$

In terms of the partitioned matrix and partitioned vectors above, the two-dimensional quantity: $[(1+z_2^\dagger \boldsymbol{S}_{22}^{-1} z_2)/(1+z^\dagger \boldsymbol{S}^{-1} z);\ (1+z_2^\dagger \boldsymbol{S}_{22}^{-1} z_2)^{-1}]$ is a maximal invariant statistic. The quantity $(z^\dagger \boldsymbol{S}^{-1} z) = (z_2^\dagger \boldsymbol{S}_{22}^{-1} z_2) + z_{1.2}^\dagger \boldsymbol{S}_{1.2}^{-1} z_{1.2})$, where $\boldsymbol{S}_{1.2} = \boldsymbol{S}_{11} - \boldsymbol{S}_{12}\boldsymbol{S}_{22}^{-1}\boldsymbol{S}_{21}$ and is the Schur complement of the block $\boldsymbol{S}_{22}$ of matrix $\boldsymbol{S}$. The $M \times 1$ vector $z_{1.2} = z_1 - \boldsymbol{S}_{12}\boldsymbol{S}_{22}^{-1} z_2$.

3.C.1 Distribution of Maximal Invariant Statistic

As described in [44], it is convenient to evaluate the joint distribution by holding the "2-components" of various vectors fixed to obtain the conditional distribution of relevant quantities. Unless stated otherwise we use the term "conditional distribution" to imply distribution of a quantity with the 2-components held constant. In the discussion below, the quantities z_1 and $\boldsymbol{Y}_1$ are random while the quantities z_2 and $\boldsymbol{Y}_2$ are fixed and are not allowed to change. Let the vectors $\boldsymbol{a}_n \in \mathcal{C}^{K\times 1};\ n = 1, 2, \ldots, K$ denote an orthonormal basis for $\mathcal{C}^{K\times 1}$. Since the columns of $\boldsymbol{Y}_2 \in \mathcal{C}^{(N-M)\times K}$ are drawn from a zero-mean, statistically independent and identically distributed, complex Gaussian distribution, each realization of the matrix $\boldsymbol{Y}_2$ has rank $(N-M)$ with probability 1. For a fixed $\boldsymbol{Y}_2$, let the vectors $\boldsymbol{a}_n \in \mathcal{C}^{K\times 1};\ n = 1, 2, \ldots, (K-N+M)$ denote an orthonormal basis for the null space of $\boldsymbol{Y}_2$. Thus, $\boldsymbol{Y}_2 \boldsymbol{a}_n = \boldsymbol{0}_{(N-M)\times 1};\ n = 1, 2, \ldots, (K-N+M)$. The remaining vectors $\boldsymbol{a}_n \in \mathcal{C}^{K\times 1};\ n = (K-N+M+1), \ldots, K$ therefore denote the orthonormal basis for the column space of $\boldsymbol{Y}_2$. Therefore,

$$\boldsymbol{S}_{1.2} = \boldsymbol{Y}_1 \left[\boldsymbol{I}_K - \boldsymbol{Y}_2^\dagger (\boldsymbol{Y}_2 \boldsymbol{Y}_2^\dagger)^{-1} \boldsymbol{Y}_2\right] \boldsymbol{Y}_1^\dagger = \boldsymbol{Y}_1 \left[\sum_{n=1}^{K-N+M} \boldsymbol{a}_n \boldsymbol{a}_n^\dagger\right] \boldsymbol{Y}_1^\dagger. \tag{3.70}$$

The above result follows from the observation that the matrix $\boldsymbol{Y}_2^\dagger (\boldsymbol{Y}_2 \boldsymbol{Y}_2^\dagger)^{-1} \boldsymbol{Y}_2$ is idempotent and has eigenvalue 0 with multiplicity $K-N+M$ with eigenvectors $\boldsymbol{a}_n;\ n = 1, 2, \ldots, (K-N+M)$ and eigenvalue 1 with multiplicity $N-M$ with eigenvectors

$\boldsymbol{a}_n;\ n=(K-N+M+1),\ldots,K$. Since, $\boldsymbol{a}_n \in \mathcal{C}^{K\times 1};\ n=1,2,\ldots,K$ are an orthonormal basis for $\mathcal{C}^{K\times 1}$, we have $\boldsymbol{I}_K=\sum_{n=1}^{K}\boldsymbol{a}_n\boldsymbol{a}_n^{\dagger}$, and therefore,

$$
\begin{aligned}
z_{1.2} &= z_1 - \boldsymbol{Y}_1\boldsymbol{Y}_2^{\dagger}(\boldsymbol{Y}_2\boldsymbol{Y}_2^{\dagger})^{-1}z_2 \\
&= z_1 - \boldsymbol{Y}_1\left[\sum_{n=1}^{K}\boldsymbol{a}_n\boldsymbol{a}_n^{\dagger}\right]\boldsymbol{Y}_2^{\dagger}(\boldsymbol{Y}_2\boldsymbol{Y}_2^{\dagger})^{-1}z_2 \\
&= z_1 - \boldsymbol{Y}_1\left[\sum_{n=K-N+M+1}^{K}\boldsymbol{a}_n\boldsymbol{a}_n^{\dagger}\right]\boldsymbol{Y}_2^{\dagger}(\boldsymbol{Y}_2\boldsymbol{Y}_2^{\dagger})^{-1}z_2.
\end{aligned}
\tag{3.71}
$$

The last equation above results from $\boldsymbol{Y}_2\boldsymbol{a}_n=\boldsymbol{0}_{(N-M)\times 1};\ n=1,2,\ldots,(K-N+M)$. Since $\boldsymbol{Y}_1 \sim \mathcal{CN}(\boldsymbol{0}_{M\times K},\boldsymbol{I}_M\otimes\boldsymbol{I}_K)$ for both hypotheses H_0 and H_1, (3.70) and (3.71) show that the quantities $\boldsymbol{S}_{1.2}$ and $z_{1.2}$ are constructed from statistically independent components of $\boldsymbol{Y}_1$ and are therefore statistically independent for fixed 2-components.

$$
\begin{aligned}
\mathbb{E}_2[z_{1.2}|H_0] &= \boldsymbol{0}_{M\times 1}, \\
\mathbb{E}_2[z_{1.2}|H_1] &= \boldsymbol{v}_1 \\
||\boldsymbol{v}_1||^2 &= \left(\boldsymbol{\gamma}^{\dagger}\boldsymbol{D}^{\dagger}\boldsymbol{R}^{-1}\boldsymbol{D}\boldsymbol{\gamma}\right)\cos^2(\theta)=\delta\cos^2(\theta).
\end{aligned}
\tag{3.72}
$$

The last equation above follows from the definition of the generalized cosine of the mismatch angle in (3.66) and the definition of the SINR in (3.67). Note that the original signal matrix $\boldsymbol{D}=[\boldsymbol{H}\ \boldsymbol{H}_{\perp}]$ was partitioned into two orthogonal and complimentary subspaces $<\boldsymbol{H}>$ and $<\boldsymbol{H}_{\perp}>$ in $\mathcal{C}^{N\times 1}$. And therefore, the two pre-whitened subspaces: $<\bar{\boldsymbol{H}}>$ and $<\bar{\boldsymbol{D}}_2>$ are not necessarily orthogonal subspaces. This means that components of the original signal vector in the subspace $<\boldsymbol{H}_{\perp}>$ (due to signal model mismatch) can add/subtract to the signal to be detected in the original subspace $<\boldsymbol{H}>$.

Define the $K\times 1$ matrix $\boldsymbol{W}=\boldsymbol{Y}_2^{\dagger}(\boldsymbol{Y}_2\boldsymbol{Y}_2^{\dagger})^{-1}z_2$, the conditional covariance matrix of vector $z_{1.2}=z_1-\boldsymbol{Y}_1\boldsymbol{W}$ under hypotheses H_0 and H_1 is:

$$
\begin{aligned}
\mathbb{E}_2[z_{1.2}z_{1.2}^{\dagger}|H_0] &= (1+\boldsymbol{W}^{\dagger}\boldsymbol{W})\boldsymbol{I}_M \\
&= (1+z_2^{\dagger}\boldsymbol{S}_{22}^{-1}z_2)\boldsymbol{I}_M, \\
\mathbb{E}_2[(z_{1.2}-\boldsymbol{v}_1)(z_{1.2}-\boldsymbol{v}_1)^{\dagger}|H_1] &= (1+z_2^{\dagger}\boldsymbol{S}_{22}^{-1}z_2)\boldsymbol{I}_M.
\end{aligned}
\tag{3.73}
$$

The above result follows by letting $\boldsymbol{y}_1(n)\in\mathcal{C}^{1\times K};\ n=1,2,\ldots,M$ denotes the nth row of $\boldsymbol{Y}_1$. The (n,m)th element of the conditional covariance matrix of vector $z_{1.2}$ under hypothesis H_0 for example is:

$$
\begin{aligned}
\mathbb{E}_2[z_{1.2}(n)z_{1.2}^{*}(m)|H_0] &= \delta_{n,m}+\mathbb{E}_2\left[\boldsymbol{W}^{\dagger}\boldsymbol{y}_1^{\dagger}(m)\boldsymbol{y}_1(n)\boldsymbol{W}\right] \\
&= (1+\boldsymbol{W}^{\dagger}\boldsymbol{W})\delta_{n,m} \\
&= (1+z_2^{\dagger}\boldsymbol{S}_{22}^{-1}z_2)\delta_{n,m};\quad 1\le n,m\le M.
\end{aligned}
\tag{3.74}
$$

In the above, $\delta_{n,m}$ is the Kronecker delta, which is 1 for $n=m$ and 0 otherwise.

Define the signal-to-interference-plus-noise loss factor $\rho = (1 + z_2^\dagger S_{22}^{-1} z_2)^{-1}$. Using Proposition 3.4, it follows that for a given loss factor ρ, the random variable $s = (z_{1.2}^\dagger S_{1.2}^{-1} z_{1.2})/(1 + z_2^\dagger S_{22}^{-1} z_2) = (z_{1.2}^\dagger S_{1.2}^{-1} z_{1.2})\rho$ has a non-central complex F distribution with parameters M, $K - N + 1$ and non-centrality parameter $\rho\ ||\mathbf{v_1}||_2^2 = \rho\ \delta \cos^2(\theta)$.

The distribution of the random variable s above was derived assuming that the 2-components of the relevant matrices and vectors are fixed. The distribution is valid for every realization of the 2-components and does not involve the specific realization of the 2-components themselves, the constraint can be removed and the distribution applies generally. With the signal power in the pre-whitened hypothesized signal subspace $<\bar{\boldsymbol{H}}>$ given by: $||\boldsymbol{v}_1||_2^2$, the signal power in the orthogonal complement subspace of $<\bar{\boldsymbol{H}}>$ is: $\boldsymbol{v}^\dagger \boldsymbol{R}^{-1} \boldsymbol{v} - ||\boldsymbol{v}_1||_2^2$. For the assumed distribution of the 2-components, the signal-to-interference-plus-noise loss factor ρ itself is obtained from Proposition 3.4 to be non-central beta with parameters $K - (N - M) + 1, N - M$ and non-centrality parameter $\delta_\perp \triangleq (\boldsymbol{v}^\dagger \boldsymbol{R}^{-1} \boldsymbol{v}) - ||\boldsymbol{v}_1||_2^2 = (\boldsymbol{v}^\dagger \boldsymbol{R}^{-1} \boldsymbol{v}) \sin^2(\theta)$. Using the notation in Proposition 3.3 to denote the non-central complex beta density, the joint density function of maximal invariant statistic $r = (1 + s)^{-1}$ and ρ is given by:

$$\begin{aligned} f(r|\rho, H_1) &= f_\beta(r; K - N + 1, M|\rho\delta \cos^2(\theta)), \\ f(\rho|H_1) &= f_\beta(\rho; K - (N - M) + 1, N - M|\delta \sin^2(\theta)), \\ f(r, \rho|H_1) &= f(r|\rho, H_1) f(\rho|H_1); \quad 0 \le r, \rho \le 1. \end{aligned} \tag{3.75}$$

For hypothesis H_0, the joint density function of maximal invariants is obtained from above by setting the signal vector to zero and so the quantity $(\boldsymbol{v}^\dagger \boldsymbol{R}^{-1} \boldsymbol{v}) = 0$ and so $||\boldsymbol{y}||_2^2 = 0$. With both non-centrality parameters set to zero (i.e. $\delta = \delta_\perp = 0$), the beta densities in (3.75) are central beta densities and so:

$$f(r, \rho|H_0) = f_\beta(r; K - N + 1, M) f_\beta(\rho; K - (N - M) + 1, N - M); \quad 0 \le r, \rho \le 1. \tag{3.76}$$

Appendix 3.D

A number of distributions relevant to this work are outlined in this appendix without proof. We do this mainly to facilitate the development in the rest of the paper. Detailed proofs of the finite sum representation in Propositions 3.2 and 3.3 may be found in Reference 44. The notations for random vectors and random variables are described in the introduction.

Proposition 3.1. *Suppose the random vector* $\boldsymbol{x} \in \mathcal{C}^{n\times 1}$, *is Gaussian distributed with* $\mathbb{E}[\boldsymbol{x}] = \boldsymbol{b}; \boldsymbol{b} \in \mathcal{C}^{n\times 1}$ *and* $\mathbb{E}[\boldsymbol{x}\boldsymbol{x}^\dagger] = \boldsymbol{I}_n$ *(denoted as* $\boldsymbol{x} \sim \mathcal{CN}(\boldsymbol{b}, \boldsymbol{I}_n)$*). Then the random variable* $y = ||\boldsymbol{x}||_2^2$ *has a non-central complex Chi-squared density with n complex degrees of freedom with non-centrality parameter* $c = ||\boldsymbol{b}||_2^2$. *This is denoted by* $y \sim \chi_n^2(c)$. *When* $c = 0$, *the random variable y has a central complex Chi-squared distribution:* $y \sim \chi_n^2$.

Proposition 3.2. *Given two statistically independent random vectors* $\boldsymbol{x} \in \mathcal{C}^{n\times 1}$ *and* $\boldsymbol{y} \in \mathcal{C}^{m\times 1}$, *with* $\boldsymbol{x} \sim \mathcal{CN}(\boldsymbol{b}, \boldsymbol{I}_n)$ *and* $\boldsymbol{y} \sim \mathcal{CN}(\boldsymbol{0}_{m\times 1}, \boldsymbol{I}_m)$. *The random variable* $z = ||\boldsymbol{x}||_2^2/||\boldsymbol{y}||_2^2$ *has a non-central complex F density with parameters n and m with non-centrality parameter* $c = ||\boldsymbol{b}||_2^2$.

This is denoted by $z \sim F_{n,m}(c)$. The cumulative distribution function of the random variable z is given by:

$$P[z \leq z_0] = \frac{z_0^n}{(1+z_0)^{n+m-1}} \times \sum_{k=0}^{m-1} \frac{\Gamma(n+m)}{\Gamma(k+n+1)\Gamma(m-k)} z_0^k G_{k+1}\left(\frac{c}{1+z_0}\right), \tag{3.77}$$

where the function $G_{k+1}(x)$ is related to the incomplete gamma function as:

$$G_{k+1}(x) = e^{-x} \sum_{n=0}^{k} \frac{x^n}{n!}. \tag{3.78}$$

Proposition 3.3. *Given a random variable $x \sim F_{n,m}(c)$, the random variable $y = (1+x)^{-1}$ has a non-central complex beta density $y \sim \beta_{m,n}(c)$:*

$$\begin{aligned} f_\beta(y; m, n|c) &= e^{-cy} \sum_{k=0}^{m} \binom{m}{k} \frac{\Gamma(m+n)}{\Gamma(m+n+k)} c^k f_\beta(y; m, n+k) \\ &= e^{-cy} f_\beta(y; m, n) \sum_{k=0}^{m} \binom{m}{k} \frac{\Gamma(n)}{\Gamma(n+k)} [c(1-y)]^k, \end{aligned} \tag{3.79}$$

where $f_\beta(y; m, n)$ is the central complex beta density function given by:

$$f_\beta(y; m, n) = \frac{\Gamma(m+n)}{\Gamma(m)\Gamma(n)} y^{m-1}(1-y)^{n-1} \qquad 0 \leq y \leq 1. \tag{3.80}$$

The cumulative distribution function of y is given by:

$$P[y \leq y_0] = 1 - \frac{1}{(m+n)} \sum_{k=0}^{m-1} f_\beta(y_0; m-k, n+k+1) G_{k+1}(cy_0). \tag{3.81}$$

Proposition 3.4. *Given a zero-mean complex Gaussian random matrix $\boldsymbol{Y} \in \mathcal{C}^{N \times K};\ K \geq N$ with $\boldsymbol{Y} \sim \mathcal{CN}(\boldsymbol{0}_{N \times K}, \boldsymbol{I}_N \otimes \boldsymbol{I}_K)$ and a statistically independent complex Gaussian random vector $\boldsymbol{x} \in \mathcal{C}^{N \times 1}$ with $\boldsymbol{x} \sim \mathcal{CN}(\boldsymbol{b}, \boldsymbol{I}_N)$, the random variable $z = \boldsymbol{x}^\dagger (\boldsymbol{Y}\boldsymbol{Y}^\dagger)^{-1} \boldsymbol{x}$ has a non-central complex F density with parameters N and $K-N+1$ with non-centrality parameter $c = ||\boldsymbol{b}||_2^2$. That is, $z \sim F_{N,K-N+1}(c)$. And from Proposition 3.3, the random variable $y = (1+z)^{-1}$ has a complex non-central beta density $y \sim \beta_{K-N+1,N}(c)$.*

References

[1] J. Ward, "Space–Time Adaptive Processing for Airborne Radar," Technical Report 1015, MIT Lincoln Laboratory, Lexington, MA, 1994.

[2] R. Klemm, "Principles of Space–Time Adaptive Processing," Institute of Electrical Engineering, London, UK, 2002.

[3] J. R. Guerci, "Space–Time Adaptive Processing for Radar," Artech House, Norwood, MA, 2003.

[4] J. Li, R. S. Blum, P. Stoica, A. H. Haimovich, and M. C. Wicks: Guest Editors, "Special Issue on MIMO Radar and Its Applications," IEEE Journal on Selected Topics in Signal Processing, vol. 4, no. 1, February 2010.

[5] K. W. Forsythe and D. W. Bliss, "MIMO Radar Waveform Constraints for GMTI," IEEE Journal of Special Topics in Signal Processing, vol. 4, no. 1, pp. 21–32, February 2010.

[6] J. L. Krolik and R. H. Anderson, "Maximum Likelihood Coordinate Registration for Over-the-Horizon Radar," IEEE Transactions on Signal Processing, vol. 45, no. 4, pp. 945–959, April 1997.

[7] G. Fabrizio and A. Farina, "An Adaptive Fitting Algorithm for Blind Waveform Estimation in Diffuse Multipath Channels," IET Radar Sonar Navigation, vol. 3, no. 4, pp. 384–405, 2009.

[8] G. Fabrizio, F. Colone, P. Lombardo, and A. Farina, "Adaptive Beamforming for High-Frequency Over-the-Horizon Passive Radar," IET Radar Sonar Navigation, vol. 5, no. 1, pp. 322–330, 2011.

[9] S. Bose and A. Steinhardt, "A Maximal Invariant Framework for Adaptive Detection with Structured and Unstructured Covariance Matrices," IEEE Transactions on Signal Processing, vol. SP-43, no. 9, pp. 2164–2175, September 1995.

[10] S. Bose and A. Steinhardt, "Optimum Array Detector for a Weak Signal in Unknown Noise," IEEE Transactions on Aerospace and Electronic Systems, vol. AES-32, no. 3, pp. 911–922, July 1996.

[11] E. J. Kelly, "An Adaptive Detection Algorithm," IEEE Transactions on Aerospace and Electronic Systems, vol. AES-22, no. 1, pp. 115–127, March 1986.

[12] C. G. Khatri and C. R. Rao, "Effects of Estimated Noise Covariance Matrix in Optimal Signal Detection," IEEE Transactions on Acoustics Speech and Signal Processing, vol. 35, no. 5, pp. 671–679, May 1987.

[13] F. C. Robey, D. R. Fuhrmann, E. J. Kelley, and R. Nitzberg, "A CFAR Adaptive Matched Filter Detector," IEEE Transactions on Aerospace and Electronic Systems, vol. AES-28, no. 1, pp. 208–216, January 1992.

[14] E. Conte, M. Lops, and G. Ricci, "Asymptotically Optimum Radar Detection in Compound-Gaussian Clutter," IEEE Transactions on Aerospace and Electronic Systems, vol. AES-31, no. 2, pp. 617–625, April 1995.

[15] S. Kraut and L. L. Scharf, "The CFAR Adaptive Subspace Detector Is a Scale-Invariant GLRT," IEEE Transactions on Signal Processing, vol. 47, no. 9, pp. 2538–2541, September 1999.

[16] C. D. Richmond, "Performance of the Adaptive Sidelobe Blanker Detection Algorithm in Homogeneous Environments," IEEE Transactions on Signal Processing, vol. 48, no. 5, pp. 1235–1247, May 2000.

[17] S. Kraut, L. L. Scharf, and R. W. Butler, "The Adaptive Coherence Estimator: A Uniformly Most-Powerful-Invariant Detection Statistic," IEEE Transactions on Signal Processing, vol. 53, no. 2, pp. 427–438, January 2005.

[18] S. Bidon, O. Besson, and J.-Y. Tourneret, "The Adaptive Coherence Estimator Is the Generalized Likelihood Ratio Test for a Class of Heterogeneous Environments," IEEE Signal Processing Letters, vol. 15, pp. 281–284, 2008.

[19] N. B. Pulsone and M. A. Zatman, "A Computationally Efficient Two-Step Implementation of the GLRT," IEEE Transactions on Signal Processing, vol. 48, no. 3, pp. 609–616, March 2000.

[20] N. B. Pulsone and C. M. Rader, "Adaptive Beamformer Orthogonal Rejection Test," IEEE Transactions on Signal Processing, vol. 49, no. 3, pp. 521–529, March 2001.

[21] K. Gerlach and M. J. Steiner, "Fast Converging Adaptive Detection of Doppler-Shifted, Range-Distributed Targets," IEEE Transactions on Signal Processing, vol. 48, no. 9, pp. 2686–2690, September 2000.

[22] E. Conte, A. De Maio, and G. Ricci, "Recursive Estimation of the Covariance Matrix of a Compound-Gaussian Process and Its Application to Adaptive CFAR Detection," IEEE Transactions on Signal Processing, vol. 50, no. 8, August 2002.

[23] O. Besson and L. L. Scharf, "CFAR Matched Direction Detector," IEEE Transactions on Signal Processing, vol. 54, no. 7, pp. 2840–2844, July 2006.

[24] A. De Maio, "Rao Test for Adaptive Detection in Gaussian Interference with Unknown Covariance Matrix," IEEE Transactions on Signal Processing, vol. 55, no. 7, pp. 3577–3584, July 2007.

[25] F. Bandiera, O. Besson, and G. Ricci, "Theoretical Performance Analysis of the W-ABORT Detector," IEEE Transactions on Signal Processing, vol. 56, no. 5, pp. 2117–2121, May 2008.

[26] A. De Maio, S. De Nicola, Y. Huang, S. Zhang, and A. Farina, "Adaptive Detection and Estimation in the Presence of Useful Signal and Interference Mismatches," IEEE Transactions on Signal Processing, vol. 57, no. 2, pp. 436–450, February 2009.

[27] R. S. Raghavan, N. B. Pulsone, and D. J. McLaughlin, "Performance of the GLRT for Adaptive Vector Subspace Detection," IEEE Transactions on Aerospace and Electronic Systems, vol. AES-32, no. 4, pp. 1473–1487, October 1996.

[28] S. Z. Kalson, "An Adaptive Array Detector with Mismatch Signal Rejection," IEEE Transactions on Aerospace and Electronic Systems, vol. AES-28, no. 1, pp. 195–207, January 1992.

[29] H. L. Van Trees, "Detection, Estimation and Linear Modulation Theory, Part-I," Wiley, New York, 1971.

[30] J. V. DiFranco and W. L. Rubin, "Radar Detection," Artech House, Dedham, MA, 1980.

[31] C. W. Helstrom, "Statistical Theory of Signal Detection," Pergamon Press, Oxford, 1960.

[32] D. Middleton, "An Introduction to Statistical Communication Theory," McGraw Hill, New York, 1960.

[33] A. D. Whalen, "Detection of Signals in Noise," Academic Press, Boston, 1971.

[34] L. L. Scharf, "Statistical Signal Processing: Detection, Estimation, and Time Series Analysis," Addison-Wesley, Reading, MA, 1991.

[35] S. M. Kay, "Fundamentals of Statistical Signal Processing; Detection Theory (vol. II)," Prentice Hall, NJ, 1998.

[36] E. L. Lehmann, "Testing Statistical Hypotheses," Second Edition, Wadsworth, Inc., Belmont, CA, 1991.

[37] T. W. Anderson, "An Introduction to Multivariate Statistical Analysis," Second Edition, Wiley, New York, Chapter 5, 1984.

[38] C. R. Rao, "Linear Statistical Inference and Its Applications," Wiley, New York, 1973.

[39] R. J. Muirhead, "Aspects of Multivariate Statistical Theory," Wiley, New York, Chapter 6, 1982.

[40] M. L. Eaton, "Multivariate Statistics – A Vector Space Approach," Lecture Notes – Monograph Series, vol. 53, Institute of Mathematical Statistics, Beachwood, Ohio, 2007.

[41] M. Abramowitz and I. Stegun: Editors, "Handbook of Mathematical Functions with Formulas, Graphs and Mathematical Tables," Dover Publications, Inc., New York, 1972.

[42] J. I. Marcum, "A Statistical Theory of Target Detection By Pulsed Radar," RAND Research Memo. RM-754, December, 1947.

[43] A. H. Nuttall, "Some Integrals Involving the Q_M Function," IEEE Transactions on Information Theory, pp. 95–96, January 1975.

[44] E. J. Kelly and K. M. Forsythe, "Adaptive Detection and Parameter Estimation for Multidimensional Signal Model," Technical Report # 848, MIT Lincoln Laboratory, April 1989.

[45] D. A. Shnidman, "The Calculation of the Probability of Detection and the Generalized Marcum Q-Function," IEEE Transactions on Information Theory, vol. 35, no. 2, pp. 389–400, March 1989.

[46] M. Zatman, "How Narrow is Narrowband?" IEE Proceedings on Radar, Sonar and Navigation, vol. 145, no. 2, pp. 85–91, April 1998.

[47] F. Gini and A. Farina, "Matched Subspace CFAR Detection of Hovering Helicopters,' IEEE Transactions on Aerospace and Electronic Systems, vol. AES-35, no. 4, pp. 1293–1305, October 1999.

[48] E. N. Gilbert and S. P. Morgan, "Optimum Design of Directive Antenna Arrays Subject to Random Variations," Bell System Technical Journal, pp. 637–663, May 1955.

[49] A. Papoulis, "Probability, Random Variables and Stochastic Processes," Third Edition, McGraw Hill, New York, 1991.

[50] R. J. Mailloux, "Covariance Matrix Augmentation to Produce Adaptive Array Pattern Troughs," Electronics Letters, vol. 31. no. 10, pp. 771–772, May 1995.

[51] M. Zatman, "Production of Adaptive Array Troughs by Dispersion Synthesis," Electronics Letters, vol. 31. no. 25, pp. 2141–2142, December 1995.

[52] J. R. Guerci, "Theory and Application of Covariance Matrix Tapers for Robust Adaptive Beamforming," IEEE Transactions on Signal Processing, vol. 47, no. 4, pp. 977–985, April 1999.

[53] R. S. Raghavan, "Maximal Invariants and Performance of Some Invariant Hypothesis Tests for an Adaptive Detection Problem," IEEE Transactions on Signal Processing, vol. 61, no. 14, pp. 3607–3619, July 2013.

[54] R. S. Raghavan, "Analysis of Steering Vector Mismatch on Adaptive Noncoherent Integration," IEEE Transactions on Aerospace and Electronic Systems, vol. 49, no. 4, pp. 2496–2508, October 2013.

[55] L. L. Scharf and B. Friedlander, "Matched Subspace Detectors," IEEE Transactions on Signal Processing, vol. 42, no. 8, pp. 2146–2157, August 1994.

[56] S. Kraut, L. L. Scharf, and T. McWhorter, "Adaptive Subspace Detector," IEEE Transactions on Signal Processing, vol. 49, no. 1, pp. 1–16, January 2001.

[57] R. S. Raghavan, H. E. Qiu, and D. J. McLaughlin, "CFAR Detection in Dutter with Unknown Correlation Properties," IEEE Transactions on Aerospace and Electronic Systems, vol. 31, no. 2, pp. 647–657, April 1995.

CHAPTER 4

Two-Stage Detectors for Point-Like Targets in Gaussian Interference with Unknown Spectral Properties

Antonio De Maio[1], *Chengpeng Hao*[2], *and Danilo Orlando*[3]

4.1 Introduction: Principles of Design

In recent years, the design of the so-called *tunable* receivers has raised a significant interest in the radar community. The class of tunable detectors has been shown to be an effective means to attack detection of mainlobe targets or rejection of coherent repeater interferers in the presence of clutter and/or possible noise-like interferers. As a matter of fact, a tunable receiver allows adjusting the rate at which the probability of detection, P_d say, falls down when the received signal departs from the nominal one. In this case, a mismatch between the nominal and the actual steering vector is present. In the sequel, we refer to the receiver capability of rejecting/detecting signals as *directivity*. Existing receivers can be classified according to their directivity as follows [1]:

- *robust receivers*, which provide good detection performances in the presence of echoes containing signal components not aligned with the nominal (transmitted) signal;
- *selective receivers*, which are capable of rejecting signals whose signature unlikely corresponds to the signal of interest in order to avoid false alarms.

As already stated tunable receivers can be robust or selective by setting proper design parameters. Among the several papers published in open literature, we cite here References 1–19. The expert reader would have a point concerning References 2 and 19, wherein the authors design adaptive receivers assuming that the nominal signal belongs to a preassigned subspace,

[1]Dipartimento di Ingegneria Elettrica e delle Tecnologie dell'Informazione, Università degli Studi di Napoli "Federico II", Napoli, Italy
[2]State Key Laboratory of Acoustics, Institute of Acoustics, Chinese Academy of Sciences, Beijing, China
[3]Facoltà di Ingegneria, Università degli Studi "Niccolò Cusano", Roma, Italy

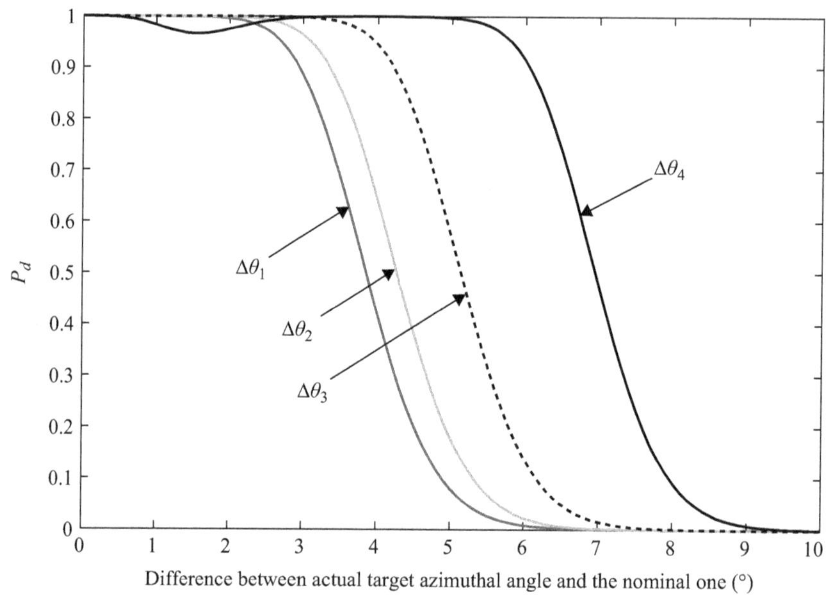

Figure 4.1 P_d versus angular difference between the actual direction of the target and the nominal one for the subspace detector.

$\mathbb{H}$ say, of the observables (see also Chapter 3). These decision schemes are typically classified as steady receivers. However, they can be tuned modifying the properties of $\mathbb{H}$. To be more definite, in the decision statistic $\mathbb{H}$ is represented by a full-column-rank matrix $\boldsymbol{H}$. The rank and/or the angular separation (measured in terms of the inner product in the Euclidean space) between the columns of $\boldsymbol{H}$ allow controlling the level of robustness to mismatched signals. To have an idea, in Figure 4.1 (parameters used in this figure can be found in Subsection 4.2.2), we plot P_d versus the difference between the actual target azimuthal angle and the nominal one for a given value of the Signal-to-Noise Ratio (SNR) and for different angular separations between the columns of $\boldsymbol{H}$. In particular, we consider four increasing values of the angular separation, i.e.,

$$\Delta\theta_1 < \Delta\theta_2 < \Delta\theta_3 < \Delta\theta_4. \tag{4.1}$$

In References 7, 8, and 14, the authors derive receivers obtained by merging the decision statistics of existing steady detectors. As a matter of fact, it is possible to exploit the similarities between the decision statistics to come up with a receiver whose directivity encompasses those of the merged stages as special cases. To give an example, consider the following statistics

$$d_1 = \frac{A}{B(1+C)}, \quad d_2 = \frac{A}{B}, \quad d_3 = \frac{A}{BC}, \tag{4.2}$$

where the random variables $A, B, C \in \mathcal{R}^+$, and combine them to obtain three new parametric decision statistics

$$d_{12}(\mu_{12}) = \frac{A}{B(1+\mu_{12}C)}, \quad \mu_{12} \in [0,\ 1], \tag{4.3}$$

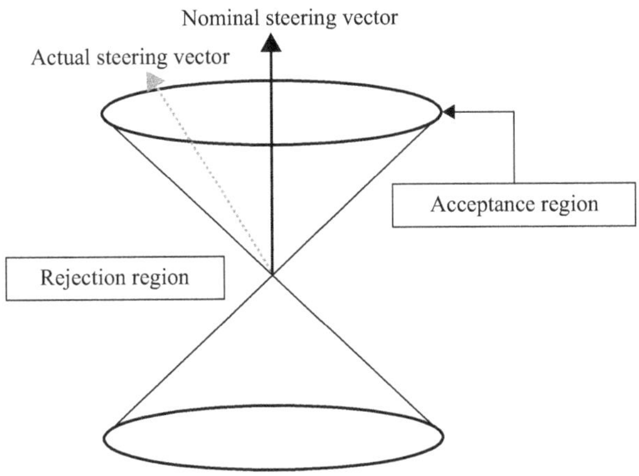

Figure 4.2 Conic acceptance/rejection idea.

$$d_{13}(\mu_{13}) = \frac{A}{B(\mu_{13} + C)}, \quad \mu_{13} \in [0,\ 1], \tag{4.4}$$

$$d_{123}(\mu_1, \mu_2) = \frac{A}{B(\mu_1 + \mu_2 C)}, \quad (\mu_1, \mu_2) \in \{[0,\ 1] \times [0,\ 1]\} \setminus (0, 0), \tag{4.5}$$

where $\times$ denotes the Cartesian product of sets. Observe that d_{12} has operating characteristics in between those of d_1 and d_2, d_{13} has operating characteristics in between those of d_1 and d_2, while d_{123} has operating characteristics in between those of d_1, d_2, and d_3.

Another family of tunable receivers can be found in References 3–6, where, at the design stage, the authors assume that the possible useful signal belongs to a proper cone with axis the nominal steering vector (see Figure 4.2). The directivity depends on the cone aperture, which is governed by a design parameter. Decision schemes described so far compare the value of a statistic with a threshold to decide for the presence of useful signal. We refer to this class of receivers as *parametric receivers*.

A tunable receiver can also be obtained by cascading two (steady and/or parametric) detectors with opposite behaviors in terms of directivity. The presence of a signal (H_1 hypothesis) is declared if and only if each stage is above the respective threshold, otherwise the overall detector decides for the null hypothesis, H_0 say. Observe that this architecture can be viewed as a logical AND between the two stages (see Figure 4.3) and, hence, the order of the stages is not significant to the decision (the truth table is given in Table 4.1). In the following, we refer to this decision schemes as *two-stage detectors*.

The two-stage paradigm is widespread in radar systems. Several examples are available in open literature. For instance, search radars, which do not provide fine range and/or angular measurements, are usually used in conjunction with tracking radars, whose resolution cell is small compared to that of a search radar [20, 21]. Another system exploiting the two-stage concept is the so-called SideLobe Blanker (SLB), which allows to suppress coherent repeater interference. The idea is that, employing an auxiliary antenna in addition to the main antenna, it is possible, by a proper choice of the antenna gains, to distinguish signals entering the sidelobes from those entering the main beam (and the former may be suppressed) [22–24]. As a matter of fact, when the ratio between the power of the signal received by the auxiliary channel in a

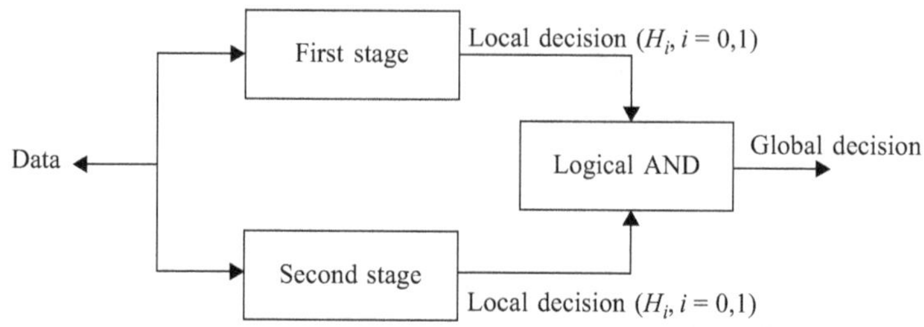

Figure 4.3 The two-stage architecture.

Table 4.1 Truth table of a two-stage detector.

Local decision	H_0	H_1
H_0	H_0	H_0
H_1	H_0	H_1

certain range cell and that of the signal received by the main channel is larger than a preassigned threshold, then the radar signal in that range cell is blanked (the received signal is interpreted as being produced by either an impulsive interference or the backscattering from discrete clutter sources impinging on the radar sidelobes). The antenna gains and the blanking threshold have to be set in order to

- maximize the probability of blanking an impulsive jammer;
- minimize the probability of blanking a true target received by the radar main beam.

However, in the presence of high-duty jammers, the SLB persists in the blanking mode for long time with the consequence that target detection is inhibited. The SideLobe Canceler (SLC) system represents a viable means to overcome this drawback. SLC uses an array of auxiliary antennas to adaptively estimate the direction of arrival and the power of the jammers and, subsequently, to modify the receiving pattern of the radar antenna placing nulls in the jammers' directions. Combinations of the SLB and SLC can be used to face with both type of interferences [25]. It is important to observe here that these two-stage systems exploit antenna-related techniques.

In the sequel, we focus on two-stage receivers based upon signal-processing techniques, which have been preliminary investigated in References 26 and 12. The considered architectures can be represented by the following equation

$$t_1(\mathbf{Z}) \underset{H_0}{\overset{H_1}{\gtrless}} \eta_1 \quad \text{and} \quad t_2(\mathbf{Z}) \underset{H_0}{\overset{H_1}{\gtrless}} \eta_2, \tag{4.6}$$

where $\mathbf{Z} \in \mathcal{C}^{N\times(K+1)}$, with $N, K \in \mathcal{N}$, $K \geq N$ (see the next section for the justification of such a constraint), is the data matrix, $t_i \in \mathcal{R}, i = 1, 2$, are the decision statistics, and (η_1, η_2) is the pair of thresholds used to ensure a preassigned value of probability of false alarm (P_{fa}). Actually, there exist infinite combinations of the above thresholds that provide the same value of P_{fa}. To give evidence of this, in Figure 4.4 we show the contours of constant P_{fa} of a

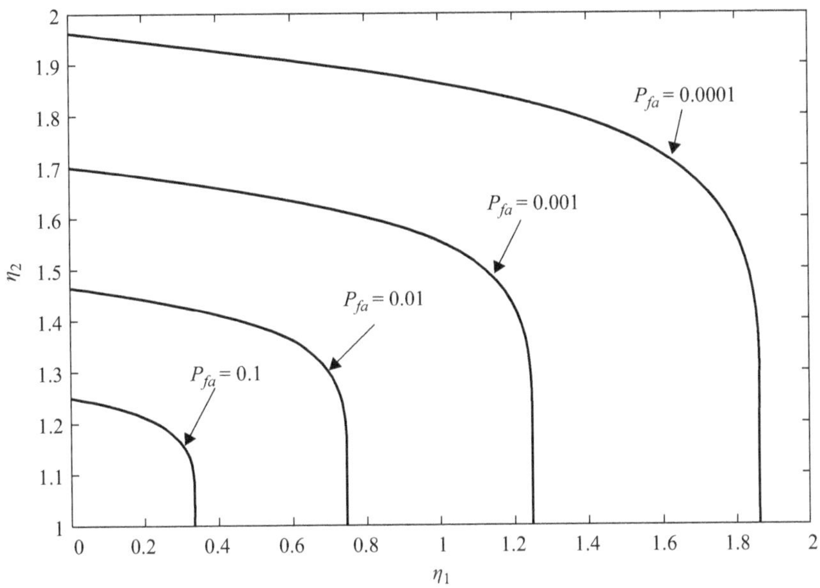

Figure 4.4 Contours of constant P_{fa} of a two-stage detector.

two-stage detector. On the other hand, the behavior in terms of directivity and P_d depends on the specific values of the thresholds (this point is further explained in the next section). Specifically, moving along a contour of constant P_{fa} yields infinite gradations between the two stages in terms of matched detection performances and directivity (see Figures 4.5 and 4.6) [13]. Observe that if $t_1 > 0$ and $t_2 > 0$, then

$$\eta_1 = 0, \quad \eta_2 > 0 \Rightarrow \text{two-stage receiver} \equiv \text{second stage}, \tag{4.7}$$

$$\eta_1 > 0, \quad \eta_2 = 0 \Rightarrow \text{two-stage receiver} \equiv \text{first stage}. \tag{4.8}$$

The range of directivity of a two-stage detector depends on the behavior of the single stages. In particular, coupling detectors with opposite behaviors, for instance, a robust receiver (as the Adaptive Matched Filter (AMF) [27] or the Subspace Detector (SD) [2]) and a selective receiver (as the whitened adaptive beamformer orthogonal rejection test (W-ABORT) [28]), leads to a decision scheme capable of ensuring a wide range of directivity. In addition, it is possible to exploit the flexibility of the two-stage paradigm to reduce the range of directivity towards the selectivity or robustness choosing detectors with similar behaviors. In Reference 1, it is shown that very selective receivers provide enhanced rejection capabilities of signals not aligned with the nominal one at the price of lower detection performances for matched signals. In order to compensate for the detection loss while retaining a high level of rejection of mismatched signals, a selective receiver can be connected with another detector having a lower selectivity but superior matched detection performance. The overall detector can guarantee enhanced rejection capabilities of sidelobe signals and, at the same time, a limited performance degradation under perfect matching conditions (*selective two-stage receivers*). Following the same line of reasoning, *robust two-stage decision schemes* can be designed.

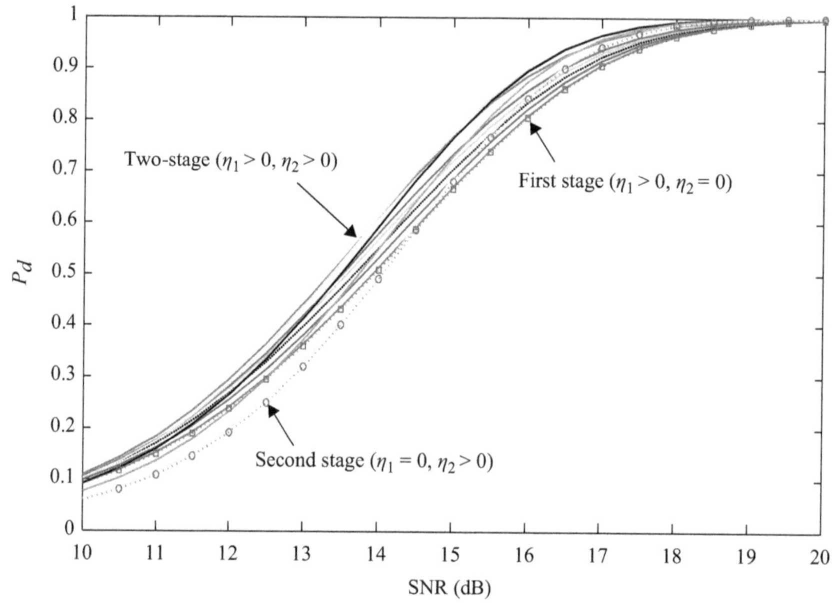

Figure 4.5 P_d versus SNR of a two-stage detector for different values of the thresholds along a contour of constant P_{fa}.

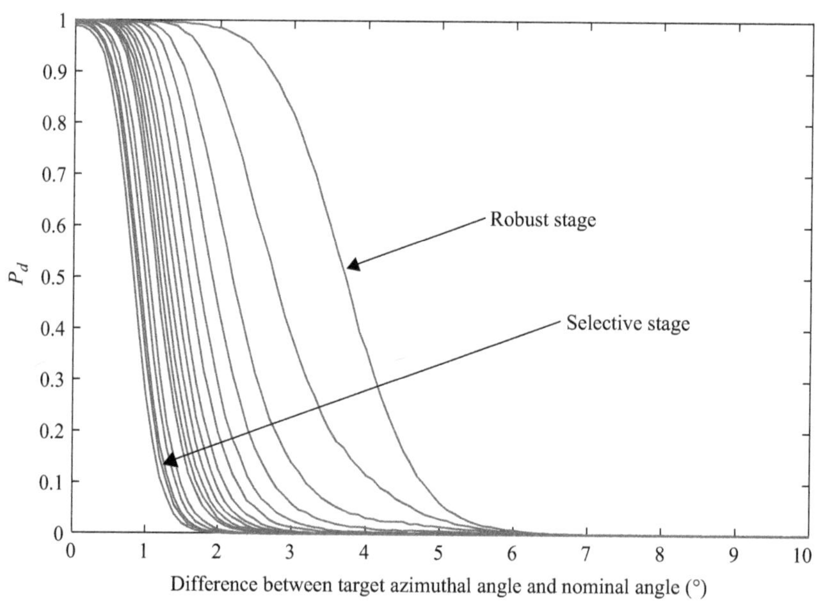

Figure 4.6 P_d versus angular separation between the actual direction of the target and the nominal one for a two-stage detector using different values of the thresholds along a contour of constant P_{fa}.

In the next section, we review most of possible architectures that can be conceived following the above design principles.

4.2 Two-Stage Architecture Description, Performance Analysis, and Comparisons

This section is devoted to the description and performance assessment of the most possible two-stage detectors that can be obtained by coupling existing decision schemes. For the reader's ease, as a preliminary step, we provide a brief description of the mathematical tools used for the performance analysis. Specifically, given the data matrix $\boldsymbol{Z} = [z\, z_1\ \ldots\ z_K]$, we formulate the detection problem to be solved as a binary hypothesis testing problem, i.e.,

$$\begin{cases} H_0 : \begin{cases} z = \boldsymbol{n}, \\ z_k = \boldsymbol{n}_k, \quad k = 1, \ldots, K, \end{cases} \\ \\ H_1 : \begin{cases} z = \alpha \boldsymbol{p} + \boldsymbol{n}, \\ z_k = \boldsymbol{n}_k, \quad k = 1, \ldots, K, \end{cases} \end{cases} \tag{4.9}$$

where

- $z \in \mathcal{C}^{N\times 1}$, with $N \in \mathcal{N}$ the dimension of the considered vectors, denotes the Cell Under Test (CUT);
- $z_k \in \mathcal{C}^{N\times 1}$, $k = 1, \ldots, K$, are the training samples (or secondary data) that are used for estimation purposes; they are usually chosen as the range cells surrounding that under test in order to preserve the homogeneity of the data. Moreover, we assume that $K \geq N$ to make the sample covariance matrix of the secondary data invertible with probability one;
- $\boldsymbol{n}$ and $\boldsymbol{n}_k$, $k = 1, \ldots, K$, are independent and identically distributed complex normal random vectors [1] with zero mean and covariance matrix $\boldsymbol{M} \in \mathcal{C}^{N\times N}$;
- $\alpha \in \mathcal{C}$ is a deterministic but unknown factor that accounts for the effects of the transmitting antenna gain, the radiation pattern of the array sensors, the two-way path loss, the radar cross section of the target, etc.;
- $\boldsymbol{p} \in \mathcal{C}^{N\times 1}$ is the actual steering vector.

Two remarks are now in order. First, notice that we assume the so-called *homogeneous environment*, where CUT and secondary data share the same spectral properties of the disturbance. However, the homogeneous environment might not be met in realistic situations: see, for example, Reference 29 and the references therein. Second, target returns may lie along a direction different from that of the nominal steering vector, $\boldsymbol{v}$ say. For this reason, in the sequel, we distinguish the actual steering vector from the nominal one and define the level of mismatch between $\boldsymbol{v}$ and $\boldsymbol{p}$ as follows [30]

$$\cos^2\theta = \frac{|\boldsymbol{v}^\dagger \boldsymbol{M}^{-1}\boldsymbol{p}|^2}{(\boldsymbol{p}^\dagger \boldsymbol{M}^{-1}\boldsymbol{p})(\boldsymbol{v}^\dagger \boldsymbol{M}^{-1}\boldsymbol{v})}, \tag{4.10}$$

where $\theta \in [0, \pi/2]$ is the mismatch angle between $\boldsymbol{p}$ and $\boldsymbol{v}$.

Moreover, we recall here those well-known statistics which come in handy to evaluate P_d (for matched and mismatched signals) and P_{fa} for the decision schemes described in the next subsections. More precisely, we will show that (in some situations) each stage can be expressed as function of these variables and, hence, it becomes straightforward to evaluate the probabilities of interest by means of closed form expressions. The first one is an equivalent form of Kelly's detector [31]

$$\bar{t}_{K} = \frac{t}{1-t} = \frac{|z^{\dagger}S^{-1}v|^2}{(v^{\dagger}S^{-1}v)\left[1 + z^{\dagger}S^{-1}z - \dfrac{|z^{\dagger}S^{-1}v|^2}{v^{\dagger}S^{-1}v}\right]}, \tag{4.11}$$

where

$$t = \frac{|z^{\dagger}S^{-1}v|^2}{(1 + z^{\dagger}S^{-1}z)(v^{\dagger}S^{-1}v)} \tag{4.12}$$

is the well-known statistic of the receiver derived in Reference 31 with

$$S = \sum_{k=1}^{K} zz^{\dagger} \tag{4.13}$$

K times the sample covariance matrix based upon the secondary data, while the second one is

$$\beta = \frac{1}{1 + z^{\dagger}S^{-1}z - \dfrac{|z^{\dagger}S^{-1}v|^2}{v^{\dagger}S^{-1}v}} \tag{4.14}$$

and is referred to as *loss factor* (see (4.19)). It is not difficult to prove that under the H_0 hypothesis [2, 30, 31]

- $\bar{t}_K$ is distributed according to a complex central F-distribution with $1, K-N+1$ complex degrees of freedom and it is independent of β;
- β obeys a complex central distribution with $K-N+2, N-1$ complex degrees of freedom.

On the other hand, under the H_1 hypothesis

- $\bar{t}_K$, given β, is subject to a complex noncentral F-distribution with $1, K-N+1$ complex degrees of freedom and noncentrality parameter r_t defined as follows

  $$r_t^2 = \text{SNR}\ \beta\ \cos^2\theta, \tag{4.15}$$

  where[1]

  $$\text{SNR} = |\alpha|^2 p^{\dagger} M^{-1} p \tag{4.16}$$

  and $\cos^2\theta$ is given by (4.10);

[1] It is important to note that the actual steering vector may differ from the nominal one.

- β is ruled by a complex noncentral beta distribution with $K-N+2, N-1$ complex degrees of freedom and noncentrality parameter r_β defined as

$$r_\beta = \text{SNR } \sin^2\theta, \tag{4.17}$$

where

$$\sin^2\theta = 1 - \cos^2\theta. \tag{4.18}$$

Observe that $\beta \in [0, 1]$ and hence in (4.15)

$$\text{SNR}\beta \leq \text{SNR}. \tag{4.19}$$

It is important to recall that the stochastic representations used in the next subsections are aimed at writing the test statistics of the two-stage architectures in terms of the same random variables. In Chapter 3, the reader can find alternative statistical characterizations of the most common detectors.

Finally, for the sake of completeness, we report in Table 4.2 the expressions of the probability density functions (pdfs) and Cumulative Distribution Functions (CDFs) of the aforementioned complex random variables (for further details we refer the interested reader to References 1, 2, and 32).

4.2.1 The Adaptive Sidelobe Blanker

The Adaptive Sidelobe Blanker (ASB) has been proposed in References 12 and 13 in order to mitigate the high number of false alarms of the AMF [27] due to the presence of clutter inhomogeneities. The ASB is obtained cascading the AMF, given by

Table 4.2 Pdfs and CDFs of useful complex distributions (the noncentrality parameter is denoted by δ).

Complex central F-distribution pdf with N, M complex degrees of freedom	$f(x) = \frac{(N+M-1)!}{(N-1)!(M-1)!} \frac{x^{N-1}}{(1+x)^{N+M}}, \quad x \geq 0$
Complex central F-distribution CDF with N, M complex degrees of freedom	$F(x) = \frac{x^N}{(1+x)^{N+M-1}} \sum_{k=0}^{M-1} \binom{N+M-1}{N+k} x^k, \quad x \geq 0$
Complex noncentral F-distribution pdf with N, M complex degrees of freedom	$f(x) = \frac{(N+M-1)!}{(N-1)!(M-1)!} \frac{x^{N-1} e^{-\delta^2/(1+x)}}{(1+x)^{N+M}} \sum_{k=0}^{M} \binom{M}{k} \frac{(N-1)!}{(N+k-1)!} \left(\frac{\delta^2 x}{1+x}\right)^k,$ $x \geq 0$
Complex noncentral F-distribution CDF with N, M complex degrees of freedom	$F(x) = \frac{x^N e^{-\delta^2/(1+x)}}{(1+x)^{N+M-1}} \sum_{k=0}^{M-1} \binom{N+M-1}{N+k} x^k \sum_{i=0}^{k} \left(\frac{\delta^2}{1+x}\right)^i \frac{1}{i!}, \ x \geq 0$
Complex central beta distribution pdf with N, M complex degrees of freedom	$f(x) = \frac{(N+M-1)!}{(N-1)!(M-1)!} x^{N-1}(1-x)^{M-1}, \quad 0 \leq x \leq 1$
Complex central beta distribution CDF with N, M complex degrees of freedom	$F(x) = x^{N+M-1} \sum_{k=0}^{M-1} \binom{N+M-1}{k} \left(\frac{1-x}{x}\right)^k, \quad 0 \leq x \leq 1$
Complex noncentral beta distribution pdf with N, M complex degrees of freedom	$f(x) = e^{-\delta^2 x} \sum_{k=0}^{N} \binom{N}{k} \frac{(N+M-1)!\delta^{2k} x^{N-1}(1-x)^{M+k-1}}{(M+k-1)!(N-1)!}, \ 0 \leq x \leq 1$
Complex noncentral beta distribution CDF with N, M complex degrees of freedom	$F(x) = 1 - \frac{x^{N-1} e^{-\delta^2 x}}{(1-x)^{-M}} \sum_{k=0}^{N-1} \binom{N+M-1}{k+M} \left(\frac{1-x}{x}\right)^k \sum_{i=0}^{k} \frac{(\delta^2 x)^i}{i!},$ $0 \leq x \leq 1$

$$t_{\text{AMF}} = \frac{|z^\dagger S^{-1} v|^2}{v^\dagger S^{-1} v} \underset{H_0}{\overset{H_1}{\gtrless}} \eta_{\text{AMF}} \tag{4.20}$$

and the Adaptive Coherent Estimator (ACE) [33, 34], also known as Adaptive Normalized Matched Filter (ANMF), whose expression is

$$t_{\text{ACE}} = \frac{|z^\dagger S^{-1} v|^2}{(z^\dagger S^{-1} z)(v^\dagger S^{-1} v)} \underset{H_0}{\overset{H_1}{\gtrless}} \eta_{\text{ACE}}. \tag{4.21}$$

As already stated, in order to derive closed-form expressions for the P_{fa} and the P_d, we need to recast the statistics of the two considered receivers as functions of the same quantities. To this end, we use the following equivalent test in place of that defined in (4.21)

$$\begin{aligned}
\bar{t}_{\text{ACE}} &= \frac{t_{\text{ACE}}}{1 - t_{\text{ACE}}} \\
&= \frac{|z^\dagger S^{-1} v|^2}{(v^\dagger S^{-1} v)\left[1 + z^\dagger S^{-1} z - \dfrac{|z^\dagger S^{-1} v|^2}{v^\dagger S^{-1} v}\right]} \frac{\left[1 + z^\dagger S^{-1} z - \dfrac{|z^\dagger S^{-1} v|^2}{v^\dagger S^{-1} v}\right]}{\left[1 + z^\dagger S^{-1} z - \dfrac{|z^\dagger S^{-1} v|^2}{v^\dagger S^{-1} v}\right] - 1} \\
&= \bar{t}_{\text{K}} \frac{1/\beta}{1/\beta - 1} \\
&= \bar{t}_{\text{K}} \frac{1}{1 - \beta} \underset{H_0}{\overset{H_1}{\gtrless}} \bar{\eta}_{\text{ACE}},
\end{aligned} \tag{4.22}$$

where

$$\bar{\eta}_{\text{ACE}} = \frac{\eta_{\text{ACE}}}{1 - \eta_{\text{ACE}}}. \tag{4.23}$$

In addition, observe that the decision statistic of the AMF can be recast as

$$\begin{aligned}
t_{\text{AMF}} &= \frac{|z^\dagger S^{-1} v|^2}{(v^\dagger S^{-1} v)\left[1 + z^\dagger S^{-1} z - \dfrac{|z^\dagger S^{-1} v|^2}{v^\dagger S^{-1} v}\right]} \left[1 + z^\dagger S^{-1} z - \dfrac{|z^\dagger S^{-1} v|^2}{v^\dagger S^{-1} v}\right] \\
&= \frac{\bar{t}_{\text{K}}}{\beta}.
\end{aligned} \tag{4.24}$$

Using the above results, it is possible to write the P_{fa} of the ASB as follows

$$\begin{aligned}
P_{fa}(\eta_{\text{AMF}}, \bar{\eta}_{\text{ACE}}) &= \Pr_{H_0}\left\{t_{\text{AMF}} > \eta_{\text{AMF}},\ \bar{t}_{\text{ACE}} > \bar{\eta}_{\text{ACE}}\right\} \\
&= \Pr_{H_0}\left\{\frac{\bar{t}_{\text{K}}}{\beta} > \eta_{\text{AMF}},\ \bar{t}_{\text{K}} \frac{1}{1 - \beta} > \bar{\eta}_{\text{ACE}}\right\} \\
&= \Pr_{H_0}\left\{\bar{t}_{\text{K}} > \max\left(\eta_{\text{AMF}}\beta, \bar{\eta}_{\text{ACE}}(1 - \beta)\right)\right\}
\end{aligned}$$

$$= \int_0^1 \Pr_{H_0}\left\{\bar{t}_{\text{K}} > \max\left(\eta_{\text{AMF}}\beta, \bar{\eta}_{\text{ACE}}(1-\beta)\right) | \beta = b\right\} f_\beta(b)db$$

$$= 1 - \int_0^1 \Pr_{H_0}\left\{\bar{t}_{\text{K}} \leq \max\left(\eta_{\text{AMF}}\beta, \bar{\eta}_{\text{ACE}}(1-\beta)\right) | \beta = b\right\} f_\beta(b)db$$

$$= 1 - \int_0^1 F_0(\max\left(\eta_{\text{AMF}}b, \bar{\eta}_{\text{ACE}}(1-b)\right) f_\beta(b)db$$

$$= 1 - \int_0^{\frac{\bar{\eta}_{\text{ACE}}}{\eta_{\text{AMF}}+\bar{\eta}_{\text{ACE}}}} F_0(\bar{\eta}_{\text{ACE}}(1-b)) f_\beta(b)db - \int_{\frac{\bar{\eta}_{\text{ACE}}}{\eta_{\text{AMF}}+\bar{\eta}_{\text{ACE}}}}^1 F_0(\eta_{\text{AMF}}b) f_\beta(b)db, \quad (4.25)$$

where $\Pr_{H_0}\{\cdot\}$ denotes the probability of the event argument assuming that H_0 is in force, $\Pr_{H_0}\{A|B\}$ is the probability of the event A conditioned to the event B, $f_\beta(\cdot)$ is the pdf of β under H_0, and $F_0(\cdot)$ is the CDF of $\bar{t}_{\text{K}}$ under H_0 (see also Table 4.2). Observe that the fourth equality comes from the total probability theorem.

An important remark is now in order. The ASB possesses the invariance property with respect to the group of transformations defined by [35]

$$G = \left\{g : \begin{bmatrix} z & Z_s \end{bmatrix} \to \begin{bmatrix} Tz & TZ_sB^\dagger \end{bmatrix}\right\}, \quad (4.26)$$

where $Z_s = [z_1 \ \dots \ z_K]$, $B \in \mathcal{C}^{K\times K}$ is a unitary matrix, and $T \in \mathcal{C}^{N\times N}$ is a full-rank matrix such that the range of Tv coincides with the range of v. As a consequence, it ensures the CFAR property with respect to the unknown covariance matrix of the interference.

Following the same line of reasoning, it is not difficult to show that the P_d of the ASB is given by

$$P_d(\eta_{\text{AMF}}, \bar{\eta}_{\text{ACE}}, \text{SNR}, \cos^2\theta)$$
$$= \Pr_{H_1}\{t_{\text{AMF}} > \eta_{\text{AMF}}, \bar{t}_{\text{ACE}} > \bar{\eta}_{\text{ACE}}\}$$
$$= 1 - \int_0^1 \Pr_{H_1}\left\{\bar{t}_{\text{K}} \leq \max\left(\eta_{\text{AMF}}\beta, \bar{\eta}_{\text{ACE}}(1-\beta)\right) | \beta = b\right\} f_{\beta, r_\beta}(b)db$$
$$= 1 - \int_0^{\frac{\bar{\eta}_{\text{ACE}}}{\eta_{\text{AMF}}+\bar{\eta}_{\text{ACE}}}} F_1(\bar{\eta}_{\text{ACE}}(1-b)) f_{\beta, r_\beta}(b)db - \int_{\frac{\bar{\eta}_{\text{ACE}}}{\eta_{\text{AMF}}+\bar{\eta}_{\text{ACE}}}}^1 F_1(\eta_{\text{AMF}}b) f_{\beta, r_\beta}(b)db, \quad (4.27)$$

where $F_1(\cdot)$ and $f_{\beta, r_\beta}(\cdot)$ are the CDF of $\bar{t}_{\text{K}}$ and the pdf of β, respectively, under H_1.

Equations (4.25) and (4.27) encompass the following special cases

$$(4.25) = \begin{cases} P_{fa} \text{ of the AMF}, & \text{if } \bar{\eta}_{\text{ACE}} = 0 \text{ and } \eta_{\text{AMF}} > 0, \\ P_{fa} \text{ of the ACE}, & \text{if } \bar{\eta}_{\text{ACE}} > 0 \text{ and } \eta_{\text{AMF}} = 0, \\ 1 & \text{if } \bar{\eta}_{\text{ACE}} = 0 \text{ and } \eta_{\text{AMF}} = 0, \end{cases} \quad (4.28)$$

and

$$(4.27) = \begin{cases} P_d \text{ of the AMF}, & \text{if } \bar{\eta}_{\text{ACE}} = 0 \text{ and } \eta_{\text{AMF}} > 0, \\ P_d \text{ of the ACE}, & \text{if } \bar{\eta}_{\text{ACE}} > 0 \text{ and } \eta_{\text{AMF}} = 0, \\ 1 & \text{if } \bar{\eta}_{\text{ACE}} = 0 \text{ and } \eta_{\text{AMF}} = 0. \end{cases} \quad (4.29)$$

In Figure 4.7, we show the contour plots corresponding to different values of P_{fa} in the plane $\eta_{\text{AMF}}\bar{\eta}_{\text{ACE}}$ assuming $N = 16, K = 32$. In Figure 4.8 we focus on the matched detection

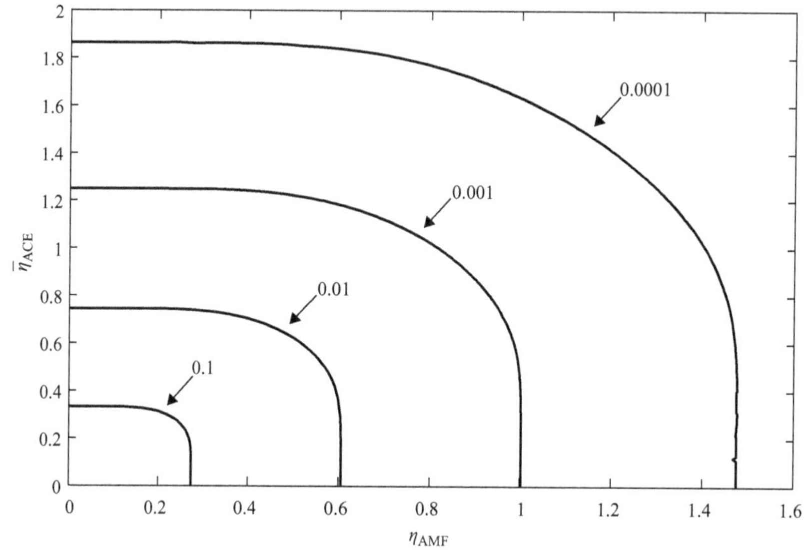

Figure 4.7 Contours of constant P_{fa} for the ASB assuming $N = 16$ and $K = 32$.

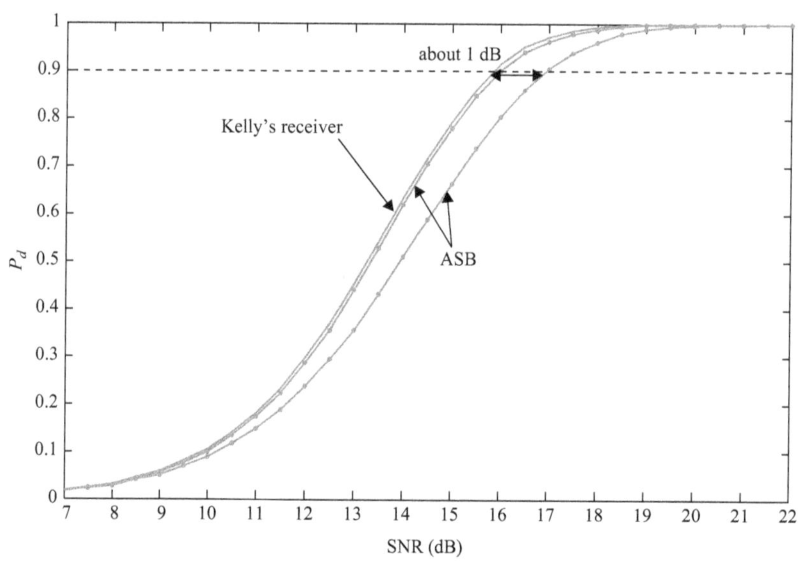

Figure 4.8 P_d versus SNR for the ASB (solid line with dot marker) and Kelly's receiver (solid line and no marker) assuming $N = 16$, $K = 32$, and $P_{fa} = 0.0001$.

performance ($\cos^2 \theta = 0$) and plot the P_d versus the SNR assuming $P_{fa} = 10^{-4}$, $N = 16$, and $K = 32$. In particular, we choose those threshold pairs that ensure the minimum and the maximum loss at $P_d = 0.9$ with respect to Kelly's detector usually regarded as benchmark detector in case of matched signals. Observe that the maximum loss with respect to Kelly's receiver at $P_d = 0.9$ is about 1 dB.

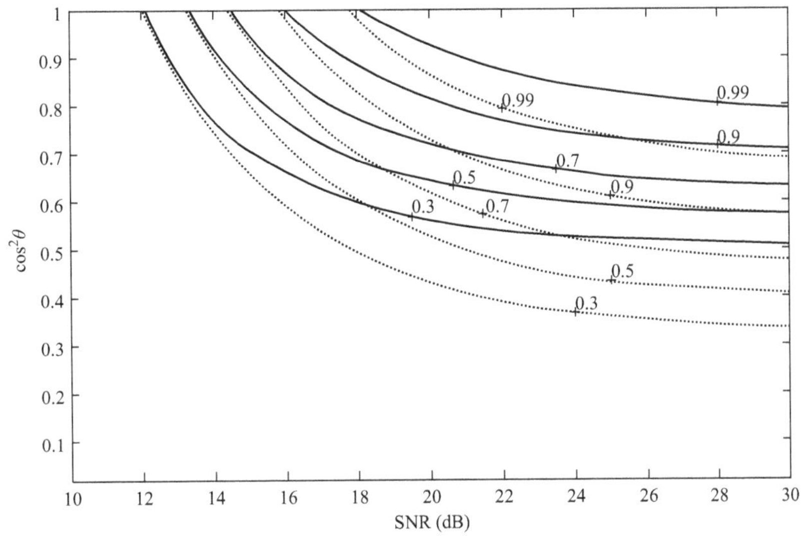

Figure 4.9 Contours of constant P_d for the ASB (solid line) and Kelly's receiver (dotted line) with $N = 16$, $K = 32$, $P_{fa} = 0.0001$, and threshold pair corresponding to the minimum loss with respect to Kelly's receiver at $P_d = 0.9$.

The performance of the ASB in the presence of mismatched signals are analyzed inspecting the contours of constant P_d represented as a function of $\cos^2\theta$, plotted vertically, and of the SNR, plotted horizontally. Such plots were introduced in Reference 10 and are referred to as *mesa plots*. Observe that reading the values of P_d on horizontal lines of a mesa plot returns the performance of the receiver in terms of P_d versus SNR for a preassigned value of the mismatch angle θ. In Figures 4.9–4.11, we analyze the behavior of the ASB in the case of mismatched signals in comparison with Kelly's receiver. Specifically, in Figure 4.9 we show that the ASB can provide better selectivity than Kelly's receiver with a negligible loss in case of matched signals. The threshold pair is chosen to obtain the minimum loss with respect to Kelly's receiver. In Figure 4.10, we consider the threshold configuration yielding the most robust behavior, which coincides to that of the AMF. Finally, in Figure 4.11 the contours of constant P_d refer to the most selective case with the constraint of a loss of about 1 dB with respect to Kelly's receiver at $P_d = 0.9$ and for matched signals. In this case, the ASB exhibits the same directivity as the ACE.

4.2.2 Modifications of the ASB towards Robustness: The Subspace-Based ASB

The Subspace-based ASB (S-ASB) is a two-stage detector proposed in Reference 16 to increase the range of directivity of the ASB towards the robustness. To this end, the AMF has been replaced with the SD [2], given by

$$t_{\text{SD}} = \frac{\boldsymbol{z}^\dagger \boldsymbol{S}^{-1}\boldsymbol{H}(\boldsymbol{H}^\dagger \boldsymbol{S}^{-1}\boldsymbol{H})^{-1}\boldsymbol{H}^\dagger \boldsymbol{S}^{-1}\boldsymbol{z}}{1+\boldsymbol{z}^\dagger \boldsymbol{S}^{-1}\boldsymbol{z}} \underset{H_0}{\overset{H_1}{\gtrless}} \eta_{\text{SD}}, \tag{4.30}$$

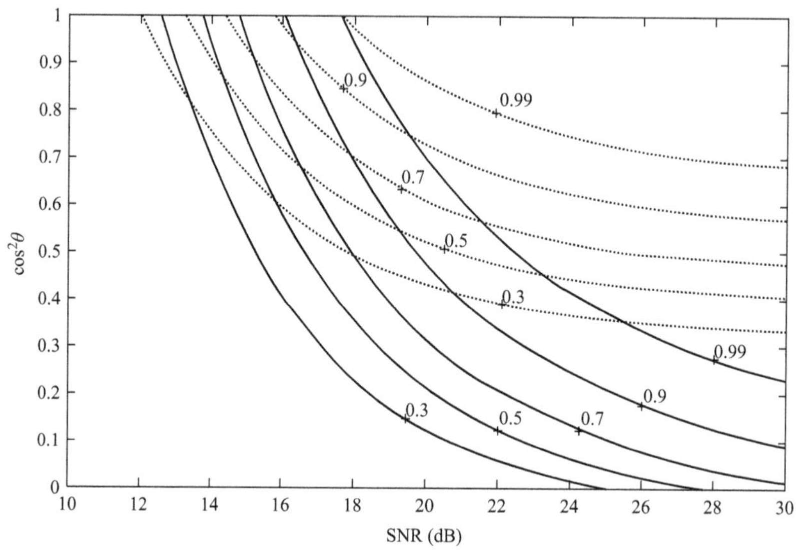

Figure 4.10 Contours of constant P_d for the ASB (solid line) and Kelly's receiver (dotted line) with $N = 16$, $K = 32$, $P_{fa} = 0.0001$, and threshold pair corresponding to the most robust behavior.

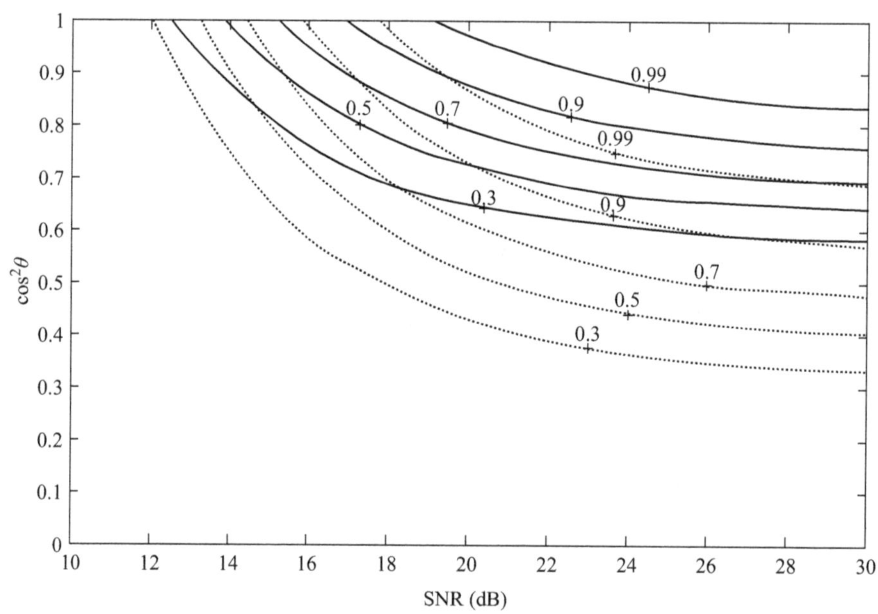

Figure 4.11 Contours of constant P_d for the ASB (solid line) and Kelly's receiver (dotted line) with $N = 16$, $K = 32$, $P_{fa} = 0.0001$, and threshold pair corresponding to the most selective behavior.

where

$$\boldsymbol{H} = \left[\boldsymbol{h}_1\, \boldsymbol{h}_2\, \ldots\, \boldsymbol{h}_q\right] \in \mathcal{C}^{N\times q}, \tag{4.31}$$

is a full-column-rank matrix. Hereafter, we assume that the nominal steering vector belongs to the subspace, $\mathbb{H}$ say, spanned by the columns of $\boldsymbol{H}$. It follows that $\mathbb{H}$ can be represented by

$$\boldsymbol{H} = \left[\boldsymbol{v}\, \boldsymbol{h}_2 \, \cdots \, \boldsymbol{h}_q\right]. \tag{4.32}$$

Notice that if $q = 1$ the SD coincides with Kelly's receiver and achieves the highest detection performances for matched signals. However, it is no longer robust to considerably mismatched signals. For this reason, in the two-stage architecture, we set $q > 1$. In particular, as shown in Reference 16, the choice

$$\boldsymbol{H} = [\boldsymbol{v}\, \boldsymbol{h}_2], \tag{4.33}$$

with $\boldsymbol{h}_2$ a vector slightly mismatched with respect to $\boldsymbol{v}$, guarantees an enhanced robustness with respect to the ASB (see also Figure 4.1 that is obtained assuming $N = 16$, $K = 32$, $P_{fa} = 10^{-4}$, and $q = 2$). In the sequel, we choose $\boldsymbol{h}_2$ such that

$$\frac{|\boldsymbol{v}^{\dagger}\boldsymbol{M}^{-1}\boldsymbol{h}_2|^2}{(\boldsymbol{v}^{\dagger}\boldsymbol{M}^{-1}\boldsymbol{v})(\boldsymbol{h}_2^{\dagger}\boldsymbol{M}^{-1}\boldsymbol{h}_2)} = 0.8949, \tag{4.34}$$

where the (i, j)th element of the covariance matrix $\boldsymbol{M}$ is given by $\rho^{|i-j|}$, $i, j = 1, \ldots, N$, with $\rho = 0.95$. Closed-form expressions for the P_d and P_{fa} can be obtained finding those stochastic representations for the SD and the ACE that are functions of the same random variables. To this end, we consider the following equivalent decision statistic for the SD

$$\bar{t}_{\text{SD}} = \frac{1}{1 - t_{\text{SD}}} \tag{4.35}$$

$$= \frac{1 + \boldsymbol{z}^{\dagger}\boldsymbol{S}^{-1}\boldsymbol{z}}{1 + \boldsymbol{z}^{\dagger}\boldsymbol{S}^{-1}\boldsymbol{z} - \boldsymbol{z}^{\dagger}\boldsymbol{S}^{-1}\boldsymbol{H}(\boldsymbol{H}^{\dagger}\boldsymbol{S}^{-1}\boldsymbol{H})^{-1}\boldsymbol{H}^{\dagger}\boldsymbol{S}^{-1}\boldsymbol{z}} \tag{4.36}$$

$$= (1 + \beta_2)(\bar{t}_{\text{K}} + 1), \tag{4.37}$$

and for the ACE

$$\bar{t}_{\text{ACE}} = \bar{t}_{\text{K}}\left[1 + \frac{1}{\beta_2(1 + \beta_1) + \beta_1}\right], \tag{4.38}$$

where random variables β_1 and β_2 have the following statistical characterization

- under the H_0 hypothesis
  - β_1 is a complex central F-distributed random variable with $(N-q, K-N+q+1)$ complex degrees of freedom;
  - β_2 is ruled by a complex central F-distribution with $(q-1, K-N+2)$ complex degrees of freedom and it is statistically independent of β_1;
- under the H_1 hypothesis
  - β_1 is a complex noncentral F-distributed random variable with $(N-q, K-N+q+1)$ complex degrees of freedom and noncentrality parameter r_1 defined by

$$r_1^2 = \text{SNR} \sin^2\theta \|\boldsymbol{v}_2\|^2; \tag{4.39}$$

- given β_1, β_2 is ruled by a complex noncentral F-distribution with $(q-1, K-N+2)$ complex degrees of freedom and noncentrality parameter r_2 defined by

$$r_2^2 = \frac{\text{SNR}\sin^2\theta\|\boldsymbol{v}_1\|^2}{1+\beta_1}. \tag{4.40}$$

In (4.39) and (4.40), $\boldsymbol{v}_1 \in \mathcal{C}^{(q-1)\times 1}$ and $\boldsymbol{v}_2 \in \mathcal{C}^{(N-q)\times 1}$ are defined as follows

$$\boldsymbol{U}\boldsymbol{M}^{-1/2}\boldsymbol{p} = \sqrt{\boldsymbol{p}^\dagger\boldsymbol{M}^{-1}\boldsymbol{p}}\begin{bmatrix} v\cos\theta \\ \boldsymbol{v}_1\sin\theta \\ \boldsymbol{v}_2\sin\theta \end{bmatrix}, \tag{4.41}$$

where $v \in \mathcal{C}$ and $\boldsymbol{U} \in \mathcal{C}^{N\times N}$ is the unitary matrix such that

$$\boldsymbol{U}\boldsymbol{H}_{qr} = \begin{bmatrix} 1\,0\ldots 0 \\ 0\,1\ldots 0 \\ 0\,0\ldots 1 \\ 0\,0\ldots 0 \\ \vdots\ \vdots\ \ddots\ \vdots \\ 0\,0\ldots 0 \end{bmatrix} \in \mathcal{R}^{N\times q} \tag{4.42}$$

with $\boldsymbol{H}_{qr} \in \mathcal{C}^{N\times q}$ being a slice of unitary matrix (i.e., $\boldsymbol{H}_{qr}^\dagger\boldsymbol{H}_{qr} = \boldsymbol{I}_q$) obtained by means of the QR factorization of $\boldsymbol{H}$, namely

$$\boldsymbol{H} = \boldsymbol{H}_{qr}\boldsymbol{H}_T, \tag{4.43}$$

where $\boldsymbol{H}_T \in \mathcal{C}^{q\times q}$ is an invertible upper triangular matrix. Moreover, the noncentrality parameter of the distribution of $\bar{t}_K$ under H_1 (given by (4.15)) can be recast as

$$r_t^2 = \frac{\text{SNR}\cos^2\theta}{(1+\beta_1)(1+\beta_2)}. \tag{4.44}$$

As a final remark, it is worth noticing that also the loss factor can be written as function of β_1 and β_2

$$\beta = \frac{1}{(1+\beta_1)(1+\beta_2)}. \tag{4.45}$$

Further details on the derivation of the statistical characterization of the S-ASB can be found in References 1 and 16.

Now, gathering the above results, the P_{fa} of the S-ASB is given by

$$\begin{aligned}
P_{fa}(\bar{\eta}_{\mathrm{SD}}, \bar{\eta}_{\mathrm{ACE}}) &= \Pr_{H_0}\left\{\bar{t}_{\mathrm{SD}} > \bar{\eta}_{\mathrm{SD}},\ \bar{t}_{\mathrm{ACE}} > \bar{\eta}_{\mathrm{ACE}}\right\} \\
&= \Pr_{H_0}\left\{(1+\beta_2)(\bar{t}_{\mathrm{K}}+1) > \bar{\eta}_{\mathrm{SD}},\ \bar{t}_{\mathrm{K}}\left[1+\frac{1}{\beta_2(1+\beta_1)+\beta_1}\right] > \bar{\eta}_{\mathrm{ACE}}\right\} \\
&= 1 - \Pr_{H_0}\left\{\bar{t}_{\mathrm{K}} \le \max\left(\frac{\bar{\eta}_{\mathrm{SD}}}{1+\beta_2} - 1, \bar{\eta}_{\mathrm{ACE}}\frac{\beta_2(1+\beta_1)+\beta_1}{(1+\beta_1)(1+\beta_2)}\right)\right\} \\
&= 1 - \int_0^{+\infty}\int_0^{+\infty} \Pr_{H_0}\left\{\bar{t}_{\mathrm{K}} \le \max\left(\frac{\bar{\eta}_{\mathrm{SD}}}{1+\beta_2} - 1, \bar{\eta}_{\mathrm{ACE}}\frac{\beta_2(1+\beta_1)+\beta_1}{(1+\beta_1)(1+\beta_2)}\right)\right| \\
&\qquad \beta_1 = b_1, \beta_2 = b_2\Bigg\} f_{\beta_1}(b_1) f_{\beta_2}(b_2) db_1 db_2 \\
&= 1 - \int_0^{+\infty}\int_0^{+\infty} F_0\left(\max\left(\frac{\bar{\eta}_{\mathrm{SD}}}{1+b_2} - 1, \bar{\eta}_{\mathrm{ACE}}\frac{b_2(1+b_1)+b_1}{(1+b_1)(1+b_2)}\right)\right) \\
&\qquad \times f_{\beta_1}(b_1) f_{\beta_2}(b_2) db_1 db_2,
\end{aligned} \tag{4.46}$$

where $f_{\beta_1}(\cdot)$ and $f_{\beta_2}(\cdot)$ denote the pdfs of β_1 and β_2 under H_0, respectively, $F_0(\cdot)$ is the CDF of $\bar{t}_{\mathrm{K}}$ under the H_0 hypothesis, and

$$\bar{\eta}_{\mathrm{SD}} = \frac{1}{1-\eta_{\mathrm{SD}}} \ge 1. \tag{4.47}$$

In Figure 4.12 we plot the contours of constant P_{fa} as functions of the two thresholds. It is clear that there exist an infinite number of threshold pairs that provide the same value of P_{fa}.

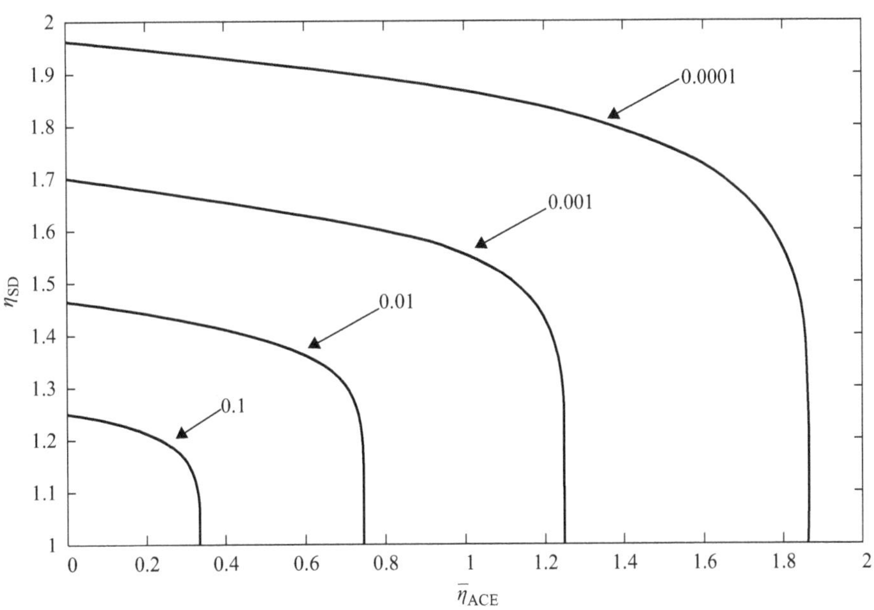

Figure 4.12 Contours of constant P_{fa} for the S-ASB assuming $N = 16$, $K = 32$, and $q = 2$.

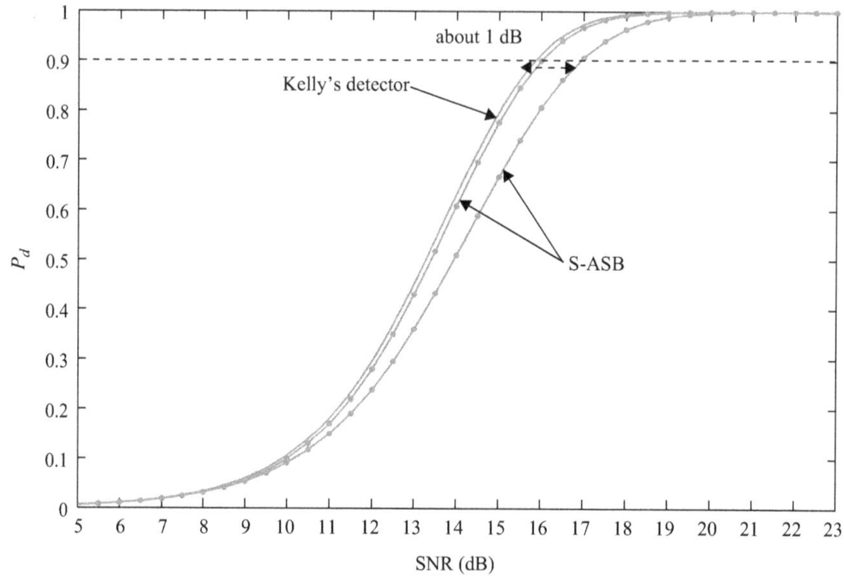

Figure 4.13 P_d versus SNR for the S-ASB (solid line with dot marker) and Kelly's receiver (solid line and no marker) assuming $N = 16$, $K = 32$, $q = 2$, and $P_{fa} = 0.0001$.

More importantly, (4.46) does not depend on $\boldsymbol{H}$, but for q (see the degrees of freedom of the considered distributions), and on $\boldsymbol{M}$. It follows that the S-ASB guarantees the CFAR property with respect to the covariance matrix of the interference.

As to the P_d, it is not difficult to show that

$$P_d(\bar{\eta}_{\text{SD}}, \bar{\eta}_{\text{ACE}}, \text{SNR}, \boldsymbol{H}, \cos^2\theta) = \Pr_{H_1}\left\{\bar{t}_{\text{SD}} > \bar{\eta}_{\text{SD}},\ \bar{t}_{\text{ACE}} > \bar{\eta}_{\text{ACE}}\right\}$$

$$= 1 - \int_0^{+\infty}\int_0^{+\infty} \Pr_{H_1}\left\{\bar{t}_{\text{K}} \le \max\left(\frac{\bar{\eta}_{\text{SD}}}{1+\beta_2} - 1, \bar{\eta}_{\text{ACE}}\frac{\beta_2(1+\beta_1)+\beta_1}{(1+\beta_1)(1+\beta_2)}\right)\middle|\ \beta_1 = b_1, \beta_2 = b_2\right\} f_{\beta_2|\beta_1}(b_2|\beta_1 = b_1) f_{\beta_1,r_1}(b_1)\,db_1\,db_2$$

$$= 1 - \int_0^{+\infty}\int_0^{+\infty} F_1\left(\max\left(\frac{\bar{\eta}_{\text{SD}}}{1+b_2} - 1, \bar{\eta}_{\text{ACE}}\frac{b_2(1+b_1)+b_1}{(1+b_1)(1+b_2)}\right)\right) \times f_{\beta_2|\beta_1}(b_2|\beta_1 = b_1) f_{\beta_1,r_1}(b_1)\,db_1\,db_2, \tag{4.48}$$

where $F_1(\,\cdot\,)$ is the CDF of $\bar{t}_{\text{K}}$ given β_1 and β_2 (and under H_1), whereas $f_{\beta_1,r_1}(\,\cdot\,)$ is the pdf of β_1 and $f_{\beta_2|\beta_1}(\,\cdot\,|\,\cdot\,)$ is the pdf of β_2 given β_1 both under H_1.

In Figure 4.13, we show the limiting behaviors of the S-ASB obtained moving along the contour of constant $P_{fa} = 0.0001$ assuming $N = 16$, $K = 32$, $q = 2$, and the columns of $\boldsymbol{H}$ are chosen as described at the beginning of this section. The detection performances of Kelly's receiver are also reported for comparison purposes. Observe that the maximum loss with respect to the benchmark at $P_d = 0.9$ is about 1 dB, while the minimum loss is about 0.2 dB. The mismatched detection performances are shown in Figures 4.14 and 4.15, where we plot the contours of constant P_d (as functions of the SNR and $\cos^2\theta$) for the S-ASB and Kelly's

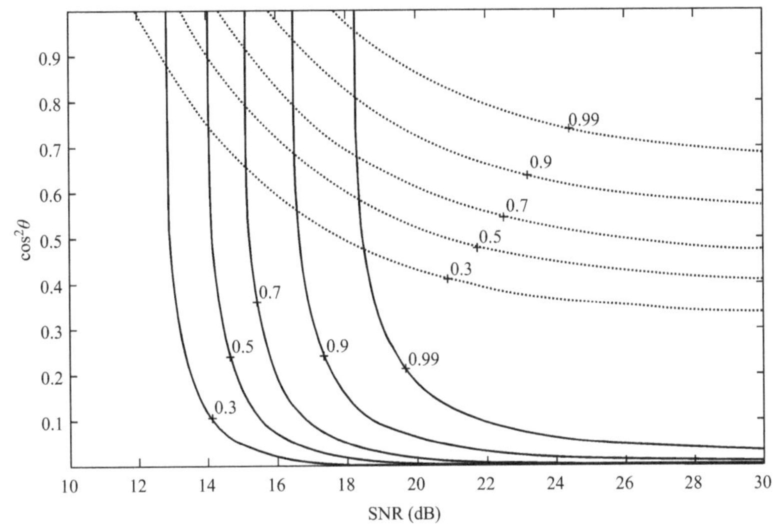

Figure 4.14 Contours of constant P_d for the S-ASB (solid line) and Kelly's receiver (dotted line) with $N = 16$, $K = 32$, $q = 2$, and threshold pair corresponding to the most robust case.

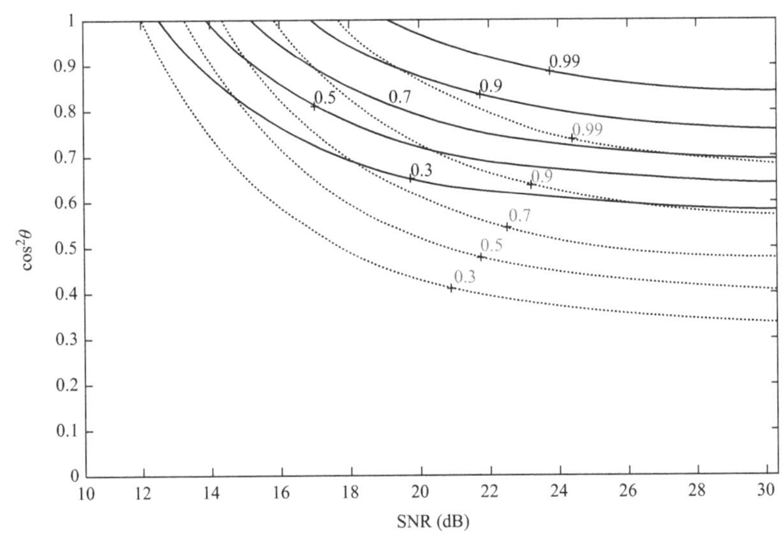

Figure 4.15 Contours of constant P_d for the S-ASB (solid line) and Kelly's receiver (dotted line) with $N = 16$, $K = 32$, $q = 2$, and threshold pair corresponding to the most selective case.

receiver. The curves shown in Figure 4.14 refer to that threshold pair that ensures the most robust behavior for the S-ASB, while the contours of Figure 4.15 are obtained choosing the threshold pair corresponding to the most selective case. Both figures assume $N = 16$, $K = 32$, $q = 2$, and $P_{fa} = 0.0001$. Finally, in Figure 4.16, we compare the most robust behavior of the S-ASB to that of the ASB. It is clear the enhanced robustness provided by the S-ASB.

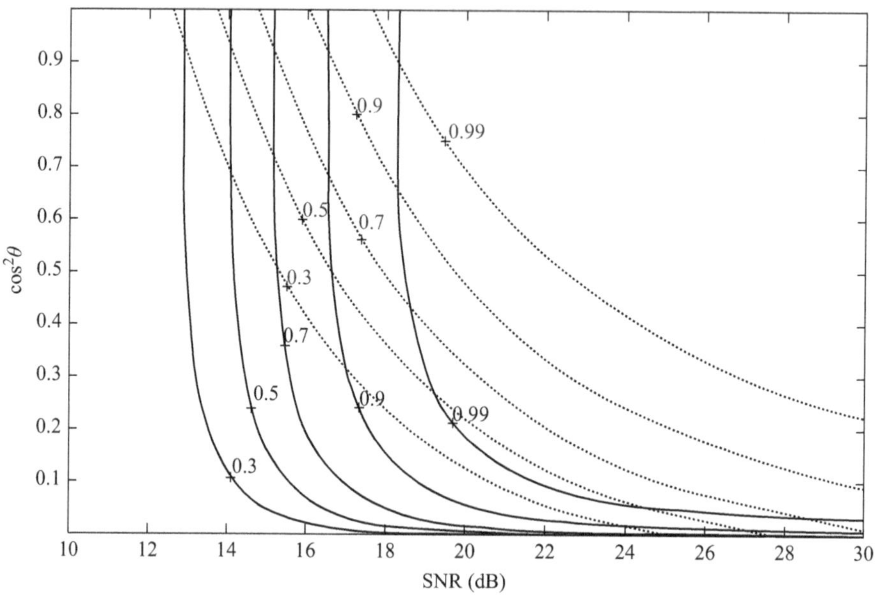

Figure 4.16 Contours of constant P_d for the S-ASB (solid line) and the ASB (dotted line) with $N = 16$, $K = 32$, $q = 2$, and threshold pair corresponding to the most robust case.

4.2.3 Modifications of the ASB towards Selectivity

In this subsection, we present two-stage solutions that increase the selectivity of the ASB. More precisely, such architectures exploit receivers that are more selective than the ACE, namely

- the Adaptive Beamformer Orthogonal Rejection Test (ABORT) devised in Reference 10 modifying the conventional noise-only hypothesis so that data possibly contain a fictitious signal orthogonal to the nominal one in the *quasi-whitened* observation space, namely after whitening data with the sample covariance matrix;
- the W-ABORT [28] obtained assuming the presence of a fictitious signal under H_0 that is orthogonal to the nominal one in the whitened observation space, namely after whitening data with the true noise covariance matrix;
- the Rao detector [36] derived by means of the Rao test design criterion;
- the parametric receiver obtained by merging the test statistic of Kelly's detector and that of the W-ABORT (WA); this architecture is referred to in the following as KWA [14].

Observe that most of selective detectors provide excellent sidelobe rejection capabilities at the price of lower matched detection performances (with respect to Kelly's receiver). However, as stated in Section 4.1, using selective receivers in a two-stage scheme allows to compensate for the detection loss while retaining a high level of selectivity. In other words, two-stage architectures provide a good trade off between matched detection performance and rejection of unwanted signals.

As a final remark, it is important to notice that two-stage detectors herein considered possess the invariance property with respect to the group of transformations (4.26). Thus, they ensure the CFAR property with respect to the unknown covariance matrix of the interference.

4.2.3.1 AMF-ABORT

The ABORT has been derived in Reference 10 by means of the Generalized Likelihood Ratio Test (GLRT) [37] applied to the following binary hypothesis testing problem

$$\begin{cases} H_0 : \begin{cases} z = \alpha_\perp \boldsymbol{v}_\perp + \boldsymbol{n}, \\ z_k = \boldsymbol{n}_k, \qquad k = 1, \ldots, K, \end{cases} \\ \\ H_1 : \begin{cases} z = \alpha \boldsymbol{v} + \boldsymbol{n}, \\ z_k = \boldsymbol{n}_k, \qquad k = 1, \ldots, K, \end{cases} \end{cases} \tag{4.49}$$

where $\alpha_\perp \in \mathcal{C}$ and

$$\boldsymbol{v}_\perp^\dagger \boldsymbol{S}^{-1} \boldsymbol{v} = 0. \tag{4.50}$$

The ABORT detector has the following form

$$t_{\text{ABORT}} = \frac{1 + t_{\text{AMF}}}{2 + z^\dagger \boldsymbol{S}^{-1} z} \tag{4.51}$$

$$= \frac{1 + \bar{t}_{\text{K}}/\beta}{1/\beta + 1 + \bar{t}_{\text{K}}/\beta} \underset{H_0}{\overset{H_1}{\gtrless}} \eta_{\text{ABORT}}. \tag{4.52}$$

In Reference 10, the authors propose a two-stage receiver obtained by cascading the AMF and the ABORT as efficient implementation of the latter. In order to obtain closed-form expression for the P_{fa} and P_d of this two-stage receiver, recast the ABORT as

$$\bar{t}_{\text{ABORT}} = \frac{t_{\text{ABORT}}}{1 - t_{\text{ABORT}}} \tag{4.53}$$

$$= \bar{t}_{\text{K}} + \beta \underset{H_0}{\overset{H_1}{\gtrless}} \bar{\eta}_{\text{ABORT}} = \frac{\eta_{\text{ABORT}}}{1 - \eta_{\text{ABORT}}}. \tag{4.54}$$

It follows that the P_{fa} of the AMF-ABORT can be written as

$$\begin{aligned} P_{fa}(\eta_{\text{AMF}}, \bar{\eta}_{\text{ABORT}}) &= \Pr_{H_0}\{t_{\text{AMF}} > \eta_{\text{AMF}}, \bar{t}_{\text{ABORT}} > \bar{\eta}_{\text{ABORT}}\} \\ &= 1 - \Pr_{H_0}\{\bar{t}_{\text{K}} \le \max(\eta_{\text{AMF}}\beta, \bar{\eta}_{\text{ABORT}} - \beta)\} \\ &= 1 - \int_0^1 F_0(\max(\eta_{\text{AMF}} b, \bar{\eta}_{\text{ABORT}} - b)) f_\beta(b) db \\ &\begin{cases} = 1 - \left[\int_0^{\bar{\eta}_{\text{ABORT}}/(1+\eta_{\text{AMF}})} F_0(\bar{\eta}_{\text{ABORT}} - b) f_\beta(b) db + \int_{\bar{\eta}_{\text{ABORT}}/(1+\eta_{\text{AMF}})}^1 F_0(\eta_{\text{AMF}} b) f_\beta(b) db \right] \\ \text{if } \bar{\eta}_{\text{ABORT}}/(1 + \eta_{\text{AMF}}) < 1, \\ = 1 - \int_0^1 F_0(\bar{\eta}_{\text{ABORT}} - b) f_\beta(b) db \\ \text{otherwise,} \end{cases} \end{aligned} \tag{4.55}$$

where $F_0(\cdot)$ is the CDF of $\bar{t}_{\text{K}}$ under the H_0 hypothesis and $f_\beta(\cdot)$ is the pdf of β under the H_0 hypothesis. In Figure 4.17, we plot the contours of constant P_{fa} for the AMF-ABORT assuming

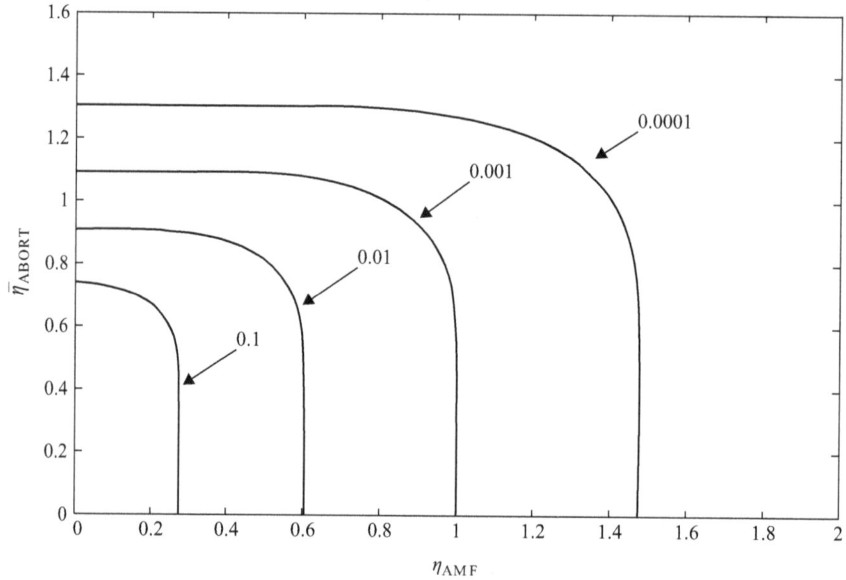

Figure 4.17 Contours of constant P_{fa} for the AMF-ABORT assuming $N = 16$ and $K = 32$.

$N = 16$ and $K = 32$. Observe that if the threshold value of one stage is given, then the P_{fa} is constant within an interval of threshold values of the other stage, e.g., for $P_{fa} = 0.0001$

$$\begin{array}{rcl} \bar{\eta}_{\text{ABORT}} \approx 1.3 & \text{and} & 0 \leq \eta_{\text{AMF}} \leq 0.8, \\ 0 \leq \bar{\eta}_{\text{ABORT}} \leq 0.6 & \text{and} & \eta_{\text{AMF}} \approx 1.5. \end{array} \tag{4.56}$$

On the other hand, the P_d has the following expression

$$P_d(\eta_{\text{AMF}}, \bar{\eta}_{\text{ABORT}}, \text{SNR}, \cos^2\theta) = \Pr_{H_1}\{t_{\text{AMF}} > \eta_{\text{AMF}}, \bar{t}_{\text{ABORT}} > \bar{\eta}_{\text{ABORT}}\}$$

$$\begin{cases} = 1 - \left[\int_0^{\bar{\eta}_{\text{ABORT}}/(1+\eta_{\text{AMF}})} F_1(\bar{\eta}_{\text{ABORT}} - b) f_{\beta,r_\beta}(b)db + \int_{\bar{\eta}_{\text{ABORT}}/(1+\eta_{\text{AMF}})}^{1} F_1(\eta_{\text{AMF}} b) f_{\beta,r_\beta}(b)db\right] \\ \text{if } \bar{\eta}_{\text{ABORT}}/(1+\eta_{\text{AMF}}) < 1, \\ = 1 - \int_0^1 F_1(\bar{\eta}_{\text{ABORT}} - b) f_{\beta,r_\beta}(b)db \\ \text{otherwise,} \end{cases} \tag{4.57}$$

where $F_1(\cdot)$ is the CDF of $\bar{t}_K$ under the H_1 hypothesis and $f_{\beta,r_\beta}(\cdot)$ is the pdf of β under the H_1 hypothesis. In Figure 4.18 we plot P_d versus SNR for the AMF-ABORT assuming $N = 16$, $K = 32$, and $P_{fa} = 0.0001$. More precisely, we show the curves corresponding to the minimum and the maximum loss with respect to Kelly's receiver at $P_d = 0.9$. Observe that the maximum detection loss of this two-stage receiver is about 0.4 dB while that of the ASB is about 1 dB for the chosen parameter values. However, the AMF-ABORT exhibits a lower selectivity with respect to the ASB (and hence to the S-ASB) as shown in Figure 4.19, where we assume a maximum loss with respect to Kelly's detector of about 1 dB at $P_d = 0.9$ and for matched signals. On the other hand, the AMF-ABORT guarantees the same robustness to mismatched signals as the ASB, namely that of the AMF as shown in Figure 4.20, where the contours of constant P_d are perfectly overlapped.

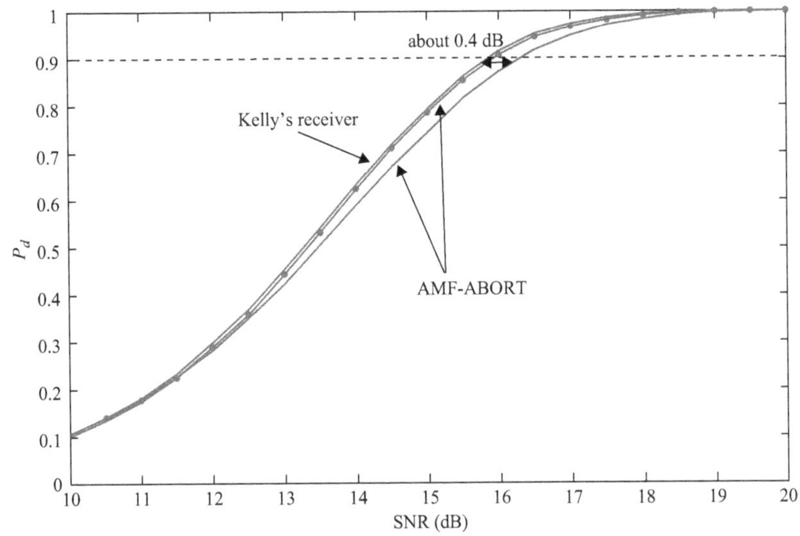

Figure 4.18 P_d versus SNR for the AMF-ABORT (solid line with dot marker) and Kelly's receiver (solid line and no marker) assuming $N = 16$, $K = 32$, and $P_{fa} = 0.0001$.

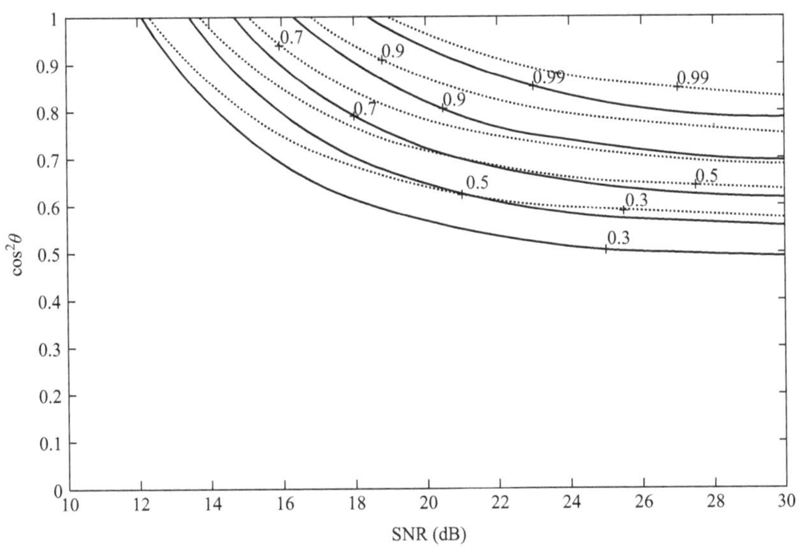

Figure 4.19 Contours of constant P_d for the AMF-ABORT (solid line) and the ASB (dotted line) with $N = 16$, $K = 32$, and threshold pair corresponding to the most selective case and under the constraint of a maximum loss with respect to Kelly's receiver of about 1 dB at $P_d = 0.9$ and for matched signals.

4.2.3.2 AMF-Rao

The Rao detector [36] is very selective for low values of N and K, but it exhibits poor matched detection performance. In order to face with this drawback, the Rao detector is coupled with

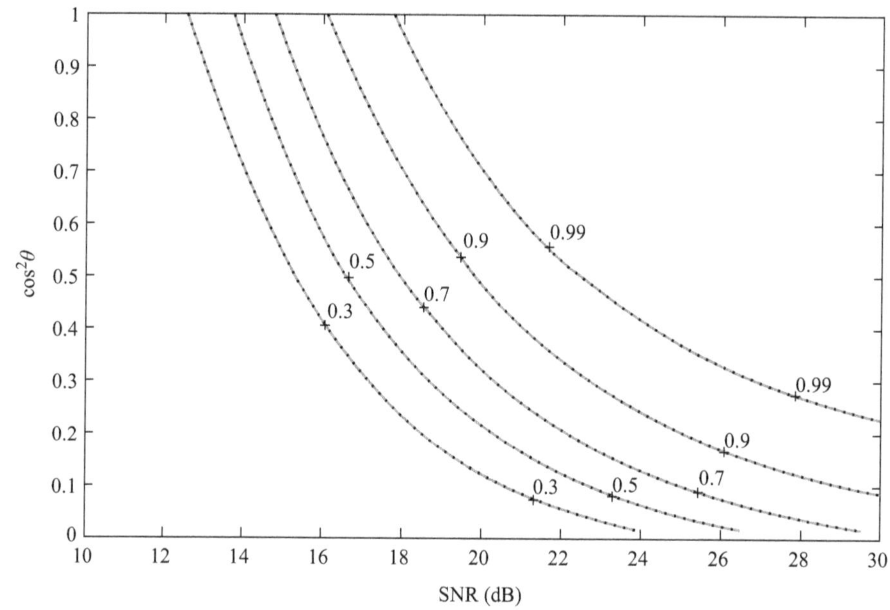

Figure 4.20 Contours of constant P_d for the AMF-ABORT (solid line) and the ASB (dotted line) with $N = 16$, $K = 32$, and threshold pair corresponding to the most robust case.

the AMF to obtain a two-stage detector capable of guaranteeing both enhanced rejection performances of unwanted signals and good matched detection performance [36]. The P_{fa} and the P_d can be easily evaluated after recasting the Rao detector as follows[2]

$$t_{\text{RAO}} = \frac{|z^\dagger (S + zz^\dagger)^{-1} v|^2}{v^\dagger S^{-1} v} \tag{4.58}$$

$$= t\beta \tag{4.59}$$

$$= \frac{\bar{t}_{\text{K}}}{1+\bar{t}_{\text{K}}}\beta \underset{H_0}{\overset{H_1}{\gtrless}} \eta_{\text{RAO}}. \tag{4.60}$$

It is not difficult to show that if $\beta > \eta_{\text{RAO}}$

$$\begin{aligned} P_{fa}(\eta_{\text{AMF}}, \eta_{\text{RAO}}) &= \Pr_{H_0}\{t_{\text{AMF}} > \eta_{\text{AMF}},\ t_{\text{RAO}} > \eta_{\text{RAO}}\} \\ &= 1 - \Pr_{H_0}\{\bar{t}_{\text{K}} \le \max(\eta_{\text{AMF}}\beta, \eta_{\text{RAO}}/(\beta - \eta_{\text{RAO}}))\} \\ &= \int_{\eta_{\text{RAO}}}^{1} \left[1 - F_0(\max(\eta_{\text{AMF}} b, \eta_{\text{RAO}}/(b - \eta_{\text{RAO}})))\right] f_\beta(b)\, db \end{aligned} \tag{4.61}$$

[2]For further details on the Rao test design criterion, we refer the reader to Reference 38.

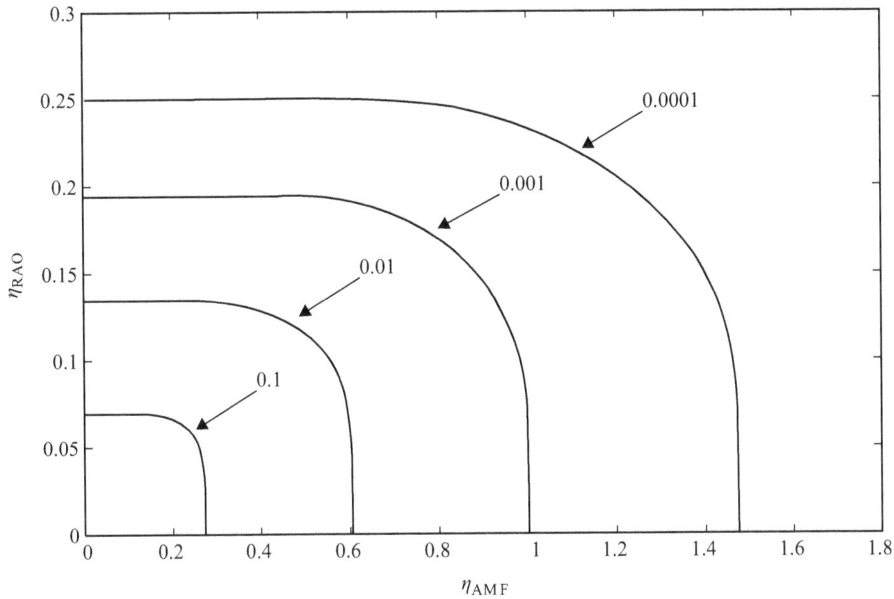

Figure 4.21 Contours of constant P_{fa} for the AMF-Rao assuming $N = 16$ and $K = 32$.

and

$$P_d(\eta_{AMF}, \eta_{RAO}, \text{SNR}, \cos^2\theta) = \Pr_{H_1}\{t_{AMF} > \eta_{AMF},\ t_{RAO} > \eta_{RAO}\}$$
$$= \int_{\eta_{RAO}}^{1} \left[1 - F_1(\max(\eta_{AMF} b, \eta_{RAO}/(b - \eta_{RAO})))\right] f_{\beta,r_\beta}(b)db, \quad (4.62)$$

where $F_i(\cdot)$, $i = 0, 1$, is the CDF of $\tilde{t}_K$ under the H_i hypothesis, $i = 0, 1$, $f_\beta(\cdot)$ and $f_{\beta,r_\beta}(\cdot)$ are the pdfs of β under H_0 and H_1, respectively. On the other hand, if $\beta < \eta_{RAO}$, $P_{fa} = P_d = 0$.

The contour plots corresponding to different values of P_{fa} are shown in Figure 4.21 for $N = 16$ and $K = 32$. Again, observe that, given a threshold value for one stage, the P_{fa} does not change within a given interval of threshold values for the other stage. In Figure 4.22, we show the matched detection performance of the AMF-Rao assuming $N = 16$, $K = 32$, and $P_{fa} = 0.0001$. In particular, notice that the maximum loss with respect to Kelly's receiver at $P_d = 0.9$ is about 2 dB. This is due to the fact that the Rao detector can guarantee an enhanced selectivity with respect to the ACE at the price of lower detection performances in case of matched signals. As a matter of fact, the maximum loss of ASB and of the S-ASB with respect to Kelly's receiver is about 1 dB for the considered parameter values. The excellent rejection capabilities of the Rao detector give rise to an increased selectivity of the overall detector as shown in Figure 4.23 where the AMF-Rao is compared to the ASB under the constraint of a maximum detection loss for matched signals with respect to Kelly's detector of about 1 dB at $P_d = 0.9$.

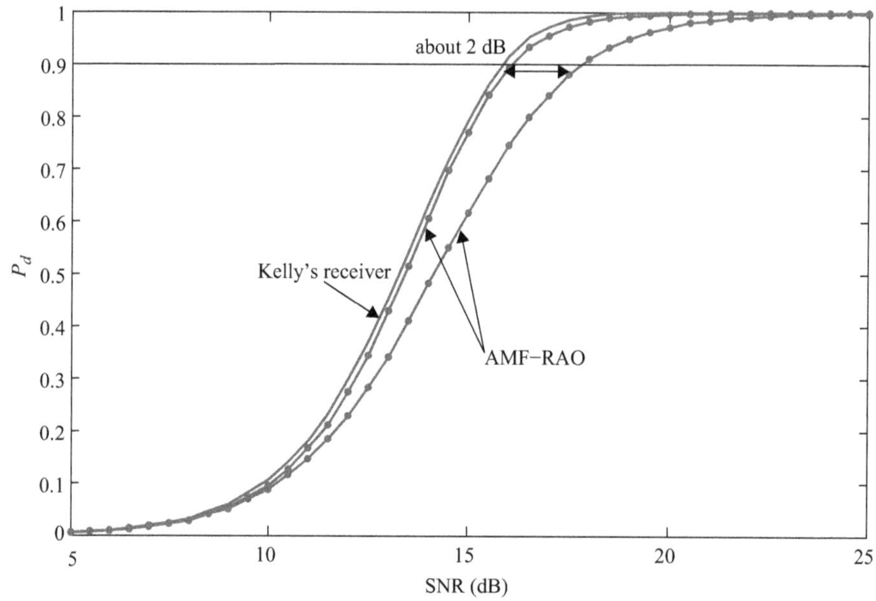

Figure 4.22 P_d versus SNR for the AMF-Rao (solid line with dot marker) and Kelly's receiver (solid line and no marker) assuming $N = 16$, $K = 32$, and $P_{fa} = 0.0001$.

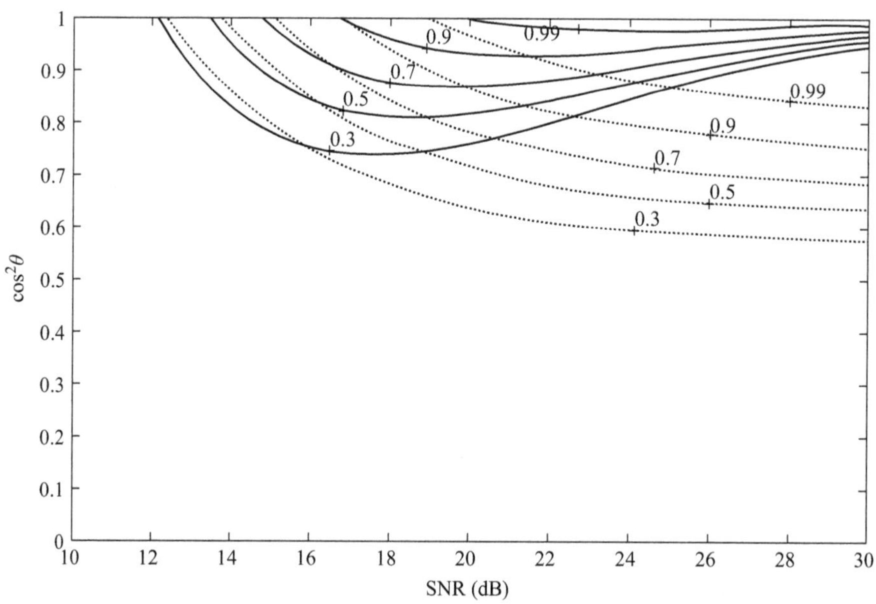

Figure 4.23 Contours of constant P_d for the AMF-Rao (solid line) and the ASB (dotted line) with $N = 16$, $K = 32$, and threshold pair corresponding to the most selective case under the constraint of a maximum loss of about 1 dB with respect to Kelly's receiver at $P_d = 0.9$ and matched signals.

4.2.3.3 AMF-WA

The W-ABORT has been proposed in Reference 28 by applying the GLRT to the following binary hypothesis testing problem

$$\begin{cases} H_0 : \begin{cases} z = \alpha_\perp v_\perp + n, \\ z_k = n_k, \qquad k = 1, \dots, K, \end{cases} \\ \\ H_1 : \begin{cases} z = \alpha v + n, \\ z_k = n_k, \qquad k = 1, \dots, K, \end{cases} \end{cases} \tag{4.63}$$

where $\alpha_\perp \in \mathcal{C}$ and

$$v_\perp^\dagger M^{-1} v = 0. \tag{4.64}$$

The decision scheme is given by

$$t_{\mathrm{WA}} = \frac{1}{\left[\dfrac{|z^\dagger S^{-1} v|^2}{(1 + z^\dagger S^{-1} z)(v^\dagger S^{-1} v)} - 1\right]^2 (1 + z^\dagger S^{-1} z)}$$

$$= \frac{1}{\left[\dfrac{\bar{t}_{\mathrm{K}}}{1 + \bar{t}_{\mathrm{K}}} - 1\right]^2 (1/\beta + \bar{t}_{\mathrm{K}}/\beta)} \tag{4.65}$$

$$= (1 + \bar{t}_{\mathrm{K}})\beta \underset{H_0}{\overset{H_1}{\gtrless}} \eta_{\mathrm{WA}}. \tag{4.66}$$

It follows that

$$\begin{aligned} P_{fa}(\eta_{\mathrm{AMF}}, \eta_{\mathrm{WA}}) &= \Pr_{H_0}\{t_{\mathrm{AMF}} > \eta_{\mathrm{AMF}},\ t_{\mathrm{WA}} > \eta_{\mathrm{WA}}\} \\ &= 1 - \Pr_{H_0}\{\bar{t}_{\mathrm{K}} \le \max(\eta_{\mathrm{AMF}}\beta, \eta_{\mathrm{WA}}/\beta - 1)\} \\ &= 1 - \int_0^1 F_0(\max(\eta_{\mathrm{AMF}} b, \eta_{\mathrm{WA}}/b - 1)) f_\beta(b)\, db \end{aligned} \tag{4.67}$$

and

$$\begin{aligned} P_d(\eta_{\mathrm{AMF}}, \eta_{\mathrm{WA}}, \mathrm{SNR}, \cos^2\theta) &= \Pr_{H_1}\{t_{\mathrm{AMF}} > \eta_{\mathrm{AMF}},\ t_{\mathrm{WA}} > \eta_{\mathrm{WA}}\} \\ &= 1 - \int_0^1 F_1(\max(\eta_{\mathrm{AMF}} b, \eta_{\mathrm{WA}}/b - 1)) f_{\beta, r_\beta}(b)\, db, \end{aligned} \tag{4.68}$$

where $F_i(\cdot)$, $i = 0, 1$, is the CDF of $\bar{t}_{\mathrm{K}}$ under the H_i hypothesis, $i = 0, 1$, $f_\beta(\cdot)$ and $f_{\beta, r_\beta}(\cdot)$ are the pdfs of β under H_0 and H_1, respectively.

The contours of constant P_{fa} are analogous to those already analyzed in previous subsections and, hence, we do not report here such contour plots. As to the matched detection performance, Figure 4.24 highlights that also for this receiver the maximum loss with respect to Kelly's detector increases to about 2 dB. In Figures 4.25 and 4.26, we compare the AMF-WA to the AMF-Rao for different values of N and K. More precisely, Figure 4.25 assumes $N = 16$ and $K = 32$, whereas in Figure 4.26 $N = 30$ and $K = 60$. The curves of both figures are obtained

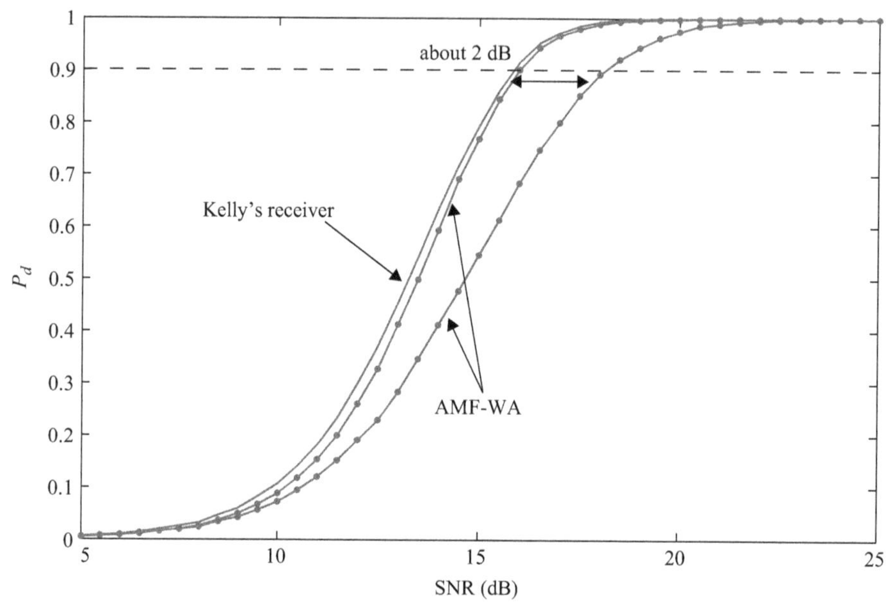

Figure 4.24 P_d versus SNR for the AMF-WA (solid line with dot marker) and Kelly's receiver (solid line and no marker) assuming $N = 16$, $K = 32$, and $P_{fa} = 0.0001$.

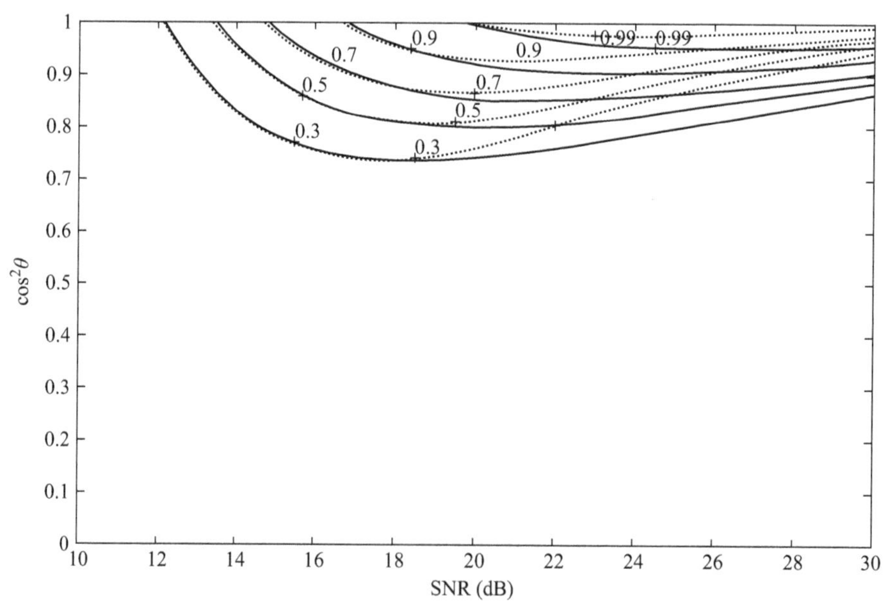

Figure 4.25 Contours of constant P_d for the AMF-WA (solid line) and the AMF-Rao (dotted line) with $N = 16$, $K = 32$, $P_{fa} = 0.0001$, and threshold pair corresponding to the most selective case under the constraint of a maximum loss of about 1 dB with respect to Kelly's receiver at $P_d = 0.9$ and matched signals.

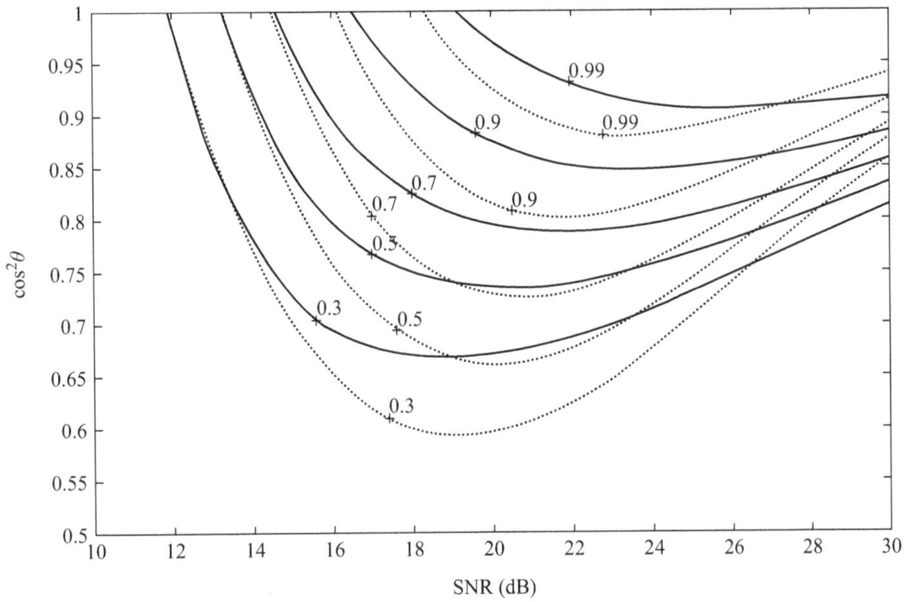

Figure 4.26 Contours of constant P_d for the AMF-WA (solid line) and the AMF-Rao (dotted line) with $N = 30$, $K = 60$, $P_{fa} = 0.0001$, and threshold pair corresponding to the most selective case under the constraint of a maximum loss of about 1 dB with respect to Kelly's receiver at $P_d = 0.9$ and matched signals.

under the constraint that the loss with respect to Kelly's detector is less than (about) 1 dB for the perfectly matched case. It is clear that the AMF-Rao is more sensitive than the AMF-WA to the system parameters N and K. More precisely, the AMF-Rao becomes less selective as N and/or K increase and the AMF-WA detector is slightly superior to the AMF-Rao in terms of selectivity for the system parameters considered in Figure 4.26.

4.2.3.4 AMF-KWA

The KWA is a receiver obtained by merging the statistic of Kelly's detector and that of the W-ABORT [14]. The decision statistic of the KWA is obtained by recasting Kelly's receiver and the W-ABORT in a proper manner. Specifically, consider the following equivalent form of Kelly's statistic

$$\tilde{t}_{\text{K}} + 1 = \frac{1 + z^\dagger S^{-1} z}{1 + z^\dagger S^{-1} z - \dfrac{|z^\dagger S^{-1} v|^2}{v^\dagger S^{-1} v}} \tag{4.69}$$

and rewrite the W-ABORT as

$$t_{\text{WA}} = \frac{1 + z^\dagger S^{-1} z}{\left[1 + z^\dagger S^{-1} z - \dfrac{|z^\dagger S^{-1} v|^2}{v^\dagger S^{-1} v}\right]^2}. \tag{4.70}$$

The KWA detector is given by

$$t_{\text{KWA}} = \frac{1 + z^\dagger S^{-1} z}{\left[1 + z^\dagger S^{-1} z - \frac{|z^\dagger S^{-1} v|^2}{v^\dagger S^{-1} v}\right]^{2\gamma}} \tag{4.71}$$

$$= (\bar{t}_{\text{K}} + 1)\beta^{2\gamma-1} \underset{H_0}{\overset{H_1}{\gtrless}} \eta_{\text{KWA}}, \tag{4.72}$$

where $\gamma \geq 0$ is the tuning parameter. Observe that it encompasses Kelly's detector and W-ABORT for $\gamma = 1/2$ and $\gamma = 1$, respectively. In addition, it is statistically equivalent to the energy detector, namely to

$$z^\dagger S^{-1} z \underset{H_0}{\overset{H_1}{\gtrless}} \eta_{\text{ED}}, \tag{4.73}$$

for $\gamma = 0$.

Gathering the above result yields

$$\begin{aligned} P_{fa}(\eta_{\text{AMF}}, \eta_{\text{KWA}}, \gamma) &= \Pr_{H_0}\{t_{\text{AMF}} > \eta_{\text{AMF}},\ t_{\text{KWA}} > \eta_{\text{KWA}}\} \\ &= 1 - \Pr_{H_0}\left\{\bar{t}_{\text{K}} \leq \max(\eta_{\text{AMF}}\beta, \eta_{\text{KWA}}/\beta^{2\gamma-1} - 1)\right\} \\ &= 1 - \int_0^1 F_0(\max(\eta_{\text{AMF}} b, \eta_{\text{KWA}}/b^{2\gamma-1} - 1)) f_\beta(b) db \end{aligned} \tag{4.74}$$

and

$$\begin{aligned} P_d(\eta_{\text{AMF}}, \eta_{\text{KWA}}, \text{SNR}, \cos^2\theta, \gamma) &= \Pr_{H_1}\{t_{\text{AMF}} > \eta_{\text{AMF}},\ t_{\text{KWA}} > \eta_{\text{KWA}}\} \\ &= 1 - \int_0^1 F_1(\max(\eta_{\text{AMF}} b, \eta_{\text{KWA}}/b^{2\gamma-1} - 1)) f_{\beta,r_\beta}(b) db, \end{aligned} \tag{4.75}$$

where $F_i(\,\cdot\,)$, $i = 0, 1$, is the CDF of $\bar{t}_{\text{K}}$ under the H_i hypothesis, $i = 0, 1$, $f_\beta(\,\cdot\,)$ and $f_{\beta,r_\beta}(\,\cdot\,)$ are the pdfs of β under H_0 and H_1, respectively.

For the sake of brevity, we do not report here the curves of constant P_{fa}, which highlight that in order to preserve a preassigned value of P_{fa}, any variation of η_{AMF} requires a significant modification of $\eta_{\text{KWA}}(> 0.1)$. The matched detection performance (assuming $\gamma = 1.3$) is shown in Figure 4.27, where the maximum loss with respect to Kelly's receiver is greater than that of the AMF-WA and AMF-Rao and increases to about 4 dB. As to the performance in case of mismatched signals, we compare the AMF-KWA to the AMF-WA and the AMF-Rao assuming $N = 16, K = 32$ and $N = 30, K = 60$. In particular, in Figures 4.28 and 4.29, we show that the AMF-KWA with $\gamma = 1.3$ is slightly more selective than the AMF-WA and provides rejection capabilities comparable to those of the AMF-Rao. The contours of constant P_d reported in Figures 4.30 and 4.31 refer to $N = 30$ and $K = 60$. In this case, the AMF-KWA with $\gamma = 1.3$ is slightly superior in terms of selectivity to both the AMF-WA and the AMF-Rao.

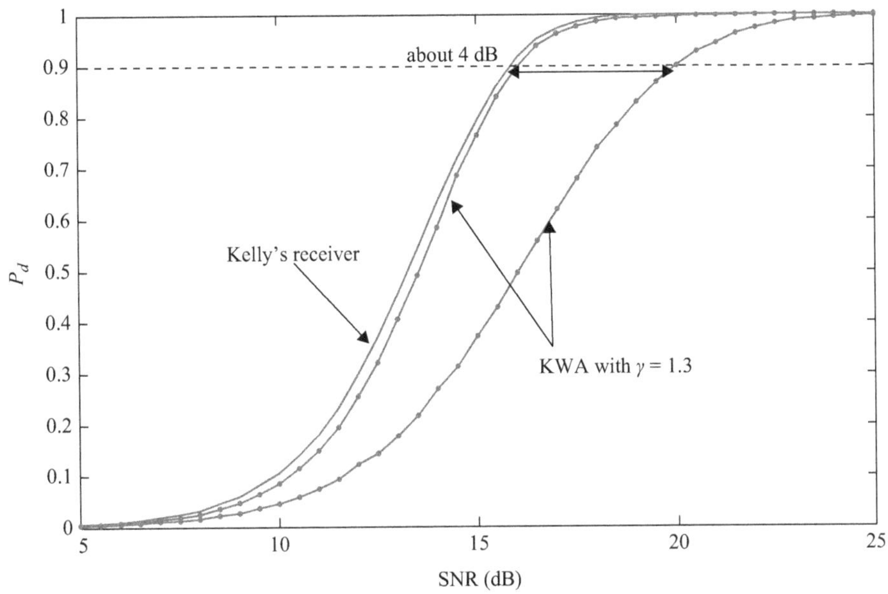

Figure 4.27 P_d versus SNR for the AMF-KWA (solid line with dot marker) and Kelly's receiver (solid line and no marker) assuming $N = 16$, $K = 32$, and $P_{fa} = 0.0001$.

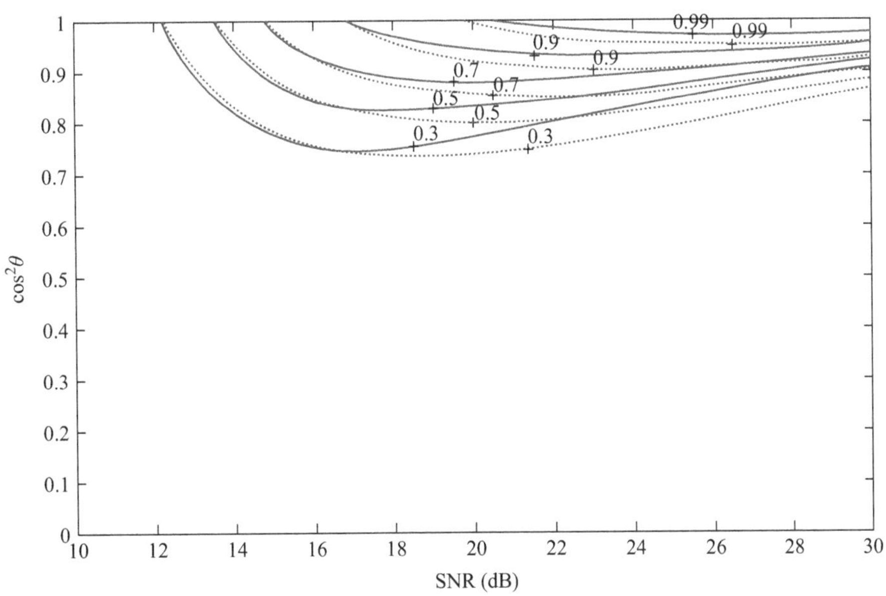

Figure 4.28 Contours of constant P_d for the AMF-KWA (solid line) and the AMF-WA (dotted line) with $N = 16, K = 32, P_{fa} = 0.0001, \gamma = 1.3$, and threshold pair corresponding to the most selective case under the constraint of a maximum loss of about 1 dB with respect to Kelly's receiver at $P_d = 0.9$ and matched signals.

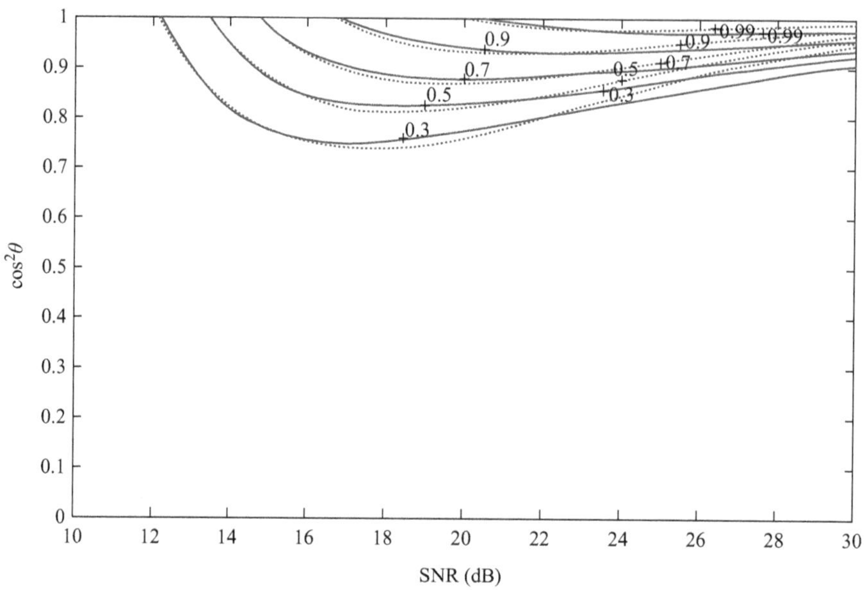

Figure 4.29 Contours of constant P_d for the AMF-KWA (solid line) and the AMF-Rao (dotted line) with $N = 16, K = 32, P_{fa} = 0.0001, \gamma = 1.3$, and threshold pair corresponding to the most selective case under the constraint of a maximum loss of about 1 dB with respect to Kelly's receiver at $P_d = 0.9$ and matched signals.

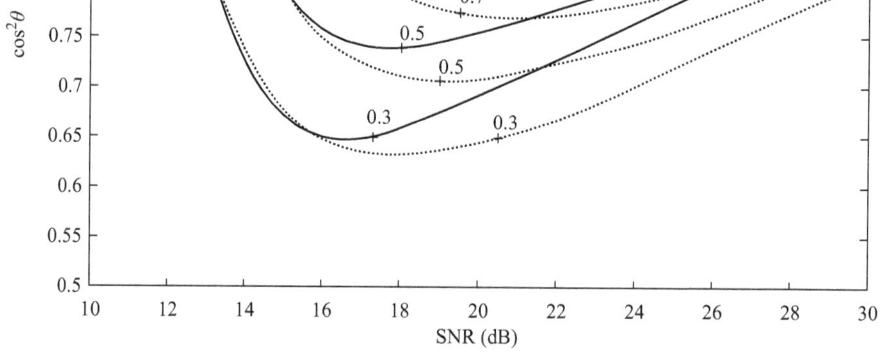

Figure 4.30 Contours of constant P_d for the AMF-KWA (solid line) and the AMF-WA (dotted line) with $N = 30, K = 60, P_{fa} = 0.0001, \gamma = 1.3$, and threshold pair corresponding to the most selective case under the constraint of a maximum loss of about 1 dB with respect to Kelly's receiver at $P_d = 0.9$ and matched signals.

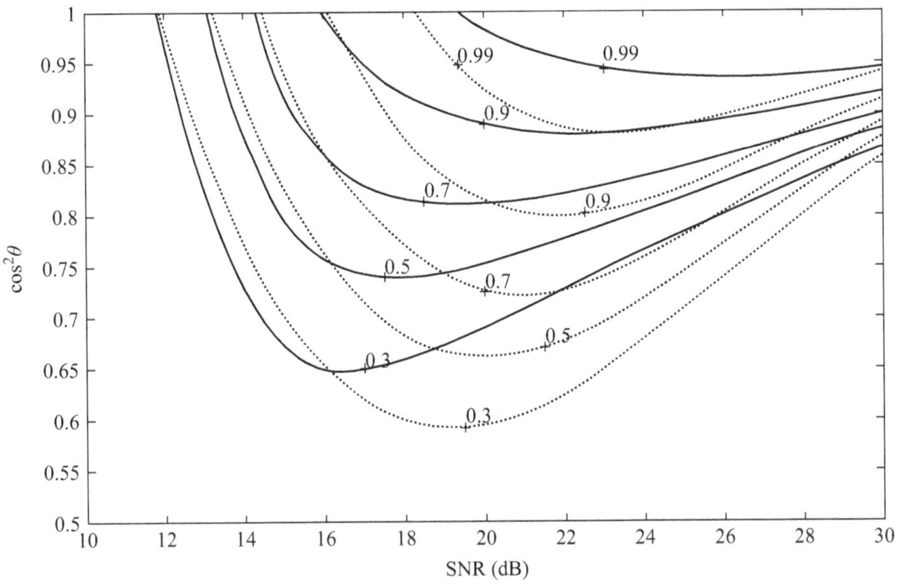

Figure 4.31 Contours of constant P_d for the AMF-KWA (solid line) and the AMF-Rao (dotted line) with $N = 30$, $K = 60$, $P_{fa} = 0.0001$, $\gamma = 1.3$, and threshold pair corresponding to the most selective case under the constraint of a maximum loss of about 1 dB with respect to Kelly's receiver at $P_d = 0.9$ and matched signals.

4.2.4 Modifications of the ASB towards both Selectivity and Robustness

This subsection is devoted to the analysis of those two-stage architectures that extend the range of directivity of the ASB in both robustness and selectivity. To this end, the robust stage of the ASB, namely the AMF, is replaced by the SD (see also Subsection 4.2.2), while in place of the ACE, which is the selective stage of the ASB, more recent selective solutions (Rao detector, W-ABORT, and KWA) are adopted (see also Subsection 4.2.3).

4.2.4.1 WAS-ASB

Cascading the SD and the W-ABORT yields a two-stage architecture that increases the selectivity of the S-ASB. In the following, we refer to this receiver as WAS-ASB. In order to derive closed-form expression for the P_{fa} and P_d we make use of stochastic representation (4.37) for the SD and observe that the decision statistic of W-ABORT admits the following expression

$$t_{\mathrm{WA}} = \frac{\bar{t}_{\mathrm{K}} + 1}{(\beta_1 + 1)(\beta_2 + 1)}. \tag{4.76}$$

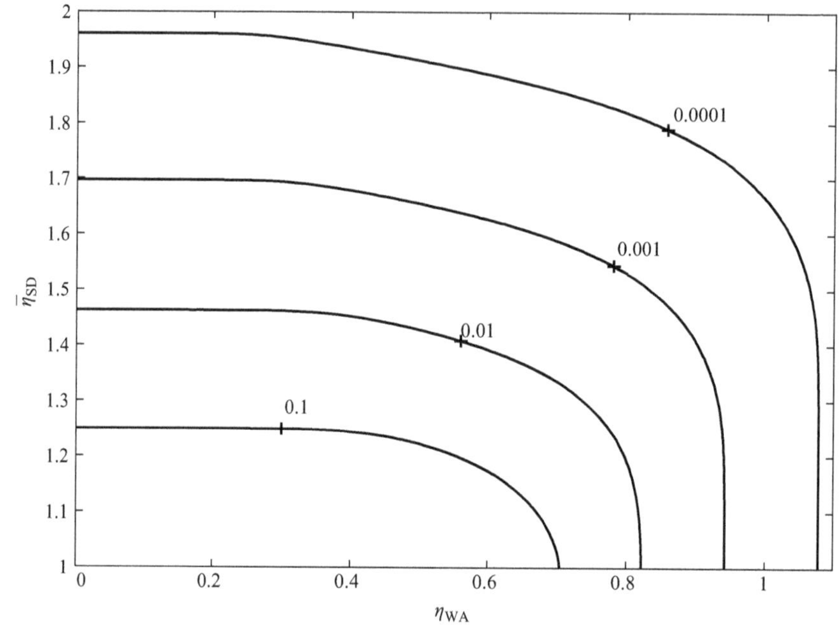

Figure 4.32 Contours of constant P_{fa} for the WAS-ASB assuming $N = 16$, $K = 32$, and $q = 2$.

Thus, we can write

$$\begin{aligned} P_{fa}(\bar{\eta}_{SD}, \eta_{WA}) &= \Pr_{H_0}\{\bar{t}_{SD} > \bar{\eta}_{SD}, t_{WA} > \eta_{WA}\} \\ &= 1 - \Pr_{H_0}\{\bar{t}_K \le \max(\bar{\eta}_{SD}/(1+\beta_2) - 1, \eta_{WA}(1+\beta_1)(1+\beta_2) - 1)\} \\ &= 1 - \int_0^{+\infty}\int_0^{+\infty} F_0\left(\max\left(\frac{\bar{\eta}_{SD}}{1+b_2} - 1, \eta_{WA}(1+b_1)(1+b_2) - 1\right)\right) \\ &\quad \times f_{\beta_1}(b_1) f_{\beta_2}(b_2) db_1 db_2 \end{aligned} \tag{4.77}$$

and

$$\begin{aligned} P_d(\bar{\eta}_{SD}, \eta_{WA}, \cos^2\theta, \text{SNR}, \boldsymbol{H}) &= \Pr_{H_1}\{\bar{t}_{SD} > \bar{\eta}_{SD}, t_{WA} > \eta_{WA}\} \\ = 1 - \int_0^{+\infty}\int_0^{+\infty} F_1\left(\max\left(\frac{\bar{\eta}_{SD}}{1+b_2} - 1, \eta_{WA}(1+b_1)(1+b_2) - 1\right)\right) \\ \times f_{\beta_2|\beta_1}(b_2|\beta_1 = b_1) f_{\beta_1, r_1}(b_1) db_1 db_2, \end{aligned} \tag{4.78}$$

where $F_i(\cdot)$, $i = 0, 1$, is the CDF of $\bar{t}_K$ under the H_i hypothesis, $i = 0, 1$, $f_{\beta_1}(\cdot)$ and $f_{\beta_2}(\cdot)$ are the pdfs of β_1 and β_2 under H_0, respectively, finally $f_{\beta_1, r_1}(\cdot)$ is the pdf of β_1 and $f_{\beta_2|\beta_1}(\cdot|\cdot)$ is the pdf of β_2 given β_1 both under H_1.

Inspection of (4.77) highlights that the WAS-ASB possesses the CFAR property with respect to $\boldsymbol{M}$. For the reader's ease, Figure 4.32 shows the contour plots for the WAS-ASB corresponding to different values of P_{fa}, as functions of the threshold pairs, $N = 16$, $K = 32$, and $r = 2$. Note that, for a preassigned value of P_{fa}, $\bar{\eta}_{SD}$ does not approximately change for values of η_{WA} ranging from 0 to 0.2. As to the matched detection performances, in Figure 4.33 we show the

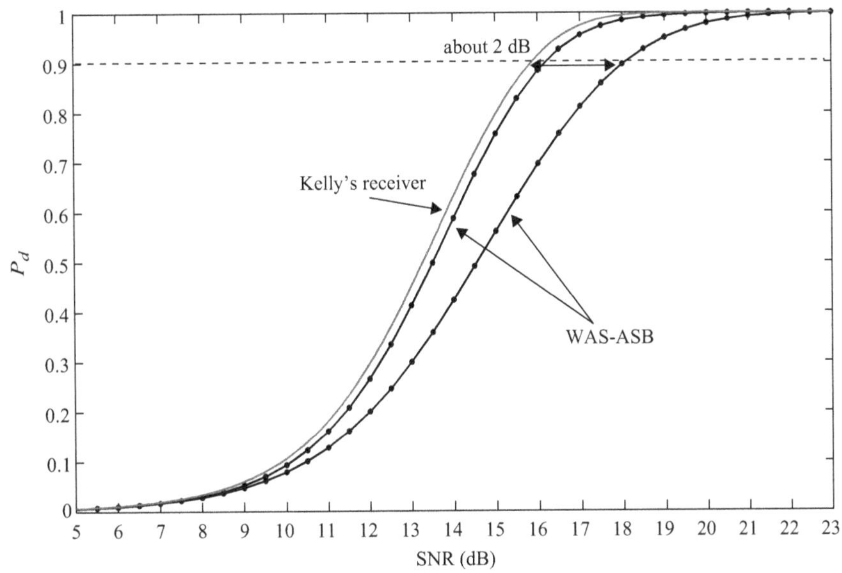

Figure 4.33 P_d versus SNR for the WAS-ASB (point marker) and Kelly's receiver (no marker) assuming $N = 16$, $K = 32$, $q = 2$, and $P_{fa} = 0.0001$.

limiting behaviors of the WAS-ASB assuming $P_{fa} = 0.0001$, $N = 16$, $K = 32$, $q = 2$, and the columns of $\boldsymbol{H}$ are chosen as described at the beginning of Section 4.2.2. The maximum detection loss with respect to Kelly's receiver at $P_d = 0.9$ is about 2 dB, while the minimum loss is about 0.35 dB.

The performance of the WAS-ASB in the presence of mismatched signals is analyzed in Figures 4.34 and 4.35, where we plot the contours of constant P_d as functions of the SNR and $\cos^2\theta$. More precisely, in Figure 4.34 we select the threshold pair that returns the most robust behavior, while in Figure 4.35 the most selective behavior of the WAS-ASB is compared with that of the S-ASB. Both figures assume that the maximum admissible loss with respect to Kelly's receiver is about 1 dB at $P_d = 0.9$ and refer to matched signals. It is clear that the WAS-ASB provides a wider range of directivity than the S-ASB.

4.2.4.2 KWAS-ASB

The KWA described in Subsection 4.2.3.4 can be used in place of the W-ABORT as selective stage of the WAS-ASB. The new resulting two-stage architecture further extends the range of directivity of the WAS-ASB [14]. In the sequel, we refer to this detector as KWAS-ASB. Exploiting the results contained in the previous section, it is easy to show that

$$\begin{aligned} P_{fa}(\bar{\eta}_{\text{SD}}, \eta_{\text{KWA}}, \gamma) &= \Pr_{H_0}\{\bar{t}_{\text{SD}} > \bar{\eta}_{\text{SD}}, t_{\text{KWA}} > \eta_{\text{KWA}}\} \\ &= 1 - \Pr_{H_0}\{\bar{t}_{\text{K}} \leq \max(\bar{\eta}_{\text{SD}}/(1+\beta_2) - 1, \eta_{\text{KWA}}[(1+\beta_1)(1+\beta_2)]^{2\gamma-1} - 1)\} \\ &= 1 - \int_0^{+\infty}\int_0^{+\infty} F_0\left(\max\left(\frac{\bar{\eta}_{\text{SD}}}{1+b_2} - 1, \eta_{\text{KWA}}[(1+b_1)(1+b_2)]^{2\gamma-1} - 1\right)\right) \\ &\quad \times f_{\beta_1}(b_1) f_{\beta_2}(b_2) db_1 db_2 \end{aligned} \tag{4.79}$$

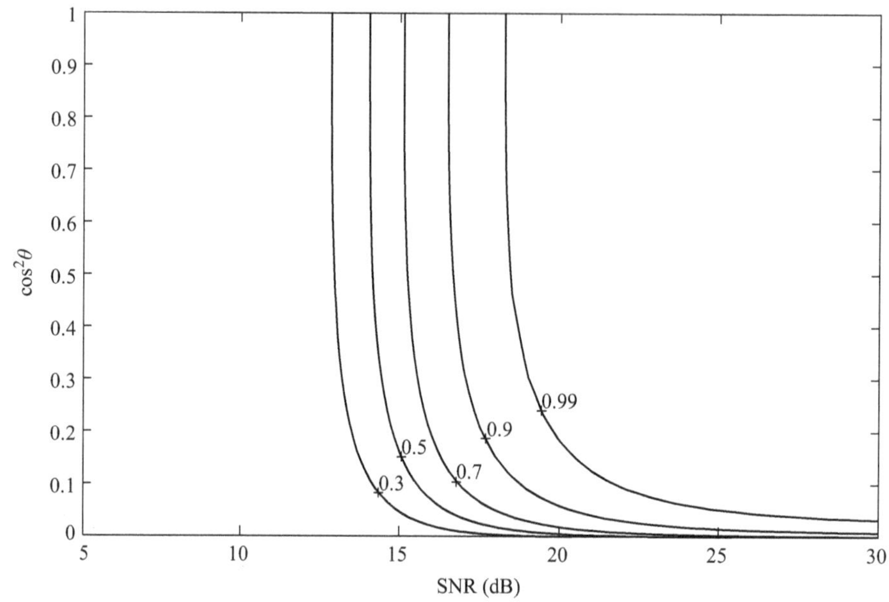

Figure 4.34 Contours of constant P_d for the WAS-ASB assuming $N = 16$, $K = 32$, $q = 2$, $P_{fa} = 0.0001$, and threshold pair that ensures the most robust behavior with the constraint of a maximum loss with respect to Kelly's receiver of about 1 dB.

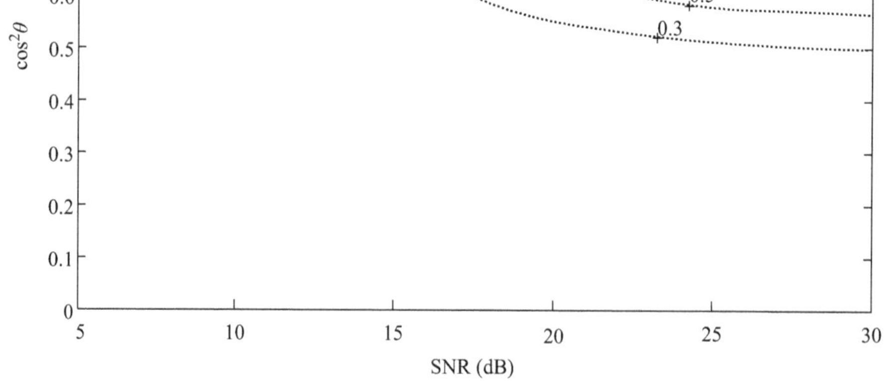

Figure 4.35 Contours of constant P_d for the WAS-ASB (solid line) and the S-ASB (dotted line) assuming $N = 16$, $K = 32$, $q = 2$, $P_{fa} = 0.0001$, and threshold pairs that ensure the most selective behavior with the constraint of a maximum loss with respect to Kelly's receiver of about 1 dB.

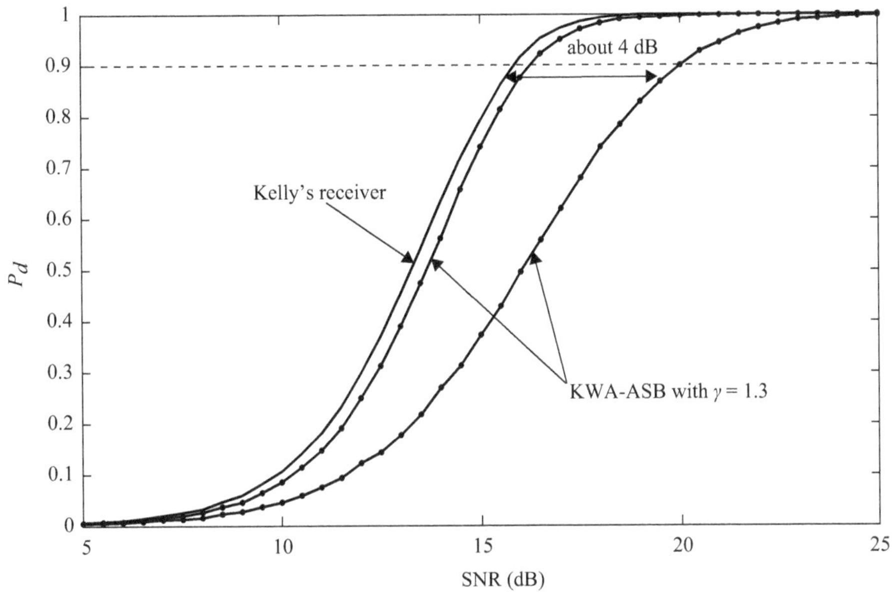

Figure 4.36 P_d versus SNR for the KWAS-ASB with $\gamma = 1.3$ (point marker) and Kelly's receiver (no marker) assuming $N = 16$, $K = 32$, $q = 2$, and $P_{fa} = 0.0001$.

and

$$
\begin{aligned}
P_d(\bar{\eta}_{\text{SD}}, \eta_{\text{KWA}}, \cos^2\theta, \text{SNR}, \boldsymbol{H}, \gamma) &= \Pr_{H_1}\{\bar{t}_{\text{SD}} > \bar{\eta}_{\text{SD}}, t_{\text{KWA}} > \eta_{\text{KWA}}\} \\
&= 1 - \int_0^{+\infty}\int_0^{+\infty} F_1\left(\max\left(\frac{\bar{\eta}_{\text{SD}}}{1+b_2} - 1, \eta_{\text{KWA}}[(1+b_1)(1+b_2)]^{2\gamma-1} - 1\right)\right) \\
&\quad \times f_{\beta_2|\beta_1}(b_2|\beta_1 = b_1) f_{\beta_1,r_1}(b_1) db_1 db_2,
\end{aligned} \tag{4.80}
$$

where $F_i(\cdot)$, $i = 0, 1$, is the CDF of $\bar{t}_{\text{K}}$ under the H_i hypothesis, $i = 0, 1$, $f_{\beta_1}(\cdot)$ and $f_{\beta_2}(\cdot)$ are the pdfs of β_1 and β_2 under H_0, respectively, finally $f_{\beta_1,r_1}(\cdot)$ is the pdf of β_1 and $f_{\beta_2|\beta_1}(\cdot\,|\,\cdot)$ is the pdf of β_2 given β_1 both under H_1.

The contours of constant P_{fa} are analogous to those of the WAS-ASB and, for this reason, are omitted. The matched detection performance is shown in Figure 4.36: the maximum loss with respect to Kelly's receiver at $P_d = 0.9$ is about 4 dB (see also the limiting behaviors of the WAS-ASB). As to the performance in the presence of mismatched signals, in Figure 4.37 we compare the KWAS-ASB and the WAS-ASB assuming $N = 16$, $K = 32$, and $P_{fa} = 10^{-4}$. Again, the threshold pairs are chosen to ensure a loss with respect to Kelly's receiver of about 1 dB at $P_d = 0.9$ and for matched signals. As expected, choosing values of γ greater than 1 increases the selectivity of the overall detector. This trend is confirmed by inspecting Figure 4.38, where $N = 30$ and $K = 60$. Finally, observe that the KWAS-ASB shares the SD with the WAS-ASB and, as a consequence, they exhibit more or less the same behavior in terms of robustness.

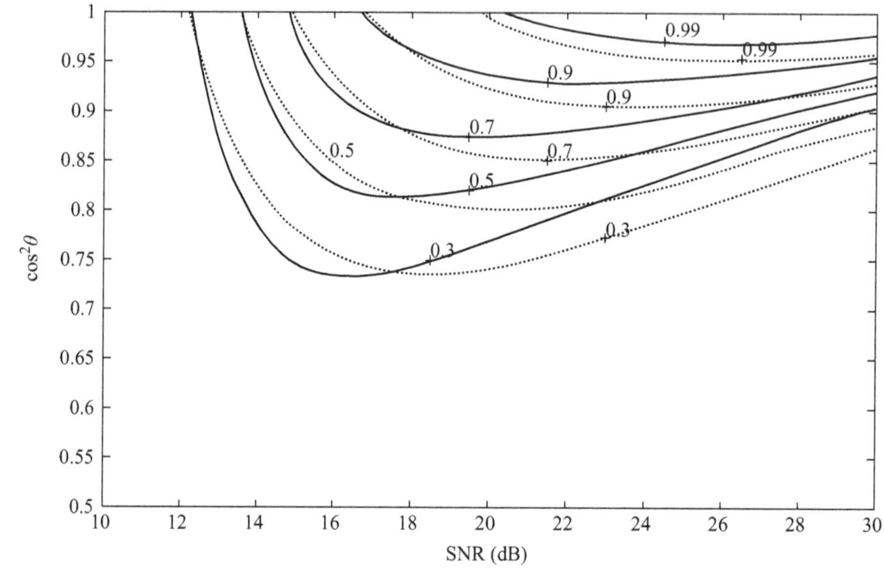

Figure 4.37 Contours of constant P_d for the KWAS-ASB with $\gamma = 1.3$ (solid line) and the WAS-ASB (dotted line) assuming $N = 16$, $K = 32$, $q = 2$, $P_{fa} = 0.0001$, and threshold pairs that ensure the most selective behavior with the constraint of a maximum loss with respect to Kelly's receiver of about 1 dB.

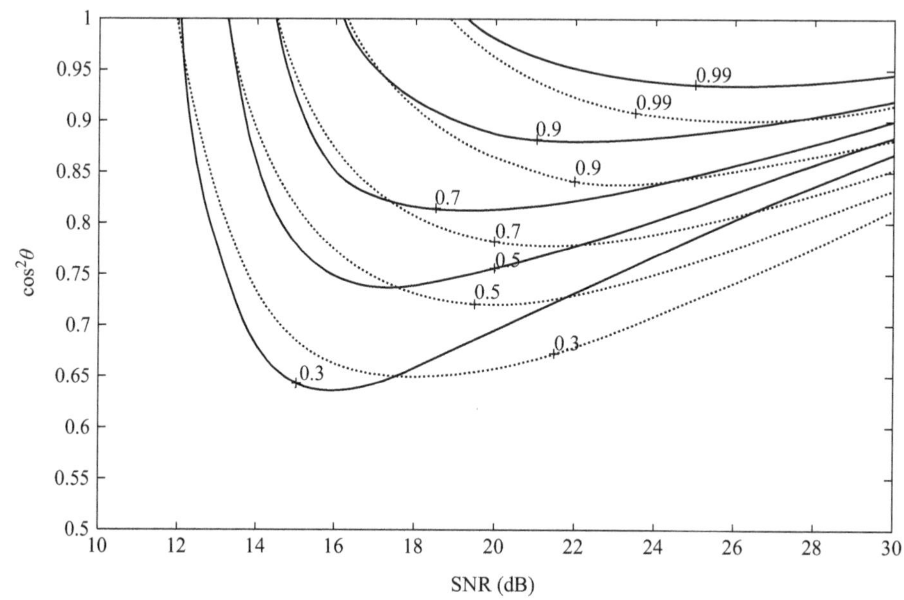

Figure 4.38 Contours of constant P_d for the KWAS-ASB with $\gamma = 1.3$ (solid line) and the WAS-ASB (dotted line) assuming $N = 30$, $K = 60$, $q = 2$, $P_{fa} = 0.0001$, and threshold pairs that ensure the most selective behavior with the constraint of a maximum loss with respect to Kelly's receiver of about 1 dB.

4.2.4.3 SRao-ASB

Another modification of the S-ASB consists in substituting the ACE with the Rao detector. The overall structure provides better detection performance than the S-ASB and, for low values of N and K, than the KWAS-ASB. We refer to this receiver as SRao-ASB. The corresponding expressions of the P_{fa} and P_d can be obtained using results of Sections 4.2.2 and 4.2.3.2. More precisely, under the assumption that

$$\frac{1}{(1+\beta_1)(1+\beta_2)} > \eta_{\text{RAO}}, \tag{4.81}$$

the P_{fa} is given by

$$\begin{aligned} P_{fa}(\bar{\eta}_{\text{SD}}, \eta_{\text{RAO}}) &= \Pr_{H_0}\{\bar{t}_{\text{SD}} > \bar{\eta}_{\text{SD}}, t_{\text{RAO}} > \eta_{\text{RAO}}\} \\ &= 1 - \Pr_{H_0}\left\{\bar{t}_{\text{K}} \le \max\left(\frac{\bar{\eta}_{\text{SD}}}{1+\beta_2} - 1, \frac{\eta_{\text{RAO}}}{\frac{1}{(1+\beta_1)(1+\beta_2)} - \eta_{\text{RAO}}}\right)\right\} \\ &= \int_0^{\frac{1}{\eta_{\text{RAO}}}-1} \int_0^{\frac{1}{\eta_{\text{RAO}}(1+b_2)}-1} \\ &\quad \times \left[1 - F_0\left(\max\left(\frac{\bar{\eta}_{\text{SD}}}{1+b_2} - 1, \frac{\eta_{\text{RAO}}}{[(1+b_1)(1+b_2)]^{-1} - \eta_{\text{RAO}}}\right)\right)\right] \\ &\quad \times f_{\beta_1}(b_1) f_{\beta_2}(b_2) db_1 db_2, \end{aligned} \tag{4.82}$$

while P_d has the following expression

$$\begin{aligned} P_d(\bar{\eta}_{\text{SD}}, \eta_{\text{RAO}}, \text{SNR}, \cos^2\theta) &= \Pr_{H_1}\{\bar{t}_{\text{SD}} > \bar{\eta}_{\text{SD}}, t_{\text{RAO}} > \eta_{\text{RAO}}\} \\ &= 1 - \Pr_{H_1}\left\{\bar{t}_{\text{K}} \le \max\left(\frac{\bar{\eta}_{\text{SD}}}{1+\beta_2} - 1, \frac{\eta_{\text{RAO}}}{\frac{1}{(1+\beta_1)(1+\beta_2)} - \eta_{\text{RAO}}}\right)\right\} \\ &= \int_0^{\frac{1}{\eta_{\text{RAO}}}-1} \int_0^{\frac{1}{\eta_{\text{RAO}}(1+b_2)}-1} \\ &\quad \times \left[1 - F_1\left(\max\left(\frac{\bar{\eta}_{\text{SD}}}{1+b_2} - 1, \frac{\eta_{\text{RAO}}}{[(1+b_1)(1+b_2)]^{-1} - \eta_{\text{RAO}}}\right)\right)\right] \\ &\quad \times f_{\beta_2|\beta_1}(b_2|\beta_1 = b_1) f_{\beta_1,r_1}(b_1) db_1 db_2, \end{aligned} \tag{4.83}$$

where $F_1(\cdot), f_{\beta_1}(\cdot), f_{\beta_2}(\cdot), f_{\beta_1,r_1}(\cdot)$, and $f_{\beta_2|\beta_1}(\cdot\,|\,\cdot)$ have been already defined. On the other hand, if (4.81) is not valid, then $P_{fa} = P_d = 0$.

In Figure 4.39, we show the limiting behaviors of the SRao-ASB in comparison with Kelly's receiver. Observe that the maximum loss with respect to Kelly's receiver exhibited by the SRao-ASB is about 2 dB at $P_d = 0.9$ and is similar to that of the AMF-Rao. As to mismatched detection performance, since the most robust behavior of the SRao-ASB is similar to that of the KWAS-ASB, we focus on the rejection of unwanted signals. In Figures 4.40 and 4.41, we show that for low values of N and K the selectivity of the SRao-ASB is comparable to that of the KWAS-ASB (see Figure 4.40), while for high values of N and K the SRao-ASB

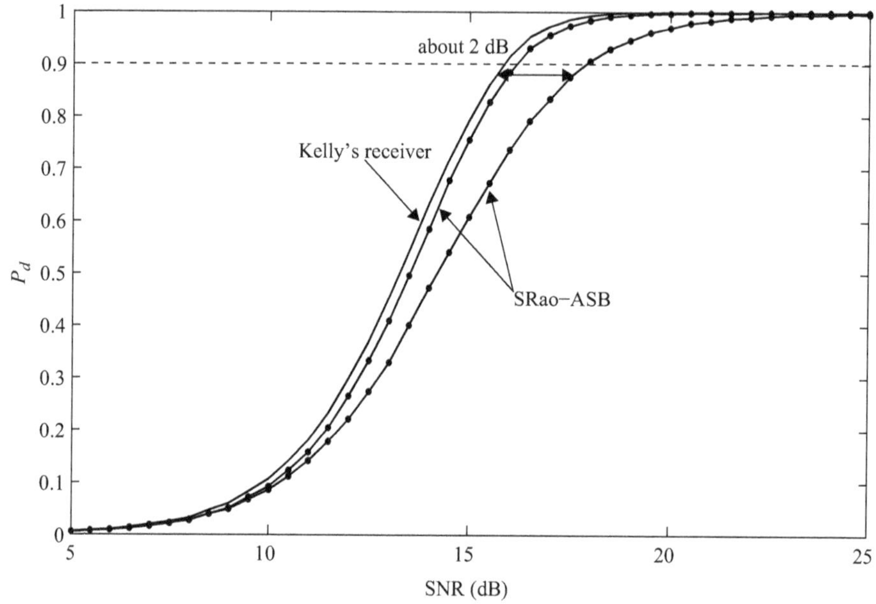

Figure 4.39 P_d versus SNR for the SRao-ASB with (point marker) and Kelly's receiver (no marker) assuming $N = 16$, $K = 32$, $q = 2$, and $P_{fa} = 0.0001$.

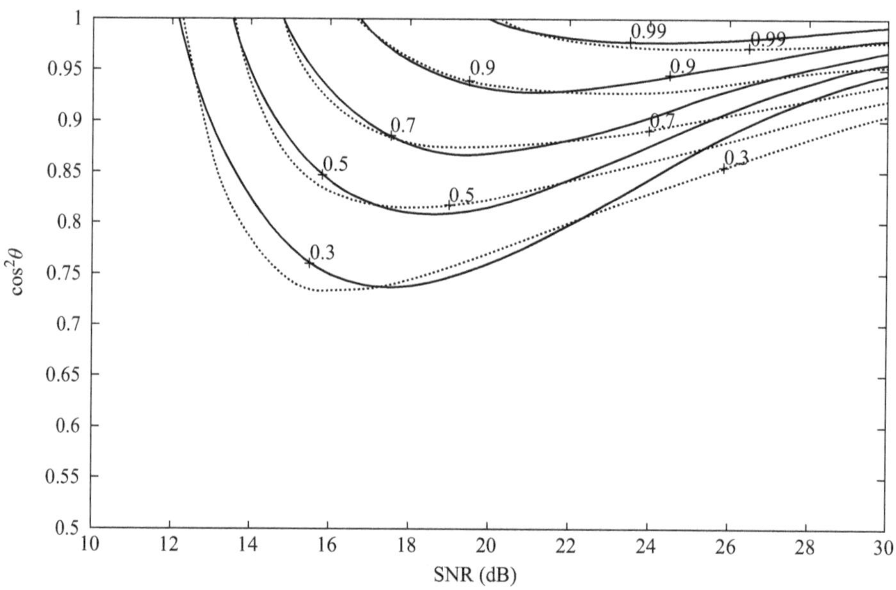

Figure 4.40 Contours of constant P_d for the SRao-ASB (solid line) and the KWAS-ASB with $\gamma = 1.3$ (dotted line) assuming $N = 16, K = 32, q = 2, P_{fa} = 0.0001$, and threshold pairs that ensure the most selective behavior with the constraint of a maximum loss with respect to Kelly's receiver of about 1 dB.

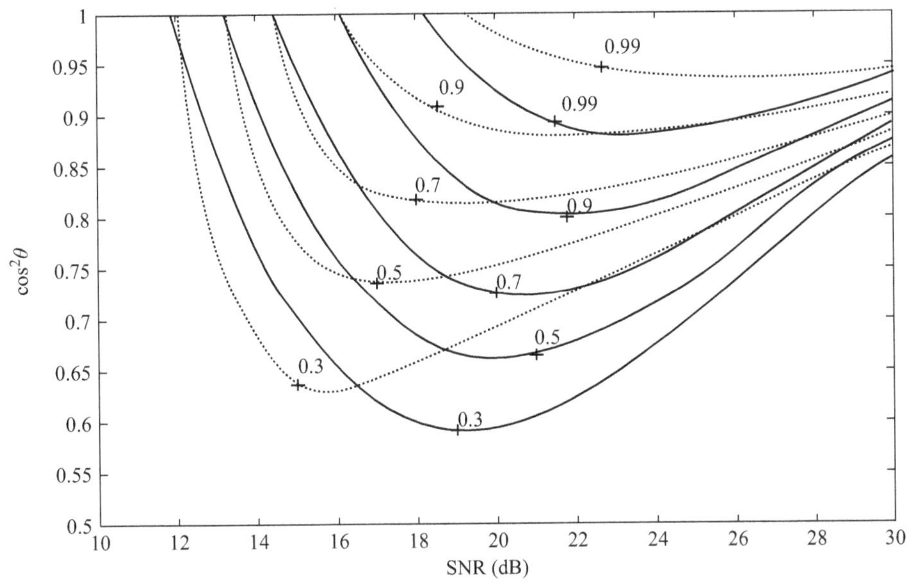

Figure 4.41 Contours of constant P_d for the SRao-ASB (solid line) and the KWAS-ASB with $\gamma = 1.3$ (dotted line) assuming $N = 30, K = 60, q = 2, P_{fa} = 0.0001$, and threshold pairs that ensure the most selective behavior with the constraint of a maximum loss with respect to Kelly's receiver of about 1 dB.

becomes less selective than the KWAS-ASB (see Figure 4.41). The above trend is typical of two-stage receivers based upon the Rao detector (see Subsection 4.2.3.2), and it is due to the fact that the Rao detector for high values of N and/or K becomes similar to the AMF.

4.2.5 Selective Two-Stage Detectors

In this subsection, the two-stage idea is exploited to design tunable receivers capable of ensuring (practically) the same performance of Kelly's detector under perfectly matching conditions (namely, with a negligible detection loss with respect to Kelly's detector) and enhanced rejection capabilities of unwanted signals. This is accomplished cascading Kelly's detector, which provides excellent detection performance for matched signals, and a selective receiver. More precisely, we consider two different pairs

(1) Kelly's receiver and the KWA (referred to as K-KWA);

(2) Kelly's receiver and the Rao detector (referred to as K-Rao).

Remarkably, this two-stage decision schemes are invariant with respect to the transformation group defined in Reference 35 and, hence, ensure the CFAR property with respect to the covariance matrix of the interference. The derivations of the P_{fa} and P_d rely on the results contained in previous sections (Figure 4.42).

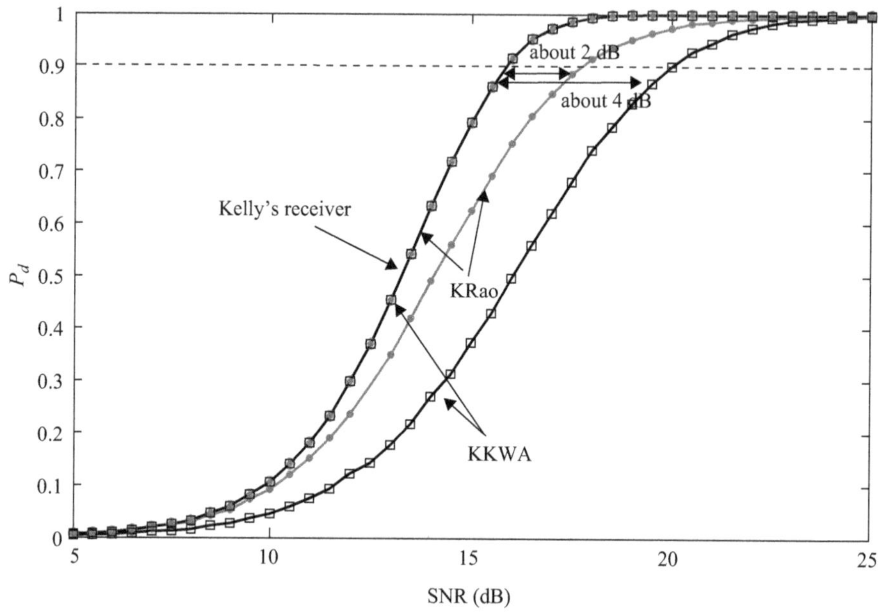

Figure 4.42 P_d versus SNR for the K-KWA with $\gamma = 1.3$ (solid line with square marker), the K-Rao (solid line with dot marker), and Kelly's receiver (solid line and circle marker) assuming $N = 16$, $K = 32$, and $P_{fa} = 0.0001$.

More precisely, the P_{fa} and P_d for the K-KWA are given by

$$\begin{aligned} P_{fa}(\bar{\eta}_{\text{K}}, \eta_{\text{KWA}}, \gamma) &= \Pr_{H_0}\{\bar{t}_{\text{K}} > \bar{\eta}_{\text{K}},\ t_{\text{KWA}} > \eta_{\text{KWA}}\} \\ &= 1 - \Pr_{H_0}\{\bar{t}_{\text{K}} \leq \max(\bar{\eta}_{\text{K}}, \eta_{\text{KWA}}/\beta^{2\gamma-1} - 1)\} \\ &= 1 - \int_0^1 F_0(\max(\bar{\eta}_{\text{K}}, \eta_{\text{KWA}}/b^{2\gamma-1} - 1)) f_\beta(b) db \end{aligned} \tag{4.84}$$

and

$$\begin{aligned} P_d(\bar{\eta}_{\text{K}}, \eta_{\text{KWA}}, \gamma, \text{SNR}, \cos^2\theta) &= \Pr_{H_1}\{\bar{t}_{\text{K}} > \bar{\eta}_{\text{K}},\ t_{\text{KWA}} > \eta_{\text{KWA}}\} \\ &= 1 - \Pr_{H_1}\{\bar{t}_{\text{K}} \leq \max(\bar{\eta}_{\text{K}}, \eta_{\text{KWA}}/\beta^{2\gamma-1} - 1)\} \\ &= 1 - \int_0^1 F_1(\max(\bar{\eta}_{\text{K}}, \eta_{\text{KWA}}/b^{2\gamma-1} - 1)) f_{\beta,r_\beta}(b) db, \end{aligned} \tag{4.85}$$

respectively, where $F_i(\cdot)$, $i = 0, 1$, is the CDF of $\bar{t}_{\text{K}}$ under the H_i hypothesis, $i = 0, 1$, $f_\beta(\cdot)$ and $f_{\beta,r_\beta}(\cdot)$ are the pdfs of β under H_0 and H_1, respectively.

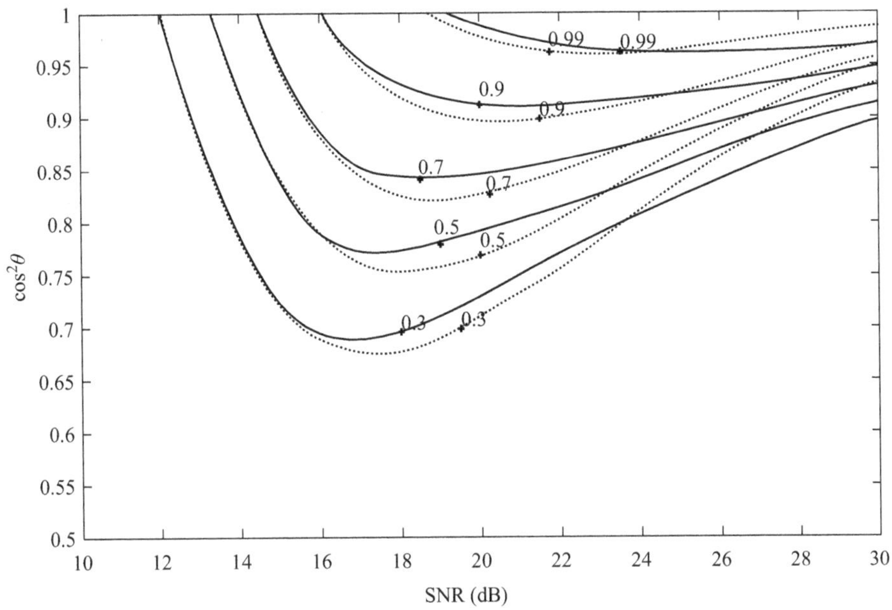

Figure 4.43 Contours of constant P_d for the K-KWA with $\gamma = 1.3$ (solid line) and the K-Rao (dotted line) with $N = 16$, $K = 32$, and threshold pair corresponding to the most selective case under the constraint of a maximum loss of about 0.4 dB with respect to Kelly's receiver at $P_d = 0.9$ and matched signals.

On the other hand, assuming that $\beta > \eta_{\mathrm{RAO}}$, the performance of the K-Rao can be evaluated as follows

$$\begin{aligned} P_{fa}(\bar{\eta}_{\mathrm{K}}, \eta_{\mathrm{RAO}}) &= \Pr_{H_0}\{\bar{t}_{\mathrm{K}} > \bar{\eta}_{\mathrm{K}},\ t_{\mathrm{RAO}} > \eta_{\mathrm{RAO}}\} \\ &= 1 - \Pr_{H_0}\{\bar{t}_{\mathrm{K}} \leq \max(\bar{\eta}_{\mathrm{K}}, \eta_{\mathrm{RAO}}/(\beta - \eta_{\mathrm{RAO}}))\} \\ &= \int_{\eta_{\mathrm{RAO}}}^{1} \left[1 - F_0(\max(\bar{\eta}_{\mathrm{K}}, \eta_{\mathrm{RAO}}/(b - \eta_{\mathrm{RAO}})))\right] f_\beta(b)db \end{aligned} \tag{4.86}$$

and

$$\begin{aligned} P_d(\bar{\eta}_{\mathrm{K}}, \eta_{\mathrm{RAO}}, \gamma, \mathrm{SNR}, \cos^2\theta) &= \Pr_{H_1}\{\bar{t}_{\mathrm{K}} > \bar{\eta}_{\mathrm{K}},\ t_{\mathrm{RAO}} > \eta_{\mathrm{RAO}}\} \\ &= 1 - \Pr_{H_1}\{\bar{t}_{\mathrm{K}} \leq \max(\bar{\eta}_{\mathrm{K}}, \eta_{\mathrm{RAO}}/(\beta - \eta_{\mathrm{RAO}}))\} \\ &= \int_{\eta_{\mathrm{RAO}}}^{1} \left[1 - F_1(\max(\bar{\eta}_{\mathrm{K}}, \eta_{\mathrm{RAO}}/(b - \eta_{\mathrm{RAO}})))\right] f_{\beta, r_\beta}(b)db. \end{aligned} \tag{4.87}$$

If $\beta < \eta_{\mathrm{RAO}}$, then $P_{fa} = P_d = 0$. The contours of constant P_{fa} for the K-KWA and the K-Rao are similar to those of the AMF-KWA and AMF-Rao, respectively, and are not reported here for the sake of brevity. As to the matched detection performance, it is worth noticing that the K-KWA exhibits a maximum loss with respect to Kelly's receiver of about 4 dB (as seen for the AMF-KWA), while the K-Rao has a maximum loss at $P_d = 0.9$ of about 2 dB (see also the AMF-Rao). In Figure 4.43, we plot the contours of constant P_d using those threshold pairs

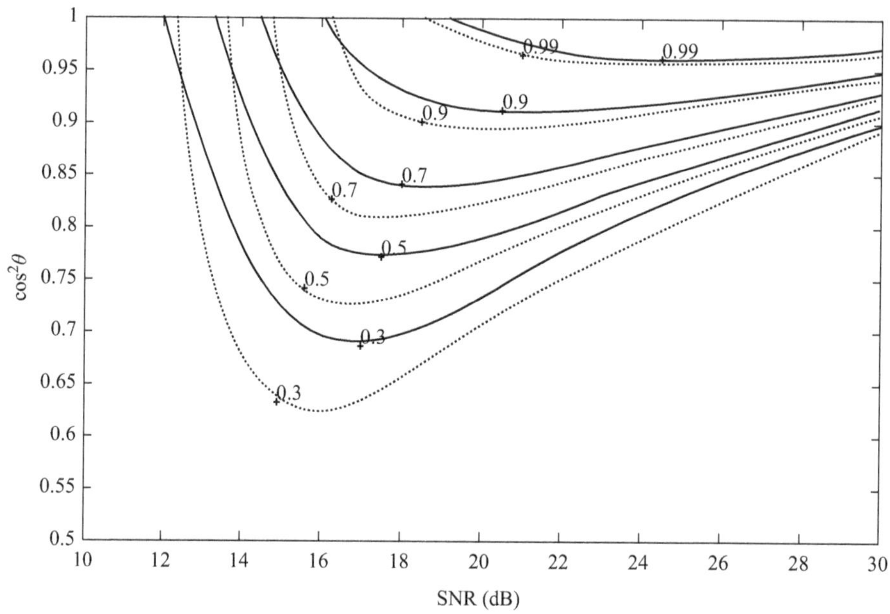

Figure 4.44 Contours of constant P_d for the K-KWA (solid line) and the KWAS-ASB (dotted line) with $N = 16$, $K = 32$, $\gamma = 1.3$, and threshold pair corresponding to the most selective case under the constraint of a maximum loss of about 0.4 dB with respect to Kelly's receiver at $P_d = 0.9$ and matched signals.

that guarantee a matched detection loss of about 0.4 dB with respect to Kelly's receiver. The mismatched detection performances are similar. More precisely, for SNR values lower than 24 dB, the K-KWA is slightly more selective than the K-Rao; conversely, for SNR $>$ 24 dB, the K-Rao provides slightly better rejection capabilities than the K-KWA. Finally, inspection of Figure 4.44 highlights that the K-KWA can guarantee a better trade off between selectivity and matched detection performance than KWAS-ASB. As a matter of fact, the K-KWA is slightly more selective than the KWAS-ASB under the constraint of a matched detection loss of 0.4 dB with respect to Kelly's receiver.

Before concluding remarks, it is worth giving a qualitative summary of the detection algorithms analyzed up to this point. To this end, a classification in terms of range of directivity is given in Table 4.3, where S_i, $i = 0, 1, 2, \ldots$, is the level of selectivity with $S_i < S_{i+1}$, and R_i, $i = 0, 1, 2, \ldots$, denotes the level of robustness with $R_i < R_{i+1}$.

4.3 Conclusions

This chapter has provided a survey on the two-stage detection of point-like targets embedded in homogeneous Gaussian disturbance. The family of two-stage detectors belongs to the more general class of tunable receivers, which allow to modify their directivity tuning proper parameters. They are obtained cascading two detectors with opposite behaviors in terms of directivity. The presence of a signal is declared if and only if each stage is above the respective threshold.

Table 4.3 Reviewed two-stage detectors.

Receiver	Directivity range		Maximum loss (dB) with respect to the benchmark, $N = 16$, $K = 32$, $P_{fa} = 10^{-4}$
ASB	R_0	S_1	1
S-ASB	R_1	S_1	1
AMF-ABORT	R_0	S_0	0.4
AMF-Rao	R_0	S_3 for low values of N, K; S_1 for high values of N, K	2
AMF-WA	R_0	S_2	2
AMF-KWA ($\gamma = 1.3$)	R_0	S_3	4
WAS-ASB	R_1	S_2	2
KWAS-ASB	R_1	S_3	4
SRao-ASB	R_1	S_3 for low values of N, K; S_1 for high values of N, K	2
K-KWA ($\gamma = 1.3$)	–	S_3	4
K-Rao	–	S_3 for low values of N, K; S_1 for high values of N, K	2

Several two-stage solutions are reviewed highlighting that it is possible to extend the range of directivity towards selectivity and/or robustness. Closed-form expressions for the P_{fa} and the P_d (for both matched and mismatched signals) are derived, and the performance analysis, carried out by means of numerical integration techniques, is aimed at showing that these flexible solutions can guarantee a good trade off between rejection of sidelobe interferers and detection of mainlobe targets.

It is important to note that the considered two-stage detectors ensure the CFAR property with respect to the disturbance covariance matrix. Moreover, from the perspective of the invariance theory [35], the subspace-based detectors, which offer the widest range of directivity, do not share the same invariance property as those detectors based upon the AMF or Kelly's receiver.

Finally, plans for future work include several extensions of the considered detection strategies; for instance, it would be of interest to address mismatch problems concerning extended targets. In this case, new models are to be defined and analyzed. Another issue that needs attention is the design and the analysis of multistage detection structures, i.e., formed by cascading more than two stages.

References

[1] F. Bandiera, D. Orlando, and G. Ricci, "Advanced Radar Detection Schemes under Mismatched Signal Models," *Synthesis Lectures on Signal Processing No. 8*, Morgan & Claypool Publishers, San Rafael, CA, 2009.

[2] E. J. Kelly and K. Forsythe, "Adaptive Detection and Parameter Estimation for Multidimensional Signal Models," Lincoln Lab, MIT, Lexington, Tech. Rep. No. 848, April 19, 1989.

[3] A. De Maio, "Robust Adaptive Radar Detection in the Presence of Steering Vector Mismatches," *IEEE Transactions on Aerospace and Electronic Systems*, Vol. 41, No. 4, pp. 1322–1337, October 2005.

[4] F. Bandiera, A. De Maio, and G. Ricci, "Adaptive CFAR Radar Detection with Conic Rejection," *IEEE Transactions on Signal Processing*, Vol. 55, No. 6, pp. 2533–2541, June 2007.

[5] F. Bandiera, D. Orlando, and G. Ricci, "CFAR Detection Strategies for Distributed Targets under Conic Constraints," *IEEE Transactions on Signal Processing*, Vol. 57, No. 9, pp. 3305–3316, September 2009.

[6] C. Hao, F. Bandiera, J. Yang, and D. Orlando, "Adaptive Detection of Multiple Point-Like Targets Under Conic Constraints," *Progress in Electromagnetic Research*, Vol. 129, pp. 231–250, 2012.

[7] S. Z. Kalson, "An Adaptive Array Detector with Mismatched Signal Rejection," *IEEE Transactions on Aerospace and Electronic Systems*, Vol. 28, No. 1, pp. 195–207, January 1992.

[8] C. Hao, B. Liu, S. Yan, and L. Cai, "Parametric Adaptive Radar Detector with Enhanced Mismatched Signals Rejection Capabilities," *EURASIP Journal on Advances in Signal Processing*, Vol. 2010, Article ID 375136, 11 pages.

[9] A. De Maio, "Rao Test for Adaptive Detection in Gaussian Interference with Unknown Covariance Matrix," *IEEE Transactions on Signal Processing*, Vol. 55, No. 7, pp. 3577–3584, July 2007.

[10] N. B. Pulsone and C. M. Rader, "Adaptive Beamformer Orthogonal Rejection Test," *IEEE Transactions on Signal Processing*, Vol. 49, No. 3, pp. 521–529, March 2001.

[11] N. B. Pulsone and M. A. Zatman, "A Computationally Efficient Two-Step Implementation of the GLRT," *IEEE Transactions on Signal Processing*, Vol. 48, No. 3, pp. 609–616, March 2000.

[12] C. D. Richmond, "Statistical Performance Analysis of the Adaptive Sidelobe Blanker Detection Algorithm," *Proceedings of 31st Annual Asilomar Conference on Signals, Systems, and Computers*, Pacific Grove, CA, USA, November 1997.

[13] C. D. Richmond, "Performance of the Adaptive Sidelobe Blanker Detection Algorithm in Homogeneous Environments," *IEEE Transactions on Signal Processing*, Vol. 48, No. 5, pp. 1235–1247, May 2000.

[14] F. Bandiera, D. Orlando, and G. Ricci, "One-Stage and Two-Stage Tunable Receivers," *IEEE Transactions on Signal Processing*, Vol. 57, No. 8, pp. 3264–3273, August 2009.

[15] C. D. Richmond, "The Theoretical Performance of a Class of Space–Time Adaptive Detection and Training Strategies for Airborne Radar," *Proceedings of 32nd Annual Asilomar Conference on Signals, Systems, and Computers*, Pacific Grove, CA, USA, November 1998.

[16] F. Bandiera, D. Orlando, and G. Ricci, "A Subspace-Based Adaptive Sidelobe Blanker," *IEEE Transactions on Signal Processing*, Vol. 56, No. 9, pp. 4141–4151, September 2008.

[17] F. Bandiera, O. Besson, D. Orlando, and G. Ricci, "An Improved Adaptive Sidelobe Blanker", *IEEE Transactions on Signal Processing*, Vol. 56, No. 9, pp. 4152–4161, September 2008.

[18] C. Hao, B. Liu, and L. Cai, "Performance Analysis of a Two-Stage Rao Detector," *Signal Processing*, Vol. 91, No. 8, pp. 2141–2146, August 2011.

[19] G. A. Fabrizio, A. Farina, and M. D. Turley, "Spatial Adaptive Subspace Detection in OTH Radar," *IEEE Transactions on Aerospace and Electronic Systems*, Vol. 39, No. 4, pp. 1407–1427, October 2003.

[20] M. I. Skolnik, *Introduction to Radar Systems*, 3rd ed., McGraw-Hill, New York, NY, 2001.

[21] A. Farina and F. A. Studer, *Radar Data Processing. Introduction and Tracking (Vol. I)*, John Wiley & Sons, New York, NY, 1985.

[22] A. Farina, *Antenna-Based Signal Processing Techniques for Radar Systems*, Artech House, Boston, MA, 1992.

[23] A. De Maio, A. Farina, and F. Gini, "Performance Analysis of the Sidelobe Blanking System for Two Fluctuating Jammer Models," *IEEE Transactions on Aerospace and Electronic Systems*, Vol. 41, No. 3, pp. 1082–1091, July 2005

[24] G. Cui, A. De Maio, A. Aubry, A. Farina, and L. Kong, "Advanced SLB Architectures with Invariant Receivers," *IEEE Transactions on Aerospace and Electronic Systems*, Vol. 49, No. 2, pp. 798–818, April 2013.

[25] A. Farina, L. Timmoneri, and R. Tosini, "Cascading SLB and SLC Devices," *Signal Processing*, Vol. 45, No. 2, pp. 261–266, August 1995.

[26] D. E. Kreithen and A. O. Steinhardt, "Target Detection in Post-STAP Undernulled Clutter," *Proceedings of 29th Annual Asilomar Conference on Signals, Systems, and Computers*, Vol. 2, pp. 1203–1207, November 1995.

[27] F. C. Robey, D. L. Fuhrman, E. J. Kelly, and R. Nitzberg, "A CFAR Adaptive Matched Filter Detector," *IEEE Transactions on Aerospace and Electronic Systems*, Vol. 29, No. 1, pp. 208–216, January 1992.

[28] F. Bandiera, O. Besson, and G. Ricci, "An ABORT-Like Detector With Improved Mismatched Signals Rejection Capabilities," *IEEE Transactions on Signal Processing*, Vol. 56, No. 1, pp. 14–25, January 2008.

[29] G. T. Capraro, A. Farina, H. Griffiths, and M. C. Wicks, "Knowledge-Based Radar Signal and Data Processing (A Tutorial Review)," *IEEE Signal Processing Magazine*, Vol. 23, No. 1, pp. 18–29, January 2006.

[30] E. J. Kelly, "Performance of an Adaptive Detection Algorithm; Rejection of Unwanted Signals," *IEEE Transactions on Aerospace and Electronics Systems*, Vol. 25, No. 2, pp. 122–123, March 1989.

[31] E. J. Kelly, "An Adaptive Detection Algorithm," *IEEE Transactions on Aerospace and Electronic Systems*, Vol. 22, No. 2, pp. 115–127, March 1986.

[32] N. R. Goodman, "Statistical Analysis Based on a Certain Multivariate Complex Gaussian Distribution (An Introduction)," *The Annals of Mathematical Statistics*, Vol. 34, No. 1, pp. 152–177, March 1963.

[33] E. Conte, M. Lops, and G. Ricci, "Asymptotically Optimum Radar Detection in Compound Gaussian Noise," *IEEE Transactions on Aerospace and Electronic Systems*, Vol. 31, No. 2, pp. 617–625, April 1995.

[34] S. Kraut and L. L. Scharf, "The CFAR Adaptive Subspace Detector Is a Scale-Invariant GLRT," *IEEE Transactions Signal Processing*, Vol. 47, No. 9, pp. 2538–2541, September 1999.

[35] S. Bose and A. O. Steinhardt, "A Maximal Invariant Framework for Adaptive Detection with Structured and Unstructured Covariance Matrices," *IEEE Transactions on Signal Processing*, Vol. 43, No. 9, pp. 2164–2175, September 1995.

[36] A. De Maio, "Rao Test for Adaptive Detection in Gaussian Interference with Unknown Covariance Matrix," *IEEE Transactions on Signal Processing*, Vol. 55, No. 7, pp. 3577–3584, July 2007.

[37] J. Neyman and E. S. Pearson, "On the Use and Interpretation of Certain Test Criteria for Purpose of Statistical Inference," *Biometrika*, Vol. 20, pp. 175–240, 1928.

[38] S. M. Kay, *Fundamentals of Statistical Signal Processing, Detection Theory*, (Vol. II) Prentice-Hall, Englewood Cliffs, NJ, 1998.

CHAPTER 5

Bayesian Radar Detection in Interference

Pu Wang[1], *Hongbin Li*[2], *and Braham Himed*[3]

5.1 Introduction

Knowledge-aided space–time adaptive processing (KA-STAP) advocates an intelligent utilization of *a priori* knowledge from various sources, such as previous measurements, digital geographic maps, and real-time radar platform parameters. A natural and systematic way for incorporating such *a priori* knowledge for detection is to employ a Bayesian inference framework. This is appealing for KA-STAP because the Bayesian method not only allows for the formal and systematic use of prior information (on the interference covariance matrix) but also quantifies uncertainties presented in the prior knowledge through hyperparameters. Moreover, by treating the interference covariance matrix as a random quantity, it provides an additional flexibility of modeling data heterogeneity with computationally manageable detection strategies.

This chapter aims to discuss recent advances on Bayesian KA-STAP techniques. It unfolds as the classical STAP signal model in Section 5.2 evolves into a framework of KA-STAP model including a knowledge-aided homogeneous model, a knowledge-aided partially homogeneous model, and a knowledge-aided compound Gaussian model in Section 5.3. Then in Section 5.4, a hierarchical two-layered STAP model is discussed, which provides a new way to describe the non-homogeneity between the test and training data. Section 5.5 is devoted to parametric Bayesian detectors that integrate structural space–time information, i.e., a multi-channel autoregressive (AR) process, for the interference model and the consequent Bayesian estimation. The resulting Bayesian parametric detectors allow a fast implementation and further reduction in the amount of training data needed for reliable detection.

---

[1] Schlumberger-Doll Research, Cambridge, MA, USA.
[2] Stevens Institute of Technology, Hoboken, NJ, USA.
[3] AFRL/RYMD, Dayton, OH, USA.

5.2 General STAP Signal Model

In STAP, $J > 1$ antennas transmit a coherent burst of N pulses at a constant pulse repetition frequency $f_r = 1/T_r$, where T_r is the pulse repetition interval (PRI). The transmitter frequency is $f_c = c/\lambda$, where c is the propagation velocity and λ is the wavelength. The waveform returns are collected over a time interval referred to as the coherent processing interval (see also Chapter 1). For each PRI, $K + 1$ fast-time (range) samples are collected to cover the range interval. Detection is performed one at a time over the range bins of interest. For each detection, there is a test range bin, and the neighboring range bins can be employed to provide the training data. The received data are hence organized in a $J \times N \times (K+1)$ datacube, as shown in Figure 5.1 (the reader is referred to Chapter 2 for the details concerning the construction of data vectors). For each range bin, the $J \times N$ data matrix is sampled from J antennas and N pulses. Stacking the columns of each data matrix (at each range bin) on top of each other, we obtain the test signal $\boldsymbol{r}_0$ at one range bin under test, and the training data $\boldsymbol{r}_k, k = 1, 2, \ldots, K$, from K neighboring range bins. Note that $\boldsymbol{r}_k, k = 0, 1, \ldots, K$, are vectors of dimension $JN \times 1$.

With the three-dimensional datacube, the STAP problem is to detect a multi-channel signal with an *unknown* amplitude in the presence of spatially and temporally correlated interference [1,2]:

$$\begin{aligned} H_0 : &\quad \boldsymbol{r}_0 = \boldsymbol{d}_0, \quad \boldsymbol{r}_k = \boldsymbol{d}_k, \quad k = 1, 2, \ldots, K, \\ H_1 : &\quad \boldsymbol{r}_0 = \alpha \boldsymbol{p} + \boldsymbol{d}_0, \quad \boldsymbol{r}_k = \boldsymbol{d}_k, \quad k = 1, 2, \ldots, K, \end{aligned} \tag{5.1}$$

where $\boldsymbol{p}$ is the steering vector, assumed known, which depends on the array geometry and is parameterized by the location or (angular) spatial frequency ω_s and the velocity or (angular) Doppler frequency ω_d of the target. For a uniformly equi-spaced linear array, the (normalized) steering vector is given as

$$\boldsymbol{p} = \boldsymbol{p}_d \otimes \boldsymbol{p}_s, \tag{5.2}$$

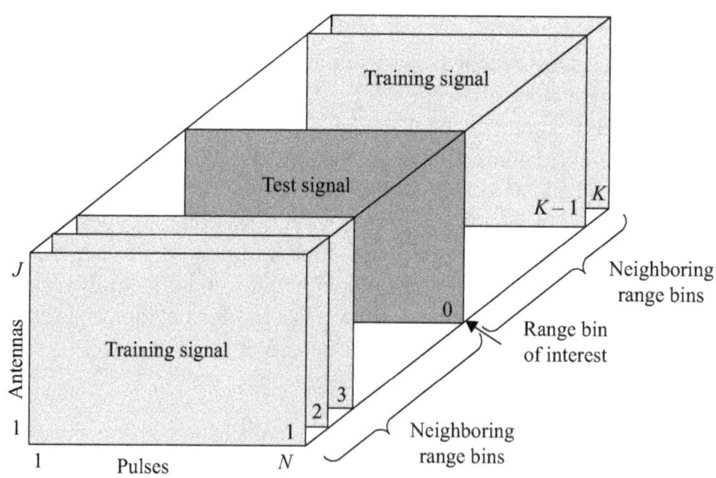

Figure 5.1 The three-dimensional STAP datacube within a coherent processing interval, where J is the number of spatial antennas, N is the number of temporal pulses, and K is the number of training range bins centered at the test range bin (indexed by 0).

where $\boldsymbol{p}_d = [1, e^{j\omega_d}, \ldots, e^{j\omega_d(N-1)}]^T/\sqrt{N}$ and $\boldsymbol{p}_s = [1, e^{j\omega_s}, \ldots, e^{j\omega_s(J-1)}]^T/\sqrt{J}$. In addition, α denotes the *unknown*, deterministic, and complex-valued signal amplitude, and the interferences $\boldsymbol{d}_0$ and $\boldsymbol{d}_k$ are assumed to follow the complex circular Gaussian distribution with zero mean and covariance matrices $\boldsymbol{M}_0$ and $\boldsymbol{M}_k$, respectively, i.e., $\boldsymbol{d}_k \sim \mathcal{CN}(\boldsymbol{0}, \boldsymbol{M}_k), k = 1, 2, \ldots, K$.

The training data are assumed to contain information about the interference in the test data. Otherwise, there is no incentive to utilize the training data. Classic STAP models provide several ways to describe such relation between the test and training data. In particular, the conventional STAP model can be used for the following scenarios:

- *Homogeneous* STAP Model ($\boldsymbol{M}_0 = \boldsymbol{M}_1 = \cdots = \boldsymbol{M}_K = \boldsymbol{\Sigma}$): The interference in the test data $\boldsymbol{d}_0$ shares the same covariance matrix $\boldsymbol{\Sigma}$ with the interference $\boldsymbol{d}_k$ in the training data.
- *Partially Homogeneous* STAP Model ($\boldsymbol{M}_0 = \lambda\boldsymbol{\Sigma}, \boldsymbol{M}_1 = \cdots = \boldsymbol{M}_K = \boldsymbol{\Sigma}$): Like the homogeneous model, the training data share the same covariance matrix, i.e., $\boldsymbol{M}_1 = \cdots = \boldsymbol{M}_K = \boldsymbol{\Sigma}$. However, there is an *unknown* power scaling factor λ between $\boldsymbol{M}_0$ and $\{\boldsymbol{M}_k\}_{k=1}^K$, i.e., $\boldsymbol{M}_0 = \lambda\boldsymbol{\Sigma}$.
- *Compound-Gaussian* STAP Model includes two kinds of heterogeneity between the test and training data and among the training data (see also Chapters 7–9). Specifically, $\boldsymbol{d}_0$ and $\boldsymbol{d}_k$, conditioned on the so-called texture components $s_k > 0$, are complex circular Gaussian distributed, i.e., $\boldsymbol{d}_k | s_k \sim \mathcal{CN}(\boldsymbol{0}, s_k^2\boldsymbol{\Sigma})$, where the texture components s_k are positive random variables, drawn from a correlated random process. Equivalently, we can write $\boldsymbol{d}_k = s_k\boldsymbol{w}_k, k = 0, 1, \ldots, K$, where the speckle components $\boldsymbol{w}_k \sim \mathcal{CN}(\boldsymbol{0}, \boldsymbol{\Sigma})$ are homogeneous between the test and training data, while the texture components s_k take into account the heterogeneity.

All the above, STAP models can be represented by the graphical model shown in Figure 5.2a, where the edges denote the conditional dependency among the corresponding nodes, the circles

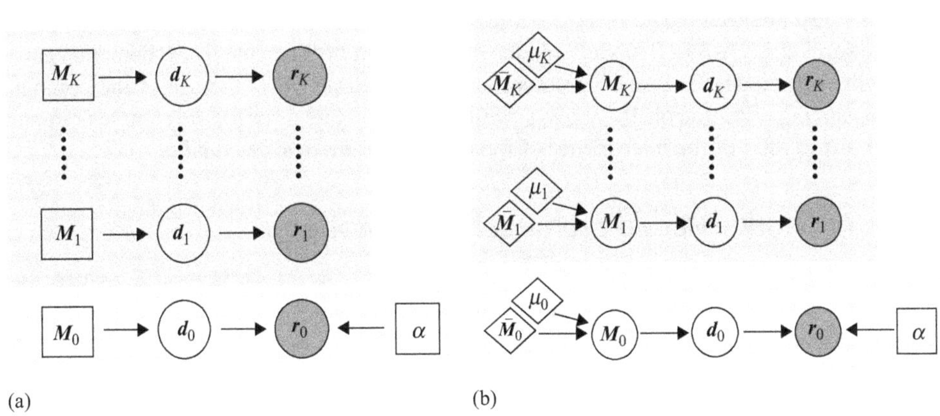

Figure 5.2 Directed graphical model representation of the STAP models (including homogeneous, partially homogeneous, and (deterministic) compound-Gaussian models). Circles denote random variables, squares denote deterministic model parameters, and diamonds denote user parameters. Shaded circles further represent observed random variables. Note the different (deterministic versus stochastic) ways of treating the interference covariance matrix $\{\boldsymbol{M}_k\}_{k=0}^K$ in the two STAP models. (a) Conventional STAP model and (b) KA-STAP model.

denote the random variables (e.g., the interference and, the observed test and training data), the squares denote the deterministic model parameters (e.g., the target reflected amplitudes and the interference covariance matrices in the conventional STAP model), and the diamonds denote user parameters (e.g., the prior covariance matrices in the KA-STAP model introduced in the following sections). Shaded circles further represent the observed random variables.

In both homogeneous and non-homogeneous scenarios, traditional STAP detectors such as the Reed–Mallett–Brennan (RMB) detector [3], Kelly's generalized likelihood ratio test (GLRT) [4], the adaptive matched filter (AMF) [5], the locally most powerful invariant test [6], the adaptive coherence estimator (ACE) [7,8], the Rao test [9], etc., involve estimating and inverting the full-dimensional space–time interference covariance matrix $\boldsymbol{M}_0$. At a minimum, $K \geq JN$ training data are needed to ensure a full-rank estimate of the $JN \times JN$ space–time covariance matrix $\boldsymbol{M}_0$. Moreover, the RMB rule [3] suggests that, on average, $K > (2JN - 3)$ training data are needed to be within 3 dB from the optimum detection performance. Such conditions may not be satisfied, especially in heterogeneous or dense-target environments that offer limited training data. Thus, it is necessary to reduce the training data requirement associated with traditional STAP detectors for practical applications.

To meet the challenge, the Bayesian approach models the covariance matrix $\{\boldsymbol{M}_k\}_{k=0}^{K}$ as a random parameter, as opposed to a deterministic parameter in the conventional STAP, along with some prior probability distribution. Referred to as KA-STAP models, the Bayesian STAP structure can be represented by the graphical model shown in Figure 5.2b, where an additional layer of nodes is employed to model the interference covariance matrices $\boldsymbol{M}_k$ with prior information embedded in the user parameters (denoted by diamonds) such as the prior covariance matrices $\bar{\boldsymbol{M}}_k$ and the hyperparameter μ_k.

5.3 KA-STAP Models

Similar to the classical STAP models, the KA-STAP models still need to establish a relation between the test and training data. This section discusses the evolution of the classical STAP models towards the KA-STAP models and the development of Bayesian STAP detection associated with each model. The key ingredient of such development is the use of the complex inverse Wishart distribution on $\boldsymbol{M}_0$ and $\boldsymbol{M}_k$, which is a conjugate prior for the complex circular Gaussian distribution of the interferences for the sake of analytical tractability.

5.3.1 Knowledge-Aided Homogeneous Model

Let us first consider the homogeneous model ($\boldsymbol{M}_0 = \boldsymbol{M}_1 = \cdots = \boldsymbol{M}_K = \boldsymbol{\Sigma}$), where the disturbance signals $\boldsymbol{d}_k, k = 1, 2, \ldots, K$, are independent and identically distributed (i.i.d.) with respect to $\boldsymbol{d}_0$, with complex circular Gaussian distribution $\boldsymbol{d}_k \sim \mathcal{CN}(\boldsymbol{0}, \boldsymbol{\Sigma})$. Accordingly, the likelihood functions under H_0 and H_1 can be expressed as

$$\begin{aligned} p(\boldsymbol{r}_0, \boldsymbol{r}_1, \ldots, \boldsymbol{r}_K | \boldsymbol{\Sigma}, H_0) &= \frac{1}{\left[\pi^{JN} \det(\boldsymbol{\Sigma})\right]^{K+1}} \exp\Big(-\mathrm{Tr}\Big(\boldsymbol{\Sigma}^{-1}\boldsymbol{\Gamma}(0)\Big)\Big) \\ p(\boldsymbol{r}_0, \boldsymbol{r}_1, \ldots, \boldsymbol{r}_K | \alpha, \boldsymbol{\Sigma}, H_1) &= \frac{1}{\left[\pi^{JN} \det(\boldsymbol{\Sigma})\right]^{K+1}} \exp\Big(-\mathrm{Tr}\Big(\boldsymbol{\Sigma}^{-1}\boldsymbol{\Gamma}(\alpha)\Big)\Big), \end{aligned} \tag{5.3}$$

$$\boldsymbol{\Gamma}(\alpha) = (\boldsymbol{r}_0 - \alpha\boldsymbol{p})(\boldsymbol{r}_0 - \alpha\boldsymbol{p})^{\dagger} + K\boldsymbol{S}, \tag{5.4}$$

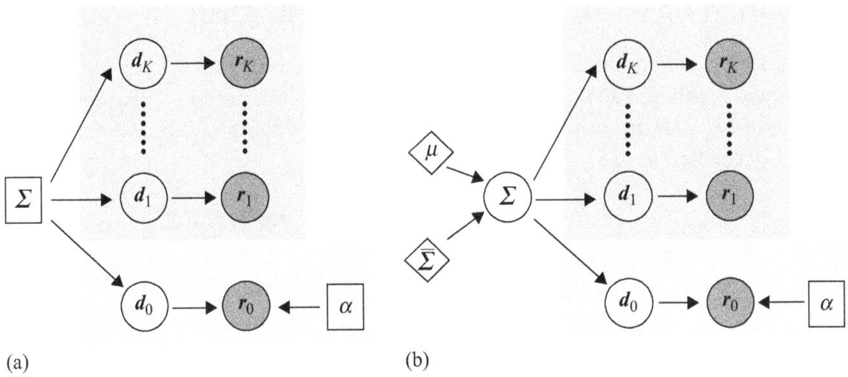

Figure 5.3 Conventional and knowledge-aided homogeneous STAP models with $\boldsymbol{M}_0 = \boldsymbol{M}_1 = \cdots = \boldsymbol{M}_K = \boldsymbol{\Sigma}$. (a) Conventional homogeneous scenario and (b) knowledge-aided homogeneous scenario.

with the sample covariance matrix given by

$$\boldsymbol{S} = \frac{1}{K}\sum_{k=1}^{K} \boldsymbol{r}_k \boldsymbol{r}_k^{\dagger}. \tag{5.5}$$

As shown in Figure 5.3, the knowledge-aided homogeneous STAP model no longer considers $\boldsymbol{\Sigma}$ as a model parameter (denoted by a square), but as a random quantity (denoted by a circle), which follows a complex inverse Wishart distribution with degrees of freedom $\mu(>JN)$ and mean $\bar{\boldsymbol{\Sigma}}$ [10–13], i.e.,

$$\boldsymbol{\Sigma} \sim \mathcal{CW}^{-1}((\mu - JN)\bar{\boldsymbol{\Sigma}}, \mu), \tag{5.6}$$

or, equivalently,

$$p(\boldsymbol{\Sigma}) = \frac{\det\left((\mu - JN)\bar{\boldsymbol{\Sigma}}\right)^{\mu}}{\tilde{\Gamma}(JN,\mu)\det(\boldsymbol{\Sigma})^{(\mu+JN)}} \exp\left(-(\mu - JN)\operatorname{Tr}\left(\boldsymbol{\Sigma}^{-1}\bar{\boldsymbol{\Sigma}}\right)\right), \tag{5.7}$$

where

$$\tilde{\Gamma}(JN,\mu) = \pi^{JN(JN-1)/2}\prod_{k=1}^{JN}\Gamma\left(\mu - JN + k\right), \tag{5.8}$$

with Γ denoting the Gamma function. The matrix $\bar{\boldsymbol{\Sigma}}$ specifies the available *a priori* knowledge about the interference, while μ quantifies the prior uncertainty. This is observed by noting that

$$E\{\boldsymbol{\Sigma}\} = \bar{\boldsymbol{\Sigma}}, \tag{5.9}$$

$$E\{(\boldsymbol{\Sigma} - \bar{\boldsymbol{\Sigma}})^2\} = \frac{\bar{\boldsymbol{\Sigma}}^2 + (\mu - JN)^2\operatorname{Tr}\{\bar{\boldsymbol{\Sigma}}\}\bar{\boldsymbol{\Sigma}}}{(\mu - JN)^2 - 1}. \tag{5.10}$$

As a result, the larger μ is, the more reliable $\bar{\boldsymbol{\Sigma}}$ is. In Figure 5.3, both parameters $\bar{\boldsymbol{\Sigma}}$ and μ are specified by the user. Nevertheless, μ can also be treated as a model parameter to be estimated, which will be discussed later in this section.

5.3.2 Bayesian GLRT (B-GLRT) and Bayesian AMF (B-AMF)

Since the interference covariance matrix $\boldsymbol{\Sigma}$ is a random variable, a GLRT for the knowledge-aided homogeneous model can be obtained by averaging out the random covariance matrix $\boldsymbol{\Sigma}$ and then finding the maximum likelihood (ML) estimate of the amplitude. Specifically, the test statistic is given by [11–13]

$$\begin{aligned} T_{\text{B-GLRT}} &= \frac{\max_{\alpha} \int p(\boldsymbol{r}_0, \boldsymbol{r}_1, \ldots, \boldsymbol{r}_K | \alpha, \boldsymbol{\Sigma}, H_1) p(\boldsymbol{\Sigma}) d\boldsymbol{\Sigma}}{\int p(\boldsymbol{r}_0, \boldsymbol{r}_1, \ldots, \boldsymbol{r}_K | \boldsymbol{\Sigma}, H_0) p(\boldsymbol{\Sigma}) d\boldsymbol{\Sigma}} \\ &= \frac{\det\left(\boldsymbol{\Gamma}(0) + (\mu - JN)\bar{\boldsymbol{\Sigma}}\right)}{\min_{\alpha} \det\left(\boldsymbol{\Gamma}(\alpha) + (\mu - JN)\bar{\boldsymbol{\Sigma}}\right)} \\ &= \frac{\left|\boldsymbol{p}^\dagger \hat{\boldsymbol{\Sigma}}^{-1} \boldsymbol{r}_0\right|^2}{\left(\boldsymbol{r}_0^\dagger \hat{\boldsymbol{\Sigma}}^{-1} \boldsymbol{r}_0 + 1\right)\left(\boldsymbol{p}^\dagger \hat{\boldsymbol{\Sigma}}^{-1} \boldsymbol{p}\right)}, \end{aligned} \tag{5.11}$$

where the integrals are evaluated as the complex inverse Wishart normalization factor and

$$\hat{\boldsymbol{\Sigma}} = K\boldsymbol{S} + (\mu - JN)\bar{\boldsymbol{\Sigma}}, \tag{5.12}$$

which linearly combines the sample covariance matrix $\boldsymbol{S}$ and the prior matrix $\bar{\boldsymbol{\Sigma}}$.

Compared with Kelly's GLRT for the conventional homogeneous STAP model

$$T_{\text{GLRT}} = \frac{|\boldsymbol{p}^\dagger (K\boldsymbol{S})^{-1} \boldsymbol{r}_0|^2}{(\boldsymbol{r}_0^\dagger (K\boldsymbol{S})^{-1} \boldsymbol{r}_0 + 1)(\boldsymbol{p}^\dagger (K\boldsymbol{S})^{-1} \boldsymbol{p})}, \tag{5.13}$$

it is clear that the B-GLRT replaces the sample covariance matrix with a regularized covariance matrix estimate $\hat{\boldsymbol{\Sigma}}$ of (5.12) which involves a simple colored loading between the sample covariance matrix $\boldsymbol{S}$ and the *a priori* $\bar{\boldsymbol{\Sigma}}$. The loading factor is linearly proportional to the prior parameter μ. It is well known that colored loading improves the performance of the Kelly's GLRT when the number of training signals is limited. In this regard, the B-GLRT provides another interpretation of the colored-loading approach for the homogeneous STAP model.

Alternatively, the B-GLRT of (5.11) can be developed from a slightly different procedure. Instead of integrating the likelihood function over $\boldsymbol{\Sigma}$, one can maximize the joint likelihood function of the signals over $\{\boldsymbol{r}_k\}_{k=0}^K$ and $\boldsymbol{\Sigma}$ [13]:

$$\begin{aligned} T_{\text{MAP-GLRT}} &= \frac{\max_{\alpha} \max_{\boldsymbol{\Sigma}} p(\boldsymbol{r}_0, \boldsymbol{r}_1, \ldots, \boldsymbol{r}_K, \boldsymbol{\Sigma} | \alpha, H_1)}{\max_{\boldsymbol{\Sigma}} p(\boldsymbol{r}_0, \boldsymbol{r}_1, \ldots, \boldsymbol{r}_K, \boldsymbol{\Sigma} | H_0)} \\ &= \frac{\max_{\alpha} \left\{ \max_{\boldsymbol{\Sigma}} p(\boldsymbol{r}_0, \boldsymbol{r}_1, \ldots, \boldsymbol{r}_K | \alpha, \boldsymbol{\Sigma}, H_1) p(\boldsymbol{\Sigma}) \right\}}{\max_{\boldsymbol{\Sigma}} p(\boldsymbol{r}_0, \boldsymbol{r}_1, \ldots, \boldsymbol{r}_K | \boldsymbol{\Sigma}, H_0) p(\boldsymbol{\Sigma})}. \end{aligned} \tag{5.14}$$

It can be seen that the above estimates of $\boldsymbol{\Sigma}$ under H_0 and H_1 are the maximum *a posteriori* (MAP) estimate. Similarly, one can replace the MAP estimate with the minimum mean square error (MMSE) estimate in the joint likelihood function and obtain the MMSE-GLRT detector that coincides with the B-GLRT [13].

Different from the one-step GLRT principle, one can use a two-step approach, similar to the classic AMF detector [5], to develop a B-AMF detector. First, assuming $\boldsymbol{\Sigma}$ is known, the GLRT or the matched filter is given as [5]

$$\begin{aligned} T(\boldsymbol{\Sigma}) &= \frac{\max_{\alpha} p(\boldsymbol{r}_0, \boldsymbol{r}_1, \ldots, \boldsymbol{r}_K | \alpha, \boldsymbol{\Sigma}, H_1)}{p(\boldsymbol{r}_0, \boldsymbol{r}_1, \ldots, \boldsymbol{r}_K | \boldsymbol{\Sigma}, H_0)} \\ &= \frac{|\boldsymbol{p}^\dagger \boldsymbol{\Sigma}^{-1} \boldsymbol{r}_0|^2}{\boldsymbol{p}^\dagger \boldsymbol{\Sigma}^{-1} \boldsymbol{p}}. \end{aligned} \tag{5.15}$$

Then, in the second step, the MAP estimate of $\boldsymbol{\Sigma}$ from training signals only is obtained as

$$\begin{aligned} \hat{\boldsymbol{\Sigma}}_{\text{MAP},K} &= \max_{\boldsymbol{\Sigma}} p(\boldsymbol{\Sigma} | \boldsymbol{r}_1, \ldots, \boldsymbol{r}_K) \\ &= \max_{\boldsymbol{\Sigma}} p(\boldsymbol{r}_0, \boldsymbol{r}_1, \ldots, \boldsymbol{r}_K | \boldsymbol{\Sigma}) p(\boldsymbol{\Sigma}) \\ &= \frac{\hat{\boldsymbol{\Sigma}}}{JN + K + \mu}. \end{aligned} \tag{5.16}$$

Finally, the Bayesian-MAP detector is obtained by substituting the MAP estimate from the training signals back to (5.15) [13]

$$T_{\text{B-AMF}} = T(\boldsymbol{\Sigma})|_{\boldsymbol{\Sigma} = \hat{\boldsymbol{\Sigma}}_{\text{MAP},K}} = \frac{|\boldsymbol{p}^\dagger \hat{\boldsymbol{\Sigma}}^{-1} \boldsymbol{r}_0|^2}{\boldsymbol{p}^\dagger \hat{\boldsymbol{\Sigma}}^{-1} \boldsymbol{p}}, \tag{5.17}$$

which extends the conventional AMF [5]

$$T_{\text{AMF}} = \frac{|\boldsymbol{p}^\dagger \boldsymbol{S}^{-1} \boldsymbol{r}_0|^2}{\boldsymbol{p}^\dagger \boldsymbol{S}^{-1} \boldsymbol{p}}, \tag{5.18}$$

by colored loading.

The B-GLRT (5.11) and the B-AMF of (5.17) were numerically studied and compared with the conventional AMF and GLRT detectors in a knowledge-aided homogeneous STAP scenario, where the disturbances $\boldsymbol{d}_k, k = 0, 1, \ldots, K$, are i.i.d. generated according to a random covariance matrix $\boldsymbol{\Sigma} \sim \mathcal{CW}^{-1}((\mu - JN)\bar{\boldsymbol{\Sigma}}, \mu)$, and the prior matrix $\bar{\boldsymbol{\Sigma}}$ has a Toeplitz-block-Toeplitz structure corresponding to a multi-channel AR process. The space–time steering vector $\boldsymbol{p}$ of (5.2) is generated using a uniform linear array of $J = 4$ antennas and $N = 8$ pulses with a normalized spatial frequency $\omega_s = 0.4\pi$ and a normalized Doppler frequency $\omega_d = 0.4\pi$. The signal-to-interference-plus-noise ratio (SINR) is defined as

$$\text{SINR} = |\alpha|^2 \boldsymbol{p}^\dagger \bar{\boldsymbol{\Sigma}}^{-1} \boldsymbol{p}. \tag{5.19}$$

The performance is evaluated in terms of probability of detection P_d for a range of SINRs subject to a probability of false alarm $P_{fa} = 0.01$.

Figure 5.4 shows the detection performance in the knowledge-aided homogeneous scenario with different values of K and μ. It is evident in Figure 5.4a and b that the B-GLRT and B-AMF detectors show significant performance improvement over the conventional GLRT

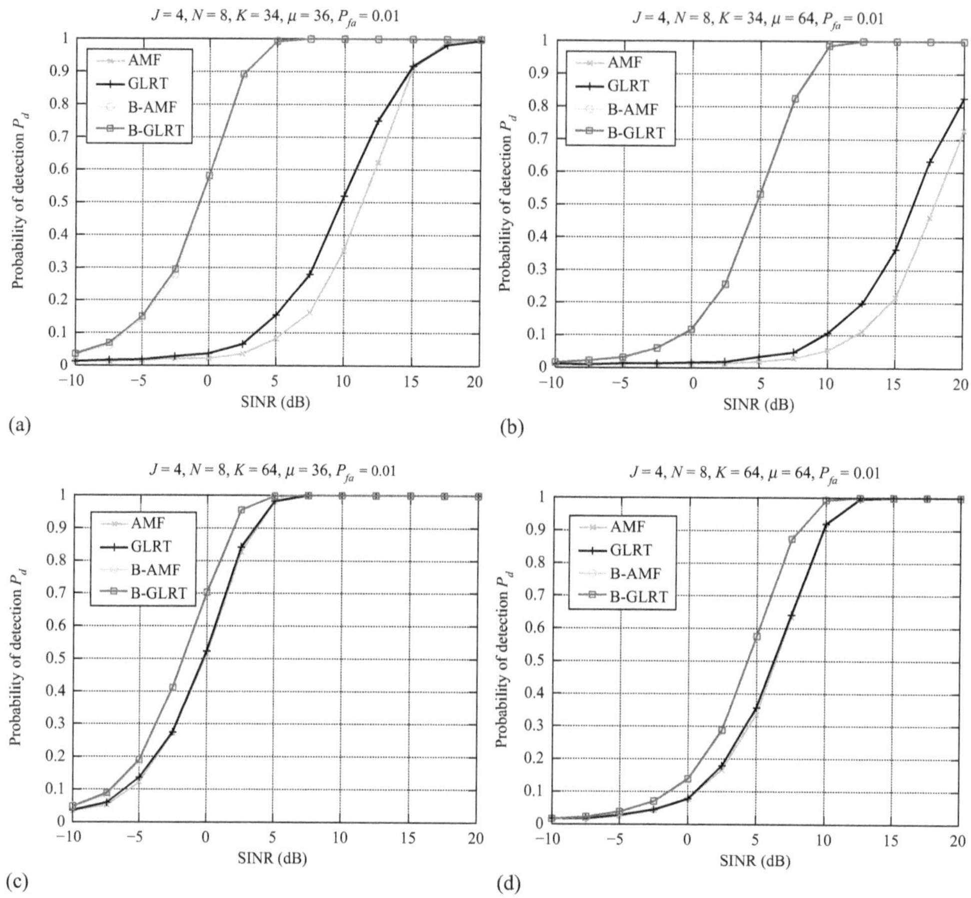

Figure 5.4 Conventional STAP detectors versus Bayesian STAP detectors in a knowledge-aided homogeneous STAP model with $J = 4, N = 8$, and $P_{fa} = 0.01$. (a) $K = 34$ and $\mu = 36$; (b) $K = 34$ and $\mu = 64$; (c) $K = 64$ and $\mu = 36$; and (d) $K = 64$ and $\mu = 64$.

and AMF detectors, when the number of training data is comparable to the space–time dimension $K \approx JN$. With sufficient training data, e.g., $K = 64$ in Figure 5.4c and d, the conventional GLRT and AMF detectors improve drastically, but the utilization of the prior knowledge still leads to some further improvement. In the four considered scenarios, the B-AMF and B-GLRT detectors show close detection performance, while the conventional GLRT is slightly better than the AMF in the case of small training data.

5.3.3 Selection of Hyperparameter

It should be noted that the hyperparameter μ is assumed to be known for the B-AMF and B-GLRT detectors. The resulting B-GLRT detector of (5.11) and the B-AMF detector of (5.17) are, hence, both functions of μ through the computation of $\hat{\boldsymbol{\Sigma}}$ in (5.12). When μ is unknown,

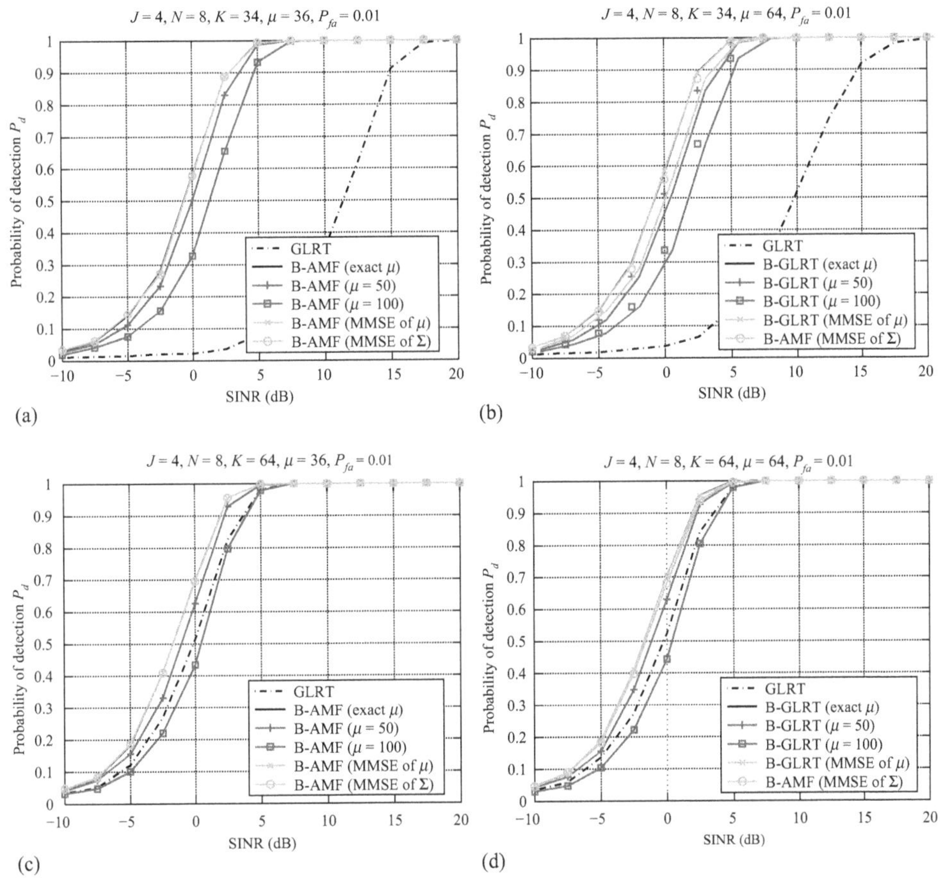

Figure 5.5 Selection of the hyperparameter μ: fixed and adaptive approaches. (a) B-AMF with varied hyperparameters μ; (b) B-GLRT with varied hyperparameters μ; (c) B-AMF with varied hyperparameters μ; and (d) B-GLRT with varied hyperparameters μ.

one may choose a large value of μ if the prior covariance matrix $\bar{\boldsymbol{\Sigma}}$ is believed to be close to the true $\boldsymbol{\Sigma}$, according to the prior variance of (5.10). However, this subjective determination of μ may not be sufficient to guide the users to assign a proper μ, if little information is available regarding the prior uncertainty.

The impact of an over-estimated hyperparameter on the detection performance is illustrated in Figure 5.5a for the B-AMF and Figure 5.5b for the B-GLRT, where the simulation configuration is the same as in Figure 5.4. Two over-estimated $\mu = 50$ and $\mu = 100$ (versus the true $\mu = 36$) were used. Notable performance degradation was observed if the assigned μ is away from the true μ. The larger the mismatch, the worse the performance.

This problem can be addressed by introducing a hierarchical stochastic homogeneous model [14], i.e., μ is modeled as a *discrete* uniform random variable over an interval, i.e.,

$$\mu \sim \text{unif}(\mu_m, \mu_M), \tag{5.20}$$

where $\mu_m(>JN)$ and μ_M are, respectively, the lower and upper bounds of μ. Then, the MMSE estimate of μ can be derived by first computing the posterior distribution of $\mu|\boldsymbol{r}_1,\ldots,\boldsymbol{r}_K$,

$$p(\mu|\boldsymbol{r}_1,\ldots,\boldsymbol{r}_K) = \int p(\mu,\boldsymbol{\Sigma}|\boldsymbol{r}_1,\ldots,\boldsymbol{r}_K)d\boldsymbol{\Sigma}$$
$$= \frac{\int p(\boldsymbol{r}_1,\ldots,\boldsymbol{r}_K|\boldsymbol{\Sigma})p(\boldsymbol{\Sigma}|\mu)p(\mu)d\boldsymbol{\Sigma}}{p(\boldsymbol{r}_1,\ldots,\boldsymbol{r}_K)}$$
$$= \bar{c}\frac{\det\big((\mu-JN)\bar{\boldsymbol{\Sigma}}\big)^{\mu}}{\det\big(K\boldsymbol{S}+(\mu-JN)\bar{\boldsymbol{\Sigma}}\big)^{K+\mu}}\frac{\tilde{\Gamma}(JN,K+\mu)}{\tilde{\Gamma}(JN,\mu)}\boldsymbol{1}_{\mu_m,\mu_M}(\mu), \tag{5.21}$$

where $\boldsymbol{1}_{\mu_m,\mu_M}(\mu)$ is the indicator function and $\bar{c}$ is a normalization factor such that $\int p(\mu|\boldsymbol{r}_1,\ldots,\boldsymbol{r}_K)d\mu = 1$. Therefore, the MMSE estimate of μ is given by the posterior mean of $\mu|\boldsymbol{r}_1,\ldots,\boldsymbol{r}_K$

$$\hat{\mu}_{\text{MMSE}} = \frac{\sum_{\mu=\mu_m}^{\mu_M}\mu h(\mu)}{\sum_{\mu=\mu_m}^{\mu_M}h(\mu)}, \tag{5.22}$$

where

$$h(\mu) = \frac{\det\big((\mu-JN)\bar{\boldsymbol{\Sigma}}\big)^{\mu}}{\det\big(K\boldsymbol{S}+(\mu-JN)\bar{\boldsymbol{\Sigma}}\big)^{K+\mu}}\frac{\tilde{\Gamma}(JN,K+\mu)}{\tilde{\Gamma}(JN,\mu)}. \tag{5.23}$$

In (5.22), the MMSE estimate of μ can be fully determined from the training signals $\{\boldsymbol{r}_k\}_{k=1}^K$ via $\boldsymbol{S}$, the prior covariance matrix $\bar{\boldsymbol{\Sigma}}$, and the hyperparameter range (μ_m,μ_M). As a result, one can directly use the MMSE estimate of μ to replace μ in (5.11) and (5.17), which yields the automatic versions of the B-AMF and B-GLRT:

$$T_{\text{B-GLRT}} = \frac{|\boldsymbol{p}^{\dagger}\hat{\boldsymbol{\Sigma}}^{-1}(\hat{\mu}_{\text{MMSE}})\boldsymbol{r}_0|^2}{(\boldsymbol{r}_0^{\dagger}\hat{\boldsymbol{\Sigma}}^{-1}(\hat{\mu}_{\text{MMSE}})\boldsymbol{r}_0+1)(\boldsymbol{p}^{\dagger}\hat{\boldsymbol{\Sigma}}^{-1}(\hat{\mu}_{\text{MMSE}})\boldsymbol{p})}, \tag{5.24}$$

$$T_{\text{B-AMF}} = \frac{|\boldsymbol{p}^{\dagger}\hat{\boldsymbol{\Sigma}}^{-1}(\hat{\mu}_{\text{MMSE}})\boldsymbol{r}_0|^2}{\boldsymbol{p}^{\dagger}\hat{\boldsymbol{\Sigma}}^{-1}(\hat{\mu}_{\text{MMSE}})\boldsymbol{p}}, \tag{5.25}$$

where

$$\hat{\boldsymbol{\Sigma}}(\hat{\mu}_{\text{MMSE}}) = K\boldsymbol{S}+(\hat{\mu}_{\text{MMSE}}-JN)\bar{\boldsymbol{\Sigma}}, \tag{5.26}$$

with $\hat{\mu}_{\text{MMSE}}$ given by (5.22). We refer to (5.24) and (5.25) as the B-GLRT and B-AMF with the MMSE estimate of μ, respectively.

Another approach is to update the MMSE estimate of $\boldsymbol{\Sigma}$ based on the hierarchical Bayesian model [14]. First, recognize that

$$\begin{aligned} p(\boldsymbol{\Sigma}|\boldsymbol{r}_1,\ldots,\boldsymbol{r}_K,\mu) &\propto p(\boldsymbol{r}_1,\ldots,\boldsymbol{r}_K|\boldsymbol{\Sigma})p(\boldsymbol{\Sigma}|\mu), \\ \boldsymbol{\Sigma}|\boldsymbol{r}_1,\ldots,\boldsymbol{r}_K,\mu &\sim \mathcal{CW}^{-1}(K\boldsymbol{S}+(\mu-JN)\bar{\boldsymbol{\Sigma}},K+\mu), \end{aligned} \tag{5.27}$$

which results in

$$p(\boldsymbol{\Sigma}|\boldsymbol{r}_1,\ldots,\boldsymbol{r}_K)=\sum_{\mu=\mu_m}^{\mu_M}p(\boldsymbol{\Sigma}|\boldsymbol{r}_1,\ldots,\boldsymbol{r}_K,\mu)p(\mu|\boldsymbol{r}_1,\ldots,\boldsymbol{r}_K), \tag{5.28}$$

with $p(\mu|\boldsymbol{r}_1,\ldots,\boldsymbol{r}_K)$ given by (5.21). The MMSE estimate of $\boldsymbol{\Sigma}$ is given by

$$\begin{aligned}\hat{\boldsymbol{\Sigma}}_{\text{MMSE}}&=\int\boldsymbol{\Sigma}p(\boldsymbol{\Sigma}|\boldsymbol{r}_1,\ldots,\boldsymbol{r}_K)d\boldsymbol{\Sigma}\\&=\int\sum_{\mu=\mu_m}^{\mu_M}\boldsymbol{\Sigma}p(\boldsymbol{\Sigma}|\boldsymbol{r}_1,\ldots,\boldsymbol{r}_K,\mu)p(\mu|\boldsymbol{r}_1,\ldots,\boldsymbol{r}_K)d\boldsymbol{\Sigma}\\&=\sum_{\mu=\mu_m}^{\mu_M}E\{\boldsymbol{\Sigma}|\boldsymbol{r}_1,\ldots,\boldsymbol{r}_K,\mu\}p(\mu|\boldsymbol{r}_1,\ldots,\boldsymbol{r}_K)\\&\overset{(a)}{=}\sum_{\mu=\mu_m}^{\mu_M}\frac{K\boldsymbol{S}+(\mu-JN)\bar{\boldsymbol{\Sigma}}}{K+\mu-JN}p(\mu|\boldsymbol{r}_1,\ldots,\boldsymbol{r}_K),\end{aligned} \tag{5.29}$$

where (a) holds since $\boldsymbol{\Sigma}|\boldsymbol{r}_1,\ldots,\boldsymbol{r}_K,\mu$ is complex inverse Wishart distributed as shown in (5.27). The above equation can be rewritten as

$$\hat{\boldsymbol{\Sigma}}_{\text{MMSE}}=\beta\boldsymbol{S}+(1-\beta)\bar{\boldsymbol{\Sigma}}, \tag{5.30}$$

where $\beta=\sum_{\mu=\mu_m}^{\mu_M}Kp(\mu|\boldsymbol{r}_1,\ldots,\boldsymbol{r}_K)/(K+\mu-JN)$. This is equivalent to a fully adaptive colored loading for the covariance matrix estimation and hence, provides a fully adaptive Bayesian detection by replacing $\hat{\boldsymbol{\Sigma}}$ in the B-GLRT and B-AMF detectors with $\hat{\boldsymbol{\Sigma}}_{\text{MMSE}}$:

$$T_{\text{B-GLRT}}=\frac{|\boldsymbol{p}^{\dagger}\hat{\boldsymbol{\Sigma}}_{\text{MMSE}}^{-1}\boldsymbol{r}_0|^2}{(\boldsymbol{r}_0^{\dagger}\hat{\boldsymbol{\Sigma}}_{\text{MMSE}}^{-1}\boldsymbol{r}_0+1)(\boldsymbol{p}^{\dagger}\hat{\boldsymbol{\Sigma}}_{\text{MMSE}}^{-1}\boldsymbol{p})}, \tag{5.31}$$

$$T_{\text{B-AMF}}=\frac{|\boldsymbol{p}^{\dagger}\hat{\boldsymbol{\Sigma}}_{\text{MMSE}}^{-1}\boldsymbol{r}_0|^2}{\boldsymbol{p}^{\dagger}\hat{\boldsymbol{\Sigma}}_{\text{MMSE}}^{-1}\boldsymbol{p}}, \tag{5.32}$$

which are referred to as, respectively, the B-GLRT and B-AMF detectors with the MMSE estimate of $\boldsymbol{\Sigma}$.

With the same configuration in Figure 5.4, we evaluate the fully adaptive B-GLRT detectors of (5.24) and (5.31) and the fully adaptive B-AMF detectors of (5.25) and (5.32). In Figure 5.5a and b, two over-estimated values of $\mu=50$ and $\mu=100$ are used, when the number of training data is fixed to $K=34$. Notable performance losses were observed by comparing their performance to the B-AMF and B-GLRT detectors with the true value of $\mu=36$. On the other hand, the two fully adaptive solutions with the MMSE estimates of μ and $\boldsymbol{\Sigma}$ show negligible performance loss for the B-AMF detector in Figure 5.5a and the B-GLRT detector in Figure 5.5b. By increasing the number of training data from $K=34$ to $K=64$, the results shown in Figure 5.5c and d reveal that the over-estimated hyperparameter $\mu=100$ makes the B-AMF and B-GLRT detectors even worse than the conventional GLRT and AMF detectors without using any prior knowledge.

5.3.4 Extensions to Partially Homogeneous and Compound-Gaussian Models

The Bayesian detection in the knowledge-aided homogeneous model has inspired the development of KA-STAP detection for other conventional STAP models. Here, we briefly discuss two extensions along this direction. The first model considered in this section is the knowledge-aided partially homogeneous model and the Bayesian ACE detector [15]. One motivation to consider the partially homogeneous model stems from the use of guard cells in radar signal processing [8, 16–18]. In STAP, a number of guard cells are often used to mitigate the sidelobe effects and hence separate the test signal and training signals, which may lead to a power difference between the test and training signals [2].

First, recall the conventional partially homogeneous model ($\boldsymbol{M}_0 = \lambda\boldsymbol{\Sigma}$ and $\boldsymbol{M}_1 = \cdots = \boldsymbol{M}_K = \boldsymbol{\Sigma}$) in (5.1) with the following assumptions

$$\boldsymbol{d}_0 \sim \mathcal{CN}(\boldsymbol{0}, \lambda\boldsymbol{\Sigma}), \quad \boldsymbol{d}_k \sim \mathcal{CN}(\boldsymbol{0}, \boldsymbol{\Sigma}), \tag{5.33}$$

with λ denoting a deterministic but unknown scaling parameter. Similar to the knowledge-aided homogeneous model, the knowledge-aided partially homogeneous model treats the interference covariance matrix $\boldsymbol{\Sigma}$ as a random quantity with the following distribution

$$\boldsymbol{\Sigma} \sim \mathcal{CW}^{-1}((\mu - JN)\bar{\boldsymbol{\Sigma}}, \mu). \tag{5.34}$$

The graphical model representation of the knowledge-aided partially homogeneous model is shown in Figure 5.6, where the interference covariance matrix is considered to be a random variable and the power scaling factor λ is considered as the model parameter to be estimated.

Compared with the B-AMF and B-GLRT in the knowledge-aided homogeneous model, the GLRT here takes one additional step to find the ML estimate of the scaling factor λ, in addition

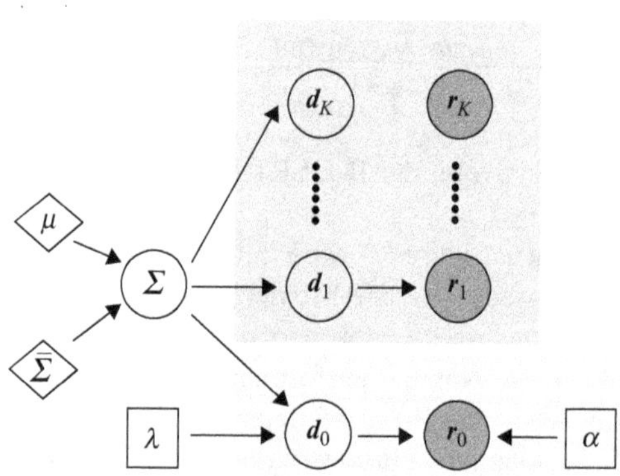

Figure 5.6 The knowledge-aided partially homogeneous STAP model with $\boldsymbol{M}_0 = \lambda\boldsymbol{\Sigma}$ and $\boldsymbol{M}_1 = \cdots = \boldsymbol{M}_K = \boldsymbol{\Sigma}$ and the power scaling factor λ is considered as an unknown model parameter.

to the integration over the random $\boldsymbol{\Sigma}$ and the ML estimate of α [15]; see Appendix 5.A for the derivation. As a result, it is not surprising to see that the resulting GLRT

$$T = \frac{\left|\boldsymbol{p}^{\dagger}\hat{\boldsymbol{\Sigma}}^{-1}\boldsymbol{r}_0\right|^2}{\left(\boldsymbol{p}^{\dagger}\hat{\boldsymbol{\Sigma}}^{-1}\boldsymbol{p}\right)\left(\boldsymbol{r}_0^{\dagger}\hat{\boldsymbol{\Sigma}}^{-1}\boldsymbol{r}_0\right)}, \tag{5.35}$$

where $\hat{\boldsymbol{\Sigma}} = K\boldsymbol{S} + (\mu - N)\bar{\boldsymbol{\Sigma}}$, is a Bayesian version of the ACE [8], which is the GLRT for the conventional partially homogeneous model

$$T_{\text{ACE}} = \frac{\left|\boldsymbol{p}^{\dagger}\boldsymbol{S}^{-1}\boldsymbol{r}_0\right|^2}{\left(\boldsymbol{p}^{\dagger}\boldsymbol{S}^{-1}\boldsymbol{p}\right)\left(\boldsymbol{r}_0^{\dagger}\boldsymbol{S}^{-1}\boldsymbol{r}_0\right)}. \tag{5.36}$$

The GLRT of (5.35), referred to as the KA-ACE detector, can also be derived using the MAP and MMSE estimates of $\boldsymbol{\Sigma}$ instead of integrating it out in the likelihood function [15].

Next, we consider the compound-Gaussian model that explicitly takes into account the power fluctuation (or the texture) across range bins, especially for heavy-tailed clutter distributions often seen in sea clutter. Similar to the knowledge-aided homogeneous model, the knowledge-aided compound-Gaussian model further assumes the speckle component $\boldsymbol{\Sigma}$ as a random variable. In particular, the interferences in both the test and the training range bins consist of two components, i.e.,

$$\boldsymbol{d}_k = s_k \boldsymbol{w}_k, \quad k = 0, 1, \ldots, K, \tag{5.37}$$

where the speckle component $\boldsymbol{w}_k$ and the texture components s_k^2 are specified as

$$\boldsymbol{w}_k \sim \mathcal{CN}(\boldsymbol{0}, \boldsymbol{\Sigma}), \tag{5.38}$$

$$\boldsymbol{\Sigma} \sim \mathcal{CW}^{-1}((\mu - JN)\bar{\boldsymbol{\Sigma}}, \mu), \tag{5.39}$$

$$s_k^2 \sim \mathcal{IG}(q_k, \beta_k), \tag{5.40}$$

with $\mathcal{IG}(q, \beta)$ denoting the inverse Gamma distribution with parameters $q > 2$ and $\beta > 0$

$$p(s^2) = \frac{\beta^q}{\Gamma(q)(s^2)^{q+1}} e^{-\beta/s^2}, \quad s^2 > 0. \tag{5.41}$$

Note that the disturbances in all cells no longer follow the Gaussian assumption, after marginalizing the randomness of the covariance matrix and the texture components.

The Bayesian detection for this knowledge-aided compound-Gaussian model is developed in a two-step approach [19]. First, assuming the speckle covariance matrix $\boldsymbol{\Sigma}$ is known, the GLRT on the test signal $\boldsymbol{r}_0$ only is derived (see Appendix 5.B)

$$T = \frac{|\boldsymbol{p}^{\dagger}\boldsymbol{\Sigma}^{-1}\boldsymbol{r}_0|^2}{(\beta_0 + \boldsymbol{r}_0^{\dagger}\boldsymbol{\Sigma}^{-1}\boldsymbol{r}_0)(\boldsymbol{p}^{\dagger}\boldsymbol{\Sigma}^{-1}\boldsymbol{p})}. \tag{5.42}$$

Then the second step is to replace the assumed $\boldsymbol{\Sigma}$ in the above GLRT by an estimate of $\boldsymbol{\Sigma}$ from the training data $\{\boldsymbol{r}_k\}_{k=1}^K$. Depending on the estimate of $\boldsymbol{\Sigma}$ from the training data, one can derive three Bayesian detectors:

- *The Marginal MAP Estimate of* $\boldsymbol{\Sigma}$ maximizes the posterior distribution $p(\boldsymbol{\Sigma}|\boldsymbol{r}_1, \ldots, \boldsymbol{r}_K)$ (see Appendix 5.B) which results in solving a fixed-point equation iteratively

$$(\mu + K + JN)\boldsymbol{\Sigma} = (\mu - JN)\bar{\boldsymbol{\Sigma}} + \sum_{k=1}^{K} \frac{(q_k + JN)\boldsymbol{r}_k\boldsymbol{r}_k^\dagger}{\beta_k + \boldsymbol{r}_k^\dagger \boldsymbol{\Sigma}^{-1}\boldsymbol{r}_k}. \tag{5.43}$$

  Note that q_k and β_k are the assumed hyperparameters related to the texture components.

- *The Joint MAP Estimate of* $\boldsymbol{\Sigma}$ jointly maximizes the posterior distribution $p(\boldsymbol{\Sigma}, s_1^2, \ldots, s_K^2|\boldsymbol{r}_1, \ldots, \boldsymbol{r}_K)$ (see Appendix 5.B) first over s_k^2 and then $\boldsymbol{\Sigma}$. Similarly, the joint MAP estimate solves a slightly different fixed-point equation:

$$(\mu + K + JN)\boldsymbol{\Sigma} = (\mu - JN)\bar{\boldsymbol{\Sigma}} + \sum_{k=1}^{K} \frac{(q_k + JN + 1)\boldsymbol{r}_k\boldsymbol{r}_k^\dagger}{\beta_k + \boldsymbol{r}_k^\dagger \boldsymbol{\Sigma}^{-1}\boldsymbol{r}_k}. \tag{5.44}$$

- *The MMSE Estimate of* $\boldsymbol{\Sigma}$ is given by the posterior mean

$$\hat{\boldsymbol{\Sigma}}_{\text{MMSE}} = \mathbb{E}\{\boldsymbol{\Sigma}|\boldsymbol{r}_1, \ldots, \boldsymbol{r}_K\} = \int \boldsymbol{\Sigma} p(\boldsymbol{\Sigma}|\boldsymbol{r}_1, \ldots, \boldsymbol{r}_K) d\boldsymbol{\Sigma}, \tag{5.45}$$

  which cannot be computed analytically. A two-step Gibbs sampling utilizing known distributions greatly facilitates the numerical computation:

  (1) numerically generate texture samples $\boldsymbol{s}^i$ according to an inverse Gamma distribution as

$$s_k^2|\boldsymbol{\Sigma}^{i-1}, \boldsymbol{r}_k \sim \mathcal{IG}(q_k + JN, \beta_k + \boldsymbol{r}_k^\dagger(\boldsymbol{\Sigma}^{i-1})^{-1}\boldsymbol{r}_k), \tag{5.46}$$

  with $\boldsymbol{\Sigma}^{i-1}$ denoting $\boldsymbol{\Sigma}$ from the $(i-1)$-th iteration.

  (2) numerically generate $\boldsymbol{\Sigma}^i$ according to an inverse complex Wishart distribution as

$$\boldsymbol{\Sigma}|\boldsymbol{s}^i, \boldsymbol{r}_1, \ldots, \boldsymbol{r}_K \sim \mathcal{CW}^{-1}\left((\mu - JN)\bar{\boldsymbol{\Sigma}} + \sum_{k=1}^{K} \boldsymbol{r}_k^\dagger \boldsymbol{r}_k/\{s_k^2\}^i, \mu + K\right). \tag{5.47}$$

  The MMSE estimate of $\boldsymbol{\Sigma}$ is given as the average of the last N_r values of the numerically generated samples $\boldsymbol{\Sigma}^i$

$$\hat{\boldsymbol{\Sigma}}_{\text{MMSE}} \approx \frac{1}{N_r} \sum_{i=N_{bi}+1}^{N_{bi}+N_r} \boldsymbol{\Sigma}^i, \tag{5.48}$$

  where N_{bi} and N_r stand for, respectively, the number of burn-in iterations and the number of Gibbs samples of $\boldsymbol{\Sigma}^i$ used in the MMSE estimate.

A closely related extension is a Bayesian detector of distributed targets in knowledge-aided compound-Gaussian noise. Interested readers may refer to [20] for more details.

5.4 Knowledge-Aided Two-Layered STAP Model

In the above section, the Bayesian approach has been applied to classical STAP models including the homogeneous, the partially homogeneous, and the compound-Gaussian models, by treating the interference covariance matrix as a random parameter and embedding the prior knowledge through its prior distribution. The resulting Bayesian STAP detectors have been observed to have a colored-loading form of the conventional STAP detectors, except the one in the knowledge-aided compound-Gaussian model (Figure 5.7).

Distinct from the knowledge-aided models in the above section, a knowledge-aided two-layered STAP model [21] that makes use of the randomness of the interference covariance matrix is introduced within the Bayesian framework. The key concept of this two-layered STAP model is to model the covariance mismatch $\boldsymbol{M}_0 \neq \boldsymbol{M}_t (\boldsymbol{M}_1 = \cdots = \boldsymbol{M}_K)$ between the test and training data in a *stochastic* way. This is achieved by assuming that the conditional distribution of $\boldsymbol{M}_t|\boldsymbol{M}_0$ is a complex inverse Wishart distribution with $\nu(>JN)$ degrees of freedom and mean $\boldsymbol{M}_0$, i.e.,

$$\boldsymbol{M}_t|\boldsymbol{M}_0 \sim \mathcal{CW}^{-1}((\nu - JN)\boldsymbol{M}_0, \mu). \tag{5.49}$$

On the one hand, the above conditional prior distribution guarantees that $\boldsymbol{M}_t$ in realizations is different from $\boldsymbol{M}_0$ with probability one. On the other hand, they are not completely irrelevant; otherwise, the training data are meaningless for the inference of $\boldsymbol{M}_0$. The conditional distribution establishes the relationship between $\boldsymbol{M}_0$ and $\boldsymbol{M}_k$ via the statistic characteristics,

$$\mathbb{E}\{\boldsymbol{M}_t|\boldsymbol{M}_0\} = \boldsymbol{M}_0, \tag{5.50}$$

$$\operatorname{cov}\{\boldsymbol{M}_t|\boldsymbol{M}_0\} = \frac{\boldsymbol{M}_0^2 + (\nu - JN)\operatorname{Tr}\{\boldsymbol{M}_0\}\boldsymbol{M}_0}{(\nu - JN)^2 - 1}, \tag{5.51}$$

where (5.50) states that, conditioned on $\boldsymbol{M}_0$, the training data are still homogeneous in average to the interference in the test data, and (5.51) reflects that the heterogeneity in average is controlled by the hyperparameter ν. The larger the hyperparameter ν, the less heterogeneity between the test and training data. Meanwhile, the second layer $\boldsymbol{M}_0$ is assumed to be complex

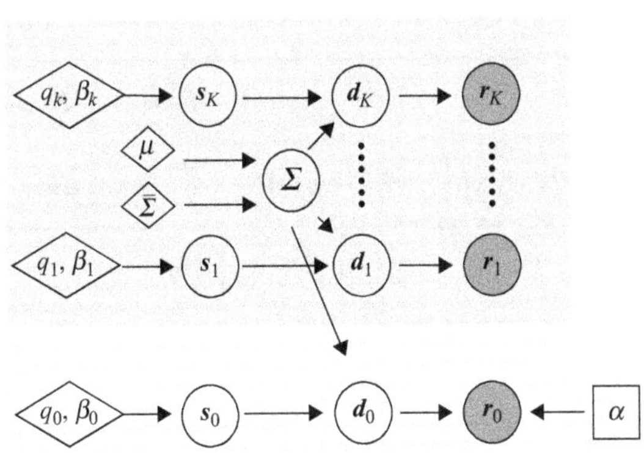

Figure 5.7 The knowledge-aided compound-Gaussian STAP model with $\boldsymbol{M}_0 = s_0^2\boldsymbol{\Sigma}$ and $\boldsymbol{M}_k = s_k^2\boldsymbol{\Sigma}, k = 1, \ldots, K$.

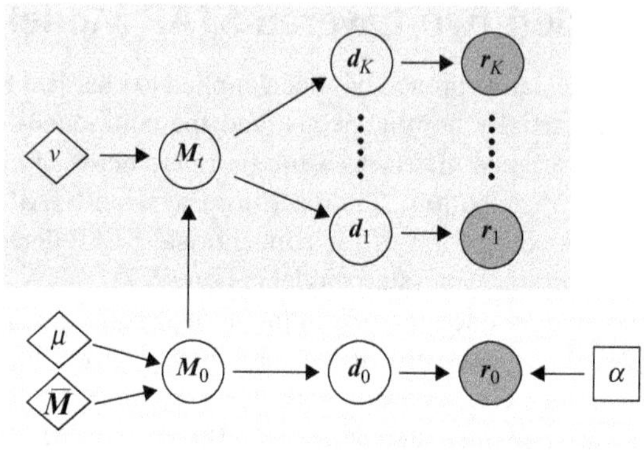

Figure 5.8 The knowledge-aided two-layered STAP model, where the covariance matrix of training data is different from that of the test data through an edge, i.e., a conditional stochastic way.

Wishart distributed with μ degrees of freedom and mean $\bar{M}$, i.e.,

$$M_0 \sim \mathcal{CW}(\mu^{-1}\bar{M}, \mu), \tag{5.52}$$

where $\bar{M}$ is again the prior knowledge. In a short summary, the stochastic non-homogeneous model is characterized by two-layered stochastic modeling: one is the knowledge-aided modeling of M_0 via its prior distribution, similar to the knowledge-aided homogeneous model; the other is the heterogeneity modeling between M_0 and M_t via the conditional distribution of $M_t|M_0$. For each layer, a hyperparameter (μ or ν) is used to control the prior uncertainty or the (average) heterogeneity. The graphical representation of the two-layered STAP model is illustrated in Figure 5.8. Note that, in this knowledge-aided two-layered STAP model, the dependence of the training data on the test data is reflected by the conditional distribution in (5.49), as denoted by the edge from M_0 to M_t in Figure 5.8.

Together with (5.1), (5.49), and (5.52), the GLRT utilizing the test and training signals simultaneously takes the form of

$$T = \frac{\max_{\alpha} \int p(r_0, r_1, \ldots, r_K|\alpha, M_0, M_t, H_1)p(M_t|M_0)p(M_0)dM_0}{\int p(r_0, r_1, \ldots, r_K|M_0, M_t, H_0)p(M_t|M_0)p(M_0)dM_0}$$
$$\overset{(a)}{\propto} \frac{\max_{\alpha} \int q(M_0)\exp\left\{-(r_0 - \alpha p)^\dagger M_0^{-1}(r_0 - \alpha p)\right\} dM_0}{\int q(M_0)\exp\left\{-r_0^\dagger M_0^{-1} r_0\right\} dM_0}, \tag{5.53}$$

where

$$q(M_0) = \frac{\det(M_0)^{\nu-1}}{\det(KS + (\nu - JN)M_0)^{\nu-K}} p(M_0). \tag{5.54}$$

Unfortunately, the ML estimate of α under H_0 has no closed-form solution. An approximate ML estimate of α was proposed in [21] by realizing that the numerator is upper bounded

$$\max_{\alpha} \int q(\boldsymbol{M}_0)\exp\left\{-(\boldsymbol{r}_0-\alpha\boldsymbol{p})^{\dagger}\boldsymbol{M}_0^{-1}(\boldsymbol{r}_0-\alpha\boldsymbol{p})\right\} d\boldsymbol{M}_0$$
$$\leq \int q(\boldsymbol{M}_0)\exp\left\{-\boldsymbol{r}_0^{\dagger}\boldsymbol{M}_0^{-1}\boldsymbol{r}_0\right\}\exp\left\{\frac{|\boldsymbol{p}^{\dagger}\boldsymbol{M}_0^{-1}\boldsymbol{r}_0|^2}{\boldsymbol{p}^{\dagger}\boldsymbol{M}_0^{-1}\boldsymbol{p}}\right\} d\boldsymbol{M}_0, \tag{5.55}$$

since

$$\max_{\alpha}(\boldsymbol{r}_0-\alpha\boldsymbol{p})^{\dagger}\boldsymbol{M}_0^{-1}(\boldsymbol{r}_0-\alpha\boldsymbol{p}) = \boldsymbol{r}_0^{\dagger}\boldsymbol{M}_0^{-1}\boldsymbol{r}_0 - \frac{|\boldsymbol{p}^{\dagger}\boldsymbol{M}_0^{-1}\boldsymbol{r}_0|^2}{\boldsymbol{p}^{\dagger}\boldsymbol{M}_0^{-1}\boldsymbol{p}}. \tag{5.56}$$

Expressing both the numerator and the denominator in terms of the posterior distribution $p(\boldsymbol{M}_0|\boldsymbol{r}_1,\ldots,\boldsymbol{r}_K)$ yields an approximate Bayesian GLRT (denoted as the AGLRT)

$$T = \frac{\displaystyle\int h(\boldsymbol{M}_0)\exp\left\{\frac{|\boldsymbol{p}^{\dagger}\boldsymbol{M}_0^{-1}\boldsymbol{r}_0|^2}{\boldsymbol{p}^{\dagger}\boldsymbol{M}_0^{-1}\boldsymbol{p}}\right\} p(\boldsymbol{M}_0|\boldsymbol{r}_1,\ldots,\boldsymbol{r}_K)d\boldsymbol{M}_0}{\displaystyle\int h(\boldsymbol{M}_0)p(\boldsymbol{M}_0|\boldsymbol{r}_1,\ldots,\boldsymbol{r}_K)d\boldsymbol{M}_0}, \tag{5.57}$$

where

$$h(\boldsymbol{M}_0) = \det(\boldsymbol{M}_0)^{-1}\exp\left(-\boldsymbol{r}_0^{\dagger}\boldsymbol{M}_0^{-1}\boldsymbol{r}_0\right). \tag{5.58}$$

It is seen that the Bayesian AGLRT of (5.57) is the ratio of the posterior mean of $h(\boldsymbol{M}_0)\exp\left(\frac{|\boldsymbol{p}^{\dagger}\boldsymbol{M}_0^{-1}\boldsymbol{r}_0|^2}{\boldsymbol{p}^{\dagger}\boldsymbol{M}_0^{-1}\boldsymbol{p}}\right)$ to the posterior mean of $h(\boldsymbol{M}_0)$ over the posterior distribution of $\boldsymbol{M}_0|\boldsymbol{r}_1,\ldots,\boldsymbol{r}_K$. Therefore, the numerical implementation can be obtained as

$$T = \frac{\displaystyle\sum_{i=N_{bi}+1}^{N_{bi}+N_r} h(\boldsymbol{M}_0^{(i)})\exp\left(\frac{\left|\boldsymbol{p}^{\dagger}\left(\boldsymbol{M}_0^{(i)}\right)^{-1}\boldsymbol{r}_0\right|^2}{\boldsymbol{p}^{\dagger}\left(\boldsymbol{M}_0^{(i)}\right)^{-1}\boldsymbol{p}}\right)}{\displaystyle\sum_{i=N_{bi}+1}^{N_{bi}+N_r} h(\boldsymbol{M}_0^{(i)})}, \tag{5.59}$$

where $\boldsymbol{M}_0^{(i)}$ denotes the Gibbs sampler from the posterior distribution $p(\boldsymbol{M}_0|\boldsymbol{r}_1,\ldots,\boldsymbol{r}_K)$ at the i-th iteration. The Gibbs sampling of $p(\boldsymbol{M}_0|\boldsymbol{r}_1,\ldots,\boldsymbol{r}_K)$ can be fulfilled via a two-step procedure:

- generate $\boldsymbol{M}_0^{(i+1)}$ according to the distribution of $\boldsymbol{M}_0|\boldsymbol{M}_t,\boldsymbol{r}_1,\ldots,\boldsymbol{r}_K$, which follows the complex Wishart distribution

$$\boldsymbol{M}_0|\boldsymbol{M}_t,\boldsymbol{r}_1,\ldots,\boldsymbol{r}_K \sim \mathcal{CW}\left(\left(\mu\bar{\boldsymbol{M}}^{-1} + (\nu - JN)\boldsymbol{M}_t^{-1}\right)^{-1}, \nu+\mu\right); \tag{5.60}$$

- generate $\boldsymbol{M}_t^{(i+1)}$ according to the distribution of $\boldsymbol{M}_t|\boldsymbol{M}_0,\boldsymbol{r}_1,\ldots,\boldsymbol{r}_K$, which follows the complex inverse Wishart distribution

$$\boldsymbol{M}_t|\boldsymbol{M}_0,\boldsymbol{r}_1,\ldots,\boldsymbol{r}_K \sim \mathcal{CW}^{-1}(K\boldsymbol{S} + (\nu - JN)\boldsymbol{M}_0, \nu+K). \tag{5.61}$$

The B-AMF detector for the stochastic two-layered STAP model, on the other hand, takes a simpler form as

$$T = \frac{|\boldsymbol{p}^\dagger \hat{\boldsymbol{M}}_0^{-1} \boldsymbol{r}_0|^2}{\boldsymbol{p}^\dagger \hat{\boldsymbol{M}}_0^{-1} \boldsymbol{p}}, \tag{5.62}$$

where $\hat{\boldsymbol{M}}_0$ can be either the MMSE or MAP estimate of $\boldsymbol{M}_0$ based on the posterior distribution of $\boldsymbol{M}_0|\boldsymbol{r}_1, \ldots, \boldsymbol{r}_K$. The MMSE estimate is taken as the sample mean of the last N_r Gibbs sampler of $\boldsymbol{M}_0^{(i)}$ from the above two-step Gibbs sampling procedure, i.e.,

$$\hat{\boldsymbol{M}}_{0,\text{MMSE}} = \frac{1}{N_r} \sum_{i=N_{bi}+1}^{N_{bi}+N_r} \boldsymbol{M}_0^{(i)}. \tag{5.63}$$

The MAP estimate of $\boldsymbol{M}_0$ is given by a closed-form expression:

$$\hat{\boldsymbol{M}}_{0,\text{MAP}} = \bar{\boldsymbol{M}}^{1/2} \boldsymbol{U} \boldsymbol{\Gamma} \boldsymbol{U}^\dagger \bar{\boldsymbol{M}}^{1/2}, \tag{5.64}$$

where $\boldsymbol{U}$ is the matrix whose i-th column is the eigenvector of the pre-whitened sample covariance matrix by the prior knowledge

$$\bar{\boldsymbol{S}} = \bar{\boldsymbol{M}}^{-1/2} \boldsymbol{S} \bar{\boldsymbol{M}}^{-1/2} = \boldsymbol{U} \boldsymbol{\Lambda} \boldsymbol{U}^\dagger, \tag{5.65}$$

with $\boldsymbol{\Lambda} = \text{diag}\{\lambda_i\}$ denoting a diagonal matrix consisting of the eigenvalue of $\bar{\boldsymbol{S}}$, and $\boldsymbol{\Gamma} = \text{diag}\{\gamma_i\}$ with

$$\gamma_i = \left(\frac{\mu - JN - K}{2\mu} - \frac{K\lambda_i}{2(\nu - JN)} \right) + \sqrt{\left(\frac{\mu - JN - K}{2\mu} - \frac{K\lambda_i}{2(\nu - JN)} \right)^2 + \frac{K(\nu + \mu - JN)}{\mu(\nu - JN)} \lambda_i}. \tag{5.66}$$

It was noted in [21] that the MAP estimate of $\boldsymbol{M}_0$ is computationally more appealing than the MMSE estimate. The interpretation of the MAP estimate is also interesting. From (5.66), the MAP estimator simply provides a regularization on the eigenvalues λ_i of the sample covariance matrix while leaving the eigenvectors intact. Moreover, this regularization from λ_i to γ_i appears to be nonlinear and controlled by the hyperparameters μ and ν.

The detection performance of the three Bayesian STAP detectors: (1) the Bayesian AGLRT detector of (5.57); (2) the B-AMF of (5.62) with the MMSE estimate of $\boldsymbol{M}_0$ (5.63), denoted as the BAMF-MMSE detector; and (3) the B-AMF of (5.62) with the MAP estimate of $\boldsymbol{M}_0$ (5.64), denoted as the BAMF-MAP detector, is evaluated in the knowledge-aided two-layered STAP model, where the interferences in the test and training data are generated with the covariance matrix following a multi-channel AR process. The spatial and temporal signatures of the target are the same as those used in Figure 5.4.

We first consider a scenario where the prior knowledge is relatively accurate with a large value of μ and a relatively homogeneous training data with a large value of ν, as shown in Figure 5.9a. A comparison with the conventional AMF and GLRT detectors shows the improvement of the Bayesian detectors in the stochastic non-homogeneous case. In Figure 5.9b where the heterogeneity between the test and training data is increased with a small value of ν, the performance improvement of the Bayesian detectors over the conventional detectors

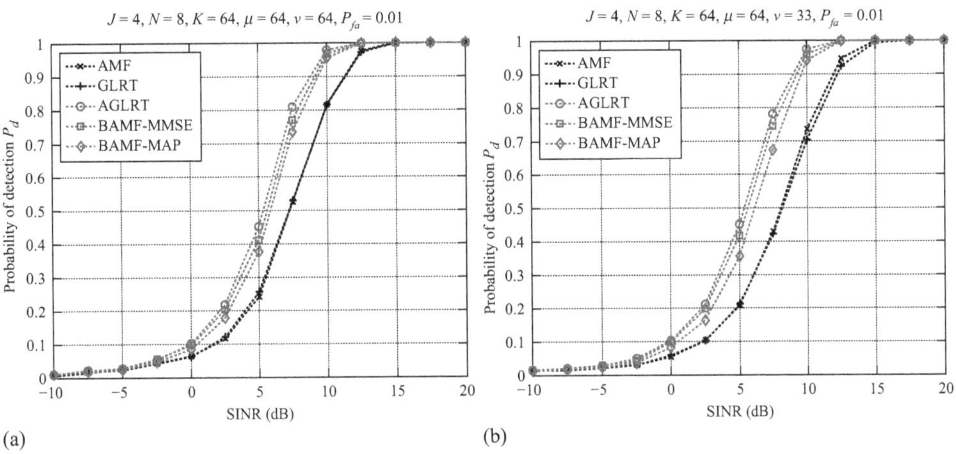

Figure 5.9 Probability of detection versus SINR in a knowledge-aided two-layered STAP model. (a) $\mu = 64$ and $\nu = 64$ and (b) $\mu = 64$ and $\nu = 33$.

is increased. Among the Bayesian detectors, the Bayesian AGLRT and the BAMF-MMSE detectors, at a cost of a higher complexity due to the Gibbs sampling, yield slightly better performance than the simple BAMF-MAP detector in both cases.

Extensions of the knowledge-aided two-layered STAP model towards the partially homogeneous model were considered in [22]. Related efforts focusing on the covariance matrix estimation and performance characterization under a similar but extended knowledge-aided heterogeneous model can be found in [23] and [24].

5.5 Knowledge-Aided Parametric STAP Model

In this section, the structural information of the interference covariance matrices $\boldsymbol{M}_k$ is exploited, together with the knowledge-aided concept, to facilitate the computation of the Bayesian detectors. The resulting knowledge-aided *parametric* STAP model employs multi-channel AR processes to describe the spatially and temporally correlated disturbances $\boldsymbol{d}_k, k = 0, 1, \ldots, K$, in the test and training data. Extensive experimental studies have proved that AR processes with a moderate model order are effective parametric models to represent the spatio-temporal correlation.

Specifically, a knowledge-aided hybrid multi-channel AR (P) model is employed involving a *stochastic* cross-channel or spatial covariance matrix $\boldsymbol{Q}$ and a *deterministic* AR correlation matrix $\boldsymbol{A}$, i.e.,

$$\boldsymbol{d}_k(n) = -\sum_{p=1}^{P} \boldsymbol{A}^{\dagger}(p)\boldsymbol{d}_k(n-p) + \boldsymbol{\varepsilon}_k(n), \quad n = 1, 2, \ldots, N, \tag{5.67}$$

where $\boldsymbol{d}_k(n)$ denote the n-th array snapshot from J spatial receivers, $\boldsymbol{A} = [\boldsymbol{A}^T(1), \boldsymbol{A}^T(2), \ldots, \boldsymbol{A}^T(P)] \in \mathbb{C}^{JP\times J}$ with $\boldsymbol{A}(p)$ denoting the AR coefficient matrix for the p-th time lag, and $\boldsymbol{\varepsilon}_k(n)$ denote the $J \times 1$ spatial noise vectors that are temporally white but spatially colored Gaussian noise: $\{\boldsymbol{\varepsilon}_k(n)\}_{k=0}^{K} \sim \mathcal{CN}(\boldsymbol{0}, \boldsymbol{Q})$, with $\boldsymbol{Q}$ denoting the *unknown* $J \times J$ spatial covariance matrix. Furthermore, the random spatial covariance matrix $\boldsymbol{Q}$ follows an inverse complex Wishart

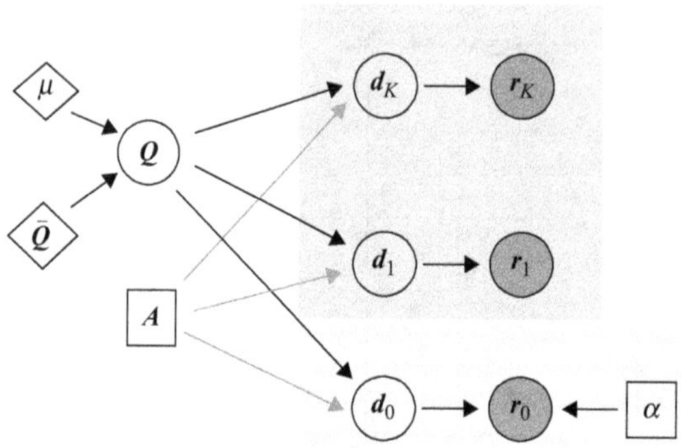

Figure 5.10 The knowledge-aided hybrid parametric STAP model, where $\boldsymbol{M}_0 = \boldsymbol{M}_1 = \cdots = \boldsymbol{M}_K = f(\boldsymbol{A}, \boldsymbol{Q})$ with the function f is characterized by a hybrid multi-channel AR process with a random spatial covariance matrix and a deterministic correlation matrix.

distribution with μ degrees of freedom and mean $\bar{\boldsymbol{Q}}$,

$$\boldsymbol{Q} \sim \mathcal{CW}^{-1}((\mu - J)\bar{\boldsymbol{Q}}, \mu), \tag{5.68}$$

similar to the knowledge-aided non-parametric STAP model but in a smaller dimension (J versus JN). Figure 5.10 illustrates the graphical model representation of the knowledge-aided hybrid parametric STAP model, where the disturbances in all cells share the same covariance matrix as a parametric form of the random matrix $\boldsymbol{Q}$ and the deterministic matrix $\boldsymbol{A}$.

With the knowledge-aided hybrid STAP model, we first consider a two-step parametric detector from the Bayesian framework, which is referred to as the knowledge-aided parametric AMF (KA-PAMF) [25]. The first step is to develop the GLRT that is partially adaptive with a given temporal AR coefficient $\boldsymbol{A}$. Then, the partial adaptive GLRT is modified to arrive at a fully adaptive detector by replacing $\boldsymbol{A}$ with the ML estimate obtained from training signals.

The partially adaptive P-GLRT with a given $\boldsymbol{A}$ is obtained by integrating out the random spatial covariance matrix $\boldsymbol{Q}$ in the likelihood ratio [25]:

$$\begin{aligned} T &= \frac{\max\limits_{\alpha} \int f_1\left(\boldsymbol{r}_0, \boldsymbol{r}_1, \ldots, \boldsymbol{r}_K \,|\alpha, \boldsymbol{Q}\right) p(\boldsymbol{Q}) d\boldsymbol{Q}}{\int f_0\left(\boldsymbol{r}_0, \boldsymbol{r}_1, \ldots, \boldsymbol{r}_K \,|\alpha = 0, \boldsymbol{Q}\right) p(\boldsymbol{Q}) d\boldsymbol{Q}} \\ &= \frac{\max\limits_{\alpha} \det(\boldsymbol{\Xi}_1)^{-L+J}}{\det(\boldsymbol{\Xi}_0)^{-L+J}} \\ &= \frac{\left|\sum\limits_{n=P}^{N-1} \tilde{\boldsymbol{p}}^{\dagger}(n)\, \boldsymbol{\Psi}^{-1} \tilde{\boldsymbol{r}}_0(n)\right|^2}{\sum\limits_{n=P}^{N-1} \tilde{\boldsymbol{p}}^{\dagger}(n)\, \boldsymbol{\Psi}^{-1} \tilde{\boldsymbol{p}}(n)}, \end{aligned} \tag{5.69}$$

where $L = (K+1)(N-P) + (\mu + J)$, and

$$\boldsymbol{\Xi}_i = \hat{\boldsymbol{R}}_{xx}(\alpha) + \hat{\boldsymbol{R}}_{yx}^{\dagger}(\alpha)\boldsymbol{A} + \boldsymbol{A}^{\dagger}\hat{\boldsymbol{R}}_{yx}(\alpha) + \boldsymbol{A}^{\dagger}\hat{\boldsymbol{R}}_{yy}(\alpha)\boldsymbol{A} + (\mu - J)\bar{\boldsymbol{Q}}, \tag{5.70}$$

where the correlation matrices in (5.70) (conditioned on α) are defined as

$$\hat{\boldsymbol{R}}_{xx}(\alpha) = \sum_{n=P}^{N-1} \left[\boldsymbol{r}_0(n) - \alpha \boldsymbol{p}(n)\right]\left[\boldsymbol{r}_0(n) - \alpha \boldsymbol{p}(n)\right]^{\dagger} + \sum_{k=1}^{K}\sum_{n=P}^{N-1} \boldsymbol{r}_k(n)\boldsymbol{r}_k^{\dagger}(n), \tag{5.71}$$

$$\hat{\boldsymbol{R}}_{yy}(\alpha) = \sum_{n=P}^{N-1} \left[\boldsymbol{y}_0(n) - \alpha \boldsymbol{t}(n)\right]\left[\boldsymbol{y}_0(n) - \alpha \boldsymbol{t}(n)\right]^{\dagger} + \sum_{k=1}^{K}\sum_{n=P}^{N-1} \boldsymbol{y}_k(n)\boldsymbol{y}_k^{\dagger}(n), \tag{5.72}$$

$$\hat{\boldsymbol{R}}_{yx}(\alpha) = \sum_{n=P}^{N-1} \left[\boldsymbol{y}_0(n) - \alpha \boldsymbol{t}(n)\right]\left[\boldsymbol{r}_0(n) - \alpha \boldsymbol{p}(n)\right]^{\dagger} + \sum_{k=1}^{K}\sum_{n=P}^{N-1} \boldsymbol{y}_k(n)\boldsymbol{r}_k^{\dagger}(n), \tag{5.73}$$

and the $JP \times 1$ vectors $\boldsymbol{y}_k(n)$ and $\boldsymbol{t}(n)$ are defined as $\boldsymbol{y}_k(n) = \left[\boldsymbol{r}_k^T(n-1), \ldots, \boldsymbol{r}_k^T(n-P)\right]^T$ and $\boldsymbol{t}(n) = \left[\boldsymbol{p}^T(n-1), \ldots, \boldsymbol{p}^T(n-P)\right]^T$. Meanwhile, the third line of (5.69) is due to the ML estimate of α that minimizes the determinant of $\boldsymbol{\Xi}_1$

$$\hat{\alpha}_{\mathrm{ML}} = \frac{\mathrm{Tr}\,(\tilde{\boldsymbol{S}}^{\dagger}\boldsymbol{\Psi}^{-1}\tilde{\boldsymbol{X}}_0)}{\mathrm{Tr}\,(\tilde{\boldsymbol{S}}^{\dagger}\boldsymbol{\Psi}^{-1}\tilde{\boldsymbol{S}})}, \tag{5.74}$$

where

$$\boldsymbol{\Psi} = \tilde{\boldsymbol{X}}_0\boldsymbol{P}^{\perp}\tilde{\boldsymbol{X}}_0^{\dagger} + \sum_{k=1}^{K}\tilde{\boldsymbol{X}}_k\tilde{\boldsymbol{X}}_k^{\dagger} + (\mu - J)\bar{\boldsymbol{Q}}, \tag{5.75}$$

$\boldsymbol{P}^{\perp}$ denotes the projection matrix operator that projects the signal to the orthogonal complement of the range of $\tilde{\boldsymbol{S}}^{\dagger}$: $\boldsymbol{P}^{\perp} = \boldsymbol{I} - \boldsymbol{P} = \boldsymbol{I} - \tilde{\boldsymbol{S}}^{\dagger}(\tilde{\boldsymbol{S}}^{\dagger})^{+}$ with $(\cdot)^{+}$ denoting the Moore–Penrose pseudoinverse, and

$$\tilde{\boldsymbol{S}} \in \mathcal{C}^{J\times(N-P)} = \left[\tilde{\boldsymbol{p}}(P), \ldots, \tilde{\boldsymbol{p}}(N-1)\right], \tag{5.76}$$

$$\tilde{\boldsymbol{X}}_k \in \mathcal{C}^{J\times(N-P)} = \left[\tilde{\boldsymbol{r}}_k(P), \ldots, \tilde{\boldsymbol{r}}_k(N-1)\right], \tag{5.77}$$

are formed from the temporally whitened signal $\tilde{\boldsymbol{r}}_k(n)$ and the temporally whitened steering vector $\tilde{\boldsymbol{p}}(n)$

$$\tilde{\boldsymbol{r}}_k(n) = \boldsymbol{r}_k(n) + \boldsymbol{A}^{\dagger}\boldsymbol{y}_k(n), \tag{5.78}$$

$$\tilde{\boldsymbol{p}}(n) = \boldsymbol{p}(n) + \boldsymbol{A}^{\dagger}\boldsymbol{t}(n). \tag{5.79}$$

To arrive at a fully adaptive P-GLRT, the ML estimate of $\boldsymbol{A}$ is obtained by using the training signals. Since the spatial covariance matrix is random, the likelihood function conditioned on $\boldsymbol{A}$ is obtained by integrating the likelihood function over the distribution of $\boldsymbol{Q}$ as

$$\begin{aligned} f(\boldsymbol{r}_1, \boldsymbol{r}_2, \ldots, \boldsymbol{r}_K | \boldsymbol{A}) &= \int f(\boldsymbol{r}_1, \boldsymbol{r}_2, \ldots, \boldsymbol{r}_K | \boldsymbol{A}, \boldsymbol{Q})\, p(\boldsymbol{Q})\, d\boldsymbol{Q} \\ &\propto \det\left(\boldsymbol{\Sigma}(\boldsymbol{A}) + (\mu - J)\bar{\boldsymbol{Q}}\right)^{-(\mu + K(N-P))}, \end{aligned} \tag{5.80}$$

where

$$\boldsymbol{\Sigma}(\boldsymbol{A}) = \sum_{k=1}^{K} \sum_{n=P}^{N-1} \boldsymbol{\varepsilon}_k(n)\, \boldsymbol{\varepsilon}_k^{\dagger}(n). \tag{5.81}$$

Therefore, the ML estimate of $\boldsymbol{A}$ is equivalent to minimizing the determinant in (5.80), which is given by

$$\hat{\boldsymbol{A}}_{\mathrm{ML}} = -\hat{\boldsymbol{R}}_{yy}^{-1} \hat{\boldsymbol{R}}_{yx}, \tag{5.82}$$

since $\boldsymbol{\Sigma}(\boldsymbol{A}) + (\mu - J)\bar{\boldsymbol{Q}} \succeq \boldsymbol{\Sigma}(\hat{\boldsymbol{A}}_{\mathrm{ML}}) + (\mu - J)\bar{\boldsymbol{Q}}$. Here, $\hat{\boldsymbol{R}}_{xx}$, $\hat{\boldsymbol{R}}_{yy}$, and $\hat{\boldsymbol{R}}_{yx}$ are defined as the second term (training signals only), respectively, in (5.71)–(5.73).

Replacing $\boldsymbol{A}$ with the ML estimate $\hat{\boldsymbol{A}}_{\mathrm{ML}}$, the fully adaptive KA-PAMF takes the following test statistic

$$T_{\text{KA-PAMF}} = \frac{\left| \sum_{n=P}^{N-1} \hat{\tilde{\boldsymbol{p}}}^{\dagger}(n)\, \hat{\boldsymbol{\Psi}}^{-1} \hat{\tilde{\boldsymbol{r}}}_0(n) \right|^2}{\sum_{n=P}^{N-1} \hat{\tilde{\boldsymbol{p}}}^{\dagger}(n)\, \hat{\boldsymbol{\Psi}}^{-1} \hat{\tilde{\boldsymbol{p}}}(n)} \underset{H_0}{\overset{H_1}{\gtrless}} \gamma_{\text{KA-PAMF}}, \tag{5.83}$$

where $\gamma_{\text{KA-PAMF}}$ is a threshold subject to a chosen probability of false alarm, the temporally whitened vectors $\hat{\tilde{\boldsymbol{r}}}_k(n)$ and $\hat{\tilde{\boldsymbol{p}}}(n)$ are similarly obtained as in (5.78) and (5.79) with $\boldsymbol{A}$ replaced by the ML estimate $\hat{\boldsymbol{A}}_{\mathrm{ML}}$, and the spatially whitening matrix $\hat{\boldsymbol{\Psi}}$ is obtained adaptively:

$$\hat{\boldsymbol{\Psi}} = \hat{\tilde{\boldsymbol{X}}}_0 \hat{\boldsymbol{P}}^{\perp} \hat{\tilde{\boldsymbol{X}}}_0^{\dagger} + \sum_{k=1}^{K} \hat{\tilde{\boldsymbol{X}}}_k \hat{\tilde{\boldsymbol{X}}}_k^{\dagger} + (\mu - J)\bar{\boldsymbol{Q}}, \tag{5.84}$$

which is similarly defined as (5.75). Compared with the Bayesian detectors in the above two sections, the KA-PAMF of (5.83) uses a temporal whitening followed by a spatial whitening process. In addition, the spatial whitening matrix $\hat{\boldsymbol{Q}}$ includes three components: one is from the training data, one is from the target-canceled test data (the first term in (5.84)), and the prior knowledge $(\mu - J)\bar{\boldsymbol{Q}}$. The inclusion of the latter allows the KA-PAMF to better handle cases with very limited training data. In terms of computational complexity, the KA-PAMF detector is in general simpler to compute than the Bayesian detectors which performs a fully spatial and temporal whitening with a computational complexity in the order of $\mathcal{O}(J^3N^3)$ to compute $\boldsymbol{M}_0^{-1}$, especially when the dimension JN is large. Specifically, the complexity of computing (5.84) is about $\mathcal{O}(J^2KNP^2) + \mathcal{O}(JN^2)$. The complexity of $\mathcal{O}(JN^2)$ for the KA-PAMF mostly comes from the calculation of the spatial whitening matrix $\hat{\boldsymbol{\Psi}}$ and its inherent Moore–Penrose pseudoinverse.

The second knowledge-aided parametric detector, referred to as the knowledge-aided parametric GLRT (KA-PGLRT), is obtained via a one-step joint estimation/optimization approach for both the signal and the interference parameters. Since the exact ML estimate of the target amplitude is intractable, we use the Schur complements and find an asymptotic but closed-form ML estimate of the amplitude. In turn, it enables us to find a simple and closed-form detection variable for the KA-PGLRT. The results show that the KA-PGLRT applies a joint spatio-and-*sub*temporal whitening with the ability to utilize the prior knowledge.

Specifically, the KA-PGLRT detector takes the following form

$$
\begin{aligned}
T &= \frac{\max\limits_{\alpha} \max\limits_{\boldsymbol{A}} \displaystyle\int f_1\left(\boldsymbol{x}_0, \boldsymbol{x}_1, \ldots, \boldsymbol{x}_K \,|\alpha, \boldsymbol{A}, \boldsymbol{Q}\right) p(\boldsymbol{Q}) d\boldsymbol{Q}}{\max\limits_{\boldsymbol{A}} \displaystyle\int f_0\left(\boldsymbol{x}_0, \boldsymbol{x}_1, \ldots, \boldsymbol{x}_K \,|\boldsymbol{A}, \boldsymbol{Q}\right) p(\boldsymbol{Q}) d\boldsymbol{Q}} \\
&= \frac{\max\limits_{\alpha} \max\limits_{\boldsymbol{A}} \det(\boldsymbol{\Xi}_1)^{-L+J}}{\max\limits_{\boldsymbol{A}} \det(\boldsymbol{\Xi}_0)^{-L+J}} \\
&= \frac{\det\left(\hat{\boldsymbol{R}}_{xx}(0) - \hat{\boldsymbol{R}}_{yx}^{\dagger}(0)\, \hat{\boldsymbol{R}}_{yy}^{-1}(0)\, \hat{\boldsymbol{R}}_{yx}(0) + (\mu - J)\,\bar{\boldsymbol{Q}}\right)}{\min\limits_{\alpha} \det\left(\hat{\boldsymbol{R}}_{xx}(\alpha) - \hat{\boldsymbol{R}}_{yx}^{\dagger}(\alpha)\, \hat{\boldsymbol{R}}_{yy}^{-1}(\alpha)\, \hat{\boldsymbol{R}}_{yx}(\alpha) + (\mu - J)\,\bar{\boldsymbol{Q}}\right)},
\end{aligned} \tag{5.85}
$$

where the second equation is obtained by integrating over the prior distribution of $\boldsymbol{Q}$, and the third equation is due to the maximization of $\det(\boldsymbol{\Xi}_i)$ with respect to $\boldsymbol{A}$, which gives

$$
\hat{\boldsymbol{A}}_{\mathrm{ML}} = -\hat{\boldsymbol{R}}_{yy}^{-1}(\alpha)\, \hat{\boldsymbol{R}}_{yx}(\alpha), \tag{5.86}
$$

where $\hat{\boldsymbol{R}}_{yy}(\alpha)$ and $\hat{\boldsymbol{R}}_{yx}(\alpha)$ are defined in (5.72) and (5.73), respectively.

As a result, the remaining step is to find the ML estimate of α, which is the solution to the following minimization problem

$$
\hat{\alpha}_{\mathrm{ML}} = \arg\min_{\alpha} \det\left(\hat{\boldsymbol{R}}_{xx}(\alpha) - \hat{\boldsymbol{R}}_{yx}^{\dagger}(\alpha)\hat{\boldsymbol{R}}_{yy}^{-1}(\alpha)\hat{\boldsymbol{R}}_{yx}(\alpha) + (\mu - J)\bar{\boldsymbol{Q}}\right). \tag{5.87}
$$

By recognizing that [26]

$$
\left[\hat{\boldsymbol{R}}_{xx}(\alpha) + (\mu - J)\bar{\boldsymbol{Q}}\right] - \hat{\boldsymbol{R}}_{yx}^{\dagger}(\alpha)\hat{\boldsymbol{R}}_{yy}^{-1}(\alpha)\hat{\boldsymbol{R}}_{yx}(\alpha) \tag{5.88}
$$

in (5.87) is the Schur complement of $\hat{\boldsymbol{R}}_{yy}(\alpha)$, which is a part of a block matrix $\hat{\boldsymbol{R}}(\alpha)$ defined as

$$
\hat{\boldsymbol{R}}(\alpha) = \begin{bmatrix} \hat{\boldsymbol{R}}_{yy}(\alpha) & \hat{\boldsymbol{R}}_{yx}(\alpha) \\ \hat{\boldsymbol{R}}_{yx}^{H}(\alpha) & \hat{\boldsymbol{R}}_{xx}(\alpha) + (\mu - J)\,\bar{\boldsymbol{Q}} \end{bmatrix}, \tag{5.89}
$$

the ML estimate of α can be obtained asymptotically as

$$
\hat{\alpha}_{\mathrm{ML}} = \frac{\mathrm{Tr}\{\boldsymbol{S}^{\dagger}\hat{\boldsymbol{R}}_{X}^{-1}\boldsymbol{X}_0\} - \mathrm{Tr}\{\boldsymbol{T}^{\dagger}\hat{\boldsymbol{R}}_{Y}^{-1}\boldsymbol{Y}_0\}}{\mathrm{Tr}\{\boldsymbol{S}^{\dagger}\hat{\boldsymbol{R}}_{X}^{-1}\boldsymbol{S}\} - \mathrm{Tr}\{\boldsymbol{T}^{\dagger}\hat{\boldsymbol{R}}_{Y}^{-1}\boldsymbol{T}\}}, \tag{5.90}
$$

where

$$
\begin{aligned}
\boldsymbol{S} \in \mathcal{C}^{J(P+1)\times(N-P)} &= \left[\boldsymbol{s}_{P+1}(P), \ldots, \boldsymbol{s}_{P+1}(N-1)\right], \\
\boldsymbol{X}_k &= \left[\boldsymbol{x}_{k,P+1}(P), \ldots, \boldsymbol{x}_{k,P+1}(N-1)\right], \\
\boldsymbol{T} \in \mathcal{C}^{JP\times(N-P)} &= \left[\boldsymbol{t}(P), \ldots, \boldsymbol{t}(N-1)\right], \\
\boldsymbol{Y}_k &= \left[\boldsymbol{y}_k(P), \ldots, \boldsymbol{y}_k(N-1)\right],
\end{aligned}
$$

with the following definitions $s_{P+1}(n) = \left[t^T(n), p^T(n)\right]^T$ and $x_{k,P+1}(n) = \left[y_k^T(n), r_k^T(n)\right]^T$, and

$$\hat{R}_X = X_0 P_S^{\perp} X_0^{\dagger} + \sum_{k=1}^{K} X_k X_k^{\dagger} + \tilde{Q}, \tag{5.91}$$

$$\hat{R}_Y = Y_0 P_T^{\perp} Y_0^{\dagger} + \sum_{k=1}^{K} Y_k Y_k^{\dagger}, \tag{5.92}$$

with

$$\tilde{Q} = \begin{bmatrix} 0 & 0 \\ 0 & (\mu - J)\bar{Q} \end{bmatrix}. \tag{5.93}$$

Taking the above ML estimate of α back into the likelihood ratio of (5.85) and followed by simplifications, the KA-PGLRT takes the final form as

$$T_{\text{KA-PGLR}} = \frac{\left| \sum_{n=P}^{N-1} s_{P+1}^{\dagger}(n) \hat{R}_X^{-1} x_{0,P+1}(n) - \sum_{n=P}^{N-1} t^{\dagger}(n) \hat{R}_Y^{-1} y_0(n) \right|^2}{\sum_{n=P}^{N-1} s_{P+1}^{\dagger}(n) \hat{R}_X^{-1} s_{P+1}(n) - \sum_{n=P}^{N-1} t^{\dagger}(n) \hat{R}_Y^{-1} t(n)} \underset{H_0}{\overset{H_1}{\gtrless}} \gamma_{\text{KA-PGLR}}, \tag{5.94}$$

where $\gamma_{\text{KA-PGLR}}$ is a threshold subject to a probability of false alarm.

It is seen from (5.94) that the KA-PGLRT detector uses a knowledge-aided colored-loading step via $\hat{R}_X$ of (5.91) to incorporate the prior $\bar{Q}$. It can further be shown that the KA-PGLRT can be interpreted as a spatio-and-subtemporal whitening across $J(P+1)$ dimensions. Compared to the KA-PAMF detector, the performance improvement of the KA-PGLRT is achieved with a slight additional complexity, due to the construction of $\tilde{Q}$ in (5.93) and the addition of $\tilde{Q}$ in (5.91). The spatio-and-subtemporal whitening of the KA-PGLRT is in between a fully adaptive STAP detector such as the B-AMF detector that employs a joint spatio-temporal whitening across all JN dimensions and the conventional P-AMF [27] that utilizes successive temporal whitening followed by spatial whitening. In terms of complexity of the whitening process, the KA-PGLRT is about $\mathcal{O}(J^2KN(P+1)^2) + \mathcal{O}(J(P+1)N^2)$, more than $\mathcal{O}(J^2KNP^2) + \mathcal{O}(JN^2)$ of the KA-PAMF, which is mainly due to the larger dimension $J(P+1) \times (N-P)$ of S in (5.91) compared with the dimension $J \times (N-P)$ of $\tilde{S}$ in (5.76).

We verify the performance of the KA-PAMF and KA-PGLRT detectors and compare these two detectors with (1) a parametric matched filter (PMF) using the prior knowledge $\bar{Q}$ as a non-adaptive estimate of Q, (2) the conventional P-AMF [27], (3) the Bayesian P-AMF (B-PAMF), and (4) the simplified P-GLRT [26]. The disturbance d_k is generated as a multi-channel AR(2) process with AR coefficients A and a spatial covariance matrix Q. The spatial covariance matrix Q is generated in each Monte-Carlo run from an inverse Wishart distribution with mean $\bar{Q}$.

We first consider cases with limited training data that are particularly challenging in practice. Specifically, when only $K = 2$ training signals are available, three scenarios are considered: (1) $N = 4J$, a preferable case for the conventional parametric detectors (i.e., the P-AMF and P-GLRT), which benefit from having many more pulses (temporal observations) than spatial channels; (2) $N = 2J$, an intermediate case; and (3) the case of $N = J + 2$, an unfavorable case for the conventional parametric detectors.

In the first scenario, the number of pulses is $N = 16$ and the number of channels is $J = 4$. It has been shown that, if $N \gg J$, the conventional parametric detectors can cope with very limited or even no range training signals [28]. Therefore, it is extremely competitive to compare the knowledge-aided parametric detectors such as the KA-PAMF and the KA-PGLRT with such parametric detectors in this case. As shown in Figure 5.11a with $\mu = 12$ and $P_{fa} = 0.01$, it is seen that all adaptive parametric detectors, except the non-adaptive PMF, achieve a close detection performance. This indicates that $K = 2$ range training signals and $N = 16$ pulses are adequate in the current case for the estimation of the unknown parameters $\boldsymbol{A}$ and $\boldsymbol{Q}$.

Then, we reduce the number of pulses to $N = 8$, a less favorable case for the conventional parametric detectors. With the same amount of range training data $K = 2$ as in the first scenario, the conventional parametric detectors are expected to experience some degradation

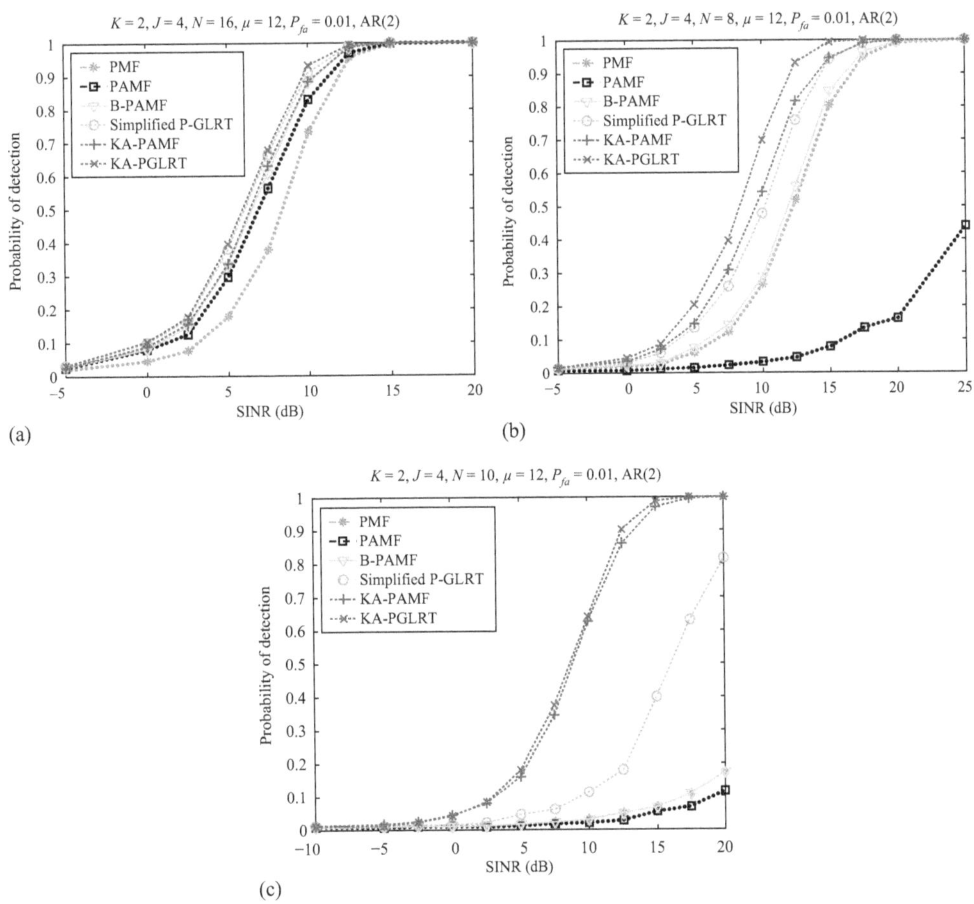

Figure 5.11 Probability of detection versus SINR with limited training data $K = 2$ when $P = 2$, $\mu = 12$, and $P_{fa} = 0.01$. (a) Adequate temporal observations ($N \gg J$); (b) moderate temporal observation ($N = 2J$); and (c) limited temporal observation ($N \approx J$).

as the number of pulses N is reduced. As confirmed in Figure 5.11b, it is evident that the KA-PAMF and KA-PGLRT detectors provide the best results among all considered parametric detectors. The conventional P-AMF, however, is unable to reliably estimate $\boldsymbol{Q}$ and $\boldsymbol{A}$ from the training data with a reduced number of temporal sampling and, hence, gives worse results. Interestingly, the simplified P-GLRT which estimates the unknown parameters from both test and training signals achieves a performance similar to that of the KA-PAMF, however with a 0.8 dB performance loss.

The most challenging case with $K = 2$ range training signals, $J = 8$ channels and $N = 10$ temporal observations, which is considered as an unfavorable case since $J \approx N$ in Figure 5.11c. It is seen that the two knowledge-aided KA-PAMF and KA-PGLRT detectors that combine knowledge learned from the test data, the training data, and the prior information are able to significantly outperform the other parametric detectors. In particular, the performance gains of the KA-PAMF and the KA-PGLRT over the simplified P-GLRT are, respectively, 6 dB and 6.6 dB. The performance gain is even larger by comparing the knowledge-aided parametric detectors with the other parametric detectors.

Finally, we consider the performance of detection as a function of the number of pulses N, when $J = 4, K = 2, P = 2, \mu = 12$, and SINR $= 10$ dB. The number of pulses increases from $N = 6$ to $N = 32$, which corresponds to a transition from a low value of N/J to a high N/J. The purpose of this simulation is to evaluate the convergence performance of the simulated parametric detectors. As shown in Figure 5.12a, it is observed that the KA-PGLRT provides the best detection performance among all considered detectors, and the KA-PAMF outperforms the competitive P-GLRT when the number of pulses is small (i.e., $N < 10$), but becomes slightly inferior when N is larger than 10. It is also observed that the KA-PAMF performs better than its counterparts, e.g., the B-PAMF and the P-AMF, for almost all values of N. When $N = 32$ or larger, all adaptive parametric detectors converge and yield similar detection performance, while the non-adaptive PMF performs the worst since it does not explore the useful information from the training signals. The convergence performance with a large number of range training data $K = 64$ is also confirmed in Figure 5.12b where $J = 4, N = 32, P = 2, K = 64, \mu = 12$,

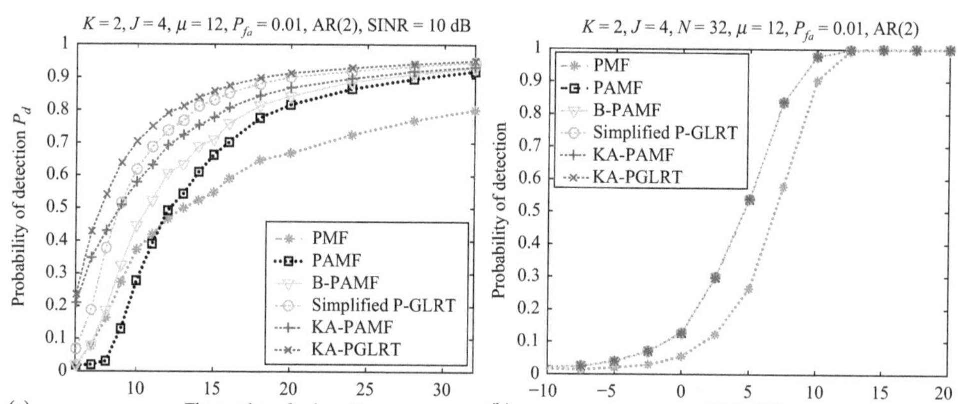

Figure 5.12 Probability of detection converges as a function of the number of temporal samples N or the number of training data K. (a) Performance convergence over N and (b) converged convergence with a sufficiently large K.

and $P_{fa} = 0.01$. It is seen that, all parametric adaptive detectors have the same performance, while the non-adaptive PMF has a performance loss of about 3 dB. The results reveal that, with sufficient training signals, the parametric adaptive detectors can fully learn the knowledge about the unknown interference covariance matrix from the training data.

5.6 Summary

This chapter discusses recent developments of KA-STAP from a Bayesian framework in three related signal models: (1) the knowledge-aided classical STAP model, (2) the knowledge-aided two-layered STAP model, and (3) the knowledge-aided parametric STAP model. In each of these models, the KA-STAP is found to yield an intuitive colored-loading solution, which outperforms the conventional STAP in homogeneous, partially homogeneous, and heterogeneous environments when training data is limited.

When there is an uncertainty with the prior knowledge, adaptive selection of the hyperparameter via a hierarchical stochastic model offers a powerful solution. In fact, the resulting detector is not only hyperparameter free but also remarkably close to the optimal performance when the true hyperparameter is known.

Finally, it is beneficial to integrate Bayesian inference with a structural model, namely the multi-channel AR model for the spatial–temporal correlation of the interference, for KA-STAP. The resulting detectors are able to take advantage of both: the colored loading from the Bayesian framework and the successive temporal-and-spatial whitening from the structural multi-channel AR modeling. This essentially allows us to tackle challenging STAP scenarios when the training data are extremely limited.

Appendix 5.A

Derivation of KA-ACE of (5.35)

Following the knowledge-aided partially homogeneous STAP model, the GLRT can be obtained as follows [15]

$$
\begin{aligned}
T &= \frac{\max\limits_{\alpha}\max\limits_{\lambda} \int p\left(\boldsymbol{r}_0, \boldsymbol{r}_1, \ldots, \boldsymbol{r}_K \,|\alpha, \lambda, \boldsymbol{\Sigma}, H_1\right) p(\boldsymbol{\Sigma}) d\boldsymbol{\Sigma}}{\max\limits_{\lambda} \int p\left(\boldsymbol{r}_0, \boldsymbol{r}_1, \ldots, \boldsymbol{r}_K \,|\lambda, \boldsymbol{\Sigma}, H_0\right) p(\boldsymbol{\Sigma}) d\boldsymbol{\Sigma}} \\
&\overset{(a)}{\propto} \frac{\max\limits_{\alpha}\max\limits_{\lambda} \lambda^{-N} \det\left(\boldsymbol{\Gamma}(\alpha, \lambda) + (\mu - N)\bar{\boldsymbol{\Sigma}}\right)^{-L}}{\max\limits_{\lambda} \lambda^{-N} \det\left(\boldsymbol{\Gamma}(0, \lambda) + (\mu - N)\bar{\boldsymbol{\Sigma}}\right)^{-L}} \\
&\overset{(b)}{\propto} \frac{\boldsymbol{r}_0^{\dagger} \hat{\boldsymbol{\Sigma}}^{-1} \boldsymbol{r}_0}{\min\limits_{\alpha} (\boldsymbol{r}_0 - \alpha \boldsymbol{p})^{\dagger}\, \hat{\boldsymbol{\Sigma}}^{-1} (\boldsymbol{r}_0 - \alpha \boldsymbol{p})} \\
&\overset{(c)}{\propto} \frac{\left|\boldsymbol{p}^{\dagger} \hat{\boldsymbol{\Sigma}}^{-1} \boldsymbol{r}_0\right|^2}{\left(\boldsymbol{p}^{\dagger} \hat{\boldsymbol{\Sigma}}^{-1} \boldsymbol{p}\right)\left(\boldsymbol{r}_0^{\dagger} \hat{\boldsymbol{\Sigma}}^{-1} \boldsymbol{r}_0\right)},
\end{aligned}
\tag{5.95}
$$

where (a) is obtained as the integrals are evaluated as complex inverse Wishart normalization factors, $\boldsymbol{\Gamma}(\alpha, \lambda) = \lambda^{-1}(\boldsymbol{r}_0 - \alpha\boldsymbol{p})(\boldsymbol{r}_0 - \alpha\boldsymbol{p})^\dagger + K\boldsymbol{S}$, (b) holds by showing that the ML estimates of λ under $H_i, i = 0, 1$, are given by

$$\hat{\lambda}(\alpha) = \frac{L - JN}{JN}(\boldsymbol{r}_0 - \alpha\boldsymbol{p})^\dagger \hat{\boldsymbol{\Sigma}}^{-1}(\boldsymbol{r}_0 - \alpha\boldsymbol{p}), \tag{5.96}$$

with $L = K + \mu + 1$, (c) holds due to the ML estimate of α given by

$$\hat{\alpha}_{\mathrm{ML}} = \frac{\boldsymbol{p}^\dagger \hat{\boldsymbol{\Sigma}}^{-1} \boldsymbol{r}_0}{\boldsymbol{p}^\dagger \hat{\boldsymbol{\Sigma}}^{-1} \boldsymbol{p}}, \tag{5.97}$$

and $\hat{\boldsymbol{\Sigma}}$ is defined in (5.12), the same as that in the homogeneous model

$$\hat{\boldsymbol{\Sigma}} = K\boldsymbol{S} + (\mu - N)\bar{\boldsymbol{\Sigma}}. \tag{5.98}$$

Appendix 5.B

Derivation of Bayesian Detectors in Knowledge-Aided Compound-Gaussian Model

Here, derivations of the three Bayesian detectors in the knowledge-aided compound-Gaussian model are provided, following the development in Reference 19. First, the GLRT with a known speckle covariance matrix $\boldsymbol{\Sigma}$ takes the form of

$$\begin{aligned} T &= \frac{\max_{\alpha} \int p(\boldsymbol{r}_0|\alpha, \boldsymbol{\Sigma}, s_0^2, H_1)p(s_0^2)ds_0^2}{\int p(\boldsymbol{r}_0|\boldsymbol{\Sigma}, s_0^2, H_0)p(s_0^2)ds_0^2} \\ &\overset{(a)}{\propto} \frac{\max_{\alpha} \left(\beta_0 + (\boldsymbol{r}_0 - \alpha\boldsymbol{p})^\dagger \boldsymbol{\Sigma}^{-1}(\boldsymbol{r}_0 - \alpha\boldsymbol{p})\right)^{-(q_0+JN)}}{\left(\beta_0 + \boldsymbol{r}_0^\dagger \boldsymbol{\Sigma}^{-1}\boldsymbol{r}_0\right)^{-(q_0+JN)}} \\ &\overset{(b)}{\propto} \frac{|\boldsymbol{p}^\dagger \boldsymbol{\Sigma}^{-1}\boldsymbol{r}_0|^2}{(\beta_0 + \boldsymbol{r}_0^\dagger \boldsymbol{\Sigma}^{-1}\boldsymbol{r}_0)(\boldsymbol{p}^\dagger \boldsymbol{\Sigma}^{-1}\boldsymbol{p})}, \end{aligned} \tag{5.99}$$

where (a) is due to the marginalization over s_k^2, and (b) holds since $\hat{\alpha} = (\boldsymbol{p}^\dagger \boldsymbol{\Sigma}^{-1}\boldsymbol{r}_0)/(\boldsymbol{p}^\dagger \boldsymbol{\Sigma}^{-1}\boldsymbol{p})$.

To find an estimate of $\boldsymbol{\Sigma}$ from the training data $\{\boldsymbol{r}_k\}_{k=1}^K$, one can consider the MAP estimate of $\boldsymbol{\Sigma}$ by maximizing the posterior distribution $p(\boldsymbol{\Sigma}|\boldsymbol{r}_1, \ldots, \boldsymbol{r}_K)$ which is given as

$$\begin{aligned} p(\boldsymbol{\Sigma}|\boldsymbol{r}_1, \ldots, \boldsymbol{r}_K) &= \int p(\boldsymbol{\Sigma}, \boldsymbol{s}|\boldsymbol{r}_1, \ldots, \boldsymbol{r}_K)d\boldsymbol{s} \\ &\propto \frac{\exp\left(-(\mu - JN)\operatorname{Tr}\{\boldsymbol{\Sigma}^{-1}\bar{\boldsymbol{\Sigma}}\}\right)}{\det(\boldsymbol{\Sigma})^{\mu+K+JN}} \prod_{k=1}^{K} \left[\beta_k + \boldsymbol{r}_k^\dagger \boldsymbol{\Sigma}^{-1}\boldsymbol{r}_k\right]^{-(q_k+JN)}, \end{aligned} \tag{5.100}$$

where the joint posterior distribution $p(\boldsymbol{\Sigma}, \boldsymbol{s}|\boldsymbol{r}_1, \ldots, \boldsymbol{r}_K)$ can be computed as

$$p(\boldsymbol{\Sigma}, \boldsymbol{s}|\boldsymbol{r}_1, \ldots, \boldsymbol{r}_K) \propto p(\boldsymbol{r}_1, \ldots, \boldsymbol{r}_K|\boldsymbol{\Sigma}, \boldsymbol{s})p(\boldsymbol{\Sigma})\left[\prod_{k=1}^{K} p(s_k^2)\right]$$
$$\propto \frac{\exp\left(-(\mu - JN)\,\mathrm{Tr}\{\boldsymbol{\Sigma}^{-1}\bar{\boldsymbol{\Sigma}}\}\right)}{\det(\boldsymbol{\Sigma})^{\mu+K+JN}} \prod_{k=1}^{K} \frac{\exp\left(-(\beta_k + \boldsymbol{r}_k^{\dagger}\boldsymbol{\Sigma}^{-1}\boldsymbol{r}_k)/s_k^2\right)}{(s_k^2)^{q_k+JN+1}}, \quad (5.101)$$

where $p(\boldsymbol{r}_1, \ldots, \boldsymbol{r}_K|\boldsymbol{\Sigma}, \boldsymbol{s}) = \prod_{k=1}^{K} p(\boldsymbol{r}_k|\boldsymbol{\Sigma}, s_k^2)$ with $\boldsymbol{r}_k|\boldsymbol{\Sigma}, s_k^2 \sim \mathcal{CN}(\boldsymbol{0}, s_k^2\boldsymbol{\Sigma})$. Taking the derivative of (5.100) with respect to $\boldsymbol{\Sigma}$ and equating to zero, the so-called *marginal* MAP estimate of $\boldsymbol{\Sigma}$ is given by the fixed-point iteration

$$(\mu + K + JN)\boldsymbol{\Sigma} = (\mu - JN)\bar{\boldsymbol{\Sigma}} + \sum_{k=1}^{K} \frac{(q_k + JN)\boldsymbol{r}_k\boldsymbol{r}_k^{\dagger}}{\beta_k + \boldsymbol{r}_k^{\dagger}\boldsymbol{\Sigma}^{-1}\boldsymbol{r}_k}, \quad (5.102)$$

which can be solved iteratively.

Alternatively, one can jointly maximize the posterior distribution $p(\boldsymbol{\Sigma}, s_1^2, \ldots, s_K^2|\boldsymbol{r}_1, \ldots, \boldsymbol{r}_K)$ of (5.101) first over s_k^2 and then $\boldsymbol{\Sigma}$. The *joint* MAP estimate of $\boldsymbol{\Sigma}$ is given as follows

$$(\mu + K + JN)\boldsymbol{\Sigma} = (\mu - JN)\bar{\boldsymbol{\Sigma}} + \sum_{k=1}^{K} \frac{(q_k + JN + 1)\boldsymbol{r}_k\boldsymbol{r}_k^{\dagger}}{\beta_k + \boldsymbol{r}_k^{\dagger}\boldsymbol{\Sigma}^{-1}\boldsymbol{r}_k}. \quad (5.103)$$

Finally, one can also consider the MMSE estimate of $\boldsymbol{\Sigma}$, which is given by the posterior mean, i.e.,

$$\hat{\boldsymbol{\Sigma}}_{\text{MMSE}} = \mathbb{E}\{\boldsymbol{\Sigma}|\boldsymbol{r}_1, \ldots, \boldsymbol{r}_K\} = \int \boldsymbol{\Sigma}p(\boldsymbol{\Sigma}|\boldsymbol{r}_1, \ldots, \boldsymbol{r}_K)d\boldsymbol{\Sigma}, \quad (5.104)$$

which can be numerically computed with the Gibbs sampling since the MMSE cannot be derived analytically. The Gibbs sampling takes a two-step loop:

- At the i-th iteration, one numerically generates samples of the texture components $\boldsymbol{s}^i$ according to the distribution $p(s_k^2|\boldsymbol{\Sigma}^{i-1}, \boldsymbol{r}_k)$

$$s_k^2|\boldsymbol{\Sigma}^{i-1}, \quad \boldsymbol{r}_k \sim \mathcal{IG}(q_k + JN, \beta_k + \boldsymbol{r}_k^{\dagger}(\boldsymbol{\Sigma}^{i-1})^{-1}\boldsymbol{r}_k), \quad (5.105)$$

  with $\boldsymbol{\Sigma}^{i-1}$ denoting $\boldsymbol{\Sigma}$ from the $(i-1)$-th iteration. (5.105) holds since

$$p(s_k^2|\boldsymbol{\Sigma}^{i-1}, \boldsymbol{r}_k) \propto p(\boldsymbol{r}_k|\boldsymbol{\Sigma}^i, s_k^2)p(s_k^2)$$
$$= \frac{\exp\left(-(\beta_k + \boldsymbol{r}_k^{\dagger}\boldsymbol{\Sigma}^{-1}\boldsymbol{r}_k)/s_k^2\right)}{(s_k^2)^{q_k+JN+1}} \quad (5.106)$$

  with $p(s_k^2)$ given by (5.40).

- With the new samples of s^i, then generate numerical samples of $\boldsymbol{\Sigma}^i$ according to the probability $p(\boldsymbol{\Sigma}|\{s^i, \boldsymbol{r}_1, \ldots, \boldsymbol{r}_K)$ which is given by

$$\boldsymbol{\Sigma}|s^i, \boldsymbol{r}_1, \ldots, \boldsymbol{r}_K \sim \mathcal{CW}^{-1}\left((\mu - JN)\bar{\boldsymbol{\Sigma}} + \sum_{k=1}^{K} \boldsymbol{r}_k^{\dagger}\boldsymbol{r}_k/\{s_k^2\}^i, \mu + K\right), \tag{5.107}$$

since, similar to (5.101),

$$\begin{aligned} p(\boldsymbol{\Sigma}|s, \boldsymbol{r}_1, \ldots, \boldsymbol{r}_K) &\propto p(s, \boldsymbol{r}_1, \ldots, \boldsymbol{r}_K|\boldsymbol{\Sigma}, s)p(s) \\ &\propto \frac{\exp\left(-\operatorname{Tr}\left\{\boldsymbol{\Sigma}^{-1}\left[(\mu - JN)\bar{\boldsymbol{\Sigma}} + \sum_{k=1}^{K} \boldsymbol{r}_k\boldsymbol{r}_k^{\dagger}/s_k^2\right]\right\}\right)}{\det(\boldsymbol{\Sigma})^{\mu+K+JN}}. \end{aligned} \tag{5.108}$$

References

[1] E. J. Kelly and K. Forsythe, "Adaptive detection and parameter estimation for multidimensional signal models," Technical Report 848, Lincoln Laboratory, MIT, 1989.

[2] J. Ward, "Space–time adaptive processing for airborne radar," Technical Report 1015, Lincoln Laboratory, MIT, December 1994.

[3] I. S. Reed, J. D. Mallett, and L. E. Brennan, "Rapid convergence rate in adaptive arrays," *IEEE Transactions on Aerospace and Electronic Systems*, vol. 10, no. 6, pp. 853–863, 1974.

[4] E. J. Kelly, "An adaptive detection algorithm," *IEEE Transactions on Aerospace and Electronic Systems*, vol. 22, no. 1, pp. 115–127, March 1986.

[5] F. C. Robey, D. R. Fuhrmann, E. J. Kelly, and R. Nitzberg, "A CFAR adaptive matched filter detector," *IEEE Transactions on Aerospace and Electronic Systems*, vol. 28, no. 1, pp. 208–216, January 1992.

[6] S. Bose and A. Steinhardt, "A maximal invariant framework for adaptive detection with structured and unstructured covariance matrices," *IEEE Transactions on Signal Processing*, vol. 43, no. 9, pp. 2164–2175, September 1995.

[7] E. Conte, M. Lops, and G. Ricci, "Asymptotically optimum radar detection in compound Gaussian clutter," *IEEE Transactions on Aerospace and Electronic Systems*, vol. 31, no. 2, pp. 617–625, April 1995.

[8] S. Kraut and L. L. Scharf, "The CFAR adaptive subspace detector is a scale-invariant GLRT," *IEEE Transactions on Signal Processing*, vol. 47, no. 9, pp. 2538–2541, September 1999.

[9] A. De Maio, "Rao test for adaptive detection in Gaussian interference with unknown covariance matrix," *IEEE Transactions on Signal Processing*, vol. 55, no. 7, pp. 3577–3584, July 2007.

[10] L. Svensson and M. Lundberg, "On posterior distributions for signals in Gaussian noise with unknown covariance matrix," *IEEE Transactions on Signal Processing*, vol. 53, no. 9, pp. 3554–3571, September 2005.

[11] A. De Maio and A. Farina, "Adaptive radar detection: a Bayesian approach," in *Proceedings of the 2006 International Radar Symposium*, Krakow, Poland, May 2006, pp. 85–88.

[12] A. De Maio, A. Farina, and G. Foglia, "Adaptive radar detection: a Bayesian approach," in *Proceedings of the 2007 IEEE International Conference on Radar*, Waltham, MA, April 2007, pp. 624–629.

[13] A. De Maio, A. Farina, and G. Foglia, "Knowledge-aided Bayesian radar detectors & their application to live data," *IEEE Transactions on Aerospace and Electronic Systems*, vol. 46, no. 1, pp. 170–183, February 2010.

[14] S. Bidon, O. Besson, and J. Y. Tourneret, "Knowledge-aided STAP in heterogeneous clutter using a hierarchical Bayesian algorithm," *IEEE Transaction on Aerospace and Electronic Systems*, vol. 47, no. 3, pp. 1863–1879, July 2011.

[15] P. Wang, Z. Sahinoglu, M. O. Pun, H. Li, and B. Himed, "Knowledge-aided adaptive coherence estimator in stochastic partially homogeneous environments," *IEEE Signal Processing Letters*, vol. 18, no. 3, pp. 193–196, March 2011.

[16] O. Besson, L. L. Scharf, and S. Kraut, "Adaptive detection of a signal known only to lie on a line in a known subspace, when primary and secondary data are partially homogeneous," *IEEE Transaction on Signal Processing*, vol. 54, no. 12, pp. 4698–4705, December 2006.

[17] M. Casillo, A. De Maio, S. Iommelli, and L. Landi, "A persymmetric GLRT for adaptive detection in partially-homogeneous environment," *IEEE Signal Processing Letters*, vol. 14, no. 12, pp. 1016–1019, December 2007.

[18] A. De Maio and S. Iommelli, "Coincidence of the Rao test, Wald test, and GLRT in partially homogeneous environment," *IEEE Signal Processing Letters*, vol. 15, no. 4, pp. 385–388, April 2008.

[19] F. Bandiera, O. Besson, and G. Ricci, "Knowledge-aided covariance matrix estimation and adaptive detection in compound-Gaussian noise," *IEEE Transaction on Signal Processing*, vol. 58, no. 10, pp. 5391–5396, October 2010.

[20] F. Bandiera, O. Besson, and G. Ricci, "Adaptive detection of distributed targets in compound-Gaussian noise without secondary data: A Bayesian approach," *IEEE Transaction on Signal Processing*, vol. 59, no. 12, pp. 5698–5708, December 2011.

[21] S. Bidon, O. Besson, and J.-Y. Tourneret, "A Bayesian approach to adaptive detection in non-homogeneous environments," *IEEE Transactions on Signal Processing*, vol. 56, no. 1, pp. 205–217, January 2008.

[22] S. Bidon, O. Besson, and J.-Y. Tourneret, "The adaptive coherence estimator is the generalized likelihood ratio test for a class of heterogeneous environments," *IEEE Signal Processing Letters*, vol. 15, no. 2, pp. 281–284, February 2008.

[23] O. Besson, S. Bidon, and J.-Y. Tourneret, "Covariance matrix estimation with heterogeneous samples," *IEEE Transactions Signal Processing*, vol. 56, no. 3, pp. 909–920, March 2008.

[24] O. Besson, S. Bidon, and J.-Y. Tourneret, "Bounds for estimation of covariance matrices from heterogeneous samples," *IEEE Transactions on Signal Processing*, vol. 56, no. 7, pp. 3357–3362, July 2008.

[25] P. Wang, H. Li, and B. Himed, "Knowledge-aided parametric tests for multichannel adaptive signal detection," *IEEE Transactions on Signal Processing*, vol. 59, no. 12, pp. 5970–5982, December 2011.

[26] P. Wang, H. Li, and B. Himed, "A Bayesian parametric test for multichannel adaptive signal detection in non-homogeneous environments," *IEEE Signal Processing Letters*, vol. 17, no. 4, pp. 351–354, April 2010.

[27] J. R. Román, M. Rangaswamy, D. W. Davis, Q. Zhang, B. Himed, and J. H. Michels, "Parametric adaptive matched filter for airborne radar applications," *IEEE Transactions on Aerospace and Electronic Systems*, vol. 36, no. 2, pp. 677–692, April 2000.

[28] K. J. Sohn, H. Li, and B. Himed, "Parametric GLRT for multichannel adaptive signal detection," *IEEE Transactions on Signal Processing*, vol. 55, no. 11, pp. 5351–5360, November 2007.

CHAPTER 6

Adaptive Radar Detection for Sample-Starved Gaussian Training Conditions*

Yuri I. Abramovich[1] *and Ben A. Johnson*[1]

6.1 Introduction

The problem of radar target detection in the background interference (plus noise) environment is the central problem in statistical radar theory. As a result, there are a number of well-established optimal (in the Neyman–Pearson sense) solutions for Gaussian signals and interference models with known interference covariance matrices, as presented in the first two chapters of this book.

The problem of adaptive detection arises when the covariance matrix of the background interference (plus noise) signal environment is not known *a priori* and instead may be represented by a certain amount of observed (secondary training) background signal data. In this case, the key issue in adaptive detector design is to specify the best way of using this secondary training data, along with the primary data that is tested for presence of a target. This design problem is twofold:

- Estimate the unknown interference (plus noise) covariance matrix.
- Use this estimate for decision making regarding the presence of a target.

Despite the key importance of this very generic problem, and the significant attention it has attracted in the last four decades, a solution that can be proved optimal in the Neyman–Pearson sense over the wide range of possible environments has not been derived.

[1]Institute for Telecommunications Research, University of South Australia, Mawson Lakes, SA, Australia

*This chapter is partially based on "Modified GLRT and AMF Framework for Adaptive Detectors" by Y. I. Abramovich, N. K. Spencer, and A. Y. Gorokhov [1], "Sample-Deficient Adaptive Detection: Adaptive Scalar Thresholding versus CFAR Detector Performance" by Y. I. Abramovich, B.A. Johnson, and N. K. Spencer [2], as well as "Band-Inverse TVAR Covariance Matrix Estimation for Adaptive Detection" by Y. I. Abramovich, N.K. Spencer, and B. A. Johnson [3], all of which appeared in IEEE Transactions on Aerospace and Electronic Systems

For this reason, the development of adaptive detectors has progressed in two major directions. The first direction is where the detector processing operations remain the same as they would have been for an optimal Neyman–Pearson detector for a known covariance matrix, but using an estimate for the unknown covariance matrix derived in some manner from the secondary training data. In most cases, the maximum likelihood (ML) criterion is used for that estimation. But, despite the well-known asymptotic efficiency of these ML estimates, it is still impossible to prove optimality (in a Neyman–Pearson sense) for adaptive detectors that use such ML covariance matrix estimates. In fact, it has been demonstrated that other estimates, including those that broadly belong to a class of estimates referred to as shrinkage estimates, provide better detection performance than ML-based ones.

Therefore, a second approach has been actively pursued, established by the pioneering Kelly paper [4], where the Neyman–Pearson criterion is replaced by a different one that is sufficiently close. Specifically, the detection problem is formulated as a binary test regarding the total dataset consisting of both secondary training samples and a single primary sample, allowing for the covariance matrix of interference to be different under the two competing hypotheses. Under some additional assumptions and restrictions, this approach allows for derivation of an optimal detector using the totality of the secondary and primary data, without assigning different functions to those two sets of data. This means that the estimate of the covariance matrix is defined by the detection test itself, and not by some "separate" (say ML) estimation criterion. In most cases (including the work by Kelly), a generalized likelihood ratio test (GLRT) supplants the adaptive Neyman–Pearson criterion solution with unknown parameters replaced by their ML estimates drawn from the secondary training data. Such detectors, as demonstrated in Chapter 2, have both good detection performance and a number of other attractive properties, discussed later in this chapter.

Yet direct comparison of the Kelly GLRT detector with one derived using the ML covariance matrix estimate (i.e. the adaptive matched filter (AMF) of Robey, Fuhrmann, Kelly, and Nitzberg (RFKN) in Reference 5) shows that neither can be considered strictly optimal in the Neyman–Pearson sense. Therefore, despite the theoretical attractiveness of the Kelly GLRT detector, the problem of the "best" adaptive detector is still under intensive exploration. In this chapter, some results of relatively recent developments (from 2007 to 2010) are introduced that specifically address the problem of adaptive detection in Gaussian signal and interference environments for which the secondary data volume is not large (i.e. the sample-starved condition). The term sample-starved is meant to generically cover conditions where the number N of independent identically distributed (i.i.d.) training samples for M-variate data (an M-element antenna array, for example) are comparable with the dimension M of the data. When $N < kM$ and k is a small multiple (say less than 2–4), one can treat the training conditions as sample-starved, mainly because in this pre-asymptotic domain, one cannot rely upon the asymptotic properties of the *ML* covariance matrix estimator. Nor, for the same reasons, is optimality in the Neyman–Pearson sense for GLRT detectors considered in these sample-starved conditions, since that also relies upon asymptotic ML principles.

Correspondingly, in Section 6.2, we discuss the distinctions between adaptive detection and adaptive beamforming and then specify assumptions and statistical properties for these two quite different (but often confused) applications of adaptive signal processing. The section details the advantages and deficiencies of the existing adaptive detection techniques, and in particular discusses the role of the GLRT/ML criterion and the role of the constant false-alarm rate (CFAR) property. Based on the well-established fact that in many practical applications of adaptive filtering, diagonally loaded covariance matrix estimates result in much better performance, we propose a modified framework for GLRT and AMF adaptive detectors. In essence, we show that regularized (loaded) covariance matrix estimates adopted within these techniques, rather than being driven to zero as suggested by ML considerations, result in a likelihood ratio which is statistically indistinguishable from the likelihood ratio produced by the true (albeit unknown) covariance matrix. This particular way of regularization (shrinkage) we call "expected likelihood" (EL), and in Section 6.2, it is shown to have superior detection performance for the GLRT and AMF variants derived over the standard algorithms for a class of important interference (plus noise) scenarios, yet retain an invariance property which is practically CFAR. Specifically, while the strict CFAR property for an arbitrary interference (plus noise) scenario cannot be guaranteed, actual deviation from the nominal false-alarm rate is small and quite acceptable. Furthermore, the EL technique results in a specific analytic level of suggested diagonal loading, rather than be set in an ad hoc manner.

The diagonal loading considered in Section 6.2, while widely used, is just one possible regularization (shrinkage) of the covariance matrix estimate. It is appropriate for scenarios with signal-subspace eigenvalues that are well separated from the minimal eigenvalues in the noise subspace. A typical example of such a scenario is a limited number of strong point-source interferers impinging upon the antenna array from the far-field. But in many radar applications that deal with target detection in clutter, the interference eigenvalues gradually decrease, rather than exhibiting a "cliff" structure. Even if the dynamic range between the maximum and minimum eigenvalues is quite high, the number of "significant" eigenvalues can be quite large. This makes the number of training samples needed for covariance matrix estimation also quite high, even with diagonal loading. An alternative approach in such circumstances is to find a parametric model with a small number of parameters that can describe the covariance matrix with a sufficient fidelity. Such regularization is often referred to as "shrinkage to a structure," as discussed in Chapter 3 of this book and Reference 6. In Section 6.3, we introduce an adaptive beamforming and adaptive detection methodology that exploits a time-varying autoregressive model with order m (TVAR(m)) that requires only a $2m + 1$-wide central band in the inverse of the estimated covariance matrix that is non-zero. While the autoregressive model (AR(m)) also belongs to this class, it is considerably more restrictive. Indeed, in an AR(m) model with $m = M - 1$, we get an arbitrary positive definite Toeplitz covariance matrix, while in the TVAR(m) case, we get an arbitrary p.d. Hermitian matrix, which allows for modeling of a wider range of interference scenarios and array structures. This model is specified by a number of $(m + 1)$-variate (sample) covariance matrices, and therefore the minimal number of K-variate i.i.d. training samples required for this approach is equal to $(m + 1)$, the same number as requires for m point-source interferers in noise, but with much different eigenspectra. In Section 6.3, we demonstrate that once again, the EL approach can be used to select the order m of the TVAR(m) model to make the covariance matrix "as likely" as the true (actual) covariance matrix, and that this model order leads to enhanced performance in a number of important "clutter-limited" scenarios. These two regularization schemes (loading and TVAR(m) modeling) are not exhaustive by any means – available *a priori* information about the covariance matrix structure can also be exploited to achieve better estimation accuracy and ultimately better detection performance.

For both of these regularization approaches, proper optimization using EL leads to detectors that are "practically" CFAR under certain conditions, though the very elegant property of the GLRT and AMF detectors that adopt the unrestricted ML (sample) covariance matrix estimate disappears. This CFAR property, whereby the output signal under hypothesis H_0 (no target) has a probability density function (pdf) which does not depend on the actual covariance matrix and therefore can be precalculated, is so attractive that it has become a defacto prerequisite for design of adaptive detectors in many studies. But, as discussed later in this chapter, this approach is not practical in many circumstances. For example, when the adaptive beamformer training is performed using data that contain a co-channel interference and adaptive detection is performed downstream after coherent integration and clutter mitigation, the training data used for the adaptive beamformer have completely different statistical properties than the clutter and noise residuals at the output of the coherent processing. Moreover, in realistic environments, the statistical description of the actual radar signal within the tails of the pdf that is responsible for false alarms is rarely accurately described by theoretical pdfs. So while CFAR properties are theoretically attractive, detectors possessing CFAR frameworks are rarely sufficiently reliable for practical applications without some augmentation via adaptive threshold determination (see Chapters 3 and 4).

This circumstance generates an interesting question. With a finite number of i.i.d. training samples N, is better to use all N samples in a strictly CFAR GLRT or AMF detector, or as an alternative, use N_1 training samples for highly efficient regularized or structured covariance matrix estimates and the remaining N_2 training samples for an estimate of the power at the output of the adaptive system (beamformer) for false-alarm threshold control? It is the question that is captured in the final section of this chapter and is a natural extension of the work presented in Sections 6.2 and 6.3. We demonstrate that in practically important scenarios, the "two-stage" adaptive detection approach can outperform the "one-stage" strictly CFAR GLRT or AMF detector.

Overall, this chapter presents an alternative adaptive detector framework that heavily relies on more efficient covariance matrix estimates in the sample-deficient environment, using the EL principle to help select those regularization parameters. The detectors are then combined within a two-stage approach to allow the use of highly effective but non-CFAR designs, and under these sample-starved conditions, the resultant detectors exhibit superior behavior to the classical detectors designed using asymptotically efficient ML-based techniques.

6.2 Improving Adaptive Detection Using EL-Selected Loading

Techniques for adaptive signal processing for radar target detection in an unknown interference environment stem from the pioneering work of Reed, Mallet, and Brennan (RMB) [7], followed by Kelly's seminal paper [4], and nowadays embrace various scenarios (as already discussed in Chapters 2–4) with both Gaussian and non-Gaussian (spherically invariant) interference (see Chapters 7–9 of this book, as well as References 8–10). Depending on the practical application, two rather different formulations are considered for the adaptive detection problem.

6.2.1 Single Adaptive Filter Formed with Secondary Data, Followed by Adaptive Thresholding Using Primary Data

The first formulation, addressed in RMB [7], is considered when the secondary (training) data are used to design the adaptive filter that is then used to process the entire set of primary data (range cells, say). For example, in adaptive antenna external-noise suppression applications, a limited number of range cells (or even an "inter-dwell gap" that is free of clutter and targets) is typically used to estimate the external-noise covariance matrix and design the adaptive antenna. Since the external noise is supposed to be homogeneous over the set of range cells, this adaptive antenna weight vector is then used to process all operational range cells that usually contain clutter and possible targets. Final target detection is performed downstream after clutter suppression (moving-target indicator (MTI) or Doppler filtering) and non-coherent integration. Adaptive threshold calculation is done at the output of this signal processing chain using primary range cells. This (adaptive) detection threshold is therefore calculated for this specific (adaptive) antenna weight vector; for Gaussian primary data, the antenna output is also Gaussian with, possibly, unknown output noise power. For this (conditional) Gaussian model, a typical way to calculate adaptive thresholds was introduced by Finn and Johnson [11]. In such schemes, detection performance degradations with respect to the clairvoyant (exact) detector are due to two statistically independent factors. One factor is associated with the adaptive threshold calculations (using primary data), while the other is associated with the signal-to-noise ratio (SNR) degradation in the adaptive antenna (using the secondary data). Therefore, for such applications, analysis of the SNR pdf at the output of the adaptive antenna gives a *complete* description of the adaptive antenna (filter) design performance. Specifically, receiver operating characteristics (ROCs), derived for example in Finn and Johnson's paper [11] for a particular adaptive threshold design and a (fluctuating) target with a certain SNR, must now be additionally averaged over the SNR pdf for the given adaptive antenna algorithm. In RMB [7], this pdf for the normalized antenna output SNR was accurately calculated for the case when the unconditional ML covariance matrix estimate (sample covariance matrix) is used for adaptive antenna filter design, drawn from the training (secondary) data. This famous β-distribution has an extremely important invariance property with respect to the observed scenario, namely, it is fully specified by only two parameters: the training sample size (N) and the adaptive antenna (filter) dimension (M). The sample-support requirement

$$N \simeq 2M \tag{6.1}$$

that ensures approximately 3-dB average SNR losses compared with the clairvoyant solution has become the most quoted requirement in studies on adaptive antenna filters.

Subsequently, in the early 1980s, considerable research was focused on a specific class of interferences whose covariance matrix has a distinct difference between the magnitude of the m ($<M$) signal-subspace eigenvalues and the $n \equiv M - m$ noise-subspace eigenvalues. The prototypical model that results in this covariance matrix structure is a mixture of m powerful external-noise point sources with (internal) white noise. The minimum eigenvalue here is equal to the noise power, while the sum of the signal eigenvalues is almost equal to the overall power of the external interferences.

Interference-to-noise ratio (INR) values of 20–40 dB are not uncommon for practical adaptive antenna applications. In fact, adaptive filter (antenna) design is efficient only for scenarios with a significant ratio of maximum to minimum eigenvalue (λ_1/λ_M). Indeed, if $\boldsymbol{s}$ is the M-variate normalized useful signal (target) array-signal (steering) vector, and $\boldsymbol{R}_0$ is the M-variate interference covariance matrix, then the clairvoyant optimum filter

$$\boldsymbol{w}_{opt} \equiv \boldsymbol{R}_0^{-1}\boldsymbol{s}, \qquad \boldsymbol{s}^{\dagger}\boldsymbol{s} = 1 \tag{6.2}$$

has an SNR improvement over the "white-noise optimum" non-adaptive beamformer $\boldsymbol{w}_{wn} \equiv \boldsymbol{s}$ of

$$\eta \equiv (\boldsymbol{s}^{\dagger}\boldsymbol{R}_0^{-1}\boldsymbol{s})(\boldsymbol{s}^{\dagger}\boldsymbol{R}_0\boldsymbol{s}). \tag{6.3}$$

According to the so-called Kantorovich inequality [12]

$$(s^{\dagger}R_0^{-1}s)(s^{\dagger}R_0 s) \leq \frac{(\lambda_1 + \lambda_M)^2}{4\lambda_1\lambda_M} \qquad \text{if } s^{\dagger}s = 1. \tag{6.4}$$

Even an improvement of, say, $\eta = 10$ means that $\lambda_1/\lambda_M \gtrsim 40$ for the interference covariance matrix R_0.

Of course, the clairvoyant covariance matrix R_0 is unknown and must be estimated for practical adaptive detectors. The ML estimate given secondary training data is the sample covariance matrix

$$\hat{R} = \frac{1}{N} X_N X_N^{\dagger} \tag{6.5}$$

where $X_N \equiv [x_1, \ldots, x_N] \in \mathcal{C}^{M \times N} \sim \mathcal{CN}_N(0, R_0)$ is the N-sample secondary (training) data with i.i.d. samples x_j $(j = 1, \ldots, N)$ described by the M-variate complex (circular) Gaussian distribution with covariance matrix R_0. As per the formulation used in RMB, this sample covariance matrix can be directly substituted into (6.2.1) to form a practical adaptive filter, with the performance detailed in [7].

For the subclass of strong point-source interferences, whose eigenvalues (sorted in descending order) are such that

$$\lambda_1 > \cdots > \lambda_m \gg \lambda_{m+1} = \cdots \lambda_M, \tag{6.6}$$

it is known that augmenting the sample covariance matrix with additive white (diagonal) noise which exceeds the minimum sample eigenvalue (a.k.a. diagonal loading) prior to substitution into (6.2) leads to a quite dramatic SNR improvement over the unloaded ML covariance matrix estimate and subsequent inversion (sample matrix inversion or SMI) considered by RMB [7]. This improvement was described in References 13 and 14, where it was shown that for these scenarios the normalized output SNR does not depend on the loading factor, provided it is chosen to be within a certain range. The SNR loss factor, that in RMB was a function of M and N only, was then specified by m and N, irrespective of M.

Specifically, the RMB requirement (6.1) for the diagonally loaded SMI (LSMI) algorithm was supplanted by the condition

$$N \simeq 2m \tag{6.7}$$

which means a significant performance improvement for scenarios with $m \ll M$. Since the early 1980s, the properties of the LSMI algorithm have been widely explored and validated in practice for various radar and sonar applications [15, 16]. It is known that diagonal loading offers many other important features that make adaptive antennas robust against numerous inaccuracies in the scenario model [17–19].

6.2.2 Different Adaptive Process per Test Cell with Combined Adaptive Filtering and Detection Using Secondary Data

As outlined in Section 6.1, the superior SNR performance of LSMI over the ML-based RMB SMI technique raises concerns about the optimality of the ML criterion for covariance matrix estimation in adaptive antenna/filter design. Nevertheless, the same ML principle was exploited by Kelly [4], and subsequently by RFKN [5], when the significantly different problem of *adaptive detection* was addressed. This is the second formulation we consider for the adaptive detection problem. This problem is formulated as a hypothesis test, where a decision regarding the presence or absence of a target in a single primary datum (one range cell, say) is made based only on this datum and the auxiliary training (secondary) data. In essence, the training data here must be used to provide *all* missing information required for decision making, including adaptive antenna/filter design and adaptive thresholding. Since the same training data are used for these two purposes, adaptive antenna SNR losses (as in RMB, for example) and adaptive threshold losses (as in Finn) cannot be treated as independent random values, and specific analysis of adaptive detection ROCs is required. This was performed by Kelly [4] for the GLRT detector and by RFKN [5] for the "CFAR AMF" detector. Indeed, when the pdf of the scenario-free output statistics is known, detection can be completed without adaptive threshold calculations (over a set of homogeneous range cells).

Both Kelly's GLRT detector and the CFAR AMF detector [5] that use the ML covariance matrix estimate enjoy this important invariance property. Due to its importance, the invariance (for at least CFAR) property became a frequent prerequisite for adaptive detector design in the modern adaptive detector developments such as those of Scharf *et al.* [20–22]. Nevertheless, for some models invariant solutions may not exist (e.g. adaptive detection for different INRs in primary and secondary data [23]), or come at an excessive performance cost. It is important to emphasize that when (as is very often the case) this invariance cannot be assured, single CFAR primary data detection based on secondary data is still technically impossible. However in practice, a number of homogeneous ranges cells (even processed by different but statistically identical filters) can often be specified and again used for adaptive threshold calculations. Naturally, the additional losses of this method must be taken into consideration.

6.2.2.1 Differences between AMF and GLRT detectors

Various modifications of the adaptive detection problem have been considered since Kelly's paper [8–10, 24–26]. However, there is frequently confusion of these two quite different problems: adaptive filtering (the same filter for all primary data) and adaptive detection (a different filter for each primary datum). In Reference 27, Lekhovytskiy demonstrates that the application of non-normalized adaptive filters (in fact, randomly normalized) $W_j = \hat{\boldsymbol{R}}_j^{-1} s$ (6.5) with independent $\hat{\boldsymbol{R}}_j$ for each cell leads to a dramatic degradation in detection performance. Indeed, a single threshold for the various $\hat{\boldsymbol{R}}_j$ has to account for significant output power fluctuations that are due to variations in the norm of the adaptive filter. Clearly, this problem does not exist for single adaptive antenna applications. Confusion also occurs when the famous RMB 3-dB SNR average loss factor for $N \simeq 2M$ with the ML covariance matrix estimate is compared with, say, the 6-dB SNR degradation in ROC of the CFAR AMF detector that uses the same estimate [5].

In this section, the adaptive detection problem is considered in Kelly's context where each range cell is processed by an individually tailored filter, and the (unconditional) output statistics are averaged over a (random) set of adaptive solutions. The goal is two-fold. First, based on the well-known superior properties of the LSMI algorithm for *adaptive filter* performance, we want to investigate the performance of LSMI or loaded AMF (LAMF) *adaptive detectors*. We hope and expect that the superiority of LSMI *adaptive filters* will lead to a superiority of LAMF *adaptive detectors* for the class of interference scenarios (6.6), an expectation that turns out to be true. The other properties of LAMF detectors (such as CFAR) also need to be specified. Second, this superiority of LAMF over ML-based AMF means that the theoretical framework for adaptive detection (GLRT and AMF) needs to be modified so as to include LAMF detectors. Specifically, we expect that a modified framework would either directly lead to LAMF detectors with some theoretically specified loading factor or produce different adaptive detectors whose performance is at least as good as the LAMF performance for the class of interference scenarios (6.6).

In accordance with most works on adaptive detection [4,7], it is assumed that the training data contains only interference that is statistically mostly or completely the same as the interference within the primary data. In what follows, we consider the cases where the interferences are exactly the same (homogeneous), or just have a different power (non-homogeneous). This model comprises "supervised training conditions" since the secondary data do not contain targets nor other interference sources. The selection of such a homogeneous training dataset (via inhomogeneity detection) is an important research topic in adaptive radar studies [28–32]. However, here we reconsider the traditional GLRT and AMF solutions under their original assumptions. Recall that the AMF detector in RFKN [5] is derived as a GLRT where the covariance matrix is known. After the test statistics are derived, the ML estimate of the covariance matrix based on the secondary data is substituted for the known covariance matrix. Kelly's GLRT method is to treat the primary and secondary data as a single dataset, then for both of the hypotheses regarding the presence or absence of a target signal, ML estimates of *all* parameters (including the interference covariance matrix) are used to construct the decision rule. Therefore, both the AMF and the Kelly GLRT techniques heavily rely on the same ML principle for interference parameter estimation.

In this regard, it is quite instructive to recall the discussion in RFKN [5] comparing the performance of AMF and GLRT detectors. The AMF detection rule was considered to be inferior to GLRT simply because "the AMF test makes no use of the primary vector to estimate the covariance, therefore poorer detection performance might be expected" compared with the GLRT that "uses all the data (primary and secondary) in the likelihood optimization under each hypothesis." This argument could have been easily justified by comparing the detection performance of Kelly's GLRT technique with N secondary samples against the AMF performance with just $(N+1)$ secondary samples. If those expectations were correct, then the $(N+1)$-variate AMF should *always* outperform the N-variate Kelly GLRT. But that is not the case. Specifically, Kelly's GLRT is effectively based on two ML interference covariance matrix estimates constructed for the hypotheses H_0 and H_1 [4]. Thus for H_0,

$$\text{const}\,\hat{\boldsymbol{R}}|H_0 = \boldsymbol{X}_N\boldsymbol{X}_N^{\dagger} + \boldsymbol{Y}\boldsymbol{Y}^{\dagger} \tag{6.8}$$

where $\boldsymbol{X}_N \equiv [\boldsymbol{x}_1, \ldots, \boldsymbol{x}_N] \in \mathcal{C}^{M\times N} \sim \mathcal{CN}_N(\boldsymbol{0}, \boldsymbol{R}_0)$ are the N-sample secondary (training) data with i.i.d. samples $\boldsymbol{x}_j$ $(j = 1, \ldots, N)$ described by the M-variate complex (circular) Gaussian distribution with covariance matrix $\boldsymbol{R}_0$. For H_1,

$$\text{const}\,\hat{\boldsymbol{R}}|H_1 = \boldsymbol{X}_N\boldsymbol{X}_N^{\dagger} + \left[\boldsymbol{I}_M - \frac{\boldsymbol{s}\boldsymbol{s}^{\dagger}(\boldsymbol{X}_N\boldsymbol{X}_N^{\dagger})^{-1}}{\boldsymbol{s}^{\dagger}(\boldsymbol{X}_N\boldsymbol{X}_N^{\dagger})^{-1}\boldsymbol{s}}\right]\boldsymbol{Y}\boldsymbol{Y}^{\dagger}\left[\boldsymbol{I}_M - \frac{(\boldsymbol{X}_N\boldsymbol{X}_N^{\dagger})^{-1}\boldsymbol{s}\boldsymbol{s}^{\dagger}}{\boldsymbol{s}^{\dagger}(\boldsymbol{X}_N\boldsymbol{X}_N^{\dagger})^{-1}\boldsymbol{s}}\right]. \tag{6.9}$$

The primary sample is

$$\boldsymbol{Y} = \begin{cases} \boldsymbol{X}_0 \sim \mathcal{CN}(\boldsymbol{0}, \boldsymbol{R}_0) & \text{for } H_0 \\ \boldsymbol{X}_0 + a\boldsymbol{s} & \text{for } H_1 \end{cases} \tag{6.10}$$

where $s \in \mathcal{C}^{M\times 1}$ is the target wavefront vector, and a is the unknown (complex) target amplitude. One can see that for H_1, the primary sample $\boldsymbol{Y}$ contribution to the covariance matrix estimate in (6.9) does not retain any interference component that corresponds to the target signal wavefront. Instead, this component is rejected by the projection matrix

$$\left[\boldsymbol{I}_M - \frac{\boldsymbol{s}\boldsymbol{s}^\dagger(\boldsymbol{X}_N\boldsymbol{X}_N^\dagger)^{-1}}{\boldsymbol{s}^\dagger(\boldsymbol{X}_N\boldsymbol{X}_N^\dagger)^{-1}\boldsymbol{s}}\right] \tag{6.11}$$

along with the possible target signal. Since only this component is essential for target detection, the primary sample-produced covariance matrix update in (6.9) with this component canceled out is non-sensical. As a result, in some situations GLRT was found to be superior to AMF, and inferior to others. In RFKN [5], the authors rightly conceded that "the generalized likelihood-ratio test is not optimal in the Neyman–Pearson sense as the AMF test has a probability of detection that is higher than that of the GLRT for some situations." For the same reason, AMF is not optimal either. As we will show, with LAMF and its *non-ML* covariance matrix estimate also outperforming AMF and GLRT in some circumstances (such as (6.6)), it becomes clear that the ML estimation principle is not a guarantee of optimality and needs to be re-examined.

Diagonal loading (as well as the fast ML (FML) technique [32]) is quite different from the numerous methods that assume a restricted class of admissible covariance matrices, but still use the ML criterion. Of course, if properly adopted, any valid *a priori* information on the interference properties that somehow restricts the (ML) search should lead to improved adaptive detection performance, though the CFAR property may not be so easy to maintain. As discussed in Section 6.1, a typical example of "shrinkage to a structure" to exploit *a priori* is restricting to the class of Toeplitz covariance matrices for uniform antenna arrays (or pulse trains). In fact, it is straight-forward to demonstrate that the ML optimization restricted to the set of diagonally loaded or finite-subspace (FML) covariance matrices results in zero loading and maximal signal subspace, which drives the optimum solution to the same unconstrained (ML) sample covariance matrix estimate with ML. Therefore, loaded and FML techniques are inherently antithetical to the ML approach.

This leads us to the important question as to whether the ML principle within the GLRT and AMF framework can be replaced by another general principle that will generate adaptive detectors such as LAMF or that are at least not inferior to (say) LAMF for some scenarios. While these considerations stem mostly from the known discrepancy between SMI and LSMI performance, there are also important theoretical considerations that raise concerns regarding the ML criterion for small sample sizes (which are typical in radar applications). This leads us to a framework we refer to as "EL" [33,34]

6.2.2.2 EL Framework

Let us consider the traditional likelihood function (LF) for the covariance matrix given $N > M$ i.i.d. training samples $\boldsymbol{X}_N$ as in (6.8) [35]

$$f(\boldsymbol{R}|\boldsymbol{X}_N) = \frac{\text{const}}{\det(\boldsymbol{R})^N}\exp\left[-\text{Tr}\left(\boldsymbol{R}^{-1}\boldsymbol{X}_N\boldsymbol{X}_N^\dagger\right)\right], \tag{6.12}$$

then with probability one:

$$\det(\boldsymbol{X}_N\boldsymbol{X}_N^\dagger) \neq 0 \tag{6.13}$$

and so the likelihood ratio [36]

$$LR(\boldsymbol{R}, \boldsymbol{X}_N) \equiv \left[f(\boldsymbol{R}|\boldsymbol{X}_N)\det(\boldsymbol{X}_N\boldsymbol{X}_N^\dagger)\right]^{\frac{1}{N}} = \frac{\det(\boldsymbol{R}^{-1}\hat{\boldsymbol{R}})\exp{(M)}}{\exp\left[\mathrm{Tr}(\boldsymbol{R}^{-1}\hat{\boldsymbol{R}}_N)\right]} \tag{6.14}$$

where $\hat{\boldsymbol{R}}$ (the sample covariance matrix (6.5)) can also be treated as the likelihood function with respect to $\boldsymbol{R}$ (e.g. [35]). One can see that the (unconstrained) ML solution $\boldsymbol{R}_{ML} = \hat{\boldsymbol{R}}$ derived by Anderson [37] via direct maximization of the LF $f(\boldsymbol{R}|\boldsymbol{X}_N)$ yields the ultimate value for the likelihood ratio of unity:

$$\max_{\boldsymbol{R}} LR(\boldsymbol{R}, \boldsymbol{X}_N) = LR(\hat{\boldsymbol{R}}, \boldsymbol{X}_N) = 1 \tag{6.15}$$

irrespective of the sample support N and filter dimension M.

At the same time, for the true (exact, and in general unknown) covariance matrix $\boldsymbol{R}_0$, the pdf of the likelihood ratio

$$LR(\boldsymbol{R}_0, \boldsymbol{X}_N) = \frac{\det\left(\boldsymbol{R}_0^{-\frac{1}{2}}\hat{\boldsymbol{R}}\boldsymbol{R}_0^{-\frac{1}{2}}\right)\exp{(M)}}{\exp\left[\mathrm{Tr}(\boldsymbol{R}_0^{-\frac{1}{2}}\hat{\boldsymbol{R}}\boldsymbol{R}_0^{-\frac{1}{2}})\right]} \tag{6.16}$$

does not depend on $\boldsymbol{R}$, since

$$\hat{\boldsymbol{C}} \equiv \boldsymbol{R}_0^{-\frac{1}{2}}\hat{\boldsymbol{R}}\boldsymbol{R}_0^{-\frac{1}{2}} \sim \mathcal{CW}(N, M, \boldsymbol{I}_M) \tag{6.17}$$

where $\hat{\boldsymbol{C}}$ is the "white-noise" sample matrix with complex Wishart distribution, is (for $N > M$) specified by two parameters only, namely N and M. Later we shall introduce moments and series representations for the pdf of this ratio derived for the complex-valued case similar to the real-valued case in Reference 38.

With the pdf of $LR(\boldsymbol{R}_0, \boldsymbol{X}_N)$ determined, either analytically or via Monte-Carlo simulations, any appropriate quantile can now be chosen for use in an EL framework. If the likelihood of any given covariance matrix estimate exceeds the chosen threshold (which statistically represents the likelihood associated with the true (unknown) parameters), then the covariance matrix estimate is considered acceptable. See an example LR pdf in Figure 6.1, shown for the values $M = 12$ and $N = 24$.

Note that according to RMB [7], the average SNR degradation for SMI is about 3 dB here, which is often considered acceptable in practical applications. Based on the pdf in Figure 6.1, it seems natural to replace the ML estimate by one that generates likelihoods consistent with what is expected for the true covariance matrix. This "EL" estimation approach, unlike the ML criterion, can be shown to inherently justify the appropriate selection of parameters (such as loading factor and interference signal-subspace dimension) based on direct likelihood matching, rather than on "external considerations" as suggested in References 15 and 16. Moreover, we shall demonstrate that this EL principle, when incorporated into the GLRT and AMF framework, leads to detection rules that outperform standard GLRT and AMF in some scenarios and are similar to LAMF with its data-independent loading-factor selection.

The issue of ML estimation within the GLRT approach is not the only one that raises concerns within Kelly's "single data" GLRT method. Indeed, Kelly's approach whereby "the decision rule will be formulated in terms of the totality of input data without the *a priori* assignment of different functions to the primary and secondary input" [4] is not self-evident.

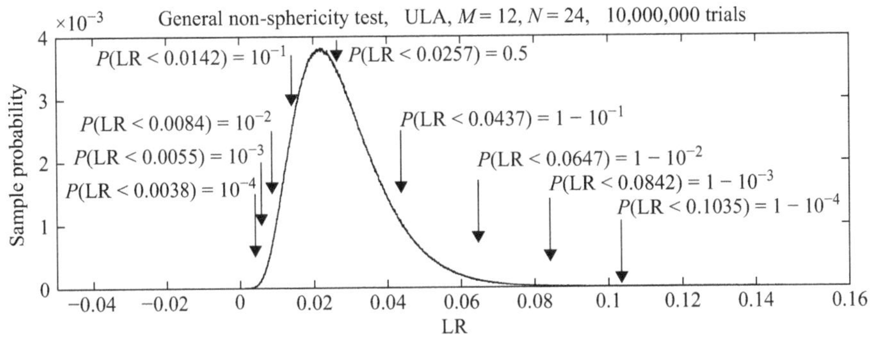

Figure 6.1 Sample pdfs for the general test, based on a 12-element array with 24 training snapshots. Note the huge distinction between the $LR(\boldsymbol{R}_0, \boldsymbol{X}_N)$ values generated by the true covariance matrix $\boldsymbol{R}_0$ and the ultimate likelihood value of unity produced by the ML estimate $\boldsymbol{R}_{ML} = \hat{\boldsymbol{R}}$. Indeed, the median LR is only 0.0257, and with probability 99.99% the LR is less than 0.1051.

It is difficult to accept two different covariance matrix estimates for the same interference, depending on the hypothesis for a single primary snapshot. It is expected that the proper methodology would search for the single interference covariance matrix that is *most supportive* to the detection problem, possibly dependent on the primary snapshot, but not on the hypothesis. This methodology should suggest both the primary and the secondary data processing governed by the detection hypothesis testing criteria on the primary data, with the *a priori* classification of the secondary data as target-free being respected. The latter means that modification of the GLRT methodology should consider two sets instead of a single-set approach using "the totality of the input data," and for the above reasons use EL rather than ML estimation. This new GLRT framework should deliver adaptive detectors that are at least as efficient as LAMF, even for the scenarios that are most favorable to LAMF (6.6). Finally, it is intuitive that if the different adaptive techniques have the same performance, the potential accuracy set by the problem formulation is approached, and the simplest technique can be favored for practical reasons.

According to the conventional single-set GLRT criterion, the decision H_1 that a target is present in a (single) snapshot $\boldsymbol{Y}$ (or absent H_0) is taken according to the rule

$$\Lambda^*(\boldsymbol{Y}) = \frac{\max\limits_{\mu\in\Omega_1} f(\boldsymbol{Y}|\mu_1, H_1)}{\max\limits_{\chi\in\Omega_0} f(\boldsymbol{Y}|\chi_1, H_0)} \underset{H_0}{\overset{H_1}{\gtrless}} h^* \tag{6.18}$$

where $f(\boldsymbol{Y}|\mu_1, H_1)$ are the (primary) data $\boldsymbol{Y}$ pdf for hypothesis H_1 (target signal present), and $\mu \in \Omega_1$ is the set of unknown (non-random) parameters that completely specify this pdf. In most cases, random parameters with an unknown (or not accurately known) marginal pdf are treated as $\mu \in \Omega_1$. Similarly, $f(\boldsymbol{Y}|\chi_1, H_0)$ are the (primary) data pdf for hypotheses H_0 (target signal absent) specified by the parameters $\chi_1 \in \Omega_0$. The threshold h^* is defined as

$$\int_{\Lambda^*(\boldsymbol{Y})>h^*} f(\boldsymbol{Y}|\chi_1, H_0)\, d\boldsymbol{Y} \le P_{FA} \qquad \forall\chi \in \Omega_0 \tag{6.19}$$

where P_{FA} is the desired probability of false alarm.

Note that for a finite-dimensional dataset $\boldsymbol{Y}$, the GLRT criterion *does not* have a rigorous theoretical justification, similar to the Bayesian rule with given marginal pdfs $f_\mu(\mu)$ and $f_\chi(\chi)$, for example. Only asymptotic considerations ($N \to \infty$) are used to justify the GLRT method, despite its obvious intuitive appeal. In practice, this means that estimates *other than the ML ones* for μ and χ in (6.18) could be employed and could result in better detection performance for finite N. This avenue should be explored whenever alternative estimates are available, especially for covariance matrix estimation.

To fit into this single-set GLRT framework, Kelly [4] considered a single total dataset $\{\boldsymbol{X}_N; \boldsymbol{Y}\}$ and introduced $\chi = \boldsymbol{R}$ and $\mu = (\boldsymbol{R}, a)$ with the two different solutions (6.8) and (6.9) for the interference covariance matrix. In our two-set GLRT (2S-GLRT) framework, we introduce

$$f(\boldsymbol{X}_N|\eta, \chi_{12}) \quad \eta \in \Omega_2 \tag{6.20}$$

$$f(\boldsymbol{Y}|\chi_0, \chi_{12}; H_0) \quad \chi_0 \in \Omega_0 \tag{6.21}$$

$$f(\boldsymbol{Y}|\mu, \chi_{12}; H_1) \quad \mu \in \Omega_1 \tag{6.22}$$

where $\chi_{12} \in \Omega_{12}$, and Ω_{12} is the set of common (interference) parameters that describe the pdf for both the primary data $\boldsymbol{Y}$ and the secondary set $\boldsymbol{X}_N$, and η, χ_0, and μ are now parameters specific to each set and hypothesis.

The *a priori* classification of the training (secondary) data $\boldsymbol{X}_N$ and the primary snapshot $\boldsymbol{Y}$ means that the estimates χ_{12} do not depend on the hypothesis H_0 versus H_1, and so is the same as appears in (6.21) and (6.22). Therefore, the following option can be considered:

$$\text{GLRT:} \qquad \Lambda_{12} = \max_{\eta,\chi_{12}} f(\boldsymbol{X}_N|\eta, \chi_{12}) \frac{\max\limits_{\mu\in\Omega_1} f(\boldsymbol{Y}|\mu, \chi_{12}; H_1)}{\max\limits_{\chi_0\in\Omega_0} f(\boldsymbol{Y}|\chi_0, \chi_{12}; H_0)} \underset{H_0}{\overset{H_1}{\gtrless}} h^*. \tag{6.23}$$

Note that this joint optimization over χ_{12} is already different from the standard AMF approach, namely:

$$\text{AMF:} \qquad \Lambda_{12}^{(2)} = \frac{\max\limits_{\mu\in\Omega_1} f(\boldsymbol{Y}|\mu, \chi_{12ML}; H_1)}{\max\limits_{\chi_0\in\Omega_0} f(\boldsymbol{Y}|\chi_0, \chi_{12ML}; H_0)} \underset{H_0}{\overset{H_1}{\gtrless}} h^* \tag{6.24}$$

where

$$\chi_{12ML} = \arg\max_{\eta,\chi_{12}} f(\boldsymbol{X}_N|\eta, \chi_{12}). \tag{6.25}$$

On the other hand, the joint optimization in (6.23) should result in a single solution in χ_{12} for both the hypotheses H_0 and H_1. It is important that this solution depends on the actual primary snapshot $\boldsymbol{Y}$, but not on the hypothesis itself regarding this primary snapshot. Therefore, at least in principle, the GLRT approach (6.23) differs from Kelly's solution and the AMF technique, even though the same ML principle is used for both GLRT (6.23) and AMF (6.24).

However, in most of the cases considered in this section, one can replace the likelihood function (LF) $f(\boldsymbol{X}_N|\eta, \chi_{12})$ by the likelihood ratio (LR)

$$LR(\boldsymbol{X}_N|\chi_{12}) = \max_{\eta} \frac{f(\boldsymbol{X}_N|\eta, \chi_{12})}{f_0(\boldsymbol{X}_N)} \in (0, 1] \tag{6.26}$$

that has the same ML solution for χ_{12ML}:

$$\arg_{\chi_{12}} \max_{\eta, \chi_{12}} f(\boldsymbol{X}_N|\eta, \chi_{12}) = \arg \max_{\chi_{12}} LR(\boldsymbol{X}_N|\chi_{12}). \tag{6.27}$$

The most important property of this LR is that, for the actual (true, exact) $\chi_{12}^{(0)}$, the pdf does not depend on $\chi_{12}^{(0)}$, i.e. scenario-free, depends only on the parameters M and N, and can be precalculated. Hence for a given probability P_0, the upper and lower bounds (α_U and α_L, respectively) can be found such that

$$\int_{\alpha_L}^{1} w[LR(\boldsymbol{X}_N|\chi_{12}^{(0)})]\, dLR = \int_{0}^{\alpha_U} w[LR(\boldsymbol{X}_N|\chi_{12}^{(0)})]\, dLR = P_0 = 1 - \varepsilon, \quad (0 < \varepsilon \ll 1) \tag{6.28}$$

so that with the high probability $(1 - 2\varepsilon)$, the exact parameters $\chi_{12}^{(0)}$ generate the LR within the specified bounds. Here $w[\,\cdot\,]$ is the scenario-free pdf for the likelihood ratio $LR(\boldsymbol{X}_N|\chi_{12}^{(0)})$.

For most cases with relatively small sample size ($N \simeq M$), Figure 6.1 shows that

$$\alpha_U \ll 1 \tag{6.29}$$

and so *the ML solution* χ_{12ML} *is far away from the true set of parameters in terms of the LR metric*. Of course, very small LRs may be generated not only by the true parameters but by a variety of completely erroneous solutions as well. For this reason, we use the "EL" approach that is based on LR matching for a certain parameterization of the estimate $\hat{\chi}_{12}(\beta)$ such that

$$\hat{\chi}_{12}(\beta_0) = \chi_{12ML} \tag{6.30}$$

and the parameterization corresponds to some valid *a priori* assumptions regarding the class of covariance matrices.

We can now propose the following 2S-GLRT techniques:

$$\text{ML-GLRT:} \qquad \Lambda_{12}^{(3)} = \max_{\beta} \frac{\max\limits_{\mu \in \Omega_1} f(\boldsymbol{Y}|\mu, \hat{\chi}_{12}(\beta); H_1)}{\max\limits_{\chi_0 \in \Omega_0} f(\boldsymbol{Y}|\chi_0, \hat{\chi}_{12}(\beta); H_0)} \underset{H_0}{\overset{H_1}{\gtrless}} h^* \tag{6.31}$$

for β such that

$$\alpha_L \leq LR(\boldsymbol{X}_N|\beta) \leq 1 \tag{6.32}$$

and

$$\text{EL-GLRT:} \qquad \Lambda_{12}^{(4)} = \max_{\beta} \frac{\max\limits_{\mu \in \Omega_1} f(\boldsymbol{Y}|\mu, \hat{\chi}_{12}(\beta); H_1)}{\max\limits_{\chi_0 \in \Omega_0} f(\boldsymbol{Y}|\chi_0, \hat{\chi}_{12}(\beta); H_0)} \underset{H_0}{\overset{H_1}{\gtrless}} h^* \tag{6.33}$$

for β such that

$$\alpha_L \le LR(\boldsymbol{X}_N|\beta) \le \alpha_U. \tag{6.34}$$

The only difference between these two methods is that ML-GLRT allows the likelihood ratio $LR(\boldsymbol{X}_N|\beta)$ generated by the parameterized estimate $\hat{\chi}_{12}(\beta)$ to arbitrarily approach the upper bound of unity, whereas EL-GLRT restricts the LR to the range of values where the exact parameters are concentrated, according to (6.28).

Similarly, we can also introduce the EL-AMF detector, where we replace ML estimates by EL ones:

$$\text{EL-AMF:} \qquad \Lambda_{12}^{(5)} = \frac{\max\limits_{\mu\in\Omega_1} f(\boldsymbol{Y}|\mu, \hat{\chi}_{12EL} : H_1)}{\max\limits_{\chi_0\in\Omega_0} f(\boldsymbol{Y}|\chi_0, \hat{\chi}_{12EL} : H_0)} \underset{H_0}{\overset{H_1}{\gtrless}} h^*. \tag{6.35}$$

The EL estimate is the estimate that, for given training data $\boldsymbol{X}_N$, generates the specific precalculated likelihood ratio LR_0, i.e.

$$LR[\boldsymbol{X}_N|\hat{\chi}_{12}(\beta_{EL})] = LR_0. \tag{6.36}$$

This value LR_0 is chosen by referring to the scenario-free pdf $LR(\boldsymbol{X}_N|\chi_{12}^{(0)})$. For example, the mean or median value of the pdf could be selected, or a specific quantile, depending on the filter designer's goals.

In the next subsection, the two-set (2S) ML-GLRT, EL-GLRT, and EL-AMF techniques are derived for the typical Gaussian model of interference and target signal. Later the detection performance of these techniques is compared with standard AMF and the clairvoyant case with known interference parameters.

6.2.2.3 2S GLRT and AMF Detectors for Gaussian Models

In most GLRT studies, the target is modeled by a vector of given structure (wavefront) with an unknown complex scaling factor that is an additional unknown deterministic parameter (see (6.10)). In this subsection, the typical target model is adopted, namely the Swerling 1 model [39], which is the Gaussian model (Rayleigh target) with uniform initial phase and Rayleigh-distributed envelope.

6.2.2.3.1 Homogeneous Interference Training Conditions; Fluctuating Target with Known Power

In this case, the only information that is assumed unavailable is the interference covariance matrix, which is identical for both the training and the primary data:

$$\Omega_0 = \emptyset, \quad \Omega_1 = \emptyset, \quad \Omega_2 = \emptyset, \quad \Omega_{12} = \{\boldsymbol{R}\} \tag{6.37}$$

and

$$f(\boldsymbol{X}_N) = \frac{1}{\pi^N \det(\boldsymbol{R})^N} \exp\left[-\text{Tr}(\boldsymbol{R}^{-1}\boldsymbol{X}_N\boldsymbol{X}_N^\dagger)\right] \tag{6.38}$$

$$f(\boldsymbol{Y}|H_1) = \frac{1}{\pi \det(\boldsymbol{R} + \sigma_s^2 \boldsymbol{s}\boldsymbol{s}^\dagger)} \exp\left[-\boldsymbol{Y}^\dagger(\boldsymbol{R} + \sigma_s^2 \boldsymbol{s}\boldsymbol{s}^\dagger)\boldsymbol{Y}\right] \tag{6.39}$$

where $\boldsymbol{s}$ is the target signal wavefront vector and σ_s^2 is the target power.

Hence, according to (6.31)–(6.34), the 2S-GLRT detection problem is

$$\Lambda_{12} = \max_{\boldsymbol{R}(\beta)} \frac{1}{1+\sigma_s^2 \boldsymbol{s}^\dagger \boldsymbol{R}^{-1}(\beta)\boldsymbol{s}} \exp\left[\frac{\sigma_s^2 |\boldsymbol{Y}^\dagger \boldsymbol{R}^{-1}(\beta)\boldsymbol{s}|^2}{1+\sigma_s^2 \boldsymbol{s}^\dagger \boldsymbol{R}^{-1}(\beta)\boldsymbol{s}}\right] \tag{6.40}$$

subject to

$$\text{ML-GLRT:} \qquad \alpha_L \le \frac{\det(\boldsymbol{R}^{-1}(\beta)\hat{\boldsymbol{R}}_N)\exp(\mathrm{M})}{\exp[\mathrm{Tr}(\boldsymbol{R}^{-1}(\beta)\hat{\boldsymbol{R}}_N)]} \le 1 \tag{6.41}$$

$$\text{EL-GLRT:} \qquad \alpha_L \le \frac{\det(\boldsymbol{R}^{-1}(\beta)\hat{\boldsymbol{R}}_N)\exp(\mathrm{M})}{\exp[\mathrm{Tr}(\boldsymbol{R}^{-1}(\beta)\hat{\boldsymbol{R}}_N)]} \le \alpha_U \tag{6.42}$$

where $\hat{\boldsymbol{R}}_N$ is the sample covariance matrix given by $\frac{1}{N}\boldsymbol{X}_N\boldsymbol{X}_N^\dagger$.

The AMF technique is based on

$$\Lambda_{12} = \frac{1}{1+\sigma_s^2 \boldsymbol{s}^\dagger \hat{\boldsymbol{R}}^{-1}(\beta)\boldsymbol{s}} \exp\left[\frac{\sigma_s^2 |\boldsymbol{Y}^\dagger \hat{\boldsymbol{R}}^{-1}(\beta)\boldsymbol{s}|^2}{1+\sigma_s^2 \boldsymbol{s}^\dagger \hat{\boldsymbol{R}}^{-1}(\beta)\boldsymbol{s}}\right] \tag{6.43}$$

subject to

$$\text{standard ML-AMF:} \qquad \hat{\boldsymbol{R}}(\beta=\beta_0) = \hat{\boldsymbol{R}}_N \tag{6.44}$$

$$\text{EL-AMF:} \qquad \frac{\det(\hat{\boldsymbol{R}}^{-1}(\beta_{EL})\hat{\boldsymbol{R}}_N)\exp(\mathrm{M})}{\exp[\mathrm{Tr}(\hat{\boldsymbol{R}}^{-1}(\beta_{EL})\hat{\boldsymbol{R}}_N)]} = LR_0. \tag{6.45}$$

In (6.41) and (6.42),

$$\gamma_0^{(1)} \equiv LR(\boldsymbol{X}_N|\boldsymbol{R}_0) = \frac{\det(N^{-1}\hat{\boldsymbol{C}})\exp(\mathrm{M})}{\exp[\mathrm{Tr}(N^{-1}\hat{\boldsymbol{C}})]} \tag{6.46}$$

where $\hat{\boldsymbol{C}} \sim \mathcal{CW}(N, M, \boldsymbol{I}_M)$, is described by a scenario-free (complex Wishart) pdf. Indeed, Appendix I of Reference 1 shows that the hth moment is given by

$$\mathbb{E}\left\{\gamma_0^{(1)h}\right\} = \left(\frac{e}{N}\right)^{Mh} \frac{(N+h)^{-M(N+h)}}{N^{-M(N+h)}} \frac{\prod_{j=1}^{M}\Gamma(N+h+1-j)}{\prod_{j=1}^{M}\Gamma(N+1-j)} \tag{6.47}$$

$$= N^{MN}e^{Mh}\frac{1}{(N+h)^{M(N+h)}} \frac{\prod_{j=1}^{M}\Gamma(N+h+1-j)}{\prod_{j=1}^{M}\Gamma(N+1-j)}. \tag{6.48}$$

The pdf for $\gamma_0^{(1)}$, $w(\gamma_0^{(1)})$ can be expressed as an infinite series by applying a Mellin transform, similar to Reference 38 (see Appendix I in Reference 1). Alternatively, the pdf can be determined via Monte-Carlo simulation. Since the pdf for $\gamma_0^{(1)}$, $w(\gamma_0^{(1)})$ depend only on M and N, simulation of *any* scenario with the same M and N (including a noise-only one) will accurately produce the pdf with sufficient trials. Bounds (both upper α_U and lower α_L) can then be determined by integrating over the pdf of $w(\gamma_0^{(1)})\,d\gamma_0^{(1)}$ to the desired level of fidelity with respect to probability of false rejection and/or false acceptance of the H_0 and H_1 hypotheses.

6.2.2.3.2 *Homogeneous Interference Training Conditions; Fluctuating Target with Unknown Power*

In this case

$$\Omega_0 = \emptyset, \quad \Omega_1 = \{\sigma_s^2\}, \quad \Omega_2 = \emptyset, \quad \Omega_{12} = \{\boldsymbol{R}\}. \tag{6.49}$$

According to (6.31), first the ML estimate of the target signal power σ_s^2 needs to be found:

$$\max_{\sigma_s^2} \frac{1}{1+\sigma_s^2 \boldsymbol{s}^\dagger \boldsymbol{R}^{-1}(\beta)\boldsymbol{s}} \exp\left[\frac{\sigma_s^2 |\boldsymbol{Y}^\dagger \boldsymbol{R}^{-1}(\beta)\boldsymbol{s}|^2}{1+\sigma_s^2 \boldsymbol{s}^\dagger \boldsymbol{R}^{-1}(\beta)\boldsymbol{s}}\right]. \tag{6.50}$$

Since $\sigma_s^2 \geq 0$, the solution is

$$\hat{\sigma}_s^2 = \begin{cases} \dfrac{|\boldsymbol{Y}^\dagger \boldsymbol{R}^{-1}(\beta)\boldsymbol{s}|^2 - \boldsymbol{s}^\dagger \boldsymbol{R}^{-1}(\beta)\boldsymbol{s}}{[\boldsymbol{s}^\dagger \boldsymbol{R}^{-1}(\beta)\boldsymbol{s}]^2} & \text{for} \dfrac{|\boldsymbol{Y}^\dagger \boldsymbol{R}^{-1}(\beta)\boldsymbol{s}|^2}{\boldsymbol{s}^\dagger \boldsymbol{R}^{-1}(\beta)\boldsymbol{s}} \geq 1 \\ 0 & \text{for} \dfrac{|\boldsymbol{Y}^\dagger \boldsymbol{R}^{-1}(\beta)\boldsymbol{s}|^2}{\boldsymbol{s}^\dagger \boldsymbol{R}^{-1}(\beta)\boldsymbol{s}} < 1. \end{cases} \tag{6.51}$$

The solution $\hat{\sigma}_s^2 = 0$ clearly means that there is no target signal present in the input data, hence our 2S-GLRT test is

$$\Lambda_{12} = \max_{\boldsymbol{R}(\beta)} \frac{\boldsymbol{s}^\dagger \boldsymbol{R}^{-1}(\beta)\boldsymbol{s}}{|\boldsymbol{Y}^\dagger \boldsymbol{R}^{-1}(\beta)\boldsymbol{s}|^2} \exp\left[\frac{|\boldsymbol{Y}^\dagger \boldsymbol{R}^{-1}(\beta)\boldsymbol{s}|^2}{\boldsymbol{s}^\dagger \boldsymbol{R}^{-1}(\beta)\boldsymbol{s}}\right] H\left(\frac{|\boldsymbol{Y}^\dagger \boldsymbol{R}^{-1}(\beta)\boldsymbol{s}|^2}{\boldsymbol{s}^\dagger \boldsymbol{R}^{-1}(\beta)\boldsymbol{s}} - 1\right) \underset{H_0}{\overset{H_1}{\gtrless}} h^* \tag{6.52}$$

where $H(x)$ is the unit step function

$$H(x) = \begin{cases} 1 & \text{for} x \geq 0 \\ 0 & \text{for} x < 0. \end{cases} \tag{6.53}$$

Observe that the function

$$f(x) = e^x/x \tag{6.54}$$

is monotonic for $x \geq 1$, and so this decision rule can be replaced by the more familiar one

$$\Lambda_{12} = \max_{\boldsymbol{R}(\beta)} \frac{|\boldsymbol{Y}^\dagger \boldsymbol{R}^{-1}(\beta)\boldsymbol{s}|^2}{\boldsymbol{s}^\dagger \boldsymbol{R}^{-1}(\beta)\boldsymbol{s}} \underset{H_0}{\overset{H_1}{\gtrless}} h^* > 1 \tag{6.55}$$

together with the same constraints on β as in (6.41) for ML-GLRT and in (6.42) for EL-GLRT. This maximization can be interpreted as the intuitively appealing maximization of the sample output signal-to-interference ratio when the adaptive filter is set to $\hat{\boldsymbol{w}}(\beta) = \hat{\boldsymbol{R}}^{-1}(\beta)\boldsymbol{s}$, and the interference output power is calculated as $\hat{\boldsymbol{w}}^\dagger(\beta)\hat{\boldsymbol{R}}(\beta)\hat{\boldsymbol{w}}(\beta)$.

Yet if we relax this restriction on $x > 1$ and consider the unconstrained (by $H(x-1)$) maximization of the Λ_{12} over $\boldsymbol{R}(\beta)$

$$\Lambda_{12} = \max_{\boldsymbol{R}(\beta)} \frac{1}{x} e^x \tag{6.56}$$

with

$$x = \frac{|\boldsymbol{Y}^\dagger \boldsymbol{R}^{-1}(\beta)\boldsymbol{s}|^2}{\boldsymbol{s}^\dagger \boldsymbol{R}^{-1}(\beta)\boldsymbol{s}} \geq 0, \tag{6.57}$$

this optimization may potentially lead to solutions with $x \to 0$, which according to (6.51) means a negative estimate of a target power and in turn, no detection. Therefore allowing for this non-constrained optimization, the detection rule should be

$$\Lambda_{12} = \frac{|\boldsymbol{Y}^\dagger \hat{\boldsymbol{R}}^{-1}(\hat{\beta})\boldsymbol{s}|^2}{\boldsymbol{s}^\dagger \hat{\boldsymbol{R}}^{-1}(\hat{\beta})\boldsymbol{s}} \underset{H_0}{\overset{H_1}{\gtrless}} h^* > 1 \tag{6.58}$$

where

$$\hat{\boldsymbol{R}}(\hat{\beta}) = \arg\max_{\boldsymbol{R}(\beta)} \frac{\boldsymbol{s}^\dagger \boldsymbol{R}^{-1}(\beta)\boldsymbol{s}}{|\boldsymbol{Y}^\dagger \boldsymbol{R}^{-1}(\beta)\boldsymbol{s}|^2} \exp\left[\frac{|\boldsymbol{Y}^\dagger \boldsymbol{R}^{-1}(\beta)\boldsymbol{s}|^2}{\boldsymbol{s}^\dagger \boldsymbol{R}^{-1}(\beta)\boldsymbol{s}}\right] \tag{6.59}$$

subject to the usual ML- (6.41) or EL-GLRT constraints (6.42).

Since the x-monotonic function (6.52) or (6.54) is replaced by the function (6.59) which is not constrained by the condition $x > 1$, one may expect different detection performance from the tests (6.55) and (6.58)–(6.59). Note also that the test (6.55) may have different performance to the unconstrained test on $\Lambda_{12} = \max_{\boldsymbol{R}(\beta)} \frac{1}{x} e^x$ even for the clairvoyant case $\boldsymbol{R} = \boldsymbol{R}_0$.

Similarly, the AMF decision rule may be introduced

$$\frac{\boldsymbol{s}^\dagger \hat{\boldsymbol{R}}^{-1}(\beta)\boldsymbol{s}}{|\boldsymbol{Y}^\dagger \hat{\boldsymbol{R}}^{-1}(\beta)\boldsymbol{s}|^2} \exp\left[\frac{|\boldsymbol{Y}^\dagger \hat{\boldsymbol{R}}^{-1}(\beta)\boldsymbol{s}|^2}{\boldsymbol{s}^\dagger \hat{\boldsymbol{R}}^{-1}(\beta)\boldsymbol{s}}\right] H\left(\frac{|\boldsymbol{Y}^\dagger \hat{\boldsymbol{R}}^{-1}(\beta)\boldsymbol{s}|^2}{\boldsymbol{s}^\dagger \hat{\boldsymbol{R}}^{-1}(\beta)\boldsymbol{s}} - 1\right) \underset{H_0}{\overset{H_1}{\gtrless}} h^* > 1 \tag{6.60}$$

or

$$\frac{|\boldsymbol{Y}^\dagger \hat{\boldsymbol{R}}^{-1}(\hat{\beta})\boldsymbol{s}|^2}{\boldsymbol{s}^\dagger \hat{\boldsymbol{R}}^{-1}(\hat{\beta})\boldsymbol{s}} \underset{H_0}{\overset{H_1}{\gtrless}} h^* > 1. \tag{6.61}$$

Here $\hat{\boldsymbol{R}}(\beta)$ should be chosen either in the standard way $\hat{\boldsymbol{R}}(\beta = \beta_0) = \hat{\boldsymbol{R}}_N$ to get the well-known ML-AMF method or by (6.45) to get the EL-AMF rule.

6.2.2.3.3 Arbitrary Scaling Factors for Interference Matrices; Fluctuating Target with Unknown Power

Here the (total) power of the interference within the training data is assumed to be different to that in the primary data, so that the interference covariance matrix is the same up to an arbitrary scaling factor [20].

More specifically,

$$\mathbb{E}\left[\boldsymbol{X}_N \boldsymbol{X}_N^\dagger\right] = c_2 N \boldsymbol{R}, \qquad \mathbb{E}\left[\boldsymbol{Y}\boldsymbol{Y}^\dagger | H_0\right] = c_1 \boldsymbol{R}. \tag{6.62}$$

$$\Omega_0 = \{c_1\}, \quad \Omega_1 = \{c_1, \sigma_s^2\}, \quad \Omega_2 = \{c_2\}, \quad \Omega_{12} = \{\boldsymbol{R}\}. \tag{6.63}$$

According to (6.26) and (6.31), one first needs to find

$$LR(\boldsymbol{X}_N|\boldsymbol{R}) = \max_{c_2} \frac{f(\boldsymbol{X}_N|c_2)}{f_0(\boldsymbol{X}_N)}. \tag{6.64}$$

Since

$$\max_{c_2} \frac{1}{\pi^N c_2^{MN} \det(\boldsymbol{R})} \exp\left[-\frac{1}{c_2}\mathrm{Tr}(\boldsymbol{R}^{-1}\boldsymbol{X}_N\boldsymbol{X}_N^\dagger)\right] \tag{6.65}$$

leads to the ML estimate

$$\hat{c}_{ML} = \frac{1}{M}\mathrm{Tr}(\boldsymbol{R}^{-1}\hat{\boldsymbol{R}}_N), \tag{6.66}$$

one ends up with the familiar "sphericity test" in (6.64):

$$LR(\boldsymbol{X}_N|\boldsymbol{R}(\beta)) = \left(\frac{\det(\boldsymbol{R}^{-1}(\beta)\hat{\boldsymbol{R}}_N)}{\left[\frac{1}{M}\mathrm{Tr}(\boldsymbol{R}^{-1}(\beta)\hat{\boldsymbol{R}}_N)\right]^M}\right)^N. \tag{6.67}$$

For the GLRT detection rule,

$$\max_{c_1} f(\boldsymbol{Y}|H_0) = \frac{\det(\boldsymbol{R}^{-1}(\beta))}{\left[\frac{1}{M}\mathrm{Tr}(\boldsymbol{R}^{-1}(\beta)\hat{\boldsymbol{R}}_N)\right]^M} \tag{6.68}$$

and

$$\max_{c_1,\sigma_s^2} f(\boldsymbol{Y}|H_1) = \max_{\substack{\sigma_s^2>0 \\ c_1>0}} \frac{\exp\left[-\boldsymbol{Y}^\dagger\boldsymbol{R}^{-1}(\beta)\boldsymbol{Y} + \frac{\sigma_s^2}{c_1}\frac{|\boldsymbol{Y}^\dagger\boldsymbol{R}^{-1}(\beta)\boldsymbol{s}|^2}{1+\frac{\sigma_s^2}{c_1}\boldsymbol{s}^\dagger\boldsymbol{R}^{-1}(\beta)\boldsymbol{s}}\right]}{\det(c_1\boldsymbol{R}(\beta))\left[1+\frac{\sigma_s^2}{c_1}\boldsymbol{s}^\dagger\boldsymbol{R}^{-1}(\beta)\boldsymbol{s}\right]}. \tag{6.69}$$

First, find σ_{sML}^2 by solving the log-likelihood equation

$$\frac{\partial}{\partial\sigma_s^2}\ln f(\boldsymbol{Y}|H_1) = 0 \tag{6.70}$$

hence

$$1 + \frac{\sigma_s^2}{c_1}\boldsymbol{s}^\dagger\boldsymbol{R}^{-1}(\beta)\boldsymbol{s} = \frac{1}{c_1}\frac{|\boldsymbol{Y}^\dagger\boldsymbol{R}^{-1}(\beta)\,\boldsymbol{s}|^2}{\boldsymbol{s}^\dagger\boldsymbol{R}^{-1}(\beta)\,\boldsymbol{s}} \tag{6.71}$$

which leads to the estimate (*cf* (6.51))

$$\hat{\sigma}_{sML}^2 = \frac{|\boldsymbol{Y}^\dagger\boldsymbol{R}^{-1}(\beta)\,\boldsymbol{s}|^2 - c_1\boldsymbol{s}^\dagger\boldsymbol{R}^{-1}(\beta)\,\boldsymbol{s}}{[\boldsymbol{s}^\dagger\boldsymbol{R}^{-1}(\beta)\,\boldsymbol{s}]^2} \tag{6.72}$$

for

$$\frac{|\boldsymbol{Y}^{\dagger}\boldsymbol{R}^{-1}(\beta)\boldsymbol{s}|^2}{\boldsymbol{s}^{\dagger}\boldsymbol{R}^{-1}(\beta)\boldsymbol{s}} \geq c_1. \tag{6.73}$$

Substituting (6.71) into (6.69) yields

$$\max_{c_1,\sigma_s^2} f(\boldsymbol{Y}|H_1) = \frac{\boldsymbol{s}^{\dagger}\boldsymbol{R}^{-1}(\beta)\boldsymbol{s}}{c_1^{M-1}\det(\boldsymbol{R}(\beta))|\boldsymbol{Y}^{\dagger}\boldsymbol{R}^{-1}(\beta)\boldsymbol{Y}|^2}\exp\left\{\frac{1}{c_1}\left[-\boldsymbol{Y}^{\dagger}\boldsymbol{R}^{-1}(\beta)\boldsymbol{Y} + \frac{|\boldsymbol{Y}^{\dagger}\boldsymbol{R}^{-1}(\beta)\boldsymbol{s}|^2}{\boldsymbol{s}^{\dagger}\boldsymbol{R}^{-1}(\beta)\boldsymbol{s}}\right]\right\}. \tag{6.74}$$

Therefore the equation

$$\frac{\partial}{\partial c_1}\ln[\max_{\sigma_s^2} f(\boldsymbol{Y}|H_1)] = 0 \tag{6.75}$$

leads to the solution

$$\hat{c}_{1ML} = \frac{1}{M-1}\left[\boldsymbol{Y}^{\dagger}\boldsymbol{R}^{-1}(\beta)\boldsymbol{Y} - \frac{|\boldsymbol{Y}^{\dagger}\boldsymbol{R}^{-1}(\beta)\boldsymbol{s}|^2}{\boldsymbol{s}^{\dagger}\boldsymbol{R}^{-1}(\beta)\boldsymbol{s}}\right]. \tag{6.76}$$

Schwarz's inequality gives us $\hat{c}_{1ML} \geq 0$, so substituting the above into (6.73) gives

$$\frac{|\boldsymbol{Y}^{\dagger}\boldsymbol{R}^{-1}(\beta)\boldsymbol{s}|^2}{\boldsymbol{s}^{\dagger}\boldsymbol{R}^{-1}(\beta)\boldsymbol{s}\boldsymbol{Y}^{\dagger}\boldsymbol{R}^{-1}(\beta)\boldsymbol{Y}} \geq \frac{1}{M} \tag{6.77}$$

so the 2S-GLRT decision rule is

$$\max_{\boldsymbol{R}(\beta)} \frac{H(\widehat{\cos}^2 - \frac{1}{M})}{\left[1-\widehat{\cos}^2\right]^{M-1}\widehat{\cos}^2} \underset{H_0}{\overset{H_1}{\gtrless}} h^* > 1 \tag{6.78}$$

where

$$\widehat{\cos}^2 \equiv \frac{|\boldsymbol{Y}^{\dagger}\boldsymbol{R}^{-1}(\beta)\boldsymbol{s}|^2}{\boldsymbol{s}^{\dagger}\boldsymbol{R}^{-1}(\beta)\boldsymbol{s}\boldsymbol{Y}^{\dagger}\boldsymbol{R}^{-1}(\beta)\boldsymbol{Y}} \tag{6.79}$$

subject to

$$\text{ML-GLRT:} \qquad \alpha_L \leq \frac{\det(\boldsymbol{R}^{-1}(\beta)\hat{\boldsymbol{R}}_N)}{\left[\frac{1}{M}\mathrm{Tr}(\boldsymbol{R}^{-1}(\beta)\hat{\boldsymbol{R}}_N)\right]^M} < 1 \tag{6.80}$$

$$\text{EL-GLRT:} \qquad \alpha_L \leq \frac{\det(\boldsymbol{R}^{-1}(\beta)\hat{\boldsymbol{R}}_N)}{\left[\frac{1}{M}\mathrm{Tr}(\boldsymbol{R}^{-1}(\beta)\hat{\boldsymbol{R}}_N)\right]^M} \leq \alpha_U. \tag{6.81}$$

It is straight-forward to show that the function

$$f(\widehat{\cos}^2) = \frac{1}{[1-\widehat{\cos}^2]^{M-1}\widehat{\cos}^2} \tag{6.82}$$

is monotonic for $\widehat{\cos}^2 \geq 1/M$, so for the same reasons as before the 2S-GLRT decision rule may be expressed in the traditional form

$$\frac{|\boldsymbol{Y}^\dagger \hat{\boldsymbol{R}}^{-1}(\beta)Y|^2}{\boldsymbol{s}^\dagger \hat{\boldsymbol{R}}^{-1}(\beta)\boldsymbol{s}\boldsymbol{Y}^\dagger \hat{\boldsymbol{R}}^{-1}(\beta)\boldsymbol{Y}} \underset{H_0}{\overset{H_1}{\gtrless}} h^* \geq \frac{1}{M} \tag{6.83}$$

where

$$\hat{\boldsymbol{R}}^{-1}(\beta) = \arg\max_{\boldsymbol{R}(\beta)} \frac{1}{[1-\widehat{\cos}^2]^{M-1}\widehat{\cos}^2} \tag{6.84}$$

subject to the same constraint (6.80) or (6.81).

Naturally, the optimized function (6.82) should not be used directly, in order to avoid the inevitable performance degradation due to the non-monotonic nature of this function over the entire interval $0 \leq \widehat{\cos}^2 \leq 1$.

We may now introduce the traditional ML-AMF solution (that is the adaptive coherence estimation [ACE] detector [10,40]) as

$$\widehat{\cos}^2_{ML} = \frac{|\boldsymbol{Y}^\dagger \hat{\boldsymbol{R}}_N^{-1}\boldsymbol{s}|^2}{\boldsymbol{s}^\dagger \hat{\boldsymbol{R}}_N^{-1}\boldsymbol{s}\boldsymbol{Y}^\dagger \hat{\boldsymbol{R}}_N^{-1}\boldsymbol{Y}} \underset{H_0}{\overset{H_1}{\gtrless}} h^* > \frac{1}{M} \tag{6.85}$$

and the EL-AMF solution as

$$\widehat{\cos}^2_{ML} = \frac{|\boldsymbol{Y}^\dagger \hat{\boldsymbol{R}}^{-1}(\beta_{EL})\boldsymbol{s}|^2}{\boldsymbol{s}^\dagger \hat{\boldsymbol{R}}^{-1}(\beta_{EL})\boldsymbol{s}\boldsymbol{Y}^\dagger \hat{\boldsymbol{R}}^{-1}(\beta_{EL})\boldsymbol{Y}} \underset{H_0}{\overset{H_1}{\gtrless}} h^* > \frac{1}{M} \tag{6.86}$$

where β_{EL} is determined by the condition

$$\frac{\det(\hat{\boldsymbol{R}}^{-1}(\beta_{EL})\hat{\boldsymbol{R}})}{\left[\frac{1}{M}\mathrm{Tr}(\hat{\boldsymbol{R}}^{-1}(\beta_{EL})\hat{\boldsymbol{R}})\right]^M} = LR_0. \tag{6.87}$$

As usual, the bounds α_L, α_U, and LR_0 are specified by the scenario-free pdf that has been derived in Reference 41 for $\gamma_0^{(2)}$:

$$\gamma_0^{(2)} = \frac{\det(\hat{\boldsymbol{C}})}{\left[\frac{1}{M}\mathrm{Tr}(\hat{\boldsymbol{C}})\right]^M} \tag{6.88}$$

$$\hat{\boldsymbol{C}} \sim \mathcal{CW}(N, M, \boldsymbol{I}_M) \tag{6.89}$$

$$w(\gamma_0^{(2)}) = C(M,N)\left[\gamma_0^{(2)}\right]^{N-M} G^{M,0}_{M,M}\left(\gamma_0^{(2)} \,\middle|\, \begin{matrix} \frac{M^2-1}{M}, \frac{M^2-2}{M}, \ldots, \frac{M^2-M}{M} \\ 0, 1, \ldots, M-1 \end{matrix}\right) \tag{6.90}$$

where

$$C(M,N) \equiv (2\pi)^{\frac{M-1}{2}} M^{\frac{1-2MN}{2}} \frac{\Gamma(MN)}{\prod_{j=1}^M \Gamma(N-j+1)} \tag{6.91}$$

and $G^{M,0}_{M,M}(\,\cdot\,)$ is Meijer's G-function [42].

Note that the above well-known models have been elucidated since they permit analytic solutions for the ML estimates $\hat{\sigma}^2_{sML}$ and $\hat{c}_{ML}$. The same methodology can be applied for more complex models where such estimates are found numerically [23].

6.2.2.4 Diagonally Loaded and FML Adaptive Detectors; "Favorable" Scenarios

According to the two-fold goal of this study, the particular parametric families are now specified for the covariance matrix estimate $\boldsymbol{R}(\beta)$ within the above 2S-GLRT and AMF detection rules as the diagonally loaded and "FML" ones. Then we consider scenarios that are known to be most favorable for LSMI and FML adaptive filter techniques. The LSMI-based LAMF detector is also specified with a data-independent (constant) loading factor. The performance of this conventional LAMF detector will be compared with that of the new 2S-GLRT and EL-AMF methods that operate with adaptive data-dependent loading-factor selection.

The scenarios best addressed by LSMI treatment (6.6) are equivalently described by a covariance matrix of the form

$$\boldsymbol{R}_0 = \mu \boldsymbol{U}_s \boldsymbol{\Lambda}_s \boldsymbol{U}_s^\dagger + \boldsymbol{U}_n \boldsymbol{U}_n^\dagger \tag{6.92}$$

where, for simplicity, the white-noise power σ_n^2 is set to unity; $\boldsymbol{U}_s$ is the $M \times m$ matrix of m "signal-subspace" eigenvectors; $\boldsymbol{U}_n$ is the $M \times n$ matrix of "noise-subspace" eigenvectors, and $\mu \boldsymbol{\Lambda}_s \gg \boldsymbol{I}_m$ is the m-variate matrix of "signal-subspace" eigenvalues. The conditions

$$m < M, \qquad \mu \gg 1 \tag{6.93}$$

(or $\text{eig}_m(\boldsymbol{R}_0) \gg \sigma_n^2$ in the general case) are "favorable," meaning that there is typically tens of dBs difference between the signal- and noise-subspace eigenvalues. For this class of interference covariance matrices, it was demonstrated [13] that the adaptive filter $\boldsymbol{w}_{LSMI}$ that arises from the LSMI technique

$$\boldsymbol{w}_{LSMI} \equiv (\beta \boldsymbol{I}_M + \hat{\boldsymbol{R}}_N)^{-1} \boldsymbol{s} \tag{6.94}$$

where $\hat{\boldsymbol{R}}_N \equiv \boldsymbol{X}_N \boldsymbol{X}_N^\dagger / N$, and the loading factor β is selected within the broad range of values

$$\mu \lambda_m \gg \beta > 1 \qquad (\text{in general, } \text{eig}_m \boldsymbol{R}_0 \gg \beta > \sigma_n^2) \tag{6.95}$$

gives the normalized output SNR (SNR loss factor)

$$\gamma_{LSMI} = \frac{[\boldsymbol{s}^\dagger (\beta \boldsymbol{I}_M + \hat{\boldsymbol{R}}_N)^{-1} \boldsymbol{s}]^2}{\boldsymbol{s}^\dagger (\beta \boldsymbol{I}_M + \hat{\boldsymbol{R}}_N)^{-1} \boldsymbol{R}_0 (\beta \boldsymbol{I}_M + \hat{\boldsymbol{R}}_N)^{-1} \boldsymbol{s}\boldsymbol{s}^\dagger \boldsymbol{R}_0^{-1} \boldsymbol{s}} \tag{6.96}$$

that is approximately described by the β-distribution

$$w(\gamma_{LSMI}) = \frac{N!}{(N-m)!(m-1)!} (1 - \gamma_{LSMI})^{m-1} \gamma_{LSMI}^{N-m}. \tag{6.97}$$

This distribution only depends on N and m, not on the loading factor or other scenario parameters. Moreover,

$$\mathbb{E}\{\gamma_{LSMI}\} \simeq 3\,\text{dB} \quad \text{for } N \simeq 2m. \tag{6.98}$$

Later in Reference 43, the same pdf was used to describe the SNR loss factor of the FML (Hung–Turner type) adaptive beamformer

$$\boldsymbol{w}_{FML} \equiv (\hat{\sigma}_n^2 \boldsymbol{I}_M + \hat{\boldsymbol{U}}_m \hat{\boldsymbol{A}}_m \hat{\boldsymbol{U}}_m^{\dagger})\boldsymbol{s} \tag{6.99}$$

where

$$\hat{\sigma}_n^2 \equiv \frac{1}{M-m} \sum_{j=1}^{M-m} \lambda_{m+j}, \qquad \hat{\boldsymbol{A}}_m \equiv \text{diag}[\hat{\lambda}_j - \hat{\sigma}_n^2] \quad \text{for } j = 1, \ldots, m \tag{6.100}$$

with $\boldsymbol{U}_j$ and λ_j ($j = 1, \ldots, M$) coming from the eigen-decomposition of the sample (ML) covariance matrix

$$\hat{\boldsymbol{R}}_N = \hat{\boldsymbol{U}} \hat{\Lambda} \hat{\boldsymbol{U}}, \qquad \hat{\boldsymbol{U}} \equiv [\hat{\boldsymbol{U}}_m, \hat{\boldsymbol{U}}_n], \quad \hat{\Lambda} \equiv \text{diag}\,[\hat{\Lambda}_m, \hat{\Lambda}_n]. \tag{6.101}$$

Note that unlike the LSMI algorithm, the FML technique requires the signal-subspace dimension (order) m to be specified. For favorable conditions (6.92)–(6.93), this order can be estimated with high accuracy by treating $\hat{\boldsymbol{R}}$ with an information-theoretic criterion (ITC) [44]. This approach is actually similar to the EL philosophy, and Subsection 6.2.2.5 shows that in many scenarios, EL matching gives as reliable an order estimate as any ITC. Therefore, for the FML method there is no practical difference between EL-FML and "conventional" FML (for (6.92)–(6.93)). For this reason, the comparative analysis is focused on the LAMF detector with constant diagonal loading:

$$\hat{\boldsymbol{R}}_{LSMI} = \beta_c \boldsymbol{I}_M + \hat{\boldsymbol{R}}_N. \tag{6.102}$$

According to the conventional AMF methodology, the LAMF detector can be derived from (6.102) being substituted into the detection test instead of the known covariance matrix [5].

The adaptive detectors that are to be compared for favorable scenarios are therefore as given below.

6.2.2.4.1 *Homogeneous Interference Training Conditions; Fluctuating Target with Unknown Power*

<u>2S-GLRT:</u>

$$\max_{\beta} \frac{|\boldsymbol{Y}^{\dagger}(\beta \boldsymbol{I}_M + \hat{\boldsymbol{R}}_N)^{-1}\boldsymbol{s}|^2}{\boldsymbol{s}^{\dagger}(\beta \boldsymbol{I}_M + \hat{\boldsymbol{R}}_N)^{-1}\boldsymbol{s}} \underset{H_0}{\overset{H_1}{\gtrless}} h^* > 1 \tag{6.103}$$

subject to

$$\alpha_L \leq \frac{\det[(\beta \boldsymbol{I}_M + \hat{\boldsymbol{R}}_N)^{-1}\hat{\boldsymbol{R}}_N] \exp(M)}{\exp \text{Tr}[(\beta \boldsymbol{I}_M + \hat{\boldsymbol{R}}_N)^{-1}\hat{\boldsymbol{R}}_N]} \leq 1 \quad \text{(ML-GLRT)} \tag{6.104}$$

or

$$\alpha_L \leq \frac{\det[(\beta \boldsymbol{I}_M + \hat{\boldsymbol{R}}_N)^{-1}\hat{\boldsymbol{R}}_N] \exp(M)}{\exp \text{Tr}[(\beta \boldsymbol{I}_M + \hat{\boldsymbol{R}}_N)^{-1}\hat{\boldsymbol{R}}_N]} \leq \alpha_U \quad \text{(EL-GLRT)} \tag{6.105}$$

ML-AMF:

$$\frac{|\boldsymbol{Y}^{\dagger}\hat{\boldsymbol{R}}_N^{-1}\boldsymbol{s}|^2}{\boldsymbol{s}^{\dagger}\hat{\boldsymbol{R}}_N^{-1}\boldsymbol{s}} \underset{H_0}{\overset{H_1}{\gtrless}} h^* > 1 \tag{6.106}$$

EL-AMF:

$$\frac{|\boldsymbol{Y}^{\dagger}(\hat{\beta}\boldsymbol{I}_M+\hat{\boldsymbol{R}}_N)^{-1}\boldsymbol{s}|^2}{\boldsymbol{s}^{\dagger}(\hat{\beta}\boldsymbol{I}_M+\hat{\boldsymbol{R}}_N)^{-1}\boldsymbol{s}} \underset{H_0}{\overset{H_1}{\gtrless}} h^* > 1 \tag{6.107}$$

where

$$\hat{\beta} \equiv \arg_{\beta}\left\{\frac{\det[(\beta\boldsymbol{I}_M+\hat{\boldsymbol{R}}_N)^{-1}\hat{\boldsymbol{R}}_N]\exp(M)}{\exp\mathrm{Tr}[(\beta\boldsymbol{I}_M+\hat{\boldsymbol{R}}_N)^{-1}\hat{\boldsymbol{R}}_N]} \equiv LR_0\right\} \tag{6.108}$$

LAMF:

$$\frac{|\boldsymbol{Y}^{\dagger}(\beta_c\boldsymbol{I}_M+\hat{\boldsymbol{R}}_N)^{-1}\boldsymbol{s}|^2}{\boldsymbol{s}^{\dagger}(\beta_c\boldsymbol{I}_M+\hat{\boldsymbol{R}}_N)^{-1}\boldsymbol{s}} \underset{H_0}{\overset{H_1}{\gtrless}} h^* > 1 \tag{6.109}$$

where the constant β_c is about two or three.

6.2.2.4.2 Non-homogeneous Interference Training Conditions; Fluctuating Target with Unknown Power

2S-GLRT:

$$\max_{\beta}\frac{|\boldsymbol{Y}^{\dagger}(\beta\boldsymbol{I}_M+\hat{\boldsymbol{R}}_N)^{-1}\boldsymbol{s}|^2}{\boldsymbol{s}^{\dagger}(\beta\boldsymbol{I}_M+\hat{\boldsymbol{R}}_N)^{-1}\boldsymbol{s}\boldsymbol{Y}^{\dagger}(\beta\boldsymbol{I}_M+\hat{\boldsymbol{R}}_N)^{-1}\boldsymbol{Y}} \underset{H_0}{\overset{H_1}{\gtrless}} h^* > \frac{1}{M} \tag{6.110}$$

subject to

$$\alpha_L \le \frac{\det[(\beta\boldsymbol{I}_M+\hat{\boldsymbol{R}}_N)^{-1}\hat{\boldsymbol{R}}_N]}{\left\{\frac{1}{M}\mathrm{Tr}[(\beta\boldsymbol{I}_M+\hat{\boldsymbol{R}}_N)^{-1}\hat{\boldsymbol{R}}_N]\right\}^M} \le 1 \qquad \text{(ML-GLRT)} \tag{6.111}$$

or

$$\alpha_L \le \frac{\det[(\beta\boldsymbol{I}_M+\hat{\boldsymbol{R}}_N)^{-1}\hat{\boldsymbol{R}}_N]}{\left\{\frac{1}{M}\mathrm{Tr}[(\beta\boldsymbol{I}_M+\hat{\boldsymbol{R}}_N)^{-1}\hat{\boldsymbol{R}}_N]\right\}^M} \le \alpha_U \qquad \text{(EL-GLRT)} \tag{6.112}$$

ML-AMF (ACE):

$$\frac{|\boldsymbol{Y}^{\dagger}\hat{\boldsymbol{R}}_N^{-1}\boldsymbol{s}|^2}{\boldsymbol{s}^{\dagger}\hat{\boldsymbol{R}}_N^{-1}\boldsymbol{s}\boldsymbol{Y}^{\dagger}\hat{\boldsymbol{R}}_N^{-1}\boldsymbol{Y}} \underset{H_0}{\overset{H_1}{\gtrless}} h^* > \frac{1}{M} \tag{6.113}$$

EL-AMF:

$$\frac{|Y^{\dagger}(\hat{\beta}I_M+\hat{R}_N)^{-1}s|^2}{s^{\dagger}(\hat{\beta}I_M+\hat{R}_N)^{-1}sY^{\dagger}(\hat{\beta}I_M+\hat{R}_N)^{-1}Y}\underset{H_0}{\overset{H_1}{\gtrless}} h^* > \frac{1}{M} \tag{6.114}$$

where

$$\hat{\beta}\equiv \arg_{\beta}\left\{\frac{\det[(\beta I_M+\hat{R}_N)^{-1}\hat{R}_N]}{\left\{\frac{1}{M}\mathrm{Tr}[(\beta I_M+\hat{R}_N)^{-1}\hat{R}_N]\right\}^M}\equiv LR_0\right\} \tag{6.115}$$

LAMF:

$$\frac{|Y^{\dagger}(\beta_c I_M+\hat{R}_N)^{-1}s|^2}{s^{\dagger}(\beta_c I_M+\hat{R}_N)^{-1}sY^{\dagger}(\beta_c I_M+\hat{R}_N)^{-1}Y}\underset{H_0}{\overset{H_1}{\gtrless}} h^* > \frac{1}{M} \tag{6.116}$$

where the constant β_c is about two or three.

Naturally the performance of these detectors compared with the FML-based ones is of interest. The FML-based detectors can be introduced in a similar way, where now the signal-subspace dimension m in (6.99) is used as a parameter in 2S-GLRT optimization. As already mentioned, EL-FML has the same performance as the FML matrix estimate using the true m for “favorable” scenarios (6.92)–(6.93).

Note that of all these detectors, only the familiar ML-AMF (ACE) detectors and standard GLRT are known to be strictly CFAR detectors [5,40]. Indeed, for signal-free primary data Y and no mismatch in the interference properties between the primary and secondary data, the above ML-AMF detectors have pdfs that are functions only of M and N. Such pdfs have been analytically derived in References 5 and 40 and can be used for false-alarm threshold calculations. For the other detectors introduced above, the strict CFAR property cannot be proven. Yet, for favorable scenarios, a certain invariance of the output signal-free statistics can be demonstrated that is sufficient for practical false-alarm threshold calculations. This invariance is specified by the following two theorems.

Theorem 6.1. *Suppose the “favorable” interference covariance matrix is of the form*

$$R_0=\mu U_s\Lambda_s U_s^{\dagger}+U_nU_n^{\dagger},\quad \mu\gg 1,\quad \Lambda_s>I_m \tag{6.117}$$

and let the loading factor β in the LSMI estimate

$$\hat{R}_{LSMI}=\beta I_M+\hat{R}_N,\quad \hat{R}_N=X_NX_N^{\dagger}/N,\quad X_N\sim \mathcal{CN}_N(0,R_0) \tag{6.118}$$

be selected within the range $\mu>\beta\gtrsim 1$, then

(a) the test statistics

$$\hat{t}_1\equiv\frac{|Y^{\dagger}(\beta I_M+\hat{R}_N)^{-1}s|^2}{s^{\dagger}(\beta I_M+\hat{R}_N)^{-1}s},\quad Y\sim\mathcal{CN}(0,R_0) \tag{6.119}$$

can be approximately (as $\mu \to \infty$) represented as

$$\hat{t}_1 \simeq \frac{\left|[\mathbf{y}_{1n}^{\dagger} - \mathbf{y}_{1s}^{\dagger}(\mathbf{Z}_s\mathbf{Z}_s^{\dagger})^{-1}\mathbf{Z}_s\mathbf{Z}_n^{\dagger}]\mathcal{Z}\mathbf{e}_1\right|^2}{\mathbf{e}_1^T\mathcal{Z}\mathbf{e}_1} \tag{6.120}$$

where

$$\mathcal{Z} \equiv \left\{\beta\mathbf{I}_n + \frac{1}{N}\mathbf{Z}_n[\mathbf{I}_N - \mathbf{Z}_s^{\dagger}(\mathbf{Z}_s\mathbf{Z}_s^{\dagger})^{-1}\mathbf{Z}_s]\mathbf{Z}_n^{\dagger}\right\}^{-1} \tag{6.121}$$

and where

$$\begin{aligned}
\mathbf{y}_{1n} &\in \mathcal{CN}^{n\times 1} \sim \mathcal{CN}(\mathbf{0}, \mathbf{I}_n)\\
\mathbf{y}_{1s} &\in \mathcal{CN}^{m\times 1} \sim \mathcal{CN}(\mathbf{0}, \mathbf{I}_m)\\
\mathbf{Z}_n &\in \mathcal{CN}^{n\times N} \sim \mathcal{CN}_N(\mathbf{0}, \mathbf{I}_n)\\
\mathbf{Z}_m &\in \mathcal{CN}^{m\times N} \sim \mathcal{CN}_N(\mathbf{0}, \mathbf{I}_m)\\
\mathbf{e}_1 &\in \mathcal{R}^{n\times 1} \equiv [1,\ 0, \ldots, 0]^T
\end{aligned} \tag{6.122}$$

and $\mathbf{y}_{1n}$, $\mathbf{y}_{1s}$, $\mathbf{Z}_n$, $\mathbf{Z}_m$ are mutually independent; and
(b) the test statistics

$$\hat{t}_2 \equiv \frac{|\mathbf{Y}^{\dagger}(\beta\mathbf{I}_M + \hat{\mathbf{R}}_N)^{-1}\mathbf{s}|^2}{\mathbf{s}^{\dagger}(\beta\mathbf{I}_M + \hat{\mathbf{R}}_N)^{-1}\mathbf{s}\mathbf{Y}^{\dagger}(\beta\mathbf{I}_M + \hat{\mathbf{R}}_N)^{-1}\mathbf{Y}}, \qquad \mathbf{Y} \sim \mathcal{CN}(\mathbf{0}, \mathbf{R}_0) \tag{6.123}$$

can be approximately (as $\mu \to \infty$) represented as

$$\hat{t}_2 \equiv \hat{t}_1 L_{LGIP}^{-1} \tag{6.124}$$

where

$$L_{LGIP} \equiv \mathbf{Y}^{\dagger}(\beta\mathbf{I}_M + \hat{\mathbf{R}}_N)^{-1}Y$$

$$\simeq \mathbf{y}_{1s}^{\dagger}\left[\frac{1}{N}\mathbf{Z}_s\mathbf{Z}_s^{\dagger} - \frac{1}{N^2}\mathbf{Z}_s\mathbf{Z}_n^{\dagger}(\beta\mathbf{I}_N + \frac{1}{N}\mathbf{Z}_n\mathbf{Z}_n^{\dagger})^{-1}\mathbf{Z}_n\mathbf{Z}_s^{\dagger}\right]^{-1}\mathbf{y}_{1s}$$

$$-2\,Re\left[\mathbf{y}_{1s}^{\dagger}(\mathbf{Z}_s\mathbf{Z}_s^{\dagger})^{-1}\mathbf{Z}_s\mathbf{Z}_n^{\dagger}\mathcal{Z}\mathbf{y}_{1n}\right] + \mathbf{y}_{1n}^{\dagger}\mathcal{Z}\mathbf{y}_{1n}. \tag{6.125}$$

The proof appears in Appendix II of Reference 1.

Despite being rather bulky, these representations mean that for scenarios satisfying (6.92)–(6.93) with sufficiently large μ, the test statistics for target-free primary data can be expressed as a function of "white-noise" i.i.d. data, and so its pdf will depend only upon the parameters N, M, m, and β. For LAMF, these representations can be used directly to calculate false-alarm thresholds, at least, by direct Monte-Carlo simulations. Of course, in this case, the order m must be specified, but this is not a problem for "favorable" scenarios (6.92)–(6.93) where ITC is quite robust. With respect to these properties, the LAMF detector can be treated as being CFAR in practice.

The derivation above provides the most accurate representations for $\hat{t}_1$ and $\hat{t}_2$, sufficient for threshold calculations even for reasonably small β. For large values, the expressions are less complicated; specifically for

$$N\beta \gg 1, \tag{6.126}$$

$$\hat{t}_1 \simeq \frac{1}{\beta} |[\mathbf{y}_{1n}^\dagger - \mathbf{y}_{1s}^\dagger (\mathbf{Z}_s \mathbf{Z}_s^\dagger)^{-1} \mathbf{Z}_s \mathbf{Z}_n^\dagger] \mathbf{e}_1|^2 \tag{6.127}$$

and

$$L_{LGIP} \simeq N \mathbf{y}_{1s}^\dagger (\mathbf{Z}_s \mathbf{Z}_s^\dagger)^{-1} \mathbf{y}_{1s} + \mathbf{y}_{1n}^\dagger \mathbf{y}_{1n} / \beta \tag{6.128}$$

and with

$$\hat{f} \equiv \mathbf{y}_{1s}^\dagger (\mathbf{Z}_s \mathbf{Z}_s^\dagger)^{-1} \mathbf{y}_{1s} \sim \frac{\hat{f}^{m-1}}{B(m, N+1-m)(1+\hat{f})^{N+1}} \tag{6.129}$$

$$\hat{g} \equiv \mathbf{y}_{1n}^\dagger \mathbf{y}_{1n} \sim \frac{\hat{g}^{M-m+1} \exp[-\hat{g}]}{\Gamma(M-m)} \tag{6.130}$$

where B is the incomplete β-function. Note that the statistics L_{LGIP} can be considered to be the loaded version of the "generalized inner product" (GIP) test that was introduced in Reference 28 for non-homogeneity detection.

In order to expand these invariance properties to the new 2S-GLRT and EL-AMF techniques, it is merely necessary to demonstrate that (under favorable interference conditions) the LRs in (6.104) and (6.111) can be approximately represented as functions of *the same* random "white-noise" variables $\mathbf{Z}_s$ and $\mathbf{Z}_n$. When dealing with GLRT optimization in (6.103) and (6.110), it is rare to get analytic expressions for the target-free thresholds. Yet, since both the optimized test statistics and constraints can be represented by "white-noise" variables, these pdfs can be defined by N, M, m, α_L, and α_U and may be precalculated using Monte-Carlo simulations.

For EL-AMF, such LR representation means that the "EL" loading factor β is a function of the same "white-noise" variables $\mathbf{Z}_s$ and $\mathbf{Z}_n$, together with the expected median LR value LR_0, and so the target-free test statistics for EL-AMF are a function of N, M, m, and LR_0 and may again be precalculated using Monte-Carlo simulations. The invariance of the LRs is formalized by the following theorem.

Theorem 6.2. *Suppose the "favorable" interference covariance matrix is of the form*

$$\mathbf{R}_0 = \mu \mathbf{U}_s \mathbf{\Lambda}_s \mathbf{U}_s^\dagger + \mathbf{U}_n \mathbf{U}_n^\dagger, \qquad \mu \gg 1 \tag{6.131}$$

and let the loading factor β in the LSMI estimate

$$\hat{\mathbf{R}}_{LSMI} = \beta \mathbf{I}_M + \hat{\mathbf{R}}_N, \qquad \hat{\mathbf{R}}_N = \mathbf{X}_N \mathbf{X}_N^\dagger / N, \quad \mathbf{X}_N \sim \mathcal{CN}_N(\mathbf{0}, \mathbf{R}_0) \tag{6.132}$$

be selected within the range $\mu > \beta \gtrsim 1$*, then the LRs*

$$\gamma_{\ell 0}^{(1)} \equiv \frac{\det(\hat{\boldsymbol{R}}_{LSMI}^{-1}\hat{\boldsymbol{R}}_N)\exp(M)}{\exp\operatorname{Tr}(\hat{\boldsymbol{R}}_{LSMI}^{-1}\hat{\boldsymbol{R}}_N)} \tag{6.133}$$

$$\gamma_{\ell 0}^{(2)} \equiv \frac{\det(\hat{\boldsymbol{R}}_{LSMI}^{-1}\hat{\boldsymbol{R}}_N)}{\left[\frac{1}{M}\operatorname{Tr}(\hat{\boldsymbol{R}}_{LSMI}^{-1}\hat{\boldsymbol{R}}_N)\right]^M} \tag{6.134}$$

may be approximately represented as

$$\gamma_{\ell 0}^{(1)} = \frac{\det[\boldsymbol{\mathcal{Z}}(\boldsymbol{\mathcal{Z}}^{-1} - \beta \boldsymbol{I}_N)]}{\exp[-\beta\operatorname{Tr}(\boldsymbol{\mathcal{Z}})]} \tag{6.135}$$

$$\gamma_{\ell 0}^{(1)}(\beta = 0) = 1 \tag{6.136}$$

$$\gamma_{\ell 0}^{(1)}(\beta \gg 1) = \beta^{-(M-m)}\det(\boldsymbol{\mathcal{Z}}^{-1} - \beta \boldsymbol{I}_N)\exp[M-m] \tag{6.137}$$

and

$$\gamma_{\ell 0}^{(2)} = \frac{\det[\boldsymbol{\mathcal{Z}}(\boldsymbol{\mathcal{Z}}^{-1} - \beta \boldsymbol{I}_N)]}{[1 - \beta\operatorname{Tr}(\boldsymbol{\mathcal{Z}}/M)]^M} \tag{6.138}$$

$$\gamma_{\ell 0}^{(2)}(\beta = 0) = 1 \tag{6.139}$$

$$\gamma_{\ell 0}^{(2)}(\beta \gg 1) = \beta^{-(M-m)}\det(\boldsymbol{\mathcal{Z}}^{-1} - \beta \boldsymbol{I}_N)\left(1 - \frac{m}{M}\right)^{-M}. \tag{6.140}$$

The proof appears in Appendix III of Reference 1.

In the next subsection, these approximations are demonstrated to be sufficiently accurate for practical false-alarm threshold calculations for scenarios that consist of strong point-source interferers (i.e. are favorable for these detectors).

6.2.2.5 Detection Performance Analysis of 2S-GLRT, AMF, and LAMF Detectors

6.2.2.5.1 "Favorable" Interference Scenario

Consider an $M = 12$-sensor uniform linear antenna array, and $m = 6$ independent Gaussian interference sources, each with 30-dB signal-to-white-noise ratio (SWNR). The interference directions-of-arrival (DOAs) were chosen to be

$$\boldsymbol{w}_6 \equiv \sin\boldsymbol{\theta}_6 = [-0.8,\ -0.4,\ 0.2,\ 0.5,\ 0.7,\ 0.9] \tag{6.141}$$

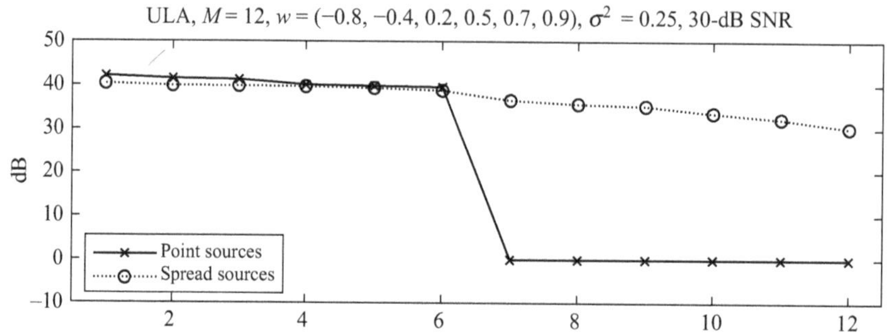

Figure 6.2 Clairvoyant eigenspectrum of the "favorable" and "unfavorable" interference covariance matrices used in the simulations. Favorable scenarios consist of strong point-source interferers, which allow for a distinct signal and noise subspace, while unfavorable scenarios include spread sources that preclude the presence of a distinct noise subspace after eigen-decomposition.

so that the eigenspectrum of the interference covariance matrix

$$\boldsymbol{R}_0 = \sigma_n^2 \boldsymbol{I}_M + \sum_{j=1}^{6} \sigma_j^2 \boldsymbol{s}(w_j)\boldsymbol{s}^\dagger(w_j) \tag{6.142}$$

where σ_n^2 is the white-noise power, and

$$\sigma_n^2 = 1, \qquad \sigma_j^2 = 1{,}000, \qquad w_j = 2\pi\frac{d}{\lambda}\sin\theta_j \tag{6.143}$$

as shown in Figure 6.2 perfectly meets the "favorable" interference conditions (6.92)–(6.93), since $\lambda_7/\lambda_8 = \lambda_7/\sigma_n^2 \simeq 35$ dB. Note that under our assumption regarding the training data consisting of N i.i.d. samples, the above adaptive detectors are applicable to any spatial, temporal, or space–time application. In fact, the eigenspectrum in Figure 6.2 has been specifically chosen to look like that of the terrain-scattered space–time covariance matrix in a side-looking airborne radar with three antenna sensors and four repetition periods [45].

Two separate target DOAs w_0 have been selected to represent two extreme cases, namely

$$\frac{s_0^\dagger[\boldsymbol{I} - s_6(s_6^\dagger s_6)^{-1}s_6^\dagger]s_0}{s_0^\dagger s_0} = \begin{cases} 0.949 & \text{for } w_0^\dagger = -0.60 \\ 0.040 & \text{for } w_0^L = 0.18. \end{cases} \tag{6.144}$$

In the first case (a "fast target" in STAP [space–time adaptive processing] terminology), total interference mitigation is not accompanied by a significant degradation in target SWNR. In the second case (a "slow target"), the target is close to at least one of the interferers and the resultant interference "nulling" leads to a dramatic signal-power reduction.

Note that for the clairvoyant detector ($\boldsymbol{R} = \boldsymbol{R}_0$), as well as for the standard GLRT and AMF detectors, this distinction does not affect the ROC if the output SNRs are identical:

$$\sigma_{sL}^2 s_0^{LH} \boldsymbol{R}_0^{-1} s_0^L = \sigma_{sH}^2 s_0^{HH} \boldsymbol{R}_0^{-1} s_0^\dagger. \tag{6.145}$$

Of course, the standard GLRT and AMF detectors are CFAR detectors and therefore their false-alarm rate (thresholds) do not depend on the particular scenario in (6.144). For the introduced 2S-GLRT, EL-AMF, and LAMF detectors, it is necessary to demonstrate that the invariance of their output target-free statistics predicted in Subsection 6.2.2.4 is sufficient for constant false-alarm thresholds.

The training sample size N in all our simulations has been chosen according to the RMB rule that ensures 3-dB average SNR losses in the SMI adaptive *filter*: $N = 2M = 24$.

6.2.2.5.2 *Homogeneous Interference Training Conditions; Fluctuating Target with Unknown Power*

First, consider the performance of the clairvoyant detector

$$\frac{|y^{\dagger} \boldsymbol{R}_0^{-1} \boldsymbol{s}|^2}{\boldsymbol{s}^{\dagger} \boldsymbol{R}_0^{-1} \boldsymbol{s}} \underset{H_0}{\overset{H_1}{\gtrless}} h > 1 \tag{6.146}$$

whose ROC has the well-known analytic expression

$$P_d = \exp\left[-\frac{h}{1+\sigma_s^2 \boldsymbol{s}_0^{\dagger} \boldsymbol{R}_0^{-1} s_0}\right] \tag{6.147}$$

where h is the threshold and $(\sigma_s^2 \boldsymbol{s}_0^{\dagger} \boldsymbol{R}_0^{-1} s_0)$ is the output SNR. For the AMF (ML-AMF) detector, the probability of false-alarm P_{FA} and of target detection can be expressed somewhat differently to RFKN [5] (see Appendix IV of Reference 1)

$$\begin{aligned} P_{FA} &= {}_2F_1(N-M+1, N-M+2, N+1; -h) \\ &= \frac{1}{(1+h)^{N-M+1}} {}_2F_1\left(N-M+1, M-1, N+1; -\frac{h}{1+h}\right) \end{aligned} \tag{6.148}$$

where ${}_2F_1(\alpha, \beta, \gamma; x)$ is the hypergeometric function [42], and

$$\begin{aligned} P_d &= \left[\frac{1+\sigma_s^2 \boldsymbol{s}^{\dagger} \boldsymbol{R}^{-1} \boldsymbol{s}}{1+\sigma_s^2 \boldsymbol{s}^{\dagger} \boldsymbol{R}^{-1} \boldsymbol{s} + h}\right]^{N-M+1} \\ &\times F_1\left(M-1, -(N-M+1), N-M+1, N+1; \frac{\sigma_s^2 \boldsymbol{s}^{\dagger} \boldsymbol{R}^{-1} \boldsymbol{s}}{1+\sigma_s^2 \boldsymbol{s}^{\dagger} \boldsymbol{R}^{-1} \boldsymbol{s}}, \frac{\sigma_s^2 \boldsymbol{s}^{\dagger} \boldsymbol{R}^{-1} \boldsymbol{s} + h}{1+\sigma_s^2 \boldsymbol{s}^{\dagger} \boldsymbol{R}^{-1} \boldsymbol{s} + h}\right) \end{aligned} \tag{6.149}$$

where $F_1(\alpha, \beta, \beta', \gamma; x, y)$ is the hypergeometric function of two variables [42]. Note that [42]

$${}_2F_1\left(M-1, 0, N+1; \frac{\sigma_s^2}{1+\sigma_s^2}\right) = 1 \tag{6.150}$$

hence for $h = 0$, $P_d = 1$, and for $\sigma_s^2 = 0$, $P_d = P_{FA}$. These analytical expressions are used to validate the results of our Monte-Carlo simulations. Specifically, we shall compare the simulated and theoretical ROC for the clairvoyant detector and use the free software routine `gsl_sf_hyperg_2F1` from the GNU Scientific Library (GSL) (http://www.gnu.org/software/gsl/) to calculate ${}_2F_1(\alpha,\ \beta,\ \gamma;\ x)$ and so find the threshold values h for false-alarm rates from 10^{-2} to 10^{-4}. Comparison of the analytically computed values with those computed

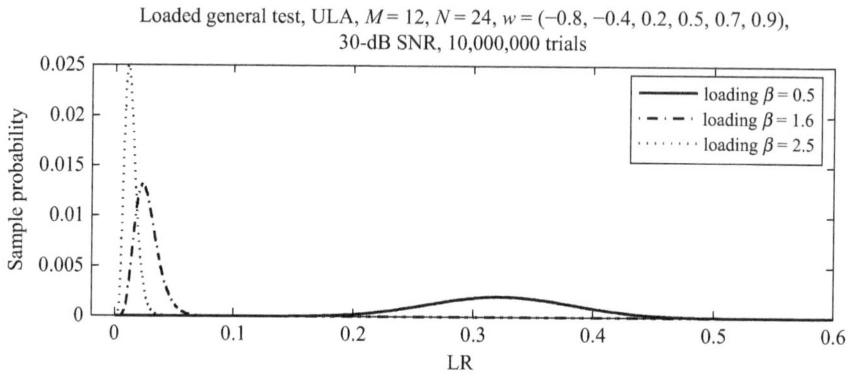

Figure 6.3 Sample pdfs for the loaded general test with different loading factors and six interferers, but with the same $M = 12$ and $N = 24$ used in Figure 6.1.

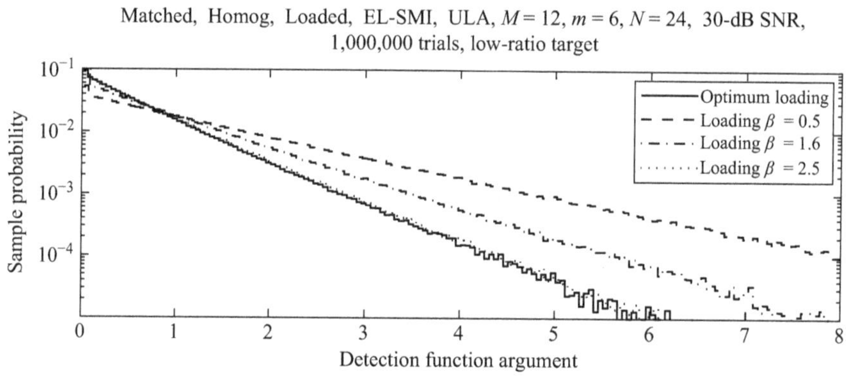

Figure 6.4 Sample pdfs for the target-free output signal of the LAMF detector, with different loading factors and a target close to one of the six interferers (a.k.a. "slow" or "low-ratio" target)

over a large number of Monte-Carlo trials demonstrates an excellent match that finally validates the accuracy of the other Monte-Carlo results.

Next we consider the performance of our new detectors, beginning with LAMF that we expect to serve as a benchmark for the 2S-GLRT and EL-AMF detectors with adaptive (data-dependent) loading factors, as suggested by the theoretical framework. Figure 6.3 shows sample pdfs calculated over 10^6 Monte-Carlo trials for the LR $\gamma_{0\ell}^{(1)}$ (6.14) and three fixed loading factors $\beta = 0.5,\ 1.6,\ 2.5$. We see that the loading factor $\beta = 1.6$ comes close to matching distribution of the LR generated by the true covariance matrix $\boldsymbol{R}_0$ shown in Figure 6.1, with a median value of 0.0257. Therefore for this scenario we expect $\beta = 1.6$ relative to a noise power of $\sigma_n^2 = 1$ to be a "sufficient" constant loading factor.

For this relatively low diagonal loading, the accuracy of our "white-noise" approximation (6.120) of the output "target-free" statistics becomes a critical issue that must be addressed first. Figure 6.4 illustrates sample pdfs for the output "target-free" statistics calculated over 10^6 Monte-Carlo trials for the "slow" target in (6.144). (The pdfs for the "fast" target are

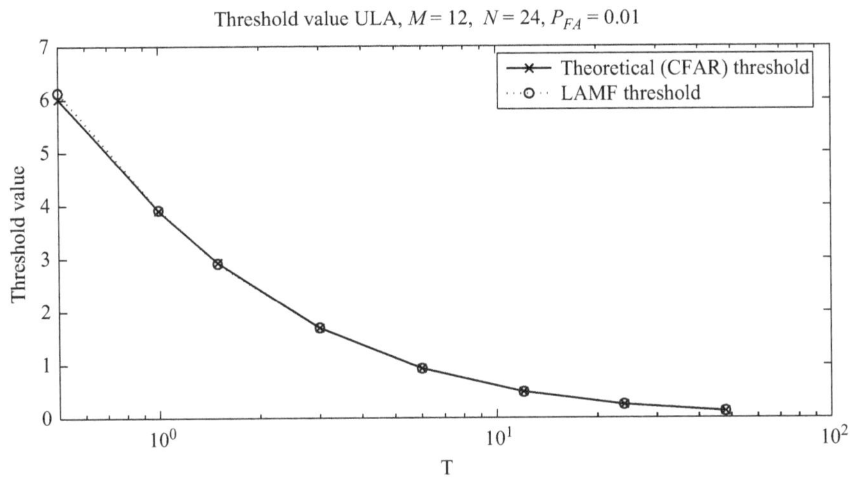

Figure 6.5 Comparison of theoretical (noise-only) and actual (target-free output signal of the LAMF detector) threshold values for $P_{FA} = 10^{-2}$ and varying loading factors. The correspondence of the curves demonstrates the CFAR nature of the LAMF detector.

not presented here, since they are indistinguishable.) The invariance of false-alarm rate with respect to the target scenario is thus demonstrated.

In order to assess the accuracy of our "white-noise" theoretical approximations for the output "target-free" statistics, we can calculate sample pdfs (10^4 trials, "slow" target scenario) of the target-free output signal for the LAMF detector for loading factors ranging from $\beta =$ 0.5 to 48 and then determine the threshold associated with the desired false-alarm rate (say $P_{FA} = 10^{-2}$). Comparing those values to the *actual threshold* values calculated (again for $P_{FA} = 10^{-2}$) using the theoretical "white-noise" representation (6.120) (10^6 trials), we find an highly accurate correspondence of threshold values, as shown in Figure 6.5. Hence for this scenario, the LAMF detector is indeed a CFAR detector, since false-alarm threshold values can be precalculated with sufficient accuracy for any given value of m (the number of dominant covariance matrix eigenvalues). This comparison also proves similar CFAR properties of EL-AMF and 2S-GLRT detectors, since they are based on the same analytical approximation (see Theorem 2).

Now we are in a position to consider the ROC of the clairvoyant, standard AMF, and LAMF detectors, which are presented in Figure 6.6 for false-alarm rate set at $P_{FA} = 10^{-4}$. The ROCs for $P_{FA} = 10^{-2}$ and 10^{-3} are not presented here, as they are little changed. (Again, the ROCs for the slow target are identical to this fast target and are also not presented.) We observe an excellent match between the simulated (ideal) and analytical (theoretical) ROCs for the clairvoyant detector (6.146), proving the accuracy of the simulations. Despite the different target model (fluctuating in our study and non-fluctuating in References 4 and 5), the standard AMF detector (ML-AMF) demonstrates performance similar to that in Reference 5. Indeed, for $P_{FA} = 10^{-4}$ and $P_d = 0.5$, the ML-AMF SNR loss factor is about 5 dB, compared with about 3 dB in Reference 7. A first-cut interpretation of this would be that AMF is incurring an additional $\sim$ 2-dB losses due to adaptive thresholding. Recall that the RMB losses are just adaptive antenna filter losses and do not include any thresholding losses.

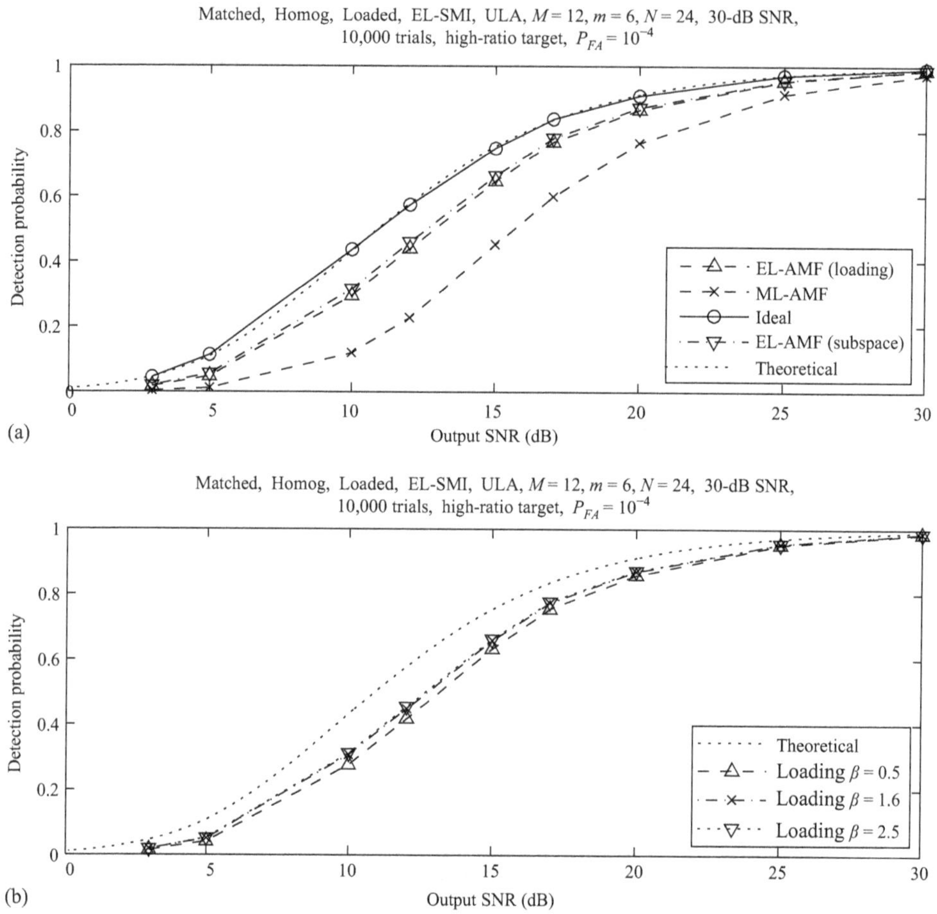

Figure 6.6 AMF ROCs for the fast target, $P_{FA} = 10^{-4}$, for various AMF variants (upper figure) and loadings (lower figure). Note that LAMF performance is not particularly sensitive to the precise loading value. Even a small diagonal loading ($\beta = 0.5$) is only marginally inferior (<0.1 dB) to LAMF for its "optimal" (EL-predicted) loading of $\beta \simeq 1.6$.

The most important result following from Figure 6.6 is that LAMF (EL-AMF) does indeed have significantly better performance. Indeed, the SNR loss factor for $P_{FA} = 10^{-4}$ and $P_D = 0.5$ is 1.6 dB, compared with 5 dB for the standard AMF detector. Less expected is the fact that a very small diagonal loading ($\beta = 0.5$) is only marginally inferior (< 0.1 dB) to LAMF for its "optimal" loading of $\beta \simeq 1.6$. The "optimality" for this fixed loading factor does not need to be very accurately specified. In fact, although not shown, the ROC performance for $\beta = 12,\ 24,\ 48$ is indistinguishable from $\beta = 1.6$. This mirrors justifications for the insensitivity in choosing loading levels for LSMI with strong point-source interferers [46,47].

Thus, our expectations regarding LAMF superiority and performance invariance with respect to the constant loading factor $1 \lesssim \beta < \mu$ are proven correct for this "favorable" scenario. Its quasi-CFAR properties obviously make LAMF especially attractive in such cases;

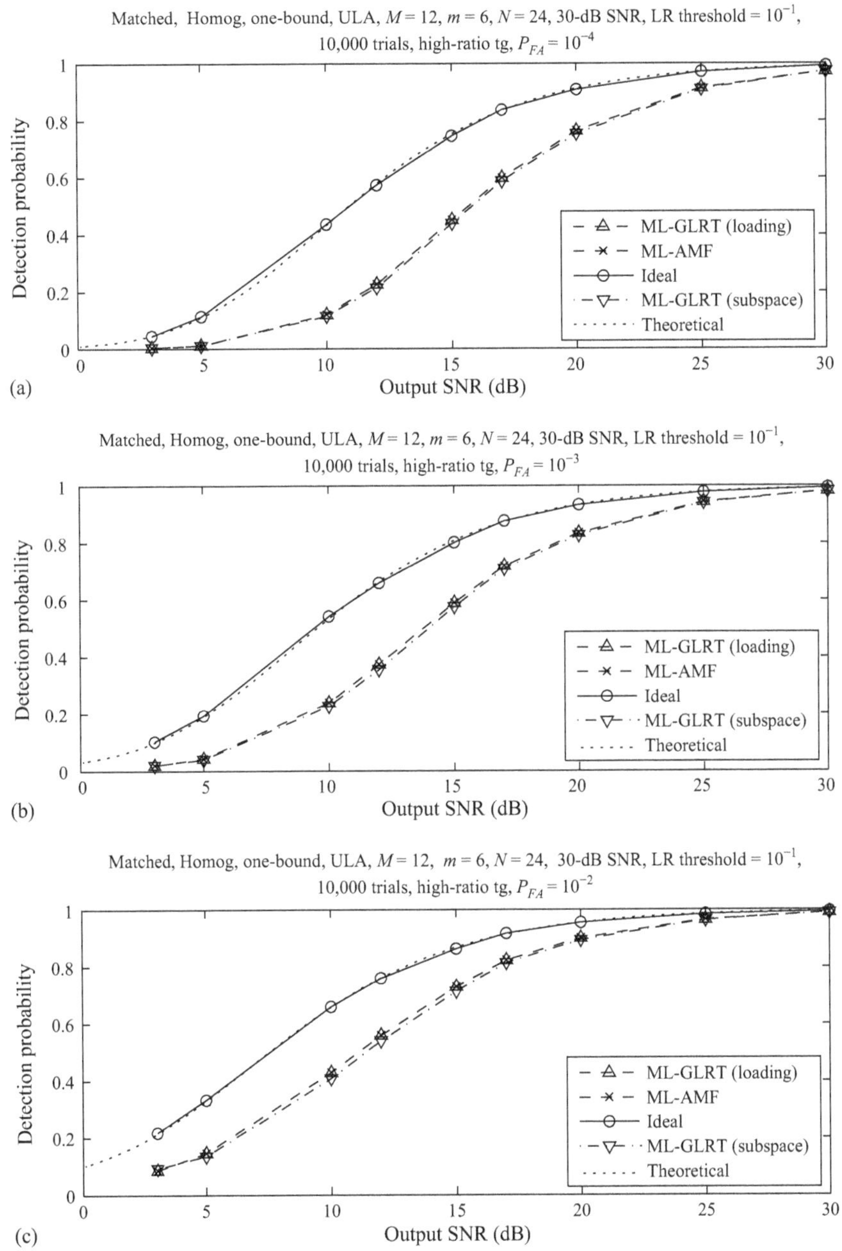

Figure 6.7 ML-GLRT ROCs for the high-ratio target for three different false-alarm probabilities.

this and its high performance sets a high bar for our "theoretically derived" 2S-GLRT and EL-AMF detectors.

Next, we consider the ROCs of the ML-GLRT method (6.103) and (6.104), where only the lower bound for the LR is introduced (that is dependent on the optimized loading factor β or interference-subspace dimension m). Figure 6.7 shows the ML-GLRT ROCs for the

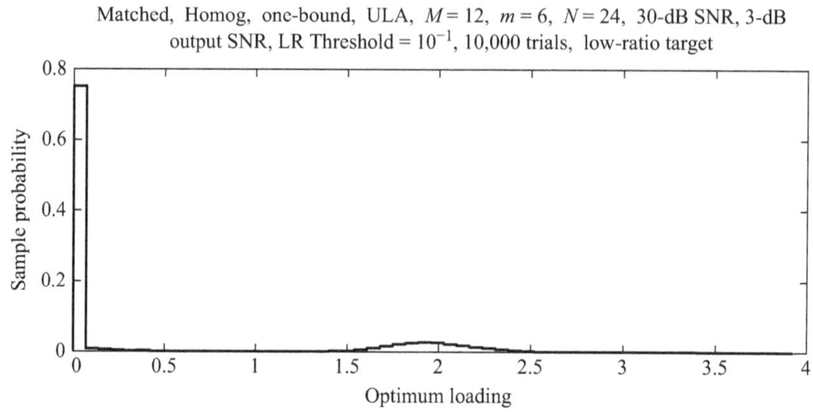

Figure 6.8 Sample pdf for the optimum loading in the ML-GLRT technique.

"high-ratio/fast" target; as before, the ROCs for the "low-ratio/slow" target are visually identical. The lower bound $\alpha_L = 0.0142$ has been chosen so that the probability of generating a LR below this threshold for the true matrix $\boldsymbol{R}_0$ is very low: $P(LR(\boldsymbol{R}_0|\boldsymbol{X}_N) < \alpha_L)$ is set to 0.1 (see Figure 6.1). As we might expect, we see that the ML-GLRT performance is practically identical to that of the (standard) ML-AMF method with its (unloaded) SMI. Again, this precise coincidence is *not* caused by a trivial zero loading-factor selection in the detection test optimization (6.103). The pdf for the selected loading factor for 3-dB output SNR is presented in Figure 6.8. We see that, despite their different nature, our 2S ML-GLRT detector has the same performance as the traditional (zero loading) ML-AMF detector. This proves our assertion that a single snapshot does not make a significant difference, even for the relatively small sample-support studies here. The difference in the covariance matrix estimates for the specific target model analyzed by Kelly [4] is responsible for the slightly better performance of ML-GLRT compared with ML-AMF, whereas for some other models these two techniques are found to be identical (e.g. [20]), or even AMF outperforming GLRT (e.g. [5]).

When both lower and upper bounds on the LR for the optimized loading are introduced in accordance with the EL-GLRT method (6.103) and (6.105), the results are completely different. Figure 6.9 shows the ROCs for the high-ratio target; the ROCs for the low-ratio target are practically identical and are not presented. Despite the broader area of admissible LR values, $P(LR < \alpha_L) = 10^{-2}$ and $P(LR > \alpha_U) = 10^{-2}$ ($\alpha_L = 0.0084$ and $\alpha_U = 0.0647$), the EL-GLRT ROCs are surprisingly close to those produced by the EL-AMF and LAMF methods. Indeed, the same improvement compared with 2S ML-GLRT and ML-AMF is observed in this case. The same practical performance for the diagonal loading and the interference-subspace dimension selection is observed, despite the fact that in this case the optimized signal-subspace dimension is far from being always correct in the 2S EL-GLRT algorithm.

This analysis of a typical scenario with fluctuating target in homogeneous interference clearly demonstrates that ML-based ML-GLRT and ML-AMF detectors share the same performance, while a significant and practically identical performance improvement is obtained for both techniques when our new EL approach is substituted for the standard ML criterion for finding the appropriate diagonal loading or interference-subspace dimension.

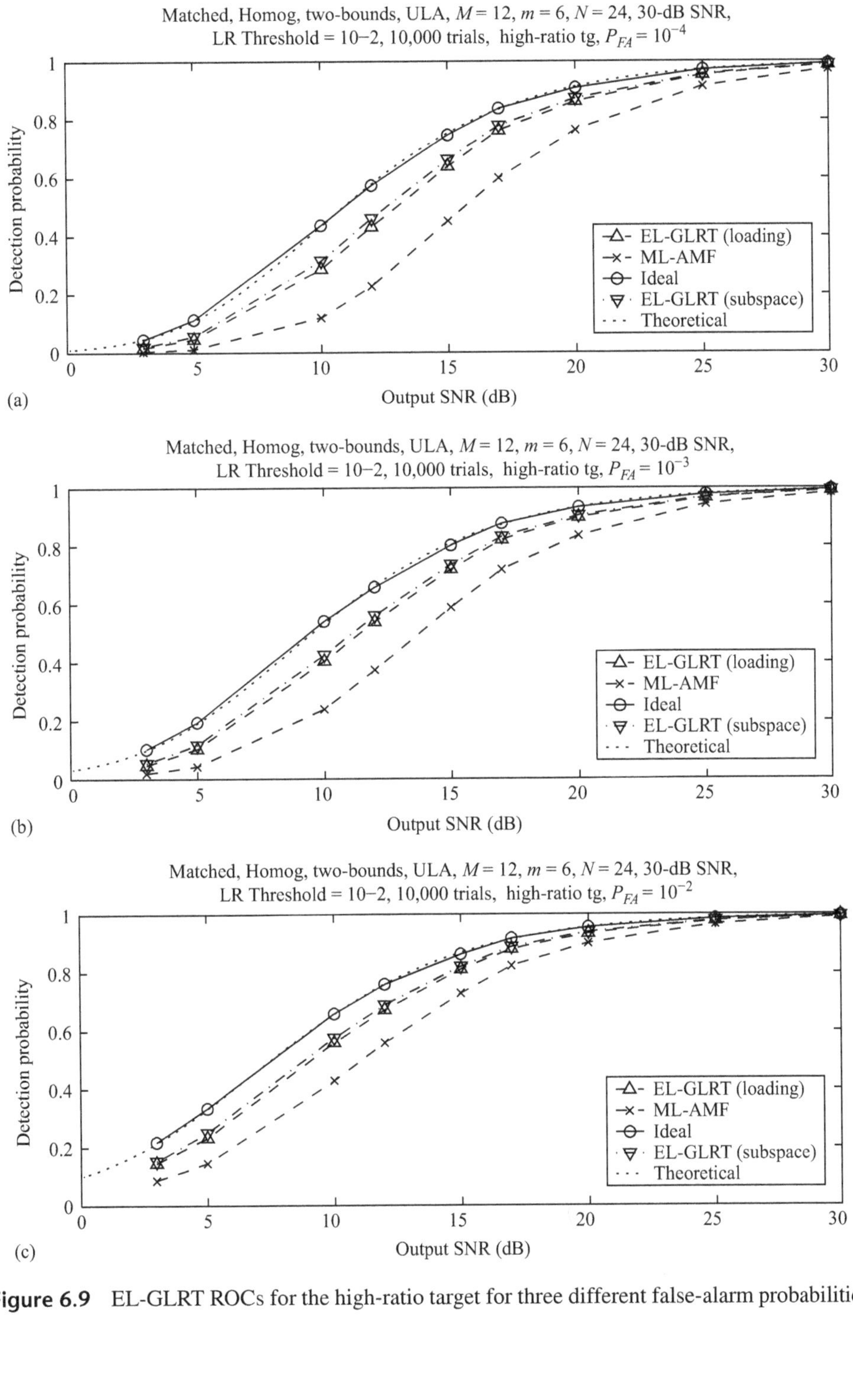

Figure 6.9 EL-GLRT ROCs for the high-ratio target for three different false-alarm probabilities.

As expected for this "favorable" scenario, the data-independent loading of LAMF (6.109) has the same performance as the "theoretically derived" EL-GLRT and EL-AMF techniques. The precise coincidence of performance for the completely different detectors EL-GLRT, EL-AMF, and LAMF most likely means that they have all approached the ultimate performance set by the adaptive detection problem formulation.

Note that the LR generated by the properly loaded ($\beta = 1.6$) matrix $\hat{\boldsymbol{R}}_{LSMI}$ is statistically close to that for the true matrix $\boldsymbol{R}_0$. This match could then be used as a new guide for selecting the diagonal loading factor, but from a theoretical viewpoint could again be treated as another justification of our EL methodology.

6.2.2.5.3 Non-homogeneous Interference Training Conditions; Fluctuating Target with Unknown Power

Here, we briefly introduce our simulation results for this alternate signal model, detection rules, and LRs (6.110)–(6.116) which demonstrate the same overall findings as for homogeneous interference training conditions.

We begin with the ROC of the clairvoyant detector:

$$\frac{|\boldsymbol{y}^{\dagger}\boldsymbol{R}_0^{-1}\boldsymbol{s}|^2}{\boldsymbol{s}^{\dagger}\boldsymbol{R}_0^{-1}\boldsymbol{s}\boldsymbol{Y}^{\dagger}\boldsymbol{R}_0^{-1}\boldsymbol{y}} \underset{H_0}{\overset{H_1}{\gtrless}} h \geq \frac{1}{M} \tag{6.151}$$

that can be analytically computed as [48]

$$P_D = \left[\frac{(1-h)(\sigma_s^2\boldsymbol{s}^{\dagger}\boldsymbol{R}^{-1}\boldsymbol{s}+1)}{(1-h)(\sigma_s^2\boldsymbol{s}^{\dagger}\boldsymbol{R}^{-1}\boldsymbol{s}+1)+h}\right]^{M-1} \tag{6.152}$$

with $P_D = P_{FA}$ for $\sigma_s^2 = 0$. This expression will again be used to validate our Monte-Carlo simulation accuracy.

As before, we start with an analysis of the traditional ML-AMF (ACE) detector (6.113) and compare it with LAMF and EL-AMF. Monte-Carlo simulation of the pdf of the "sphericity test" LR that is used for this model as the EL benchmark is for all practical purposes the same as the "general (non-sphericity) test" that we saw in Figure 6.1, and comparison of the LRs produced for various loading factors shows a correspondence with the pdf produced with a loading factor of $\beta = 2.5$.

Similar to the previous homogeneous case, we checked the accuracy of our theoretical "white-noise" approximation for the target-free output statistics (Theorem 1) and found a perfect match between these and the directly calculated false-alarm threshold values.

Figure 6.11 shows the high-ratio target sample ROCs for the clairvoyant, ML-AMF (ACE), and LAMF detectors, the latter for our three example loading factors. As expected, the theoretical and sample ROCs for the clairvoyant case match perfectly. ML-AMF (ACE) has familiar detection losses of about 3.5 dB for $P_{FA} = 10^{-4}$ and $P_D = 0.5$ [5,40].

The practically identical performance of LAMF for all three loading factors (and that is only 1 or 1.5 dB inferior to the clairvoyant detector) now comes as no surprise. Figure 6.10 also shows the sample pdf of the EL-AMF loading factor that matches $LR[\hat{\boldsymbol{R}}_{LSMI}(\beta)]$ of the loaded sample matrix to the median LR value 0.026. We see that the less accurate *a priori* assumptions on the interference covariance matrices in the primary and training data have resulted in noticeably greater admissible loading factors, from about 1.6 to 3.5. Most importantly, Figure 6.11 shows

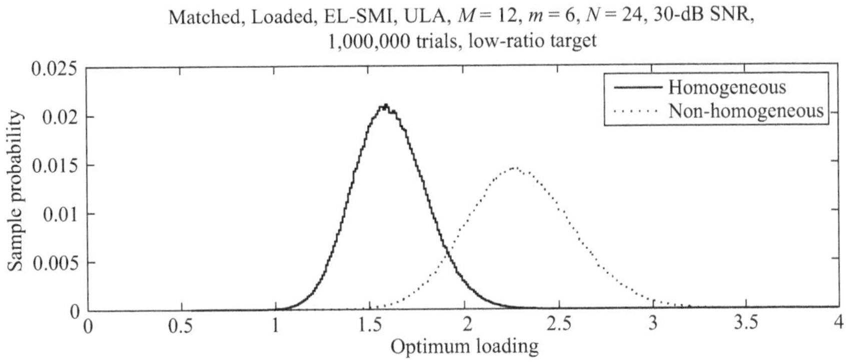

Figure 6.10 Sample pdfs for the optimum loading in the EL-AMF technique.

that EL-AMF is as good as the properly loaded LAMF. Again, the FML detector correctly found the number of sources and is as good as EL-AMF and LAMF.

Now, we consider the ML-GLRT (6.110) and (6.111) results, with its lower bound on optimized LR: $P(LR < \alpha_L) = 0.01$ ($\alpha_L = 0.0145$ according to Figure 6.3). Figure 6.12 presents the "slow" target ROCs (the "fast" ones are almost identical). Unsurprisingly, standard ML-AMF (ACE) and ML-GLRT have almost the same performance for all P_{FA} and P_D, both for diagonal loading and interference-subspace dimension selection. Once again, this coincidence does not mean the trivial equality of these routines. In fact, the ultimate interference-subspace dimension $\hat{m} = 9$ was selected in about 40% of all trials, and the true interference rank (equal to six) was selected in only 10% of trials.

Similar insight is provided by analyzing the sample pdf of the optimized ML-GLRT loading factor (not shown here). For target-free input data, this pdf is dominated by one peak at $\beta \simeq 0$ and has a second peak at $\beta \simeq 2.8$. The second peak is present for small SNRs such as 3–5 dB and disappears for sufficiently strong targets ($\gtrsim$15 dB), where zero loading dominates the selection (probability above 0.9 for SNR > 15 dB). It seems remarkable that these "random walks" in the optimized loading factor and interference-subspace dimension leave practically no trace on the ROC's behavior, as they are found to be the same as for the standard AMF (ML-AMF) test with constant zero loading.

Finally, let us consider the results of the EL-GLRT technique (6.110) and (6.112) for this scenario, where the upper and lower bounds are specified by the conditions $P(LR > \alpha_U) = 10^{-2}$, $P(LR < \alpha_L) = 10^{-2}$ (with $\alpha_L = 0.0056$ and $\alpha_U = 0.0659$ according to Figure 6.3). Figure 6.13 illustrates the EL-GLRT ROCs that repeat the trend of being practically indistinguishable from those of EL-AMF. Despite this, the EL-GLRT optimum loading factor has strikingly different behavior and is strongly dependent on the primary data SNR.

In this regard, it is instructive to analyze the sequence of sample loading-factor distributions with output SNRs varying from 3 to 30 dB (not all illustrated here). We found that for the smallest output SNR, the pdf has two distinct peaks at $\beta = 1.5$ and 3.2. As the SNR increases, the second peak decreases, until it disappears at 30-dB SNR. A detailed analysis of the maximized function $\widehat{\cos}^2(\beta)$ of (6.82) reveals the cause of the two peaks for small output SNR; Figure 6.14 shows the sample pdf $f(\widehat{\cos}^2)$ for the "slow" target and 3-dB output SNR. We see that maximization of this function in the vicinity of $\widehat{\cos}^2 = 1/M$ can "drive" the

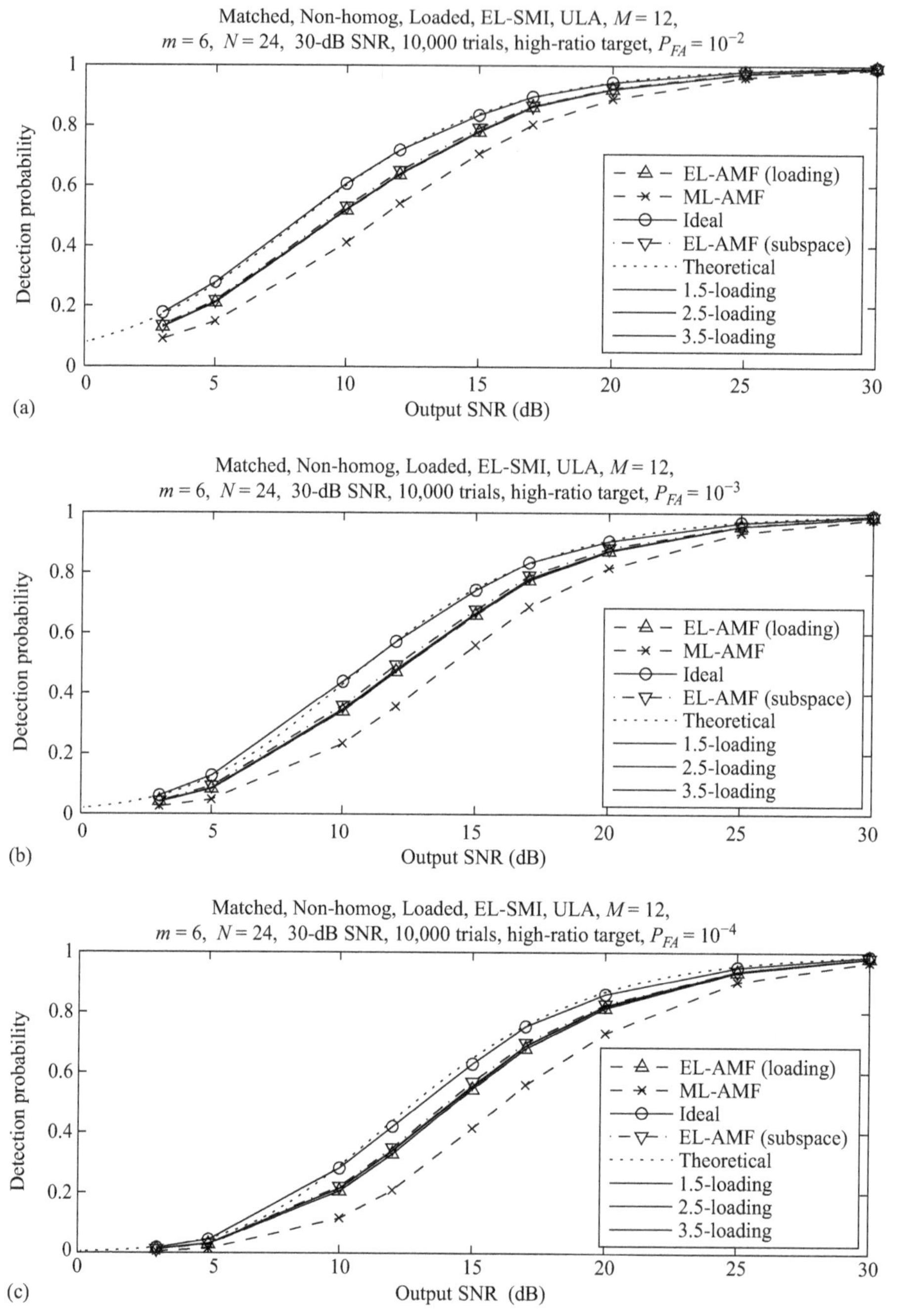

Figure 6.11 Non-homogeneous AMF ROCs for the high-ratio target for three different false-alarm probabilities.

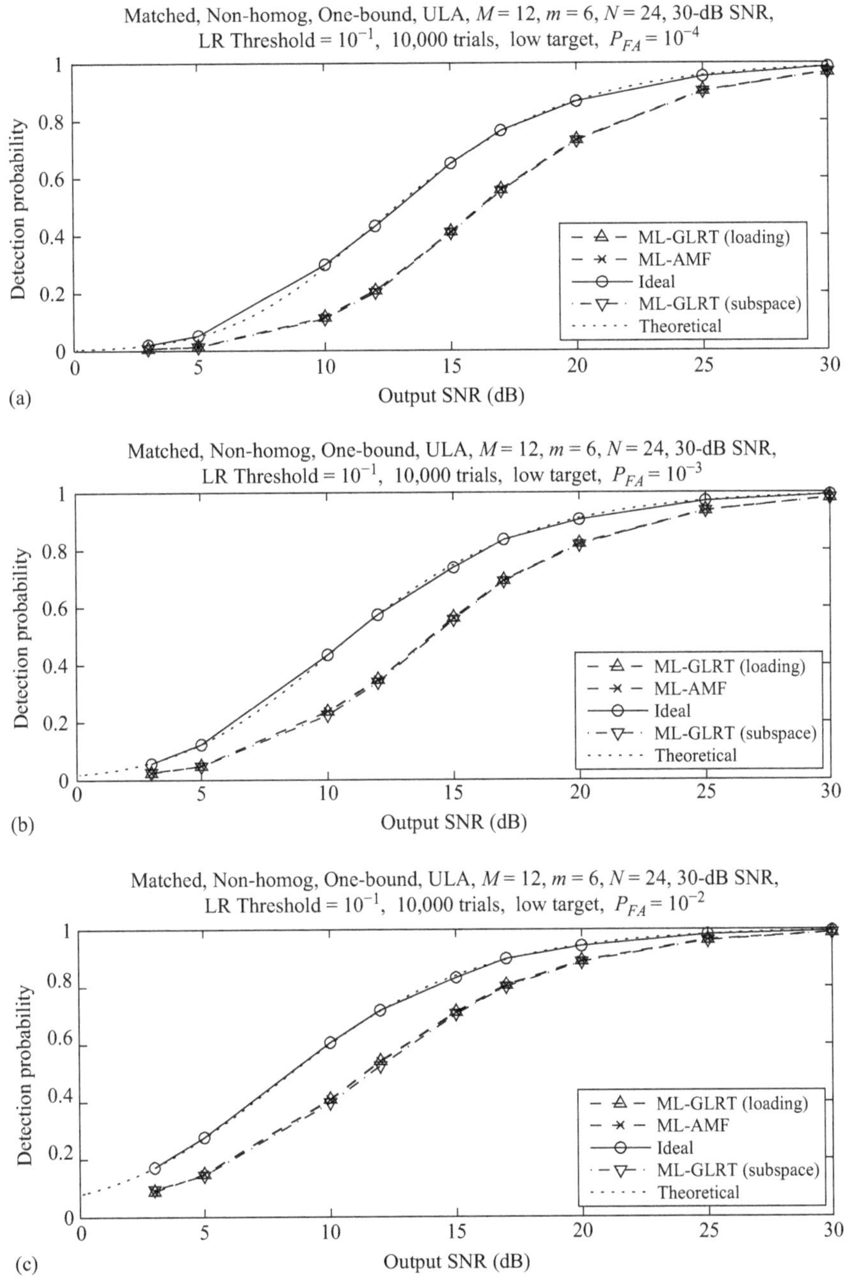

Figure 6.12 Non-homogeneous ML-GLRT ROCs for the low-ratio target for three different false-alarm probabilities.

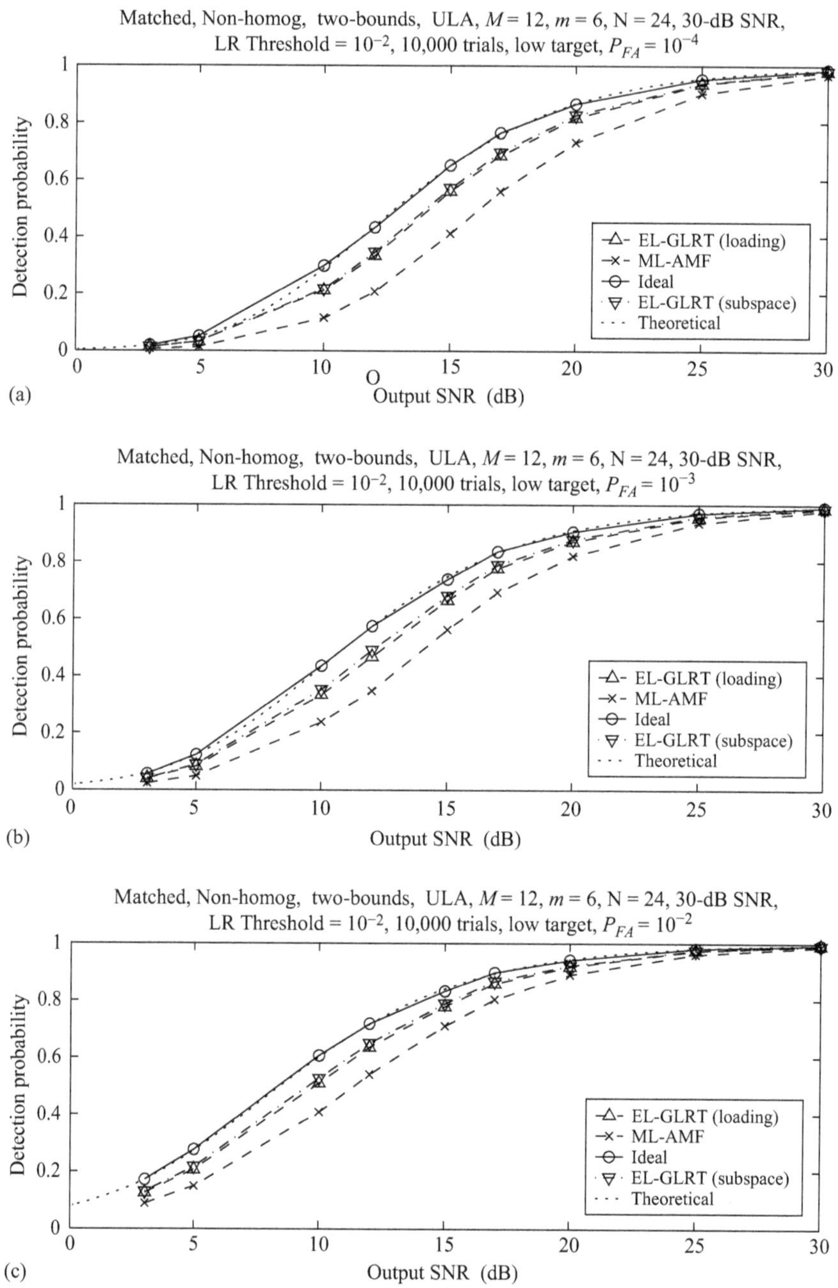

Figure 6.13 Non-homogeneous EL-GLRT ROCs for the low-ratio target for three different false-alarm probabilities.

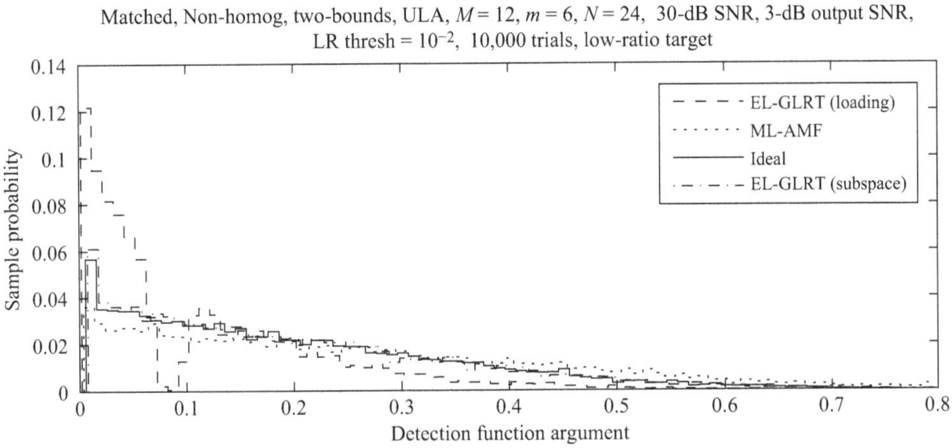

Figure 6.14 Sample pdf for $\widehat{\cos}^2$ for the "slow" target scenario and 3-dB output SNR.

argument either into the area $\widehat{\cos}^2 < 1/M$ (which means no detection) or into the admissible area with $\widehat{\cos}^2 > 1/M$. In the former case, greater loading factors (within the permitted range) are selected, while in the latter case, smaller loading factors are found to be optimal. Similar behavior is exhibited by the optimal interference-subspace dimension m. It is remarkable that, despite the significant redistribution of the optimized $\widehat{\cos}^2$ compared with the clairvoyant (and EL-AMF) cases, the overall ROCs are exactly the same as the EL-AMF ones, and only about 1 dB below the clairvoyant case.

This detailed analysis demonstrates that the accurate equivalence of EL-GLRT, EL-AMF, and LAMF performance is not due to the trivial equivalence of these routines for the given model, as in Reference 20 for example. Such an accurate performance equivalence of these three quite different algorithms suggests that in different ways both methods approach the ultimate performance set by the nature of the adaptive detection problem. It is also evident that the particular parameters used here for EL-GLRT ($P_{bound} = 10^{-2}$) and EL-AMF ($P_{median} = 0.5$) have no practical impact upon the demonstrated performance.

6.2.2.5.4 "Unfavorable" Interference Scenario

It seems quite important to conclude our study by considering the performance of our new detectors for interference models that are "unfavorable" for the LSMI and/or FML techniques, that is, scenarios with full-rank interference (no "noise subspace") and having no obvious abrupt change in the size of the (sorted) covariance matrix eigenvalues.

In fact, adaptive processing does not make sense for small ratios λ_1/λ_M. Even the clairvoyant detector has only a marginal improvement over the "white-noise-matched" receiver in this case, hence the loss associated with adaptivity can actually exceed the potential improvement. The ultimate example of such a scenario is input white noise. Of course, nobody should consider adaptive processing for internal white noise, but sometimes the external interference environment can resemble white noise, such as when the number of strong interference sources exceeds the number of antenna sensors (exceeding the degrees of freedom of the array). If the only available information about the external interference is contained in the training data, then a properly designed adaptive detector should succeed for such scenarios as well.

Let us now consider the same six-source interference scenario (6.141) and (6.143), except that each source is now equally "spread" (distributed) [49]. Here the interference covariance matrix $\boldsymbol{R}_0$ can be written as [50]

$$\boldsymbol{R}_0 = B \odot \left[\sigma_n^2 \boldsymbol{I}_M + \sum_{j=1}^{6} \sigma_j^2 \boldsymbol{s}(w_j) \boldsymbol{s}^\dagger(w_j) \right] \tag{6.153}$$

where [49]

$$B = \{\exp\left[-\nu|\ell - k|\right]\}_{\ell,k=1,\ldots,M}, \quad \beta > 0 \tag{6.154}$$

is the "spreading matrix" and $\odot$ denotes the Schur–Hadamard (element-wise) matrix product. Similar to Reference 49, we use a spreading factor of $\nu = 0.25$.

Such spreading "annihilates" the noise subspace of the original point-source covariance matrix. The eigenspectrum shown in Figure 6.2 is now characterized by a ratio $\lambda_1/\lambda_M = 10$ compared with the original $\lambda_1/\lambda_M \simeq 35$ dB. Moreover, the clairvoyant optimum filter ($W_{opt} = \boldsymbol{R}_0^{-1}\boldsymbol{s}$) has an SNR improvement over the "white-noise"-matched filter ($\boldsymbol{w}_{wn} = \boldsymbol{s}$) of only 0.40 dB for the high-ratio (fast) target and 0.27 dB for the low-ratio (slow) target, compared with the favorable scenario values of 28.5 dB and 26.7 dB, respectively.

We consider non-homogeneous training conditions with a fluctuating target of unknown power. (Based on our previous results, we expect similar behavior for the homogeneous scenario.) Since the clairvoyant and ACE ROCs do not depend on scenario and are exhaustively specified by N, M, P_{FA}, and output SNR, the ACE detector will again be about 3.5 dB inferior to the clairvoyant detector. We need to investigate whether the EL-AMF and properly loaded LAMF detectors can reduce these losses for diagonally loaded and FML sample covariance matrix estimates. Since the pdf for the sphericity test $LR(\boldsymbol{R}_0)$ is scenario-independent and so is the same as already shown in Figure 6.1, we can proceed to the pdfs of the sphericity test $LR(\hat{\boldsymbol{R}}_{LSMI})$ for the three loading factors $\beta = 3$, 5,000, and 10,000; Figure 6.15 shows the latter two.

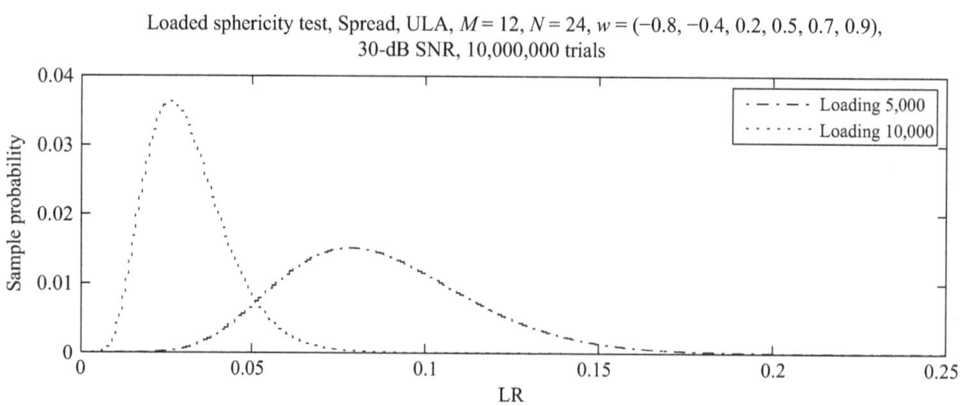

Figure 6.15 Sample pdfs for the loaded sphericity test for spread (distributed) interferences.

The traditional loading factor of $\beta = 3$ (with respect to unit internal noise power) for favorable scenarios is in this case "as good as" zero loading, as it does not affect the LR whose pdf is concentrated within the range of LRs 0.9996–1. Only for very high loading factors such as 5,000 and 10,000 (as shown in Figure 6.15) the LR pdf behaves similar to the EL benchmark. Note that $\beta = 10^4$ means that the diagonal loading here is of the same order of magnitude as the *maximum* eigenvalue of the true covariance matrix $\boldsymbol{R}_0$. In fact, such loading "drives" LAMF towards the "white-noise-matched" detector

$$\frac{|\boldsymbol{y}^\dagger \boldsymbol{s}|^2}{\boldsymbol{y}^\dagger \boldsymbol{y} \boldsymbol{s}^\dagger \boldsymbol{s}} \underset{H_0}{\overset{H_1}{\gtrless}} h > \frac{1}{M} \tag{6.155}$$

which is quite understandable given the marginal SNR improvement provided by the clairvoyant optimal filter.

Naturally, the CFAR properties for "favorable" covariance matrices that were proven in Subsection 6.2.2.4 do not hold for these scenarios and such loading factors. Indeed, for the low-ratio (slow) target and $\beta = 10^4$, the false-alarm rates $P_{FA} = 10^{-2}, 10^{-3}, 10^{-4}$ are imposed by the threshold values 0.43, 0.56, 0.66, respectively, while for the high-ratio (fast) target they are 0.22, 0.32, 0.42. Despite losing the CFAR properties, a ROC analysis of LAMF still makes sense (see Figure 6.16). Whereas the clairvoyant detector ROCs again demonstrate a perfect match between analytical calculations and simulations, we see that the LAMF performance for $\beta = 3$ is the same as for ACE, as predicted. At the same time, the "properly loaded" LAMF detector (β=5,000, 10,000) is only 0.4 dB inferior to the clairvoyant detector. The similar performance of EL-AMF now comes as no surprise.

Figure 6.17 shows the sample pdf (calculated over one million trials) of the optimum loading factor in the EL-AMF technique that was found by matching the sphericity test LR for the loaded sample covariance matrix $\hat{\boldsymbol{R}}_{LSMI}$ with the "expected" LR value $LR_0 = 0.026$ (see Figure 6.1). As we might expect, the optimum loading factor lies mostly in the range $5,000 \lesssim \beta \lesssim 10,000$. Interestingly, the optimum signal-subspace dimension histogram (not illustrated here) is not single-valued: 146 trials chose $\hat{m} = 6$, with the remainder $\hat{m} = 7$. While the "loaded" EL-AMF detector is as good as LAMF with proper loading, the EL-FML detector (where the signal-subspace dimension is found by LR matching (6.114) and (6.115)) is marginally inferior (less than 0.2 dB). The coarse discretization of the parameter m could be the reason for this small degradation.

Note that traditional recommendations regarding loading-factor selection ($\beta = 3$) or signal-subspace selection ($m = 6$), stemming from favorable conditions, are completely inappropriate here.

Whereas the properly loaded LAMF detector is statistically equivalent to EL-AMF, in practice, when only a sample covariance matrix is available, the only option seems to be to choose the "proper loading factor" (such as $\beta = 10^4$) through LR matching with the "expected" LR.

Results for the low-ratio (slow) target, as well as the 2S EL-GLRT detector, are similar: 2S ML-GLRT is as good as the conventional ACE, with 2S EL-GLRT, EL-AMF, and properly loaded LAMF being marginally inferior to the clairvoyant detector.

Finally, we wish to comment that the loss of the CFAR property for 2S EL-GLRT, EL-AMF, and LAMF is explained by the EL matching driving them close to the robust "white-noise"-matched detector, which is not a CFAR detector with respect to the covariance matrix $\boldsymbol{R}_0$ ($\neq \boldsymbol{I}_M$) and signal $\boldsymbol{s}$.

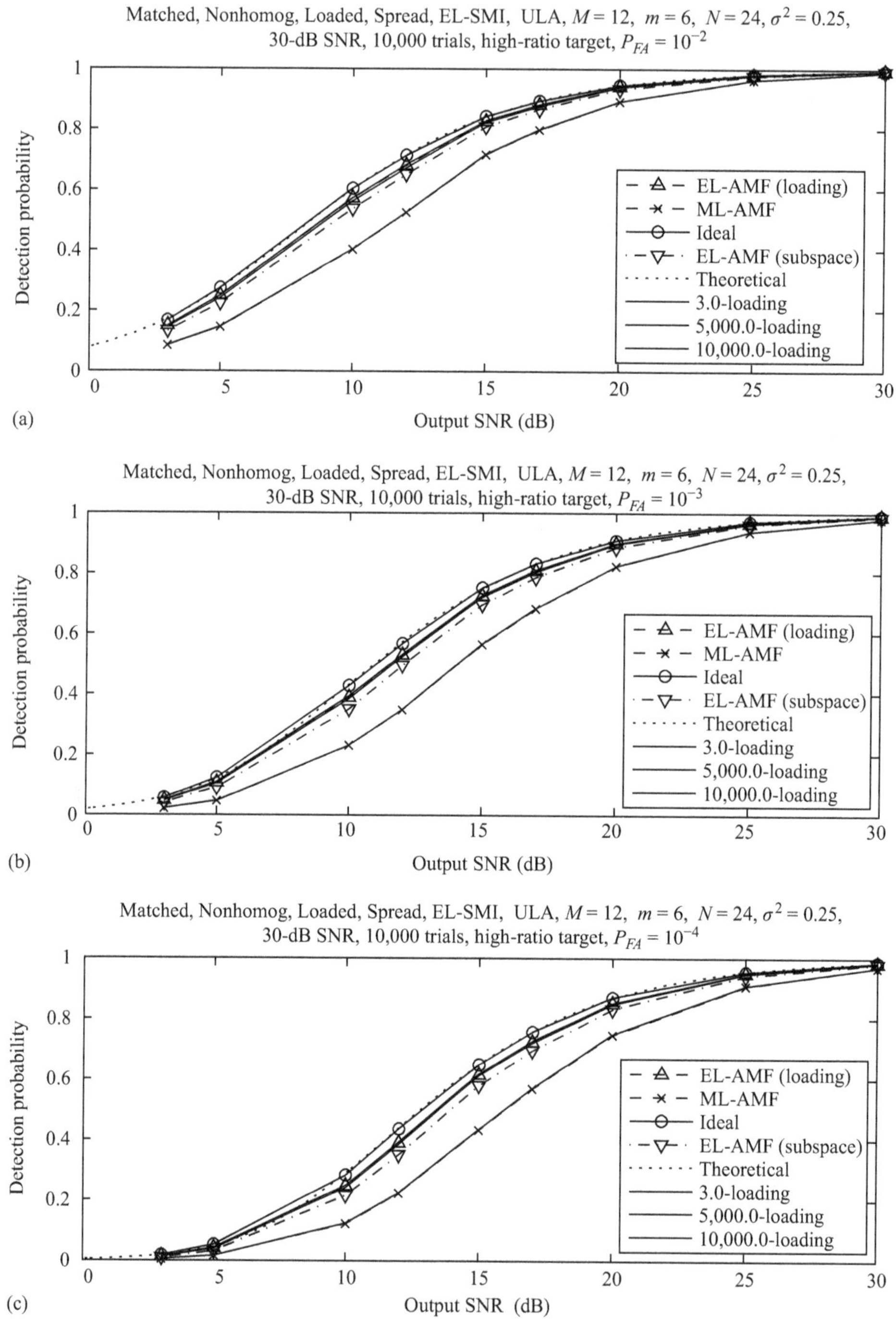

Figure 6.16 Non-homogeneous AMF ROCs for the high-ratio target for spread interferences.

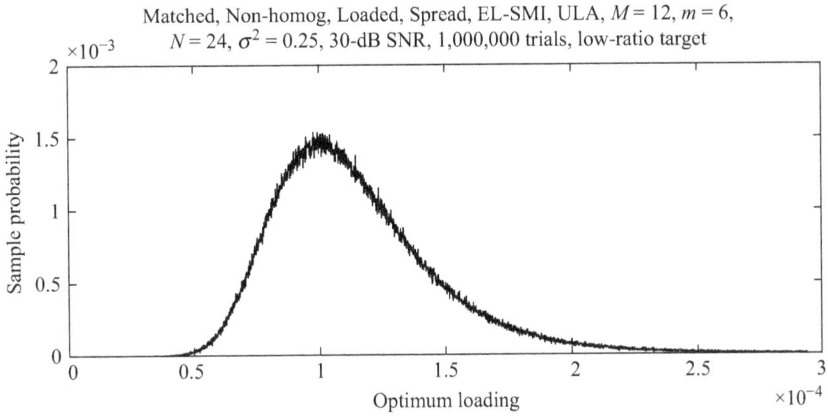

Figure 6.17 Sample pdf for the optimum loading in the EL-AMF technique for spread interferences.

6.2.3 Observations

In this section, we have diligently pursued two major goals. First, based on the well-known properties of the LSMI and FML adaptive *filter* techniques, we introduced adaptive *detectors* that use the same type of covariance matrix estimates. We did this anticipating that such detectors should have better performance than those that use the conventional ML covariance matrix estimate, at least for (favorable) interferences that have widely separated signal and noise eigensubspaces. Having achieved this, we second addressed an important theoretical issue that follows from the now-established superiority of these LSMI and FML detectors. Namely, it seemed necessary to us to suggest modification of the frameworks that were introduced for GLRT by Kelly [4] and for AMF by RFKN [5] to include these new advanced detectors. We sought a modified theoretical framework that would either justify these detectors and/or generate new ones that are at least as efficient as LAMF and FML for favorable scenarios.

More specifically, with respect to the first goal, we demonstrated that (for favorable interference) the LAMF detector that uses the diagonally loaded sample covariance matrix estimate whose loading factor is chosen from the broad range between the minimum signal and the noise eigenvalues does indeed give a significant detection performance improvement. Moreover, for such interference we demonstrated an important invariance property of the "target-free" detection statistics. We showed that these statistics are invariant with respect to the true covariance matrix and can be closely approximated by "white-noise-generated data" and that the pdf is only a function of the filter dimension, the training sample size, the signal-subspace dimension, and the loading factor. The approximation accuracy is sufficiently high to permit the precalculation of false-alarm thresholds, which means that in this case the LAMF detector has, in fact, the CFAR property.

With quite negligible losses demonstrated by LAMF compared to the clairvoyant detector, it then became a challenging problem to propose a theoretical adaptive detection framework capable of giving similar detection performance. In order to achieve this second goal of our study, we reconsidered two important issues in adaptive detection that were postulated within the traditional (Kelly's one-sample) GLRT and AMF techniques. The first concerns the arrangement made by Kelly [4] whereby "the decision rule will be formulated in terms of the totality of input data without the *a priori* assignment of different functions to the primary and secondary input." Specifically, despite the assumption that the secondary data are free of target component, "any selection rules applied to make this assumption more plausible are ignored" in his technique. We have reformulated the GLRT problem as a 2S adaptive detection problem, where the target-free status of the training data is respected, and so the interference covariance matrix estimate, while depending on the primary sample, does not depend on the hypothesis being tested.

More specifically, we have introduced a new 2S-GLRT technique whereby the single covariance matrix estimate for both hypotheses is adaptively selected from a family of admissible solutions so as to maximize the "detection function," which is the associated likelihood ratio. Unlike most published studies, we used the fluctuating Swerling I model for the target signal, which is specified by a (possibly unknown) positive power. This modeling introduced some important changes into the nature of the optimized LR; most importantly, since a non-positive ML target power estimate is inadmissible, it must be replaced by a zero estimate that corresponds to a decision that the target is absent.

The second (more important) issue that we have reconsidered here concerns the ML criterion used within the traditional GLRT and AMF techniques. It is important that only asymptotic arguments support the GLRT methodology, so there is no reason why estimates other than the ML one could not be found to be better suited for detection, especially for relatively small training sample sizes. Three straight-forward observations alerted us to challenge this dogma. The first stems from the difference in performance between Kelly's GLRT and the AMF methods; for different models, one method was found to be superior [5]. The suggestion made in Reference 5 that Kelly's technique should be better just because it involves an additional single (primary) snapshot in the covariance matrix estimation could be easily refuted, and so the performance difference should rather be attributed to the specific estimates used.

The second observation arises from the fact that the LSMI or FML sample covariance matrix estimates, which are found to be very successful in LAMF (FML) detectors, are not the ML covariance matrix estimates.

The third theoretical observation arises from a comparison of the actual LR (which is just the normalized likelihood function) produced by the exact (true) covariance matrix and the (unconstrained) ML estimate adopted by both GLRT and AMF. While the ML covariance matrix estimate (sample matrix) always delivers the ultimate value for the maximized LR, equal to one irrespective of the sample size, the exact covariance matrix yields much smaller LRs for relatively small sample support. For the specific example analyzed in this section, with a $M = 12$-sensor uniform linear antenna array and $N = 24$ samples, the median value of the LR for the exact covariance matrix is found to be only 0.025, and with probability 0.99 the LR lies in the range 0.008–0.07. Thus for relatively small sample support, the ML estimator is extremely far from the true covariance matrix, even in terms of the LR/LF metric.

For this reason, we have introduced an approach called "EL," whereby we try to find the estimate that statistically generates the same LR as the exact covariance matrix. This is feasible in practice since the pdf for the LR generated by the exact covariance matrix does not depend on the matrix itself, but only the parameters M and N, and so it can be precalculated.

We have used the above well-known families of covariance matrix estimates: the diagonally loaded sample matrix (i.e. the loaded unconstrained ML solution) and the finite-subspace interference approximation of the ML solution. For these estimates, respectively, the traditional ML criterion drives the loading factor to zero, and the interference-subspace dimension to its maximum. For 2S ML-GLRT, the loading factor and interference-subspace dimension are constrained only by the lower bound on LR, while the maximum LR value is allowed to reach the ultimate value of unity, as per the ML solution. Despite the new formulation, our Monte-Carlo simulations demonstrated that 2S ML-GLRT detection performance is almost the same as for the traditional ML-AMF (ACE) method that uses the standard ML covariance matrix estimate (zero loading, full interference rank). Thus, the detection performance is again proven to be dominated by the type of ML estimate rather than the choice of GLRT or AMF method that uses this estimate.

Our new EL method searches for the diagonal loading or interference-subspace dimension so that the modified EL estimate gives a LR value properly within the range of LRs expected for the exact covariance matrix. To be more specific, for the 2S EL-GLRT method, we maximize the detection function over the set of (loaded, finite-rank) solutions bounded by the precalculated upper and lower LR bounds. For the EL-AMF (ACE) technique, we seek the loaded solution that generates the median LR value of the exact covariance matrix. For finite-rank approximations that have only a finite number of solutions, we simply find the one that is closest to the upper LR bound, if no solution within the bounds is available.

Our extensive Monte-Carlo simulations for a scenario with signal- and noise-subspace eigenvalues separated by several tens of dBs showed that the EL-GLRT and EL-AMF methods have practically the same performance. For finite-rank approximation in this example, all solutions were of the same subspace dimension, as only the true interference rank estimate was closest to the bounded LR region, making no difference between "adaptive" and "fixed" signal-subspace dimension. While the loading factor fluctuates in the well-known range of 1.5–3 times the white-noise power, depending on the output SNR, the detection performance of EL-GLRT is the same as for the EL-AMF approach, whereby the loading factor is selected based on the training sample only. Most importantly, this performance is significantly better than that of the ML-GLRT and ML-AMF detectors (the loss factor improved to 1–1.5 dB below the clairvoyant case for $P_d = 0.5$ and $P_{FA} = 10^{-2} - 10^{-4}$ compared with 5 dB for the standard GLRT and ACE techniques) and again is practically the same in this case as for robust selection of the constant loading factor ($\beta = 3\sigma_n^2$) for the LAMF technique.

These results demonstrate that our EL criterion for the proper families (diagonally loaded, finite interference rank) gives a significant improvement in detection performance compared with the ML criterion, which for small sample support produces solutions far away from the exact ones. We emphasize that the introduced families include the standard (unconstrained) ML covariance matrix estimate, while the major distinction stems from the attempt to get a statistically close LR to that of the exact covariance matrix, rather than just the (ultimate) maximum LR value. This is an important distinction from some optimum search over a restricted set of covariance matrices, such as the class of Toeplitz covariance matrices. Any reliable *a priori* structural information on the covariance matrix should always lead to a detection improvement, however we chose the most generic families specifically to underline the difference in criteria (EL versus ML), rather than any possible difference in covariance matrix description.

More specifically, this approach allowed us to generate solutions that for favorable scenarios and LSMI/FML applications demonstrated the same detection performance as the "constant-loaded" LAMF detector. The fact that in this case the modified 2S EL-GLRT and EL-AMF framework had the same performance may be treated as an additional justification for the "conventional" LSMI/FML technique that is now used in the adaptive detectors. Indeed, we believe that the surprisingly accurate coincidence of ROCs for the three quite different techniques as 2S EL-GLRT, EL-AMF (with adaptively chosen loading), and "conventional" LAMF (with constant loading) just means that they all approach the best possible performance that is set by the adaptive detection problem formulation itself.

For favorable scenarios, it has been known since [13,14] that the loading factor can be robustly selected so that the SNR performance of the LSMI adaptive filter does not depend on any particular loading value. Now we have demonstrated that this property extends to include LAMF detection performance under the condition that the false-alarm threshold is adjusted to the selected loading factor.

The other simulation scenario that we considered was specifically selected to not have this favorable property, with its full-rank "spread" interference sources and relatively small separation between signal- and noise-subspace eigenvalues. We demonstrated that for a given sample support, a constant loading factor can still be selected that makes LAMF performance as good as EL-AMF performance with its data-dependent loading factor. Both these detectors are significantly better than the standard AMF (ACE) technique, even though at the expense of losing the CFAR property. Yet, the loading factor for this scenario must be chosen to be comparable to the *maximum* eigenvalue, unlike our previous favorable scenario whose optimum loading was comparable to the white-noise power (minimum eigenvalue). It is evident that some "scenario recognition" needs to be considered in order to avoid erroneous loading-factor selection. From this viewpoint, the EL-AMF method that has a universal LR fitting for adaptive loading-factor selection based only on the training data has an important practical advantage.

Finally, for favorable scenarios, the EL-AMF and 2S EL-GLRT detectors enjoy the CFAR property, similar to the LAMF detector. The output target-free statistics for these detectors can be approximated with high accuracy by white-noise-only data. The pdfs of these statistics are functions of sample support N, filter (antenna) dimension M, number of dominant eigenvalues m, the lower and upper LR bounds α_L, α_U (for EL-GLRT), and the median LR value (for EL-AMF). While accurate analytic expressions for these pdfs have not been derived, we demonstrated that white-noise Monte-Carlo simulations can be used to give sufficiently accurate false-alarm threshold values.

6.3 Improving Adaptive Detection Using Covariance Matrix Structure

It is shown in Section 6.2 and Reference 1 that rank reduction techniques such as loading and FML can also be applied to adaptive detectors that encounter "small-rank" covariance matrix estimates (CMEs). In cases described by a model of finite-rank interference and white noise, we demonstrated both a significant performance improvement and robust behavior of such "small-rank" techniques.

However, for applications that deal with spread/distributed/scattered/broadband interference sources (in terms of Doppler and/or spatial frequency) [51], the rank of the covariance matrix is typically not so small nor well defined, and so other parametric models are applied, such as an autoregressive model of order m, AR(m), where m is small compared to M [52,53].

We emphasize that in dealing with spread interferences (or backscattered clutter) in this section, we still only consider scenarios where the clairvoyant filter $\boldsymbol{w}_{opt}$ is significantly better than the non-adaptive "white-noise optimal" filter $\boldsymbol{w}_{WN} = \boldsymbol{s}/(\boldsymbol{s}^{\dagger}\boldsymbol{s})$, in contrast to the spread sources illustrated in Figure 6.2. In our considered cases here, the performance improvement is much greater than the totality of losses associated with adaptivity (e.g. finite-sample support), and so for spread interferences we are still dealing with a quite large ratio of largest to smallest eigenvalue of the true covariance matrix $\boldsymbol{R}$, λ_1/λ_M, in accordance with the Kantorovich inequality given in (6.2.1). Thus, instead of the "cliff-like" interference eigenspectrum for "small-rank interference" models given in (6.6) that were considered favorable in the previous Section 6.2, spread interferences require us to consider a smoother "slope-like" eigenspectrum that may be better described by a small-order AR model.

While AR(m) models and, in general, stationary models with Toeplitz covariance matrices are used in some practical applications [54], they are often considered too restrictive. Indeed, in dealing with time sequences in TAP (that is typically applied to adaptively mitigate clutter), we must consider strictly stationary clutter and strictly periodic pulse trains. If the clutter is not stationary (e.g. due to platform motion in airborne radar, or ionospheric motion in over-the-horizon radar) or the pulse train is not periodic, the clutter covariance matrix is no longer Toeplitz. For antenna array applications, the Toeplitz covariance model is even more restrictive since it implies a perfectly calibrated uniform linear antenna array.

Therefore, by removing this stationarity assumption, instead of the AR(m) model that can describe any M-variate positive definite (p.d.) Toeplitz covariance matrix (for $m = M - 1$), we now want to find a model that analogously describes (for $m = M - 1$) any M-variate Hermitian covariance matrix. One such approximation is well known and frequently used in adaptive processing: the above model of $m < M$ point (rank-one) sources in white noise. Unfortunately, for spread sources this model may require m to be equal to the number of dominant covariance matrix eigenvalues, and hence considerably exceed the number n of spread interferences. For such scenarios, we seek an "AR(m)-like" model that can describe a Hermitian covariance matrix with some number of dominant eigenvalues that significantly exceeds the model order m.

We expect that for the maximum possible order $m_{\max} = M - 1$, such an "AR(m)-like" model will accurately describe any given M-variate p.d. Hermitian covariance matrix. Such an approximation is provided by the Dym–Gohberg band-matrix extension technique [55], whereby any given M-variate p.d. Hermitian matrix $\boldsymbol{R}$ is transformed into a unique TVAR(m) approximation $\boldsymbol{R}_m$ with the remarkable property that its elements are exactly the same as those of $\boldsymbol{R}$ within the $(2m + 1)$-wide central band, while outside this band its elements are extended such that all elements of the inverse matrix R_m^{-1} outside the band are zero. (Note that the AR(m) approximation of a p.d. Toeplitz matrix has the same property.)

We recently demonstrated [56] that the Dym–Gohberg (DG) transformation of the traditional sample CME $\hat{\boldsymbol{R}}$ averaged over $N > m$ i.i.d. Gaussian training samples is the ML estimate of the TVAR(m) covariance matrix $\boldsymbol{R}_m$. Then in References 57 and 58, we introduced methods for estimating the order m of a TVAR(m) or AR(m) model, again given $N > m$ i.i.d. Gaussian data samples. These results allow TVAR(m)-based adaptive detectors to be used effectively for truly TVAR interferences, truly AR interferences, and non-TVAR interferences. In the first two cases, any study should focus on the performance degradation with respect to the clairvoyant detector caused by finite-sample support (stochastic losses). In the latter case, the study should also consider model-mismatch losses, which clearly depend on scenario. Indeed, when a TVAR(m) model serves only as an approximation to an arbitrary Hermitian covariance matrix, it may or may not be found to be the most appropriate approximation for the given matrix, steering vector, and sample size. For example, it is possible to choose between "small-rank" and "small TVAR-order" approximations and even to try to make this decision adaptively, driven only by the training data. Since a solution of the more complicated problem involving non-TVAR interference relies on the stochastic properties of TVAR(m)-based adaptive detectors, it is natural to concentrate this study on the stochastic properties of TVAR(m)-based adaptive filters and detectors, leaving the methodology for non-TVAR interference applications for separate studies [59–61].

While in References 56–58 we addressed theoretical issues of TVAR(m) model and order estimation with a few illustrative examples, this study concentrates on a performance analysis of adaptive filters and (especially) adaptive detectors for interference with TVAR(m) or AR(m) properties, limited by finite-sample support.

When considering the stochastic losses, there are two major issues regarding the efficiency of the ML TVAR CME for adaptive processing that need to be addressed. First, we need to investigate the efficiency of adaptive filters (beamformers) that use this ML estimate; specifically, we need to analyze the SNR losses associated with the sample volume N in the general case where the order of the adaptively estimated model TVAR(μ) may be overestimated, i.e. when $\mu \geq m$. Second, for TVAR(μ)-based adaptive detectors, we also need to investigate whether this ML estimate gives a satisfactory CFAR, which means that the pdf of the target-free output signal of the detector is sufficiently invariant (robust) with respect to the true TVAR(m) covariance matrix, to the extent that false-alarm thresholds can be precalculated in some way with an acceptable accuracy. (For example, in addressing the CFAR property for diagonally loaded adaptive detectors in Section 6.2 and Reference 1, we proposed certain "white-noise equivalent" models that for most important practical cases allow us to precalculate these thresholds with high accuracy.)

It is for this reason that the central issue of this section, which studies the performance of adaptive detectors based on the ML TVAR CME, is an analysis of the sensitivity of false-alarm rate on the scenario, and the search for scenario-invariant equivalent models. For adaptive detectors that are "practically CFAR," we then evaluate their performance with the standard receiver operating characteristic (ROC) curves.

This section has the following structure. Subsection 6.3.1 briefly reintroduces the TVAR(m) model from References 56 and 57 as a finite-order parametric approximation of an arbitrary Hermitian covariance matrix for adaptive estimation and describes the DG transformation that provides this approximation.

Subsection 6.3.2 shows that the SNR loss compared to the clairvoyant filter, due to finite-sample support, is not strictly independent of the true covariance matrix $\boldsymbol{R}$, nor, especially, the structure of the steering vector $\boldsymbol{s}$. A similar dependence is found for TVAR(μ)-based adaptive detectors, which are also introduced in this section, as is an equivalent model that allows the SNR loss and false-alarm rate to be precalculated with a certain accuracy that may be acceptable in many practical applications.

Subsection 6.3.3 analyzes the accuracy of this equivalent model by Monte-Carlo simulations of various TVAR(m)/AR(m) scenarios by comparing the actual and predicted performances of the TVAR(μ)-based adaptive filters/detectors. We demonstrate that the weak dependence of SNR loss and false-alarm rate on the scenario is due to a quite subtle phenomenon, namely that a certain univariate distribution of random numbers is indeed scenario-invariant, but the corresponding multivariate pdf is not. This analysis justifies the substitution of the true (correlated) TVAR(m) interference by a "white-noise" equivalent model that captures the important properties of finite-sample support and steering vector dimension. Though analytic expressions for the SNR loss and false-alarm rate are not derived for the equivalent model, false-alarm thresholds etc. can be found by Monte-Carlo simulations.

6.3.1 Background: TVAR(m) Approximation of a Hermitian Covariance Matrix, ML Model Identification and Order Estimation [56,57]

The "band-inverse" TVAR(m) approximation of an arbitrary Hermitian covariance matrix $\boldsymbol{R}$ is calculated by the Dym–Gohberg band-extension method. Let $\boldsymbol{R} \equiv \{r_{jk}\}_{jk=1\ldots M}$ be a given M-variate non-negative-definite Hermitian matrix with rank greater than m, then the TVAR(m)-model approximation of $\boldsymbol{R}$ is the p.d. Hermitian matrix $\boldsymbol{R}_m$ such that

$$\begin{cases} \{\boldsymbol{R}_m\}_{jk} = r_{jk} & \text{for } |j-k| \le m \\ \{\boldsymbol{R}_m^{-1}\}_{jk} = 0 & \text{for } |j-k| > m. \end{cases} \tag{6.156}$$

The solution to the problem of finding $\boldsymbol{R}_m$ given $\boldsymbol{R}$ is provided by the following theorem that was first proven by Dym and Gohberg [55].

Theorem 6.3. *(DG Transformation) [55] Given an M-variate Hermitian matrix $\boldsymbol{R} \equiv \{r_{jk}\}_{j,k=1,\ldots,M}$, suppose that*

$$\begin{bmatrix} r_{jj} & \cdots & r_{j,j+m} \\ \vdots & \ddots & \vdots \\ r_{j+m,j} & \cdots & r_{j+m,j+m} \end{bmatrix} > 0 \quad \text{for } j = 1, \ldots, M-m; \quad |j-k| \le m \tag{6.157}$$

i.e. all $(m+1)$-variate submatrices in the band of $\boldsymbol{R}$ are p.d. For $q = 1, \ldots, M$, let

$$\begin{bmatrix} y_{qq} \\ \vdots \\ y_{L(q),q} \end{bmatrix} = \begin{bmatrix} r_{qq} & \cdots & r_{q,L(q)} \\ \vdots & \ddots & \vdots \\ r_{L(q),q} & \cdots & r_{L(q),L(q)} \end{bmatrix}^{-1} \begin{bmatrix} 1 \\ 0 \\ \vdots \\ 0 \end{bmatrix} \tag{6.158}$$

and

$$\begin{bmatrix} z_{\Gamma(q),q} \\ \vdots \\ z_{qq} \end{bmatrix} = \begin{bmatrix} r_{\Gamma(q),\Gamma(q)} & \cdots & r_{\Gamma(q),q} \\ \vdots & \ddots & \vdots \\ r_{q,\Gamma(q)} & \cdots & r_{qq} \end{bmatrix}^{-1} \begin{bmatrix} 0 \\ \vdots \\ 0 \\ 1 \end{bmatrix} \tag{6.159}$$

where $L(q) \equiv \min\{M, q+m\}$ and $\Gamma(q) \equiv \max\{1, q-m\}$. Furthermore, let the M-variate triangular matrices $\boldsymbol{V}$ and $\boldsymbol{U}$ be defined by their elements

$$v_{jk} \equiv \begin{cases} y_{jk} y_{kk}^{-\frac{1}{2}} & \text{for } k \le j \le L(k) \\ 0 & \textit{otherwise} \end{cases} \tag{6.160}$$

$$u_{jk} \equiv \begin{cases} z_{jk} z_{kk}^{-\frac{1}{2}} & \text{for } \Gamma(k) \le j \le k \\ 0 & \textit{otherwise} \end{cases} \tag{6.161}$$

then the M-variate matrix given by

$$\boldsymbol{R}_m \equiv (\boldsymbol{V}^\dagger)^{-1} \boldsymbol{V}^{-1} = (\boldsymbol{U}^\dagger)^{-1} \boldsymbol{U}^{-1} \tag{6.162}$$

is the unique p.d. Hermitian matrix that satisfies (6.156).

By construction, $\boldsymbol{U}$ and $\boldsymbol{V}$ are band-triangular matrices with the same order (bandwidth) as the TVAR(m) model.

In Reference 62, it was proven that the p.d. submatrix condition (6.157) is the necessary and sufficient condition for the p.d. extension (6.156) to exist. Moreover, analogous to the stationary AR(m) case, among all possible extensions that exist if (6.157) is satisfied, the Dym–Gohberg extension $\boldsymbol{R}_m$ with its special band-inverse property (6.156) has the maximum possible determinant [62,63]. Therefore, analogous to the AR(m) model that is the maximum-entropy extension of the first $(m+1)$ covariance lags of a stationary process, the Dym–Gohberg extension may be treated as the generalized maximum-entropy extension of the given Hermitian band r_{jk} ($|j-k| \leq m$).

The p.d. submatrix condition (6.157) implies that, with probability one, $N > m$ i.i.d. training samples $\boldsymbol{x}_j$ ($j = 1, \ldots, N$), together with the usual sample CME

$$\hat{\boldsymbol{R}} \equiv \left\{\hat{r}_{jk}\right\}_{j,k=1,\ldots,M} = \frac{1}{N}\sum_{j=1}^{N}\boldsymbol{x}_j\boldsymbol{x}_j^{\dagger} \tag{6.163}$$

which is rank-deficient for $N < M$, are sufficient to calculate a non-degenerate ML TVAR(m) CME $\hat{\boldsymbol{R}}_m$. More particularly, if the training data have a Gaussian distribution

$$\boldsymbol{x}_j \sim \mathcal{CN}_N(\boldsymbol{0}, \boldsymbol{R}) \quad \text{for } j = 1, \ldots, N > m, \tag{6.164}$$

then we showed [56] that the DG transformation of the sample CME $\hat{\boldsymbol{R}}_m$ (denoted $DG(\hat{\boldsymbol{R}}, m)$)

$$\begin{cases} \left\{\hat{\boldsymbol{R}}_m\right\}_{jk} = \hat{r}_{jk} & \text{for } |j-k| \leq m \\ \left\{\hat{\boldsymbol{R}}_m^{-1}\right\}_{jk} = 0 & \text{for } |j-k| > m \end{cases} \tag{6.165}$$

is the exact and unique ML estimate of a TVAR(m) covariance matrix $\boldsymbol{R}_m$.

It is interesting to note that this CME $\hat{\boldsymbol{R}}_m \equiv DG(\hat{\boldsymbol{R}}, m)$ (6.165) has already been introduced in Reference 64 for $m_{\max} = N-1$ as a certain regularized estimate of the rank-deficient sample CME $\hat{\boldsymbol{R}}$, and it was observed that for a particular scenario there is an optimum order $m < m_{\max}$ in terms of adaptive filter performance.

In References 57 and 58, we also considered the problem of estimating the order of a TVAR(m) model, given the *a priori* condition

$$m \leq m_{\max} \leq N-1, \quad N \leq M. \tag{6.166}$$

The main idea here is to treat the p.d. ML CME $\hat{\boldsymbol{R}}_{m_{\max}} \equiv DG(\hat{\boldsymbol{R}}, m_{\max})$ calculated for the somehow-specified maximum possible order $m_{\max} \leq N-1$ as a sufficient statistic for a "nested" model $\hat{\boldsymbol{R}}_\mu$ with $\mu \leq m_{\max}$, since for $N < M$ the standard sample CME $\hat{\boldsymbol{R}}$ cannot serve as the TVAR model:

$$LR_{oe}(\hat{\boldsymbol{R}}_\mu) = \frac{LF[X, \hat{\boldsymbol{R}}_\mu]}{LF[X, \hat{\boldsymbol{R}}_{m_{\max}}]} \tag{6.167}$$

(*LR* is likelihood ratio, *LF* is likelihood function).

If m is the true order of the AR or TVAR training data, then for all $\mu \geq m$, the pdf of this *LR* (6.167) is scenario-invariant, i.e. $LR_{oe}(\hat{\boldsymbol{R}}_\mu)$ does not depend on $\boldsymbol{R}$, but is only a function of

M, N, and μ, similar to some other LRs that have this remarkable and practically important property [41,51]. Thus for truly TVAR(m) or AR(m) interference, we compare $LR_{oe}(\hat{\boldsymbol{R}}_\mu)$ with the threshold that has been precalculated using the scenario-invariant pdf, and we select our estimate $\hat{m}$ to be the smallest permissible value of μ that exceeds this threshold. Given M, N, $m_{\max}$, and μ, this threshold can be calculated for any desired probability of order overestimation. In this section, we mainly investigate adaptive filter and adaptive detector performance for TVAR(m) or AR(m) interference, hence this method is directly applicable.

The only remaining problem for truly TVAR(m) or AR(m) interference (that was not addressed in Reference 57) is how to select the maximum possible order $m_{\max}$ (if it is not restricted by some *a priori* considerations). Of course, if $N \le M$, then we can always select $m_{\max} = N - 1$, but this may not be the best choice. This is because for $m \ge N$ there may not be sufficient sample support available. Also, for $m \ll N$, it would be preferable to deal with $m_{\max} < N - 1$, since the "quality" of the $\hat{\boldsymbol{R}}_{N-1}$ estimate is the worst possible for N snapshots, which in turn makes the order estimation less reliable.

Therefore, the problem of selecting $m_{\max}$ must be addressed by a separate test [65]. The procedure for TVAR(m) order estimation consists of the following two steps.

Step 1 *Find the sufficient statistics* $\hat{\boldsymbol{R}}_{m_{\max}}$

Given N i.i.d. training samples ($N < M$), form the ordered sequence of ML TVAR(μ) CMEs $\hat{\boldsymbol{R}}_\mu$ ($\mu = 1, \ldots, N-1$) using the DG transformation (6.165), then find

$$m_{\max} = \arg\min_{\mu} \frac{\det(\boldsymbol{X}^\dagger \hat{\boldsymbol{R}}_\mu^{-1} \boldsymbol{X})}{\left[\mathrm{Tr}(\boldsymbol{X}^\dagger \hat{\boldsymbol{R}}_\mu^{-1} \boldsymbol{X})\right]^N} > \gamma_0 \tag{6.168}$$

where

$$\gamma_0 \equiv \arg\left\{\int_{\gamma_0}^{\infty} f(x)\, dx = P_d\right\}, \tag{6.169}$$

P_d is the probability of detection, and $f(x)$ is specified by the scenario-invariant pdf for $R_\mu = \boldsymbol{R}$

$$f(LR_{us}) = C(M,N)\, LR_{us}^{M-N}\, G_{N,N}^{N,0}\left(LR_{us} \,\middle|\, \begin{matrix} \frac{N^2-1}{N}, \frac{N^2-2}{N}, \ldots, \frac{N^2-N}{N} \\ 0,1,\ldots,N-1 \end{matrix} \right) \tag{6.170}$$

where

$$C(M,N) \equiv (2\pi)^{\frac{N-1}{2}} N^{\frac{1-2MN}{2}} \frac{\Gamma(MN)}{\prod_{j=1}^{N} \Gamma(N-j+1)} \tag{6.171}$$

and $G_{a,b}^{c,d}(\cdot)$ is Meijer's G-function [42]. If the threshold is not reached by the final $\mu = N - 1$, then the sample volume N is declared to be insufficient.

Step 2 *TVAR(μ) order estimation $\hat{m}$ given $m_{\max}$* [57]

Find $\hat{m} \equiv \min \mu$ among all μ that meet the inequality (see (6.167))

$$LR_{oe}(\hat{\boldsymbol{R}}_\mu) > \gamma_\mu \tag{6.172}$$

where

$$\gamma_\mu \equiv \arg_{\gamma} \int_{\gamma}^{1} F(x|\mu, m_{\max})\, dx \equiv 1 - P_{oo} \tag{6.173}$$

P_{oo} is the probability of order overestimation, and [57]

$$F(x|\mu, m_{\max}) \equiv C(M, N, m_{\max}, \mu) x^{(N-m_{\max}-1)} G^{(M-\mu-1),0}_{(M-\mu-1),(M-\mu-1)}\left(x \,\middle|\, \begin{matrix} m_{\max}-\mu,\ldots,m_{\max}-\mu \\ m_{\max}-\mu-1,\ldots,0,\ldots,0 \end{matrix}\right) \tag{6.174}$$

where

$$C(M, N, m_{\max}, \mu) = \prod_{j=1}^{M-\mu-1} \frac{\Gamma[N-\mu]}{\Gamma[N-\zeta_j(m_{\max})]} \tag{6.175}$$

$$\zeta_j(m_{\max}) \equiv \begin{cases} m_{\max} & \text{for } j < M - m_{\max} \\ M - j & \text{for } M - m_{\max} \le j < M - \mu. \end{cases} \tag{6.176}$$

Naturally, for an *a priori* known maximum order $m_{\max}$, Step 2 suffices.

For truly TVAR(m) models and "proper" maximum-order selection (i.e. $m < m_{\max} < N$), this method has a high accuracy of order estimation [57]; specifically, for the scenarios considered in Reference 57, the true order was estimated with a zero sample probability of order underestimation, with an overestimation probability that did not exceed P_{oo}. For this accuracy, the analysis of adaptive TVAR(μ)-based filters and detectors could have been performed for the true TVAR(m) or AR(m) order, however in what follows we analyze their performance taking into consideration possible order overestimation, the probability of which *does not* depend on scenario.

Note that the probability of order underestimation *strongly* depends on scenario, but underestimation occurs only if all the elements along the mth subdiagonal of R_m^{-1} are close to zero. For the scenarios we consider, this probability is vanishingly small, and so we ignore it in this study.

6.3.2 Performance Analysis of TVAR(m)-Based Adaptive Filters and Adaptive Detectors for TVAR(m) or AR(m) Interferences

As discussed, in this subsection, we consider the case where the true interference covariance matrix $\boldsymbol{R}$ is a p.d. TVAR(m) matrix:

$$\boldsymbol{R} = \boldsymbol{R}_m > 0, \qquad \left\{\boldsymbol{R}_m^{-1}\right\}_{jk} = 0 \quad \text{for } |j-k| > m \tag{6.177}$$

and we observe $N > m$ i.i.d. training samples, so that the ML CME $\hat{\boldsymbol{R}}_\mu$ of the TVAR(μ) model is the DG transformation/extension (6.162) of the sample matrix $\hat{\boldsymbol{R}}$ (6.163) for all $\mu < N$. The performance of the adaptive filter that uses this CME $\hat{\boldsymbol{R}}_\mu$ is measured by the SNR loss factor (6.2.1) with respect to the clairvoyant Wiener filter

$$\rho = \frac{(\boldsymbol{s}^\dagger \hat{\boldsymbol{R}}_\mu^{-1} \boldsymbol{s})^2}{(\boldsymbol{s}^\dagger \hat{\boldsymbol{R}}_\mu^{-1} \boldsymbol{R}_m \hat{\boldsymbol{R}}_\mu^{-1} \boldsymbol{s})\,(\boldsymbol{s}^\dagger \boldsymbol{R}_m^{-1} \boldsymbol{s})} < 1. \tag{6.178}$$

The TVAR(μ)-based adaptive filter performance is completely described by the pdf of this loss factor, which is in general a function of μ, N, $\boldsymbol{s}$, and $\boldsymbol{R}_m$.

The AMF adaptive detector that employs this ML CME $\hat{\boldsymbol{R}}_\mu$ to test the primary sample $\boldsymbol{y}$

$$\boldsymbol{y} = \begin{cases} \boldsymbol{x}_0 \sim \mathcal{CN}(\boldsymbol{0}, \boldsymbol{R}_m) & \text{for hypothesis } H_0 \\ \boldsymbol{x}_0 + a\boldsymbol{s}, \quad a \sim \mathcal{CN}(0, \sigma^2) & \text{for hypothesis } H_1 \end{cases} \tag{6.179}$$

where $\boldsymbol{x}_0$ is the observed M-variate interference-only data vector (snapshot) and a is the complex target amplitude (with power σ^2), is specified by the detection statistic [5]

$$f_{AMF} \equiv \frac{\left|\boldsymbol{Y}^\dagger \hat{\boldsymbol{R}}_\mu^{-1} \boldsymbol{s}\right|^2}{\boldsymbol{s}^\dagger \hat{\boldsymbol{R}}_\mu^{-1} \boldsymbol{s}} \underset{H_1}{\overset{H_0}{\lessgtr}} h_{FA} > 1. \tag{6.180}$$

Detection performance (via ROC curves) of this detector can be investigated for any desired false-alarm threshold h_{FA}, even if this depends on $\boldsymbol{R}_m$, however this analysis only shows the ultimate detection performance that is not practically attainable unless the false-alarm threshold can somehow be precalculated (to a sufficient accuracy) without *a priori* knowledge of the true covariance matrix $\boldsymbol{R}_m$. In this happy circumstance, as in Section 6.2, the detector may be called "practically CFAR," and additional detection losses associated with false-alarm threshold inaccuracies should also be analyzed.

We now similarly analyze the TVAR(m)-based AMF adaptive detector (6.180). Since the true covariance matrix has the Dym–Gohberg decomposition (6.162)

$$\boldsymbol{R}_m^{-1} = \boldsymbol{V}\boldsymbol{V}^\dagger, \tag{6.181}$$

the target-free primary data vector $\boldsymbol{x}_0$ may be presented as

$$\boldsymbol{x}_0 = (\boldsymbol{V})^{-1}\boldsymbol{\varepsilon} \quad \text{for } \boldsymbol{\varepsilon} \sim \mathcal{CN}_N(\boldsymbol{0}, \boldsymbol{I}_M). \tag{6.182}$$

By defining the matrix

$$\hat{\boldsymbol{C}}_\mu \equiv \boldsymbol{V}^\dagger \hat{\boldsymbol{R}}_\mu \boldsymbol{V} = \boldsymbol{V}^\dagger DG(\hat{\boldsymbol{R}}, \mu)\boldsymbol{V} = \boldsymbol{V}^\dagger (\hat{\boldsymbol{V}}_\mu^\dagger)^{-1} \hat{\boldsymbol{V}}_\mu^{-1} \boldsymbol{V} \tag{6.183}$$

we can express the SNR loss factor (6.178) as

$$\rho = \frac{\left\{\boldsymbol{c}^\dagger \hat{\boldsymbol{C}}_\mu^{-1} \boldsymbol{c}\right\}^2}{\left\{\boldsymbol{c}^\dagger \hat{\boldsymbol{C}}_\mu^{-2} \boldsymbol{c}\right\} \boldsymbol{c}^\dagger \boldsymbol{c}} \tag{6.184}$$

where $\boldsymbol{c} \equiv \boldsymbol{V}^\dagger \boldsymbol{s}$ and rewrite the detection statistic (6.180) for the target-free (interference-only) primary sample as

$$f_{AMF} = \frac{\left|\boldsymbol{\varepsilon}^\dagger \hat{\boldsymbol{C}}_\mu^{-1} \boldsymbol{c}\right|^2}{\boldsymbol{c}^\dagger \hat{\boldsymbol{C}}_\mu^{-1} \boldsymbol{c}} \tag{6.185}$$

where $\boldsymbol{\varepsilon} = \boldsymbol{V}^\dagger \boldsymbol{y} = \boldsymbol{V}^\dagger \boldsymbol{x}_0$. By this transformation, it is now clear that the scenario-invariance property of the loss factor and the CFARness of the adaptive detector directly depend on the invariance property of $\hat{C}_\mu$. Unfortunately, under the TVAR(m) model, the matrix $\hat{C}_\mu$ is *not* described by a spherically invariant pdf (e.g. a Wishart distribution), i.e.

$$pdf(\hat{C}_\mu) \neq pdf(Y^\dagger \hat{C}_\mu Y) \quad \text{for } Y^\dagger Y = YY^\dagger = I_M \tag{6.186}$$

where Y is some unitary matrix. Therefore, despite the fact that $\mathbb{E}\{\hat{V}_\mu\} = V$ (see (6.158) and (6.181)) and hence

$$\mathbb{E}[\hat{V}_\mu V^{-1}] = I_M, \quad \lim_{N\to\infty} \hat{C}_\mu = I_M, \tag{6.187}$$

it is not possible in general to prove the strict scenario-invariance of the SNR loss factor ρ, nor that of the detection statistic f_{AMF}, with respect to the true covariance matrix R_m.

However, in the special case of an MTI filter, where

$$s = e \equiv [1, 0, \ldots, 0]^T, \quad c^\dagger = s^T V = v_{11} e^T, \tag{6.188}$$

the loss factor in (6.184) can be transformed using the band properties of $\hat{U}$ (6.161) and $\hat{V}$ (6.160) into an expression that is well known for the $(\mu + 1)$-variate traditional SMI algorithm [7]:

$$\rho_e = \frac{\left\{e^T \left[\hat{\mathcal{R}}^{(\mu+1)}\right]^{-1} e\right\}^2}{\left\{e^T \left[\hat{\mathcal{R}}^{(\mu+1)}\right]^{-1} \mathcal{R}^{(\mu+1)} \left[\hat{\mathcal{R}}^{(\mu+1)}\right]^{-1} e\right\} e^T \left[\hat{\mathcal{R}}^{(\mu+1)}\right]^{-1} e} \tag{6.189}$$

where

$$\mathcal{R}^{(\mu+1)} \equiv \begin{bmatrix} r_{11} & \cdots & r_{1,\mu+1} \\ \vdots & \ddots & \vdots \\ r_{\mu+1,1} & \cdots & r_{\mu+1,\mu+1} \end{bmatrix} \tag{6.190}$$

and similarly for the estimated quantities. It is now clear that this SNR loss factor is exactly described by the familiar beta distribution [7]

$$f(\rho_e) = \frac{1}{B[N - \mu + 1, \mu]} (\rho_e)^{N-\mu} (1 - \rho_e)^{\mu - 1}. \tag{6.191}$$

Similarly, for this specific case, the CFAR property of the AMF detector (6.179) can be easily demonstrated.

Note that it is the spherical invariance property of the pdf $\hat{C}_\mu$ that is required to prove the strict scenario-invariance of the pdf for the loss factor ρ (6.184) (when the steering vector s is arbitrary). Moreover, if this invariance existed then this beta distribution would describe the loss-factor pdf for the general case. Unfortunately, the proof of this spherical invariance is not only intractable, but the experimental results presented below show that the loss factor *does depend* on scenario, and its pdf differs significantly from the beta distribution (6.191). In fact, we will see that the actual SNR losses exceed those predicted by this beta distribution in all of our experiments. This might be expected, since our special MTI case of $s = e_1$ involves just a single $(m + 1)$-variate block of the entire M-variate matrix, while the adaptive filter solution for the general s case involves all such blocks and hence accumulates random errors associated with the finite-sample support. For this reason, the beta distribution (6.191) can serve as a scenario-invariant *lower bound* for the loss factor.

To find a scenario-invariant *upper bound*, we may consider a specific scenario that, in terms of SNR losses (statistically at least), dominates scenarios of practical interest.

Note that the SNR loss ρ in (6.184) depends on "how close" $\hat{\boldsymbol{C}}_{\mu}^{-1}$ is to the identity matrix, since only when $\hat{\boldsymbol{C}}_{\mu}^{-1} \rightarrow \boldsymbol{I}_M$ does the loss factor tend to unity. Therefore, proximity of $\hat{\boldsymbol{C}}_{\mu}^{-1}$ to $\boldsymbol{I}_M$, and more importantly, dependence of this proximity on scenario, should be associated with the behavior of the SNR loss factor ρ and its dependence on scenario. Of course, the complete description of these properties is provided by the exact multivariate pdf of $\hat{\boldsymbol{C}}_{\mu}^{-1}$. Yet proximity of a random Hermitian matrix to the identity matrix can also be determined by a number of statistical tests (e.g. see Reference 36). In Reference 57, we chose to use the likelihood-ratio test (that is optimal for a complex Wishart distribution [36]):

$$LR_{\mu}(\hat{\boldsymbol{C}}_{\mu}) = \frac{\exp{(M)}\det(\hat{\boldsymbol{C}}_{\mu})}{\exp[\mathrm{Tr}(\hat{\boldsymbol{C}}_{\mu})]} \leq 1 \tag{6.192}$$

and proved that the pth moment of its pdf (for $\mu \geq m$) is

$$\mathcal{E}\left\{[LR_{\mu}(\hat{\boldsymbol{C}}_{\mu})]^p\right\} = \frac{N^{MN}\exp(Mp)}{(N+p)^{M(N+p)}} \prod_{j=1}^{M} \frac{\Gamma[N-L(j)-j+p]}{\Gamma[N-L(j)-j]} \tag{6.193}$$

where

$$1 \leq L(j) - j \leq \mu \tag{6.194}$$

i.e. this pdf *does not depend on the true covariance matrix* $\boldsymbol{R}_m$. Hence the manner in which $\hat{\boldsymbol{C}}_{\mu} \rightarrow \boldsymbol{I}_M$ as $N \rightarrow \infty$ does not depend on the scenario, at least within the metric of the considered LR. While the scenario-invariance of this particular LR pdf does not imply scenario-invariance of the overall $\hat{\boldsymbol{C}}_{\mu}$ pdf, we can expect that transformations (such as Hermitian forms) involving $\hat{\boldsymbol{C}}_{\mu}$ are also not very sensitive to scenario.

Unfortunately, $\hat{\boldsymbol{C}}_{\mu}^{-1}$ does not enjoy a similarly invariant LR. Indeed, according to (6.183),

$$\hat{\boldsymbol{C}}_{\mu}^{-1} = \boldsymbol{V}^{-1}\hat{\boldsymbol{V}}_{\mu}^{-1}\hat{\boldsymbol{V}}_{\mu}^{\dagger}(\boldsymbol{V}^{\dagger})^{-1} = \boldsymbol{V}^{-1}\hat{\boldsymbol{R}}_{\mu}^{-1}(\boldsymbol{V}^{\dagger})^{-1} \tag{6.195}$$

hence

$$LR_{\mu}(\hat{\boldsymbol{C}}_{\mu}^{-1}) = \frac{\exp M\det(\hat{\boldsymbol{C}}_{\mu}^{-1})}{\exp\left[\mathrm{Tr}(\hat{\boldsymbol{C}}_{\mu}^{-1})\right]} = LR[\hat{\boldsymbol{V}}_{\mu}^{\dagger}\boldsymbol{R}_m\hat{\boldsymbol{V}}_{\mu}] \equiv LR(\hat{D}). \tag{6.196}$$

In Reference 57, we showed that $\det\hat{\boldsymbol{C}}_{\mu}^{-1}$ (for $\mu \geq m$) can be presented as the product of independent variables with a pdf that is a function only of N, m, and μ. Therefore, if the diagonal elements in $\hat{D}$ were found to be *independent* variables whose pdf was also fully specified by N, m, and μ, then the invariance of the pdf for $LR_{\mu}(\hat{\boldsymbol{C}}_{\mu}^{-1})$ can be proven, once again leading to hope for the invariance of Hermitian forms of $\hat{\boldsymbol{C}}_{\mu}^{-1}$, as in the SNR loss factor

ρ (6.184). While it is straight-forward to prove that each diagonal element in $\hat{D}$ has a scenario-invariant pdf (Appendix 1 in Reference 57), the multivariate pdf for the diagonals does not, and so neither does the pdf for $LR_\mu(\hat{C}_\mu^{-1})$. Yet, dependence on scenario in $\operatorname{tr}\hat{C}_\mu^{-1}$ is rather weak, via the correlation of random variables with the invariant multivariate pdf. This means we can expect any scenario variations in the pdf for $LR_\mu(\hat{C}_\mu^{-1})$, and hence for ρ, to be quite small.

Of course, the weakest correlation for these variables occurs for uncorrelated input data, i.e. when $\boldsymbol{R}_m = \boldsymbol{I}_M$ (the "white-noise model"), and so has the biggest impact on $\operatorname{tr}\hat{C}_\mu^{-1}$ for this model in decreasing its root-mean-square error. The pdf for $LR_\mu(\hat{C}_\mu^{-1})$ should therefore be "more concentrated" for the white-noise model, with smaller "tails" expanding towards larger LR values, and correspondingly "more diagonal" properties of $\hat{C}_\mu^{-1}$. Hence we expect, on average, more stable but slightly exaggerated SNR losses ρ. In fact, our calculations in the next subsection confirm this prediction and show that the (small) differences in pdfs for the LR and SNR loss for the highly correlated TVAR model and the white-noise interference matrix $\boldsymbol{R}_m = \boldsymbol{I}_M$ meet our expectations.

We adopt the "white-noise equivalent model" in order to calculate the more realistic upper bound on the SNR loss factor (6.184):

$$\rho < \rho_{WN} = \frac{\left\{\boldsymbol{s}^\dagger \hat{\boldsymbol{F}}_\mu^{-1} \boldsymbol{s}\right\}^2}{\left\{\boldsymbol{s}^\dagger \hat{\boldsymbol{F}}_\mu^{-2} \boldsymbol{s}\right\} \boldsymbol{s}^\dagger \boldsymbol{s}} \leq 1 \tag{6.197}$$

where

$$\hat{\boldsymbol{F}}_\mu \equiv DG(\hat{\boldsymbol{R}}_{WN}, \mu), \qquad \hat{\boldsymbol{R}}_{WN} \equiv \frac{1}{N}\sum_{j=1}^{N} \boldsymbol{\varepsilon}\boldsymbol{\varepsilon}^\dagger, \quad \boldsymbol{\varepsilon} \sim \mathcal{CN}_N(\boldsymbol{0}, \boldsymbol{I}_M). \tag{6.198}$$

In fact, this approximation implies the substitution

$$\hat{\boldsymbol{C}}_\mu \equiv \boldsymbol{V}^\dagger DG(\hat{\boldsymbol{R}}, \mu)\boldsymbol{V} \rightarrow DG(\boldsymbol{V}^\dagger \hat{\boldsymbol{R}} \boldsymbol{V}, \mu) \tag{6.199}$$

i.e. we replaced the "whitened" DG transformation of the sample matrix by the DG transformation of the "whitened" sample matrix.

Despite this approximation, ρ_{WN} still depends on the steering vector; indeed, for the MTI filter $\boldsymbol{s} = \boldsymbol{e}$, this approximation coincides with (6.189), which has the β-distribution (6.191) that we proposed as a lower bound for the general case. An exact expression for the pdf of ρ_{WN} (for an arbitrary steering manifold) is not yet known, however it can be easily precalculated for any M, N, μ, and $\boldsymbol{s}$ using Monte-Carlo simulations.

The relatively weak dependence of the pdf properties of $\hat{C}_\mu^{-1}$ on scenario also suggests that the white-noise interference model $\boldsymbol{R}_m = \boldsymbol{I}_M$ can also be used to precalculate the false-alarm thresholds for the statistics f_{AMF} (6.185):

$$f_{AMF} = \frac{\left|\boldsymbol{\varepsilon}^\dagger \hat{\boldsymbol{C}}_\mu^{-1} \boldsymbol{c}\right|^2}{\boldsymbol{c}^\dagger \hat{\boldsymbol{C}}_\mu^{-1} \boldsymbol{c}} \rightarrow \frac{\left|\boldsymbol{\varepsilon}^\dagger \hat{\boldsymbol{F}}_\mu^{-1} \boldsymbol{c}\right|^2}{\boldsymbol{c}^\dagger \hat{\boldsymbol{F}}_\mu^{-1} \boldsymbol{c}}. \tag{6.200}$$

Moreover, this weak dependence suggests that a covariance matrix in any way more like $\boldsymbol{R}$ than just $\boldsymbol{I}_M$ will give better accuracy in predicting SNR losses or false-alarm thresholds.

In some practical applications, this sort of "notional" covariance matrix model can be precalculated; indeed, the ultimate "notional" model is the ML estimate $\hat{\boldsymbol{R}}_\mu$ itself. We can treat $\hat{\boldsymbol{R}}_\mu$ as the true TVAR(μ) covariance matrix in an "equivalent" scenario, then for a given steering vector $\boldsymbol{s}$ we can conduct the required number of Monte-Carlo simulations in order to generate any number of training samples that are described by the covariance matrix $\hat{\boldsymbol{R}}_\mu$. In this way, the SNR loss factor and (more importantly) false-alarm thresholds are then accurately calculated for $\hat{\boldsymbol{R}}_\mu$ (and not for $\boldsymbol{R}_m$ as we really wish). Yet, due to this weak scenario-dependence, these thresholds may be sufficiently accurate to possess the "practical CFAR" property for the detector. This technique is the essence of the bootstrap methodology, where $\hat{\boldsymbol{R}}_\mu$ is treated as the covariance matrix "within the bootstrap world" [66].

The accuracy of this bootstrap technique, and the white-noise equivalent model, is assessed numerically in the next subsection.

6.3.3 Simulation Results of TVAR(m)-Based Adaptive Detectors for TVAR(m) or AR(m) Interferences

Since accurate analytic results for TVAR(m) adaptive filter and detector performance are unavailable, the equivalent "white-noise" model introduced in Subsection 6.3.2 as an approximation must be validated by direct Monte-Carlo simulations. We remind the reader that this study concerns truly AR(m) or TVAR(m) covariance matrix models.

First consider a stationary AR(2) model that we have often used [57,58,67–73] for simple sea-clutter modeling in high-frequency over-the-horizon radars:

$$y_j = -\sum_{\kappa=1}^{2} a_\kappa y_{j-\kappa} + \sigma_0^2\,\eta_j \quad \text{for } j = 3, \ldots, M \tag{6.201}$$

where $a_1 = -1.9359$, $a_2 = 0.998$, $\sigma_0^2 = 0.009675$, and $\eta_j \sim \mathcal{CN}(0, 1)$. For the $M = 128$ length vector that corresponds to the typical number of repetition intervals (sweeps) within the coherent processing interval (dwell), we model the multiplicative Doppler-frequency (ionospheric) modulation by the diagonal matrix

$$\boldsymbol{D}_k = \operatorname{diag}\left\{\exp\left[i\frac{2\pi k}{M}\left(1 - \cos\frac{2\pi j}{M}\right)\right]\right\} \tag{6.202}$$

for $j = 1, \ldots, M$, where k is the index of the frequency modulation (FM), then

$$\boldsymbol{x} = \boldsymbol{D}_k\,\boldsymbol{y}, \qquad \boldsymbol{R}_2 \equiv \mathbb{E}\{\boldsymbol{x}\boldsymbol{x}^\dagger\} = \boldsymbol{D}_k\,\boldsymbol{N}_2\,\boldsymbol{D}_k^\dagger \tag{6.203}$$

where $\boldsymbol{N}_2 \equiv \mathcal{E}\{\boldsymbol{y}\boldsymbol{y}^\dagger\}$ is the Toeplitz covariance matrix of the AR(2) process, so the resulting process $\boldsymbol{x}$ is a TVAR(2) one. Of course, for the special stationary case $k = 0$, the TVAR(2) model collapses to an AR(2) model ($\boldsymbol{R}_2 = \boldsymbol{N}_2$).

We assume that the target signal is not affected by FM, i.e.

$$s(\nu) = \left[1, \exp(i2\pi\nu N/M), \ldots, \exp(i2\pi\nu N[M-1]/M)\right]^T \tag{6.204}$$

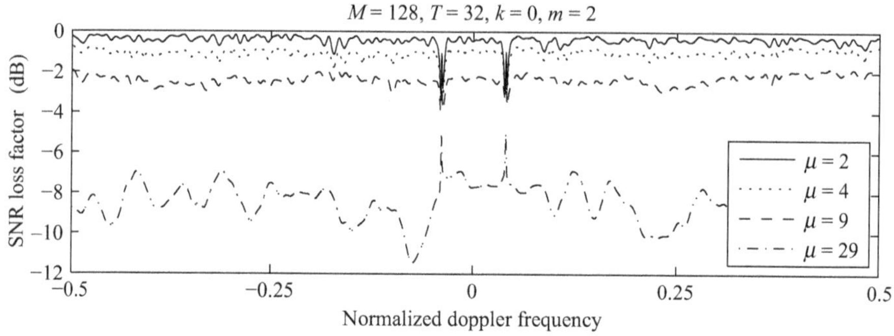

Figure 6.18 TVAR(μ)-based filter finite-sample-support SNR losses for $k = 0$ (stationary case), $N = 32$ snapshots, and various hypothesized orders μ (the true order is $m = 2$).

where ν is the Doppler frequency and is particularly interested in the SNR loss factor with respect to the clairvoyant Wiener filter (6.178):

$$\rho(\mu, \nu, N, M) = \frac{[\boldsymbol{s}^{\dagger}(\nu)\hat{\boldsymbol{R}}_{\mu}^{-1}\boldsymbol{s}(\nu)]^2}{\boldsymbol{s}^{\dagger}(\nu)\hat{\boldsymbol{R}}_{\mu}^{-1}\boldsymbol{R}_2\hat{\boldsymbol{R}}_{\mu}^{-1}\boldsymbol{s}(\nu)\boldsymbol{s}^{\dagger}(\nu)\boldsymbol{R}_2^{-1}\boldsymbol{s}(\nu)} < 1. \tag{6.205}$$

We begin this set of TVAR(2) simulations with the well-understood stationary case $k = 0$, where the Toeplitz covariance matrix $\boldsymbol{N}_2$ is estimated as a TVAR(μ) covariance matrix. The function $\boldsymbol{s}^{\dagger}(\nu)R_m^{-1}\boldsymbol{s}(\nu)$ (not illustrated) is the output SNR of the clairvoyant Wiener filter, and at the same time the inverse of the MVDR (minimum-variance distortionless response, also known as Capon) clutter spectrum. As expected, the output SNR has a minimum when the target Doppler frequency coincides with the resonant frequency of the clutter.

Figure 6.18 shows the SNR sample loss factor $\rho(\mu,\ \nu,\ N = 32,\ M = 128)$ (6.205) as a function of ν for various hypothesized orders $\mu = 2, 4, 9, 29$ (the true order is $m = 2$) calculated for a particular sample ML CME $\hat{\boldsymbol{R}}_2$. We see two sharp notches in the loss factor (i.e. relatively large losses) in the neighborhood of the resonant frequencies. Clearly the loss factor is not strictly scenario-invariant, though even for this modest sample-support case, the losses are less than 4 dB at the resonant frequencies; elsewhere the TVAR(μ) adaptive filter has a very small SNR degradation (an average of 1 dB for the correctly estimated order $\mu = 2$, rising to less than 5 dB for the significantly overestimated order $\mu = 9$) and is statistically indistinguishable across frequency.

These results show that adaptive TVAR(μ) "embedding" of a stationary AR(m) model does not incur significant degradation, and even the modest sample support $N \simeq 3\mu$ has SNR losses comparable with those of the traditional SMI technique with $N \simeq 2M$.

For a slightly non-stationary case (very small FM parameter k in (6.202)), Figure 6.19 with its $k = 5$ shows us that the departure from the corresponding stationary case in Figure 6.18 is, not surprisingly, insignificant. Here we observe the same two notches at the resonant frequencies, but already for $k = 20$ (Figure 6.20) these have disappeared, leaving the loss factor independent of Doppler frequency. At the same time, the losses away from the resonant frequencies are scenario-invariant, being practically the same for $k = 0, 5, 20$ for the fixed $N = 32$.

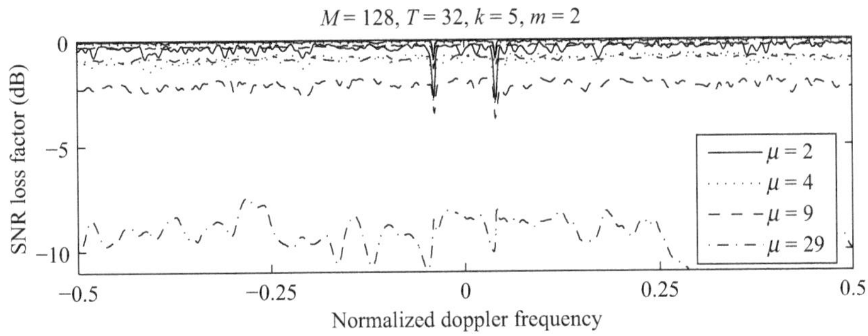

Figure 6.19 TVAR(μ)-filter finite-sample-support SNR losses for $k = 5$ (slightly non-stationary case) and $N = 32$ snapshots.

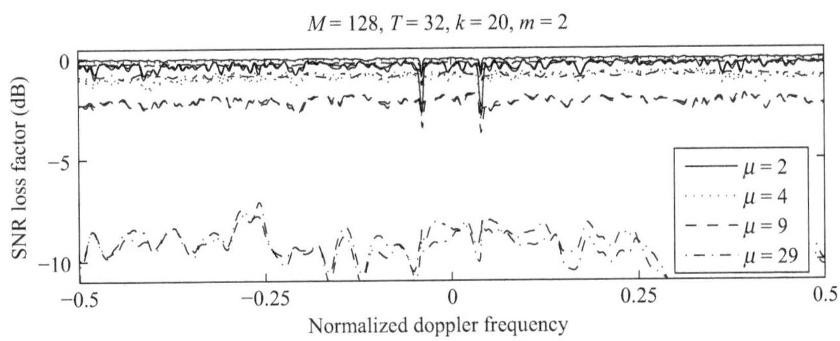

Figure 6.20 TVAR(μ)-filter finite-sample-support SNR losses for $k = 20$ (non-stationary case) and $N = 32$ snapshots.

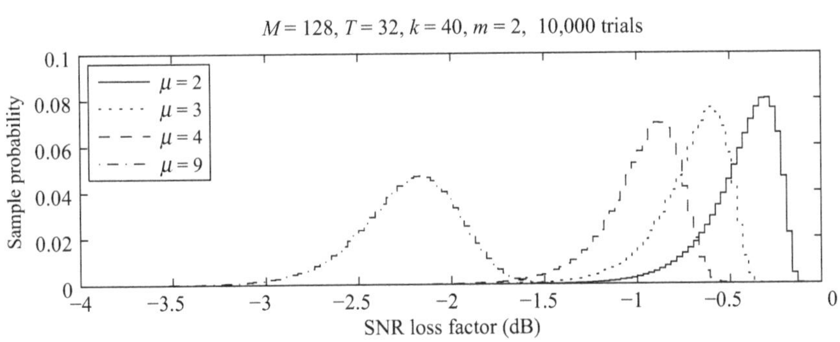

Figure 6.21 Histogram of TVAR(μ)-filter finite-sample-support SNR losses for $k = 40$ and $N = 32$ snapshots for various μ.

This ν-invariance means that we may calculate a (sample) pdf of ρ averaged over all Doppler frequencies, which we do so at Figure 6.21 for 32 snapshots and 10,000 Monte-Carlo trials. By looking at the $\mu = 9$ curve, we again see that the modest sample support $N \simeq 3(\mu + 1)$ (for $\mu \geq m$) gives rise to average losses of about 2.2 dB.

Table 6.1 Interference point-source directions for the TVAR(5) simulation experiment.

n	θ_1	θ_2	θ_3	θ_4	θ_5	θ_6
1	20°					
2	20°	23°				
4	0°	20°	23°	60°		
6	0°	10°	20°	23°	40°	60°

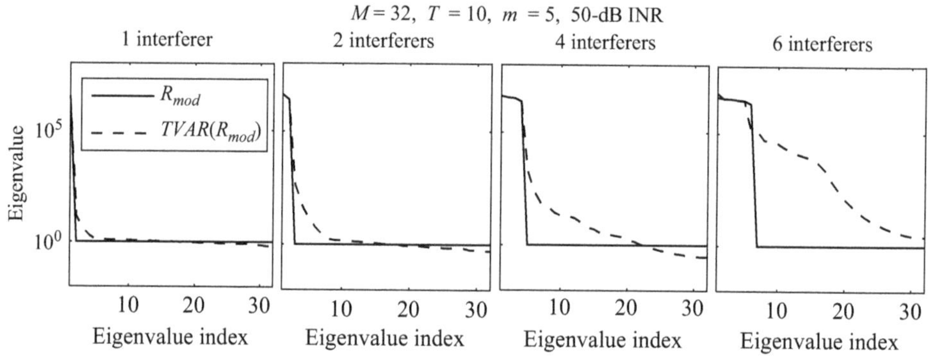

Figure 6.22 Eigenspectra of the TVAR(5) covariance matrix $N_5 = DG(R_{\text{mod}}, 5)$ for various numbers of point-source interferers n in the TVAR(5) simulation experiment.

To summarize the results so far, our TVAR(2) model shows SNR loss-factor invariance with respect to the target Doppler frequency (except for the notches in the stationary case) and to the FM index k. We now wish to demonstrate this quasi-invariance over the broader class of AR(m)/TVAR(m)-type covariance matrices.

To do so, we need to create a set of AR(m)/TVAR(m) models that, for the same order m, cover a wide range of different covariance matrix properties. We therefore consider an $M = 32$-sensor uniform linear antenna array, with an interference covariance matrix $\boldsymbol{R}_5$ that is the DG transformation $DG(R_{\text{mod}}, m = 5)$ of the covariance matrix R_{mod} calculated for various numbers of point sources in white noise $n = 1, 2, 4, 6$ (the subscript "mod" represents "model"). Naturally, for this uniform array, this transformation just results in an AR(5) model with the Toeplitz matrix N_5, but the same approach can be applied to a non-uniform (sparse) array to yield a TVAR(5) model. The target direction was chosen to be $\theta_0 = -15°$, with the interference directions in R_{mod} presented in Table 6.1.

The common INR for all sources was chosen to vary from $-\infty$ (no interference, noise-only) through -10, 10, 50 dB, with 10^5 Monte-Carlo trials undertaken for each scenario. By selecting such different signal scenarios for the same fifth-order autoregressive model, we tried to span a representative variety for our SNR loss-factor invariance investigation.

First, Figure 6.22 compares the "cliff-like" eigenspectrum of the original covariance matrix

$$\boldsymbol{R}_{\text{mod}} = \sum_{j=1}^{n} \sigma_j^2 \boldsymbol{s}(\theta_j)\boldsymbol{s}^\dagger(\theta_j) + \sigma_0^2 \boldsymbol{I}_M \tag{6.206}$$

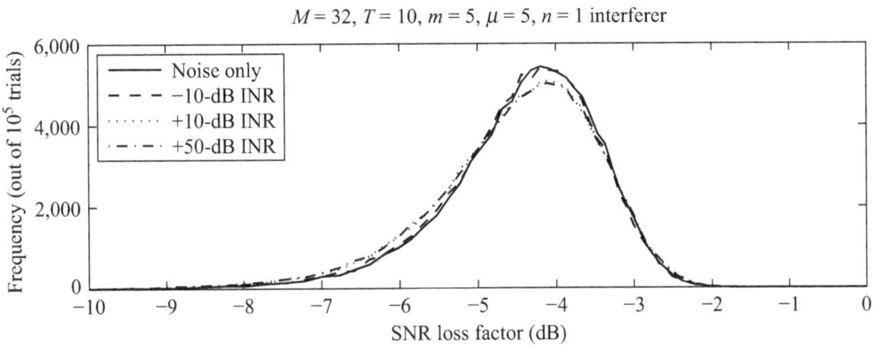

Figure 6.23 Sample pdfs of the SNR loss factor for TVAR(5) interference with $n = 1$ point source for various INRs.

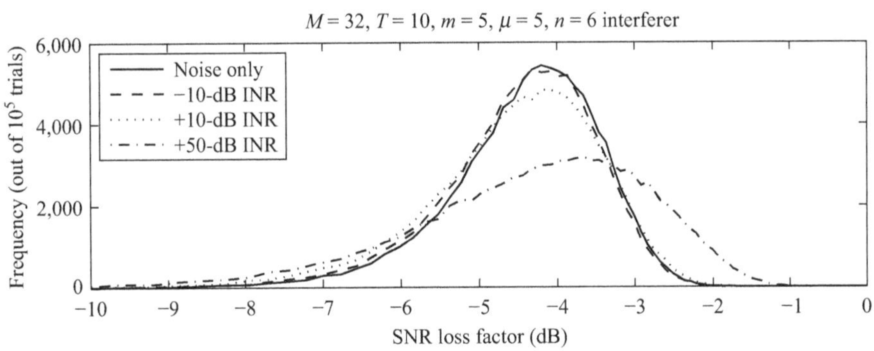

Figure 6.24 Sample pdfs of the SNR loss factor for TVAR(5) interference with $n = 6$ point sources for various INRs.

against that of its DG transformation $DG(\boldsymbol{R}, m = 5)$ for $n = 1, 2, 4, 6$ interference point sources. (As usual, the eigenvalues are sorted into decreasing magnitude.) While each case is *exactly* described by the same AR(5) model, the eigenspectra are completely different. For a single interferer, there is an almost negligible "spill over" of energy beyond the single dominant eigenvalue (both in number and magnitude), whereas the presence of four sources already induces a "heavy tail" beyond the fourth dominant eigenvalue of $\boldsymbol{R}$. For $n = 6$ (NB: $n > m$), the eigenspectrum of $\boldsymbol{R}_5 \equiv DG(\boldsymbol{R}_{\text{mod}}, 5)$ bares no resemblance to that of $\boldsymbol{R}_{\text{mod}}$, with only a few of the smallest eigenvalues approaching the AWGN power level of unity. Thus, despite the fact that all these scenarios are exactly described by the AR(5) model, the properties of the matrix $\boldsymbol{R}_5 \equiv DG(\boldsymbol{R}_{\text{mod}}, 5)$ vary dramatically.

Considering such different AR(5) eigenspectra, the practical invariance of the corresponding sample SNR loss-factor pdfs, illustrated by Figures 6.23 and 6.24, is remarkable. Recall that the AWGN-only case (no interference, "noise only") was suggested in Subsection 6.3.2 to serve as the scenario-invariant upper bound of SNR losses. Specifically, a comparison of the sample pdfs for $n = 1, 2, 4, 6$ interferers at 50-dB INR shows a gradual shift into the area of smaller losses as n increases; yet even for the worst case with $n = 6$, the difference between the true median loss factor and its "white-noise approximation" is less than 0.3 dB.

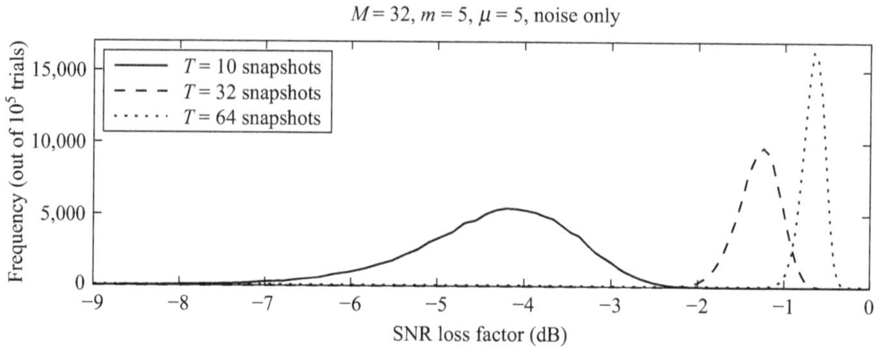

Figure 6.25 Sample pdf's of the upper bound of the SNR loss factor for TVAR(5) interference for various sample sizes N.

So while not strictly scenario-invariant, the SNR losses are shown to be quite insensitive and the white-noise equivalent model (6.197) to be a quite accurate upper bound.

Figure 6.25 investigates another dimension of the parameter space: we show sample pdfs of this upper bound for the sample sizes $N = 10, 32, 64$ to show how quickly the losses reduce as the sample support increases. Comparing this figure with the completely different scenario in Figure 6.21 reveals that the losses are dominated by N and $\mu > m$, and rather insensitive to M and the signal scenario. This very important "invariance" property of the SNR loss factor in the adaptive filter that is constructed from the TVAR(μ) ML CME is similar to that of the filter built from the LSMI CME for scenarios with "cliff-like" eigenspectra; but instead of the number of dominant eigenvalues, we here need to consider the order μ of the TVAR(μ) model.

Note that these results must not be confused with the case where TVAR(μ) modeling is applied to data that are described by a non-TVAR covariance matrix model, such as $\boldsymbol{R}_{\text{mod}}$. The application of TVAR(m) modeling for arbitrary (non-TVAR) interference is left to a separate study employing a dataset produced by the Knowledge Aided Sensor Signal Processing and Expert Reasoning (KASSPER) program, which is a phenomenological ground-clutter model [74]. We remind the reader that in all the above Monte-Carlo trials, the training data have been simulated as

$$\boldsymbol{x}_j = \boldsymbol{R}_5^{\frac{1}{2}} \boldsymbol{\varepsilon}_j, \quad \boldsymbol{\varepsilon}_j \sim \mathcal{CN}(\boldsymbol{0}, \boldsymbol{I}_M), \quad \boldsymbol{R}_5 \equiv DG(\boldsymbol{R}_{\text{mod}}, 5) \tag{6.207}$$

which means we are using an exact AR(5) model.

Now that the SNR loss factor of the TVAR(m)-based adaptive filter has been shown to be almost invariant with respect to the underlying scenario, the next subsection of experimental simulations deals with analyzing in detail the accuracy of the white-noise equivalent model and its relation to the loss-factor upper bound.

First recall from (6.181) and (6.183) that

$$\hat{\boldsymbol{C}}_\mu \equiv \boldsymbol{V}^\dagger \hat{\boldsymbol{R}}_\mu \boldsymbol{V}, \quad \boldsymbol{R}_m^{-1} = \boldsymbol{V}\boldsymbol{V}^\dagger. \tag{6.208}$$

Figure 6.26 introduces the sample pdfs for the sphericity test [36]

$$LR(\boldsymbol{C}) \equiv \frac{[\det(\boldsymbol{C})]^{(1/M)}}{\frac{1}{M}\operatorname{Tr}(\boldsymbol{C})} \tag{6.209}$$

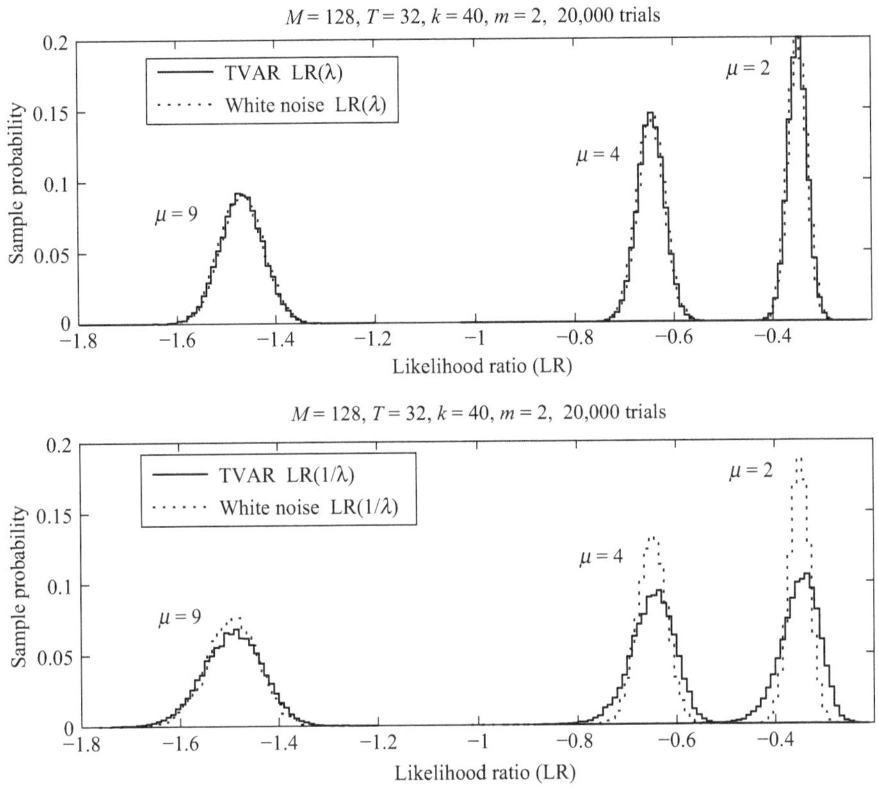

Figure 6.26 Sample pdfs of the sphericity-test likelihood ratio for various hypothesized TVAR model orders μ.

calculated for both the direct $\hat{\boldsymbol{C}}_\mu$ and the inverse $\hat{\boldsymbol{C}}_\mu^{-1}$ sample matrices (labeled "TVAR") for various $\mu > m$, with the other parameters as in Figure 6.21. Also shown are the LRs for the "white-noise equivalent" matrices $\hat{\boldsymbol{F}}_\mu$ (6.198) and $\hat{\boldsymbol{F}}_\mu^{-1}$. In terms of the proximity of the tested matrix $\hat{\boldsymbol{C}}_\mu^{-1}$ to the diagonal one ($c_0 \boldsymbol{I}_M$, $c_0 > 0$), this test is as powerful as the "general test" (6.192) [36]. In fact, due to the proven invariance of $\det \boldsymbol{C}$ for all the above-mentioned matrices, this test provides a direct insight into the impact of the scenario-dependent behavior of $\operatorname{tr} \hat{\boldsymbol{C}}_\mu^{-1}$. As expected, the sample pdfs for $LR(\hat{\boldsymbol{C}}_\mu)$ and $LR(\hat{\boldsymbol{F}}_\mu)$ (upper subfigure) are identical. Meanwhile, the pdfs for $LR(\hat{\boldsymbol{C}}_\mu^{-1})$ and $LR(\hat{\boldsymbol{F}}_\mu^{-1})$ (lower subfigure) are visibly different for $\mu = 2, 4$, but by $\mu = 9$ the difference is diminishing.

We can now demonstrate that this difference is due to a scenario-dependent correlation of the diagonal elements of $\hat{D}_\mu$, whereas the univariate pdf of these elements is scenario-invariant. Figure 6.27 presents sample pdfs of the diagonal elements of the matrices

$$\hat{\boldsymbol{C}}_\mu^{-1} \equiv \boldsymbol{V}^{-1} \hat{\boldsymbol{R}}_\mu^{-1} (\boldsymbol{V}^\dagger)^{-1}, \qquad \hat{\boldsymbol{F}}_\mu^{-1} \equiv DG(\hat{\boldsymbol{R}}_{WN}, \mu), \qquad \hat{D}_\mu \equiv \hat{\boldsymbol{V}}_\mu^\dagger \boldsymbol{R}_m \hat{\boldsymbol{V}}_\mu. \tag{6.210}$$

While the $\hat{D}_\mu$ and $\hat{\boldsymbol{F}}_\mu^{-1}$ distributions are practically the same, the $\hat{\boldsymbol{C}}_\mu^{-1}$ one is slightly different. Since $\boldsymbol{R}_m = (\boldsymbol{V}^\dagger)^{-1} \boldsymbol{V}^{-1}$, $\operatorname{Tr}(\hat{\boldsymbol{D}}_\mu) = \operatorname{Tr}(\hat{\boldsymbol{C}}_\mu^{-1})$, and $\det(\hat{\boldsymbol{D}}_\mu) = \det(\hat{\boldsymbol{C}}_\mu^{-1})$, and hence this

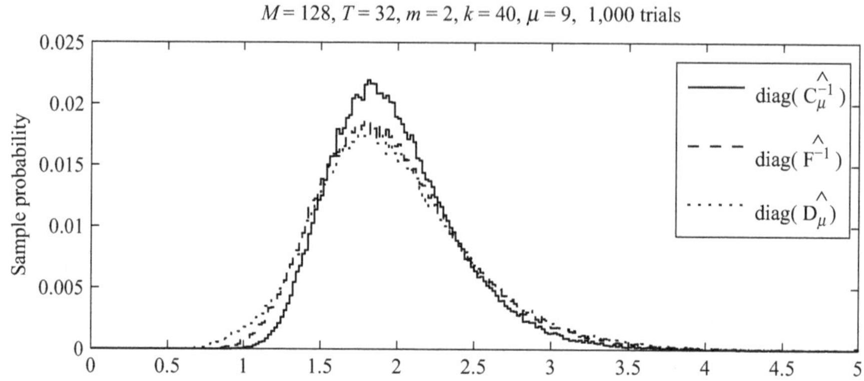

Figure 6.27 Sample pdfs of the diagonal elements in various analyzed sample matrices.

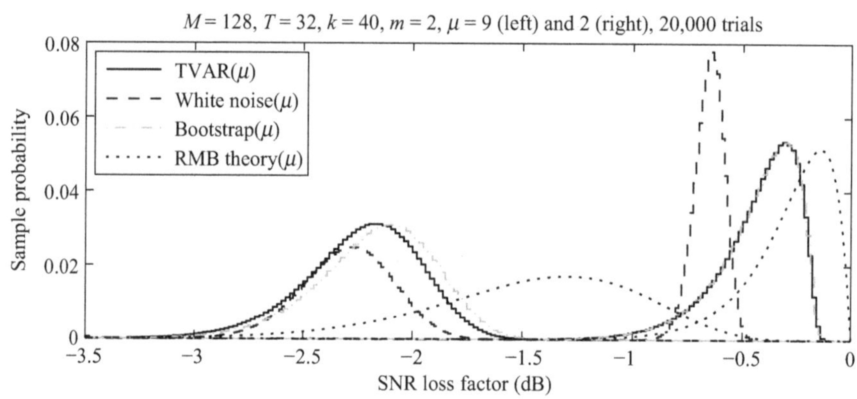

Figure 6.28 Histogram of various filter finite-sample-support SNR losses for $k = 40$ and $N = 32$ snapshots. (The set of four curves on the left refer to $\mu = 9$, the right to $\mu = 4$.)

difference can only be explained by a different (scenario-dependent) correlation of the equally distributed elements in $\hat{\boldsymbol{D}}_\mu$ and $\hat{\boldsymbol{F}}_\mu^{-1}$. Our white-noise equivalent model ignores these distinctions and therefore serves only as a close approximation.

The accuracy of this approximation is illustrated by simulation results shown in Figure 6.28. The first curve (labeled "TVAR(μ)") has already been presented in Figure 6.21, the "white-noise (μ)" curve is our equivalent-model upper bound (6.197), and the "RMB theory (μ)" curve is the lower-bound beta distribution (6.191). We see that the lower bound does indeed underestimate the actual SNR loss factor, especially for larger orders. On the contrary, the upper bound becomes more accurate for larger μ; this is not surprising since Figure 6.26 has already shown that the likelihood ratios for $\hat{\boldsymbol{C}}_\mu^{-1}$ and $\hat{\boldsymbol{F}}_\mu^{-1}$ correspond very closely. Still, even at its worst ($\mu = m = 2$), the white-noise equivalent model does not overestimate the average losses by more than 0.4 dB.

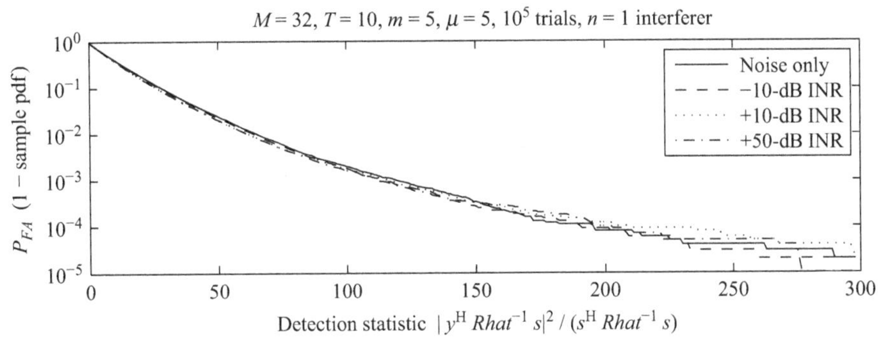

Figure 6.29 Sample false-alarm thresholds for TVAR(5) interference with $n = 1$ point source.

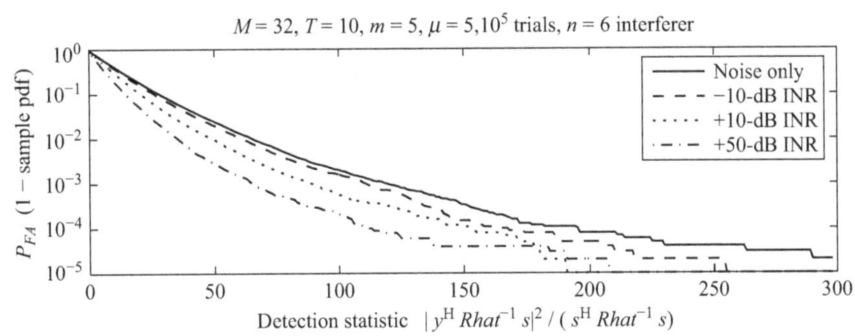

Figure 6.30 Sample false-alarm thresholds for TVAR(5) interference with $n = 6$ point sources.

To recapitulate, our analysis shows the high SNR performance of TVAR(μ)-based adaptive filters for TVAR(m) interference and the acceptable accuracy of the SNR-loss prediction provided by the white-noise equivalent model. We may now turn our attention to the performance of the adaptive detector (6.180). First, we need to validate the accuracy of the white-noise model (6.200) for predicting false-alarm thresholds.

The sample dependence of false-alarm probability P_{FA} on the detection-statistic threshold value h_{FA} (6.180) of the TVAR(m)-based AMF adaptive detector over 10^5 trials is shown in Figures 6.29 and 6.30 for the same TVAR(5) scenario as in Figures 6.22–6.24. We see that the white-noise equivalent model (6.200) mostly leads to an overestimation of the true false-alarm rate (i.e. larger than the true rate), with the worst case being observed for $n = 6$ and 50-dB INR where the white-noise-derived threshold results in $P_{FA} = 0.8 \times 10^{-4}$ instead of the nominal (i.e. desired) $P_{FA} = 10^{-3}$. The best case here is for $n = 1$; while for $n = 2$ (not shown here) we observe a very slight false-alarm underestimation, for example, the nominal $P_{FA} = 10^{-3}$ gives us $P_{FA} = 1.8 \times 10^{-3}$. Of course, for this experiment with 10^5 Monte-Carlo trials, reliable and meaningful results can only be made for $P_{FA} \gtrsim 10^{-4}$.

It is interesting that for $\mu > m$ this scenario-dependence becomes less significant (not shown here): the notable difference between the white-noise and actual thresholds for $\mu = 5$ almost disappears for $\mu = 7$.

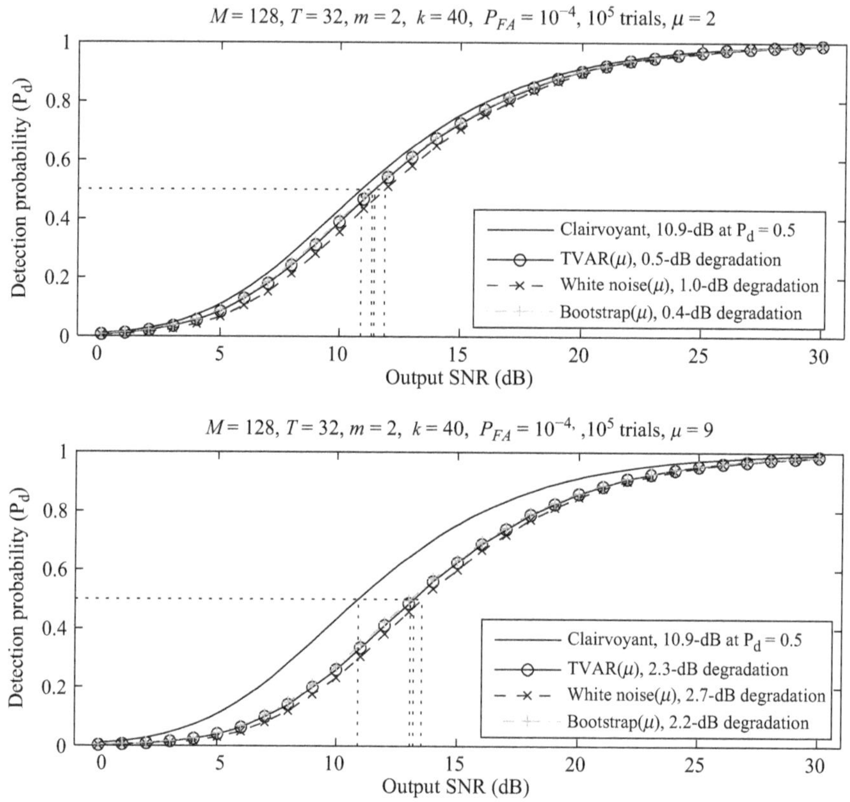

Figure 6.31 Detection performance of TVAR(2)-based adaptive detector for problem dimension $M = 128$ and $N = 32$ snapshots.

As predicted, the pdf for the "white-noise matrix" $\hat{\boldsymbol{F}}_{\mu}^{-1}$ is more concentrated around its mean value, which is almost the same for $\hat{\boldsymbol{C}}_{\mu}^{-1}$ and $\hat{\boldsymbol{F}}_{\mu}^{-1}$. From a practical viewpoint, the true false-alarm rate that is *below* the desired one (0.0008 instead of 0.0010) is acceptable provided that the SNR losses associated with the smaller-than-required false-alarm rate are appropriate. On the contrary, if the true false-alarm rate exceeds the desired one, this can overload a tracker and result in too many false tracks.

Figure 6.31 shows the ROCs for the Figure 6.21-scenario for the usual two hypothesized orders. Each subfigure presents the ROC for the clairvoyant detector, the adaptive detector using the (almost) exact false-alarm thresholds precalculated via Monte-Carlo simulations (labeled "TVAR(μ)"), for the "practical" thresholds precalculated by the white-noise model, and finally the bootstrap method. As expected, the smallest degradations (with respect to the clairvoyant detector) occur for correct order estimation $\mu = m = 2$. For the exact threshold, the loss is only 0.5 dB at 50% probability of detection, which is marginally above the median SNR losses in the TVAR(2) filter (about 0.4 dB according to Figures 6.21 and 6.28). Introducing practical thresholds adds another 0.5 dB of degradation, meaning a total SNR loss of 1 dB compared to the clairvoyant detector. Even for the significant overestimation of $\mu = 9$, practical thresholding

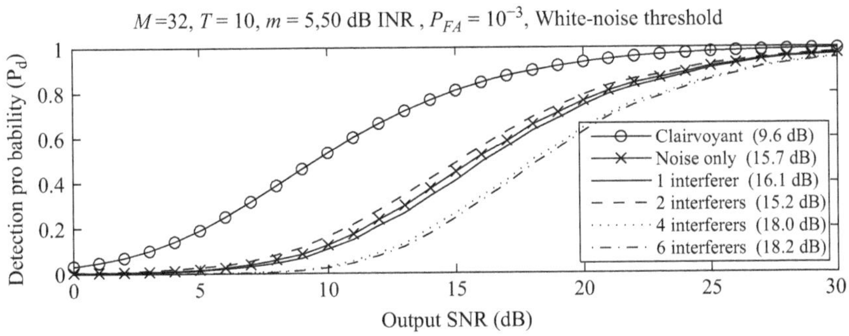

Figure 6.32 Detection performance of TVAR(5)-based adaptive detector with ideal false-alarm thresholds for 10-dB INR.

contributes only 0.4 dB in additional losses, which means acceptable accuracy for the threshold calculation based on the white-noise model. Similar observations follow from simulating the ROCs for the TVAR(5) simulation experiment, with $N = 10$ snapshots and $m = \mu = 5$ (not shown here).

Figure 6.32 shows the ROCs for thresholds calculated using the white-noise equivalent model (6.200) that for the cases $n = 0, 1, 4, 6$ lead to additional degradation, but at most equal to 2.7 dB.

Overall, this scenario with its relatively small sample support ($N = 2m$) demonstrates not only inferior detection performance compared with the previous scenario (with $N = 3.6\, m_{\max}$) but also less accurate false-alarm threshold prediction by the white-noise equivalent model, resulting in greater additional SNR losses. In this regard, it is interesting to explore the performance capabilities of the "bootstrap" approach. As discussed in Subsection 6.3.2, instead of the white-noise model, we treat the ML CME $\hat{\boldsymbol{R}}_\mu$ as the true TVAR(μ) covariance matrix and simulate enough training vectors (10^5, say) from $\hat{\boldsymbol{R}}_\mu$ to calculate false-alarm thresholds and other statistical parameters (which, though derived from $\hat{\boldsymbol{R}}_\mu$ rather than the desired $\boldsymbol{R}_m$, are weakly dependent on scenario). Figure 6.28 also shows the sample pdfs of the SNR loss factor in the adaptive filter, drawn from the "bootstrap data." We see that this approach is very successful, as it almost replicates the actual distributions, unlike the (upper-bound) white-noise model. The bootstrap-derived thresholds are more similarly accurate than the white-noise ones (not shown here). Not surprisingly, such accuracy carries over to bootstrap ROCs that are almost identical to those calculated with (almost-) exact thresholds (Figure 6.31). Of course, a real-time implementation of the bootstrap method may be difficult, unlike the white-noise model. Yet, in principle at least, the established weak dependence of adaptive detector characteristics, such as false-alarm thresholds and SNR loss factor on the true TVAR(m) covariance matrix, allows us to consider applying various CFAR techniques to "fill the gap" between the white-noise model (practical but somewhat inaccurate) and the bootstrap method (somewhat impractical but accurate).

Up to this point, we have deliberately analyzed filter and detector performance for quite high orders $\mu \geq m$ in order to demonstrate that moderate or gross order overestimation does not incur a significant loss; but of course, TVAR(μ) adaptive detector performance should include order estimation as an integral part of the entire process. For the scenario of $M = 128$, $N = 32$, $k = 40$, $m = 2$, and $m_{\max} = 5$, we have already demonstrated [73] that the order is correctly estimated ($\hat{m} = m = 2$) in all 10^3 Monte-Carlo trials for a probability of order overestimation $P_{oo} = 10^{-4}$, 10^{-3}, while for $P_{oo} = 10^{-2}$ was it overestimated in eight trials, as one might expect. With this high level of accuracy in our order estimation algorithm, it is obvious that the previously presented ROCs (for $\mu = m$) change very little when considering the entire detection process. What we also demonstrated is that the maximum permissible order $m_{\max} = 5$ can be properly selected by using Step 1 (6.168) of our order estimation technique.

This analysis concludes our performance study of TVAR(μ) or AR(μ) adaptive detection that includes (a) TVAR(μ) ML covariance matrix estimation using the DG transformation of the sample covariance matrix $\hat{\boldsymbol{R}}$, (b) TVAR($\hat{m}$) model order selection using Step 1 ($m_{\max}$ estimation) followed by Step 2 ($\hat{m}$ estimation), and (c) "practically CFAR" false-alarm threshold calculation using the "white-noise" or "bootstrap" equivalent models.

The final issue with TVAR(m)-based adaptive detection that we deal with here is a comparison of its detection performance with the efficiency of another "sample-deficient" approach that is associated with "small-rank" covariance matrix estimation. Specifically, we consider the (diagonally) loaded AMF (LAMF) detector studied in Section 6.2 and reconsider our TVAR(5) simulation experiment.

Recall that the TVAR(5) covariance matrix $\boldsymbol{N}_5$ is the DG transformation of the underlying signal scenario $DG(R_{\text{mod}}, 5)$. For such models with "slope-like" eigenspectra, as in the case $n = 4, 6$ in Figure 6.22, we cannot expect the LAMF detector to be "practically CFAR", whereas it is for "cliff-like" scenarios such as $n = 1, 2$ with their small-rank covariance matrix. (Indeed, both the false-alarm threshold and the loading factor are not easily specified for "slope-like" eigenspectra.) For this reason, we will compare the potential performance of the TVAR(m) and LAMF detectors with ideal false-alarm thresholds, bearing in mind that additional detection losses are likely to be incurred by practical false-alarm threshold calculation.

The (diagonally) loaded CME is [13,14]

$$\hat{\boldsymbol{R}}_\beta = \hat{\boldsymbol{R}} + \beta \boldsymbol{I}_M \tag{6.211}$$

where β is the loading factor. For "cliff-like" scenarios (6.6), with their m dominant interferers in white noise, the loading factor can be robustly selected within the range [46]

$$\sigma_0^2 < \beta \ll \hat{\lambda}_m \tag{6.212}$$

where $\hat{\lambda}_m$ is the minimum "signal-subspace" eigenvalue of the sample CME $\hat{\boldsymbol{R}}$ averaged over the $N > m$ snapshots.

For TVAR(m) models with "slope-like" eigenspectra, we consider two options for loading-factor selection: the first is

$$\beta = 2\sigma_0^2 \tag{6.213}$$

which is traditional for "cliff-like" scenarios, while the second is based on the recently proposed "EL" methodology [1]. Specifically, the loading factor is selected so that the undersampled LR (6.192) generated by the loaded CME $\hat{\boldsymbol{R}}_\beta$ is equal to the mean value of the LR generated by the true covariance matrix $\boldsymbol{R}$, as specified by the pdf (6.170).

Figure 6.33 shows the eigenspectrum of R_{mod} (6.206) in the $n = 4$-source case from the TVAR(5) simulation experiment. Also shown are the eigenspectra of its TVAR(5) transformation and of one particular sample $\hat{\boldsymbol{R}}$, and the level of the EL loading β_{EL}. We can immediately

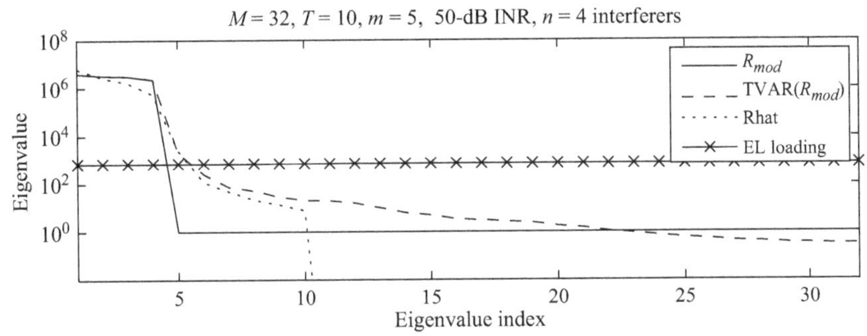

Figure 6.33 Detection performance of TVAR(5)-based adaptive detector with ideal false-alarm thresholds for 50-dB INR.

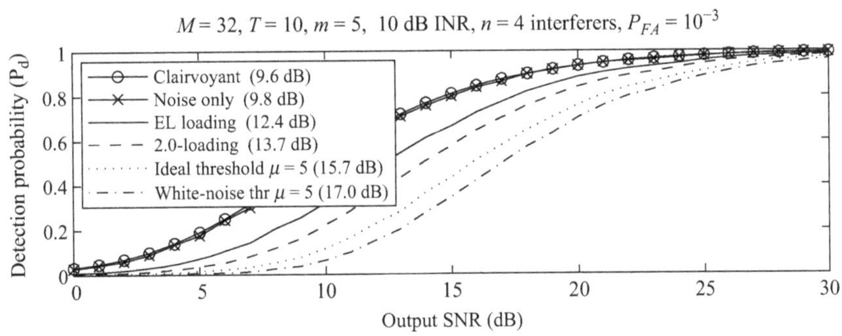

Figure 6.34 Detection performance of TVAR(5)-based adaptive detector with "white-noise" false-alarm thresholds for 50-dB INR.

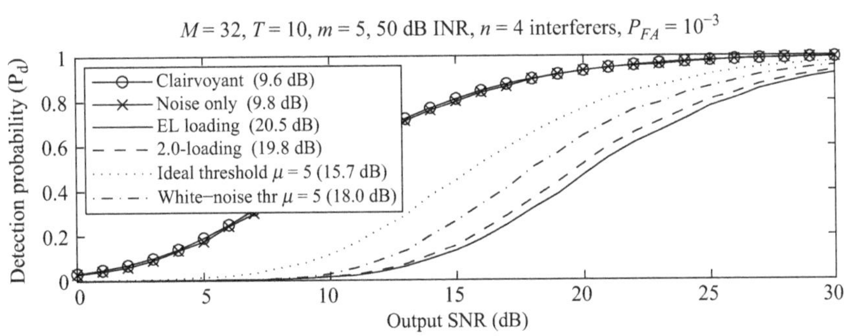

Figure 6.35 Eigenspectra of the TVAR(5) simulation true and sample covariance matrices for $n = 4$.

observe that this loading factor is almost 30 dB above the noise floor in this case, unlike the traditional $\beta = 2\sigma_0^2$.

As discussed, we compare ROCs calculated for the ideal false-alarm thresholds. Figure 6.34 shows ROCs for TVAR(5) and LAMF detectors for the scenario of 10-dB SNR in the "interim" covariance matrix model R_{mod} with n point sources in white noise (6.206). Figure 6.35 repeats

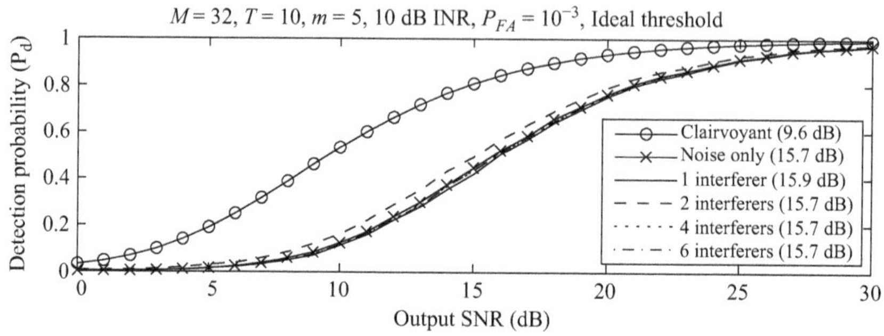

Figure 6.36 Detection performance of LAMF and TVAR(5)-based adaptive detectors for 10-dB INR.

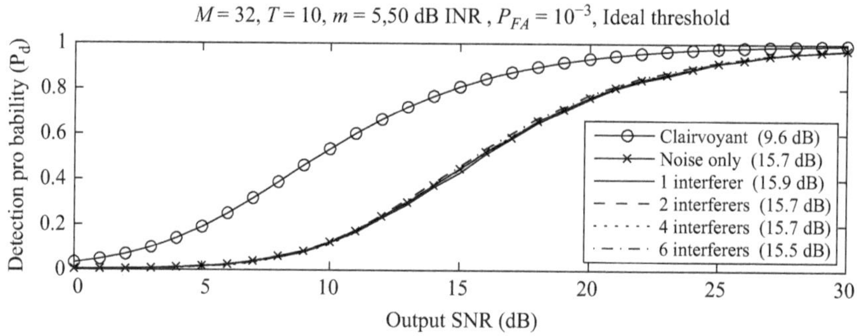

Figure 6.37 Detection performance of LAMF and TVAR(5)-based adaptive detectors for 50-dB INR.

the experiment for 50-dB SNR. The LAMF detector here adopts EL-selected loading, though conventional loading (6.213) leads to very similar performance, despite a significant difference in loading values β. Comparison of the introduced ROCs leads to some important conclusions. First, we see from Figures 6.36 and 6.37 that the performance of the TVAR(5) detector (with ideal false-alarm thresholds, as discussed above) is almost scenario-invariant (n and INR) and is characterized by 6-dB losses. On the contrary, Figures 6.34–6.38 demonstrate that the performance of the LAMF detector is strongly dependent on the underlying signal scenario.

The most important conclusion comes from comparing these ROCs with those in Figure 6.36: at this relatively low INR, LAMF is superior to TVAR(5). Specifically, for EL loading and a single source, LAMF detection losses are negligible ($10.3 - 9.6 = 0.7$ dB); even for six sources, its losses are $14.8 - 9.6 = 5.2$ dB (compared with $15.7 - 9.6 = 6.1$ dB for TVAR(5)).

Lastly, consider Figure 6.38 which are the LAMF ROCs for $n = 0, 1, 2, 4, 6$ and 50-dB INR, with EL loading (6.213). For one or two interference sources, the LAMF detector is still (potentially) better than the TVAR(5)-based one; however, for more interferers the LAMF losses increase enormously: 10 dB for $n = 4$ and 20 dB for $n = 6$, making the LAMF detector practically useless for such scenarios. Conventional loading (not shown here) demonstrates very similar results.

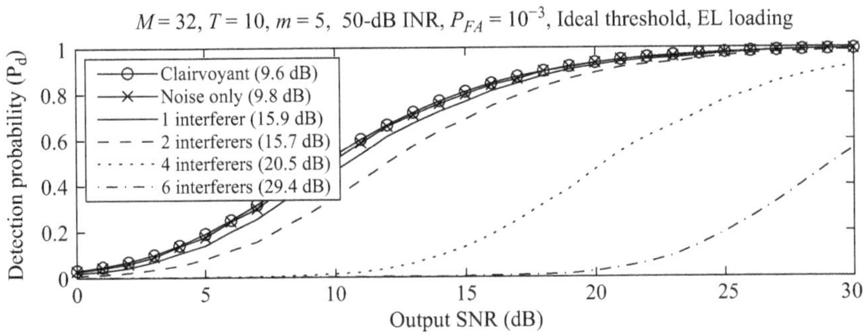

Figure 6.38 Detection performance of LAMF adaptive detector for 50-dB INR and EL loading.

An explanation of this behavior follows from considering the TVAR(5) eigenspectra for four interferers in Figure 6.22. Rather than a "cliff-like" eigenspectrum, there are a quite large number of dominating eigenvalues, which make the diagonal loading method inefficient. For 10-dB INR (not illustrated), the TVAR(5) model eigenspectrum "tail" (i.e. those eigenvalues smaller than the fifth largest one, λ_5) becomes comparable with the additive white-noise "floor," which makes LSMI more efficient than the TVAR(m) detector. On the contrary, at 50-dB INR, the "tail" as seen in Figure 6.33 is still well above the "floor," even for the last non-zero eigenvalue $\hat{\lambda}_{10}$ of $\hat{\boldsymbol{R}}$.

This analysis demonstrates that the comparative performance of diagonally loaded versus TVAR(m)-based adaptive detectors strongly depends on the eigenspectrum of the interference covariance matrix, with LAMF being superior for "cliff-like" scenarios and TVAR(m) being superior for "slope-like" scenarios. The fact that a particular interference may be *exactly* described by a TVAR(m) model does *not* necessarily mean that for limited sample support the TVAR(m) detector is always better. From a theoretical viewpoint, this observation reinforces the important statement made in Reference 1 regarding the role of the ML covariance matrix estimation criterion in the adaptive detection problem; namely that even a properly structured ML TVAR(m) ML CME $\hat{\boldsymbol{R}}_m$ may lead to a worse detection performance than the "ad hoc" diagonally loaded estimate. From a practical viewpoint, of course, it must be remembered that we have compared ROCs with ideally set false-alarm thresholds, so any practical approach may introduce additional considerations into the relative merits of the two detectors.

6.3.4 Observations

We have analyzed the performance of adaptive filters (beamformers) and adaptive detectors that exploit the recent ML estimate of an M-variate band-inverse covariance matrix, drawn from a set of N independent Gaussian training samples. Band-inverse Hermitian covariance matrices describe the class of time-varying autoregressive random processes of order m, TVAR(m), and can be used to approximate an arbitrary Hermitian covariance matrix with a "slope-like" eigenspectrum, which is typical of spread/scattered/distributed/broadband sources. The minimum sample support required for ML covariance matrix estimation is just greater than the TVAR(m) order ($N > m$), and so for all cases with $m \ll M$ there is a significant reduction in sample support, compared with the traditional SMI technique that requires $N \geq M$ training samples.

The SNR loss factor with respect to the clairvoyant filter has been analyzed for adaptive filters that are constructed from the ML TVAR(μ) covariance matrix estimate. For interferences that arise from a TVAR(m) process, we analyzed the SNR losses due to the finite sample support N involved in TVAR(μ) ($\mu \geq m$) ML covariance matrix estimation for signal scenarios with different properties of interference and target steering vector. First, we demonstrated that (unlike the conventional SMI technique), the strict invariance of the SNR loss factor with respect to the true covariance matrix and signal steering vector does not exist for TVAR-based adaptive filters.

For an adaptive MTI filter, whose M-variate steering vector has only its first element nonzero, the SNR loss factor is *exactly* described by a beta distribution with parameters N and $(\mu + 1)$. For a general steering vector, the SNR losses statistically exceed those with that $\beta(N, \mu + 1)$ distribution. This leads to our suggestion of using this beta distribution as a lower bound for SNR losses. We demonstrated that, apart from some very specific cases where the target steering vector is practically in the clutter subspace, the SNR loss factor is mainly determined by the same two parameters, being very insensitive to the true covariance matrix and the (general) target steering vector. This insensitivity allows us to propose an (approximate) upper bound, namely, the adaptive filter SNR loss factor wherein uncorrelated (white-noise) data are used in place of the true TVAR(m) covariance matrix. Though we have not derived a closed-form analytic pdf of the "white-noise" loss factor, the pdf can be precalculated for any given M, N, μ, and steering vector using Monte-Carlo simulations. Our analysis showed that this scenario-invariant upper bound in most cases only slightly overestimates the true SNR loss factor (0.2–0.5 dB) and so is suitable for the practical assessment of TVAR(μ) adaptive filter efficiency. Similar to the famous RMB rule $N \simeq 2M$ to obtain average 3-dB SNR losses, we demonstrated that a TVAR(μ) adaptive filter and general steering vector require

$$N \simeq 3\mu \quad \text{for } \mu \geq m \tag{6.214}$$

training samples, which for $m \ll M$ means a very significant sample-support reduction.

We have also analyzed adaptive detectors that are based on the ML TVAR(μ) CME, focusing on CFAR properties and ROC performance. Similar to the SNR losses in the adaptive filters, we showed that such detectors are not strictly CFAR, that is, in the absence of a target their output statistics depend on the interference scenario. Yet we found that the dependence of the false-alarm probability P_{FA} on the true TVAR(m) ($m \leq \mu$) interference covariance matrix is rather weak. We therefore suggested that, for $m \ll M$ and $\mu \ll N < M$, the false-alarm thresholds can be precalculated using a white-noise equivalent interference model with an acceptable accuracy in practical applications. In particular, for $M = 128$, $m = 2$, $N = 32$, and $P_{FA} = 10^{-4}, 10^{-3}, 10^{-2}$, we demonstrated that "white-noise"-derived thresholds produced a false-alarm rate of about 50–80% P_{FA}. This slight overestimation in false-alarm threshold induces a practically negligible ($\lesssim 0.6$ dB) additional SNR degradation.

However, we considered another scenario ($M = 32$, $m = 5$, $N = 10$) where the white-noise equivalent model provides less accurate thresholds. Among the range of these TVAR(5) scenarios, the worst false-alarm rate overestimation reached an order of magnitude ($P_{FA} = 10^{-4}$ instead of the nominal 10^{-3}), with an additional 2.7 dB of SNR losses. While these losses are still less than those associated with partitioning the sample support and using one subset for conventional scalar false-alarm threshold calculation, this example has emphasized the need for more accurate thresholds.

In this regard, we demonstrated that much more accurate false-alarm thresholds can be precalculated using "bootstrap" ideas, whereby the ML TVAR(μ) CME is treated as the true interference covariance matrix in the "bootstrap world," in order to generate any number of "bootstrap training data" that is then used for direct threshold calculation. Of course, a real-time practical implementation of this algorithm is questionable, though it is possible that a derivative of this technique can perform better than the white-noise equivalent model.

We have studied TVAR(μ)-based adaptive filters and detectors of an arbitrary order $\mu \geq m$ mainly to investigate the SNR losses associated with order overestimation. We demonstrated that modest order overestimation leads to a considerable (but not catastrophic) SNR and detection performance degradation in the adaptive filters and detectors, respectively. At the same time, the fact that a previous paper of ours demonstrates an extremely high order-estimation accuracy, such losses can be ignored in all scenarios considered in this section.

Finally, for truly TVAR(m) interferences, we conducted a comparative performance analysis of LAMF and TVAR(m) adaptive detectors. This analysis used true false-alarm thresholds to calculate ROCs, and so reflects potential detection performance rather than practical performance. Our study showed that, even for TVAR(m) interferences, the question of which detector is better strongly depends on the eigenspectrum of the interference covariance matrix. For our TVAR(5) interference scenario and $n = 1$ or 2 interferers, with its "cliff-like" eigenspectrum (where dominant eigenvalue "leakage" is comparable with the white-noise floor), LAMF is superior to the TVAR(5)-based detector. On the contrary, for the TVAR(5) scenario and $n = 4$ or 6 interferers, with a strongly "slope-like" eigenspectrum ($m = \mu = 5$, $N = 10$, where even the Nth eigenvalue is still well above the noise floor), the situation is reversed, with LAMF having 14-dB losses with respect to the TVAR(m) detector. The fact that a particular interference scenario is exactly described by a TVAR(m) model does not necessarily mean that the TVAR(m) detector is always better than the diagonally loaded one. This paves the way for a new hybrid detection scheme for practical applications that adaptively selects between the LAMF and TVAR(μ) detectors. When properly applied, both techniques yield a very significant detection performance improvement compared with the AMF detector that uses the conventional sample CME with more than M training samples.

We mention that dedicated logic circuitry for the practical implementation of the TVAR(μ)-based adaptive filter has already been developed by Lekhovytskiy *et al.* [64] as the "universal adaptive lattice filter," which ensures this technique is readily implementable.

6.4 Improving Adaptive Detection Using Data Partitioning

In many cases, the CFAR detectors introduced in Section 6.1, such as the AMF and GLRT detectors, cannot be used and/or can be achieved by some other design with potentially a significantly smaller sample-support requirement. Indeed, it was recently demonstrated [75] that for real radar clutter data, all known CFAR receivers (detectors) "do not respect their nominal probability of false alarm, namely they exhibit a false alarm rate higher than the value preassigned at the design stage." The significant power mismatch (inhomogeneity) of the training data collected over different range cells was considered in Reference 75 to be the main reason for this mismatch between predicted and actual false-alarm rate, while other "second-order factors," such as a possible statistical dependence among the secondary vectors (see also Reference 76), contribute to a further deviation of the observed false-alarm rate from the theoretical value.

Furthermore, in a number of applications, the CFAR properties of adaptive detectors cannot be used, even theoretically, when some discrepancy between the training and primary data are involved. Such discrepancies can arise, for example, when an adaptive antenna is used for external-noise (jamming) mitigation only, while the actual target detection is performed at the output of coherent Doppler processing that is applied for clutter rejection. Generally a small number of training range and/or Doppler cells that are free of clutter and targets (perhaps being collected during a time interval free of radar transmission) is used for interference CM estimation. The resultant adaptive antenna array (beamformer) is then applied across the operationally important (primary) range and Doppler cells in order to mitigate this interference. Target detection is then performed on the surface clutter background usually using non-adaptive clutter suppression/detection techniques, such as coherent Doppler processing. An important feature of such an application is that the sample data train the adaptive beamformer for external-noise mitigation but *does not* represent the entire background interference that a target must be finally discriminated against.

Another example has been reported in References 23 and 77, where a different INR (not just scaling) over the training and primary datasets leads to a considerable loss of "CFARness." Importantly, in most such cases this difference does not significantly degrade the adaptive antenna's ability to reject this interference; it is the invariance of the loss factor and the CFAR property that are lost. A final example is the non-CFAR "robust adaptive detector," that is designed to reduce the impact of target model mismatch [22]. In all these examples, the pdf "tails" of the detector output statistics are sensitive to mismatches between the model used for deriving the CFAR properties and the true parameters.

In each of the above-mentioned adaptive detectors, the loss of "CFARness" just means that the problem of adaptive interference mitigation and the problem of adaptive false-alarm threshold control must be treated separately: the secondary (training) data should be used to achieve efficient interference mitigation, while the adaptive thresholding should use the standard "cell-averaging" approach over a number of adaptively processed primary range/Doppler cells, as proposed by Kalson [78]. This very established adaptive threshold (CFAR) control was introduced by Finn and Johnson (FJ) [11] and modified in numerous subsequent studies (e.g. [79–81]). Of course, the training cells and the processed primary range/Doppler cells involved in adaptive thresholding should be sufficiently homogeneous, but the number of such cells required to maintain a certain loss factor (with respect to the clairvoyant threshold) does not depend on the dimension of the adapted system [11].

For the above-mentioned applications with different interference properties over the secondary and primary data, this two-stage adaptive processing scheme seems to be the only viable adaptive detection option. Yet a similar two-stage approach may be considered as an alternative to CFAR adaptive detectors, even for the homogeneous training conditions that are typically considered for classic one-stage CFAR detector design. In this classic approach, consider the typical scenario addressed in References 5 and 40 where the training sample and the single primary sample being tested contain the same interference (at least up to a scalar factor). Here CFAR adaptive detectors, in the absence of a target, have invariant output statistics (i.e. pdf) with respect to the interference covariance matrix. This means that different primary cells may be corrupted by interferences with different covariance matrices, but unless each primary cell has correspondingly different training data, the pdf of the output statistics (and hence the false-alarm rate) are the same for all these cells that are processed by *different adaptive filters*.

In practice this means that, for the "sliding window" approach (where a template comprising a fixed number of cells neighboring the tested cell moves across all resolution cells), the false-alarm threshold and false-alarm rate remain the same for all tested cells. Naturally, each resolution cell is processed by a different adaptive detector that is constructed from a different CME that comes from a cell-specific set of training data.

An alternative approach to this method would be to consider, for each tested resolution cell with N sufficiently homogeneous training samples, a similar two-stage adaptive process as for the above applications with severely non-homogeneous training data. Specifically, the set of N i.i.d. training samples allocated for any single primary range cell is divided into two sets, N_{CME} and N_{CFAR}, the former is used for interference CM estimation to design the adaptive beamformer/filter/antenna, then the latter set is used to estimate the signal statistics (power in the Gaussian case) at the output of this adaptive antenna (i.e. adaptive scalar CFAR design). It is important to understand that this adaptive CFAR threshold is therefore specifically designed for a particular adaptive beamformer, and if different primary cells are to be processed by different filters, then different thresholds must be used. In practice, of course, we may have an intermediate case where some number of adjacent primary cells are processed by the same filter/detector.

Clearly, any analytical derivations for receiver operating characteristics (ROCs) for such a two-stage adaptive detection scheme also describe the above-mentioned non-homogeneous training applications provided that N_{CME} is the secondary training volume used to design the adaptive beamformer, and N_{CFAR} is the number of sufficiently homogeneous primary cells used for adaptive thresholding. In these "non-homogeneous" applications, N_{CME} and N_{CFAR} cannot be traded-off against each other, since these two sets of training data contain different interference.

For homogeneous training conditions, two-stage adaptive processing design raises an important question: given a limited number N of i.i.d. training samples, what is the best type of CME to use, and what is the optimal partition $\{N_{CME}, N_{CFAR}\}$? The answer does not seem to be obvious. At one extreme, we can use all of the training data to get the conventional ML CME and rely upon the scenario-invariant properties of the GLRT/AMF/ACE detector output statistics. The difficulty here, as discussed, is that strict invariance can be achieved only when N exceeds the dimension of the antenna array (number of sensors) M and that detection losses considerably exceed the RMB [7] SNR degradation of 3 dB for $N \simeq 2M$.

Since the required N_{CFAR} does not depend on the antenna dimension M, any adaptive antenna (filter) technique that for efficient interference mitigation requires a significantly smaller sample support than the conventional ML CME (LSMI or FML as discussed in Section 6.2, for example) will lead to a more efficient two-stage detector than any of the CFAR GLRT/AMF/ACE detector (for a sufficiently large antenna/filter dimension M). On the contrary, if a particular CME gives rise to (at least) the "practical" CFAR property, then the two-stage detection scheme that uses the same estimate may be inferior to the "one-stage" (practically) CFAR detector, even for the optimal partition $\{N_{CME}, N_{CFAR}\}$. Still, if this degradation is relatively small, then even a modest reduction in N_{CME} that is provided by a more efficient (but non-CFAR) CME should result in superior detection performance. An investigation of this trade-off is presented in this subsection. A comparative analysis of the "one-stage" CFAR AMF and GLRT detectors are well described in Reference 5, so it is sufficient for us to compare just one of them with our new detectors.

We shall compare ROCs for the CFAR AMF detector, the two-stage detector that adopts the same conventional ML CME (with sample size not less than M), the "practically CFAR" (one-stage) LSMI detector for the "cliff-type" scenario (6.6), and the two-stage LSMI detector for the same scenario. We expect that a comparison of (one-stage) AMF and the two-stage LSMI detectors will demonstrate the detection performance improvement discussed above, which is associated with the more efficient loaded CME, even in the case where the output statistics (pdf) invariance is ignored or is non-existent. Comparison of (practically) CFAR and the two-stage detectors that use the same CME (conventional ML or diagonally loaded) will give an insight into the gains associated with exploiting (practical) invariance. Section 6.4.1 provides analytical expressions for the ROCs of the various detectors, then Section 6.4.2 introduces the results of calculations and Monte-Carlo simulations.

6.4.1 Analysis Performance of "One-Stage" Adaptive CFAR Detectors versus "Two-Stage" Adaptive Processing

We consider a scenario with a fluctuating target (Swerling I model) masked by interference of type (6.6). The primary sample $\boldsymbol{y}$ is described as

$$\boldsymbol{y} = \begin{cases} \boldsymbol{x}_0 \sim \mathcal{CN}(\boldsymbol{0}, \boldsymbol{R}) & \text{for hypothesis } H_0 \\ \boldsymbol{x}_0 + a\boldsymbol{s}, \quad a \sim \mathcal{CN}(0, \sigma^2) & \text{for hypothesis } H_1 \end{cases} \tag{6.215}$$

where $\boldsymbol{x}_0$ is the observed M-variate interference-only data vector (snapshot), $\mathcal{CN}(\boldsymbol{0}, \boldsymbol{R})$ is the complex (circular) Gaussian pdf, $\boldsymbol{s}$ is the M-variate normalized ($\boldsymbol{s}^\dagger \boldsymbol{s} = 1$) array-signal ("steering" or wavefront) vector, and a is the complex target amplitude (whose power is σ^2). The unknown covariance matrix $\boldsymbol{R}$ is estimated using the set of N i.i.d. training samples

$$[\boldsymbol{x}_1, \ldots, \boldsymbol{x}_N], \quad \boldsymbol{x}_j \sim \mathcal{CN}(\boldsymbol{0}, \boldsymbol{R}), \quad j = 1, \ldots, N. \tag{6.216}$$

6.4.1.1 "Benchmark" Detectors (A, B, C, D)

6.4.1.1.1 Detector A: Clairvoyant Detector

We first identify the benchmark detection performance for this model. The ultimate performance is achieved by the clairvoyant receiver (detector), which comprises the optimal/Wiener filter

$$\boldsymbol{w}_{opt} = \frac{\boldsymbol{R}^{-1}\boldsymbol{s}}{\boldsymbol{s}^\dagger \boldsymbol{R}^{-1}\boldsymbol{s}} \tag{6.217}$$

followed by the detector

$$\frac{|\boldsymbol{y}^\dagger \boldsymbol{R}^{-1}\boldsymbol{s}|^2}{\boldsymbol{s}^\dagger \boldsymbol{R}^{-1}\boldsymbol{s}} \underset{H_0}{\overset{H_1}{\gtrless}} h > 1 \tag{6.218}$$

where H_0 and H_1 are, respectively, the hypotheses that no target or a target is present, and h is the false-alarm threshold, then the target-free signal at the output of the filter has the distribution $\mathcal{CN}(0, 1)$. The ROC curve of this clairvoyant detector is described by the well-known analytic expression for the probability of detection

$$P_d = \exp\left[-\frac{|\ln P_{FA}|}{1+q^2}\right] \tag{6.219}$$

where

$$q^2 = \sigma^2 \boldsymbol{s}^\dagger \boldsymbol{R}^{-1} \boldsymbol{s} \tag{6.220}$$

is the output SNR of the filter.

6.4.1.1.2 Detectors B and C: Adaptive SMI/LSMI Filter with Ideal Thresholding
For an unknown true covariance matrix $\boldsymbol{R}$, the conventional ML CME is

$$\hat{\boldsymbol{R}} = \frac{1}{N} \sum_{j=1}^{N} \boldsymbol{x}_j \boldsymbol{x}_j^\dagger, \qquad N \geq M \tag{6.221}$$

whereas the LCME is

$$\hat{\boldsymbol{R}}_\beta \equiv \hat{\boldsymbol{R}} + \beta \boldsymbol{I}_M, \qquad \beta = (2 \sim 3) \times \lambda_M, \qquad \lambda_M = \sigma_0^2 \tag{6.222}$$

where σ_0^2 is the additive white-noise power. The detectors B (Adaptive SMI) and C (Adaptive LSMI) are then specified, respectively, by

$$\frac{|\boldsymbol{y}^\dagger \hat{\boldsymbol{R}}^{-1} \boldsymbol{s}|^2}{\boldsymbol{s}^\dagger \hat{\boldsymbol{R}}^{-1} \boldsymbol{R} \hat{\boldsymbol{R}}^{-1} \boldsymbol{s}} \underset{H_0}{\overset{H_1}{\gtrless}} h > 1, \qquad \frac{|\boldsymbol{y}^\dagger \hat{\boldsymbol{R}}_\beta^{-1} \boldsymbol{s}|^2}{\boldsymbol{s}^\dagger \hat{\boldsymbol{R}}_\beta^{-1} \boldsymbol{R} \hat{\boldsymbol{R}}_\beta^{-1} \boldsymbol{s}} \underset{H_0}{\overset{H_1}{\gtrless}} h > 1. \tag{6.223}$$

In the famous RMB paper [7], the pdf for the SNR losses ρ in the adaptive filter $\boldsymbol{w} = \hat{\boldsymbol{R}}^{-1} \boldsymbol{s}$ compared with those of the clairvoyant Wiener filter (6.217)

$$\rho = \frac{(\boldsymbol{s}^\dagger \hat{\boldsymbol{R}}^{-1} \boldsymbol{s})^2}{(\boldsymbol{s}^\dagger \hat{\boldsymbol{R}}^{-1} \boldsymbol{R} \hat{\boldsymbol{R}}^{-1} \boldsymbol{s})(\boldsymbol{s}^\dagger \boldsymbol{R}^{-1} \boldsymbol{s})} \tag{6.224}$$

was derived as

$$f(\rho) = \frac{1}{B[s,n]} \rho^{s-1} (1-\rho)^{n-1}, \tag{6.225}$$

with

$$s = N - M + 2, \qquad n = M - 1 \tag{6.226}$$

where $B[s,n]$ is the beta function [42]. In Reference 13, it was shown that for LSMI with the diagonal loading factor (6.222) and for scenarios whose covariance matrix is of the form (6.6), the SNR loss factor

$$\rho_\beta = \frac{(\boldsymbol{s}^\dagger \hat{\boldsymbol{R}}_\beta^{-1} \boldsymbol{s})^2}{(\boldsymbol{s}^\dagger \hat{\boldsymbol{R}}_\beta^{-1} \boldsymbol{R} \hat{\boldsymbol{R}}_\beta^{-1} \boldsymbol{s})(\boldsymbol{s}^\dagger \boldsymbol{R}^{-1} \boldsymbol{s})} \tag{6.227}$$

with $\boldsymbol{w}_\beta = \hat{\boldsymbol{R}}_\beta^{-1} \boldsymbol{s}$, is described with high accuracy by the same beta distribution (6.225), but with the parameters

$$s(\beta) = N - m + 2, \quad n(\beta) = m - 1. \tag{6.228}$$

If the losses associated with adaptive thresholding are ignored (i.e. the number N_{CFAR} of i.i.d. samples used for adaptive thresholding approaches infinity), which means that the target-free output power of the *specific* adaptive filter $\boldsymbol{w}$ or $\boldsymbol{w}_\beta$ is known exactly in (6.223):

$$\sigma_{out}^2(\boldsymbol{w}) = \frac{\boldsymbol{s}^\dagger \hat{\boldsymbol{R}}^{-1} \boldsymbol{R} \hat{\boldsymbol{R}}^{-1} \boldsymbol{s}}{(\boldsymbol{s}^\dagger \hat{\boldsymbol{R}}^{-1} \boldsymbol{s})^2}, \quad \text{or} \quad \sigma_{out}^2(\boldsymbol{w}_\beta) = \frac{\boldsymbol{s}^\dagger \hat{\boldsymbol{R}}_\beta^{-1} \boldsymbol{R} \hat{\boldsymbol{R}}_\beta^{-1} \boldsymbol{s}}{(\boldsymbol{s}^\dagger \hat{\boldsymbol{R}}_\beta^{-1} \boldsymbol{s})^2}, \tag{6.229}$$

then the probability of detection is

$$P_d = \frac{1}{B[s,t]} \int_0^1 \exp\left[-\frac{|\ln P_{FA}|}{1+q^2 x}\right] x^{s-1}(1-x)^{t-1}\,dx. \tag{6.230}$$

For s and t given by (6.226) (detector B), we get the detection probability for the conventional ML sample estimate-based detector, while for $s(\beta)$ and $t(\beta)$ given by (6.228) (detector C), we get the detection probability for the LSMI-based detector. Whereas expression (6.230) is well suited for numerical integration, in Appendix A of Reference 2, we derive the analytic expression for this integral:

$$P_d = \frac{P_{FA}}{(1+q^2)^s}\,\Phi_1\left(s,\ s+t,\ s+t;\ \frac{q^2}{1+q^2}, \frac{q^2|\ln P_{FA}|}{1+q^2}\right) \tag{6.231}$$

where $\Phi_1(a,b,c;w,z)$ is the confluent (degenerate) hypergeometric series in two variables [42]. This performance accounts for the two benchmark detectors when $N_{CME} = N$ and $N_{CFAR} \to \infty$.

6.4.1.1.3 Detector D: Optimal/Wiener Filter with Scalar Adaptive CFAR

Finally, another extreme case gives us another benchmark for performance comparison: when the covariance matrix is known (at least up to a scalar factor) and conventional CFAR thresholding is applied with all N samples used for output power estimation:

$$\frac{|\boldsymbol{y}^\dagger \boldsymbol{R}^{-1}\boldsymbol{s}|^2}{\boldsymbol{s}^\dagger \boldsymbol{R}^{-1}\hat{\boldsymbol{R}}\boldsymbol{R}^{-1}\boldsymbol{s}} \underset{H_0}{\overset{H_1}{\gtrless}} h > 1 \tag{6.232}$$

The probability of detection associated with the finite sample volume N here is well known [11]:

$$P_d = \left[1 + \frac{1-P_{FA}^{1/N}}{(1+q^2)P_{FA}^{1/N}}\right]^{-N}. \tag{6.233}$$

This performance accounts for the benchmark detector when $N_{CME} \to \infty$ and $N_{CFAR} = N$.

Thus we have the ROC curve expression for the clairvoyant benchmark, as well as for detection performance of adaptive detectors with ideal thresholding or ideal Wiener filtering. We next introduce ROC formulas for the practical (one-stage) AMF and "two-stage" adaptive detectors.

6.4.1.2 "One-Stage" Adaptive Detectors (E, F)

6.4.1.2.1 Detector E: (One-Stage) CFAR AMF Detector

This detector is specified by [5]

$$\frac{|y^\dagger \hat{R}^{-1} s|^2}{s^\dagger \hat{R}^{-1} s} \underset{H_0}{\overset{H_1}{\gtrless}} h > 1, \qquad \hat{R} = \frac{1}{N}\sum_{j=1}^{N} x_j x_j^\dagger. \tag{6.234}$$

Let $\tau \equiv N - M + 1$, then the false-alarm rate of this detector is [1]

$$P_{FA} = (1+h)^{-\tau}\, {}_2F_1\left(\tau,\, M-1,\, N+1;\, \frac{h}{1+h}\right) \tag{6.235}$$

where ${}_2F_1(\alpha, \beta, \gamma; x)$ is the hypergeometric function in one variable [42]. The probability of detection is [1]

$$P_d = \frac{N!}{\tau!(M-2)!}\int_0^{\frac{1}{1+q^2}} \frac{x^\tau[1-(1+q^2)x]^{M-2}}{(1+hx)^\tau(1-q^2x)^{N+1}}\,dx \tag{6.236}$$

$$= \left[\frac{1+q^2}{1+q^2+h}\right]^\tau F_1\left(M-1, \tau, -\tau, N+1; \frac{q^2+h}{1+q^2+h}, \frac{q^2}{1+q^2}\right) \tag{6.237}$$

where $F_1(\alpha, \beta, \beta', \gamma; x, y) \equiv F_1(\alpha, \beta', \beta, \gamma; y, x)$ is the hypergeometric function in two variables [42].

6.4.1.2.2 Detector F: (One-Stage) "Practically CFAR" LAMF Detector

This detector was introduced in Reference 1 similar to the above AMF detector.

$$t_F \equiv \frac{|y^\dagger R_\beta^{-1} s|^2}{s^\dagger R_\beta^{-1} s} \underset{H_0}{\overset{H_1}{\gtrless}} h > 1, \qquad R_\beta = \hat{R} + \beta I_M, \qquad \beta = (2 \sim 3) \times \sigma_0^2. \tag{6.238}$$

Unfortunately no analytic expressions have yet been derived for P_{FA} and P_d. However, in Reference 1 we demonstrated that the "practically CFAR" probability of false-alarm of this detector may be precalculated using the approximate representation of the output statistics (provided that $\lambda_m/\lambda_M \gg 1$):

$$t_F \simeq \frac{\left|[x_n^\dagger - x_m(Z_m Z_m^\dagger)^{-1} Z_m Z_n^\dagger] Z e\right|^2}{e^\dagger Z e} \tag{6.239}$$

where $n \equiv M - m$, $e \equiv [1, 0, \ldots, 0]^N$, and

$$Z \equiv \left\{\beta I_n + \frac{1}{N} Z_n \left[I_N - Z_m^\dagger (Z_m Z_m^\dagger)^{-1} Z_m\right] Z_n^\dagger\right\}^{-1} \tag{6.240}$$

$$Z_m \in C^{m\times N} \sim \mathcal{CN}_N(0, I_m), \qquad Z_n \in C^{n\times N} \sim \mathcal{CN}_N(0, I_n) \tag{6.241}$$

$$x_m \in C^{m\times 1} \sim \mathcal{CN}(0, I_m), \qquad x_n \in C^{n\times 1} \sim \mathcal{CN}(0, I_n) \tag{6.242}$$

and the Z_m, Z_n, x_m, x_n are all mutually independent.

We can see that P_{FA} is specified by "white-noise" random values and is a function only of M, N, m, and β. Actual probabilities of detection are evaluated in Reference 1 by direct Monte-Carlo simulations.

6.4.1.3 "Two-Stage" Adaptive Detectors (G, H) [78]

6.4.1.3.1 Detector G: Conventional SMI Adaptive Filter and Scalar CFAR Detector ("Two-Stage" AMF)

The two-stage detector that uses the conventional (unloaded) CME $\hat{\boldsymbol{R}}$ is

$$\frac{|\boldsymbol{y}^{\dagger}\hat{\boldsymbol{R}}_{CME}^{-1}\boldsymbol{s}|^2}{\boldsymbol{s}^{\dagger}\hat{\boldsymbol{R}}_{CME}^{-1}\hat{\boldsymbol{R}}_{CFAR}\hat{\boldsymbol{R}}_{CME}^{-1}\boldsymbol{s}} \underset{H_0}{\overset{H_1}{\gtrless}} h > 1 \tag{6.243}$$

where

$$\hat{\boldsymbol{R}}_{CME} = \frac{1}{N_{CME}}\sum_{j=1}^{N_{CME}} x_j x_j^{\dagger}, \quad \hat{\boldsymbol{R}}_{CFAR} = \frac{1}{N_{CFAR}}\sum_{j=N_{CME}+1}^{N} x_j x_j^{\dagger}. \tag{6.244}$$

(Note that when the samples are homogeneous, their ordering is arbitrary.) In Appendix B of Reference 2, we derive the following expressions:

$$P_{FA} = (1+h)^{-N_{CFAR}} \tag{6.245}$$

and

$$P_d = \frac{\Gamma(N_{CME}+1)}{\Gamma(N_{CME}-M+2)\Gamma(M-1)}\int_0^1 \left[\frac{1+q^2x}{1+q^2x+h}\right]^{N_{CFAR}} x^{N_{CME}-M+1}(1-x)^{M-2}\,dx \tag{6.246}$$

$$= \left[\frac{1+q^2}{1+q^2+h}\right]^{N_{CFAR}} F_1\left(M-1,\, N_{CFAR},\, -N_{CFAR},\, N_{CME}+1;\, \frac{q^2}{1+q^2+h},\, \frac{q^2}{1+q^2}\right). \tag{6.247}$$

Note the similarities between (6.237) and (6.247) (for $N_{CFAR} = \tau \equiv N - M + 1$) and the distinctions between (6.235) and (6.245).

6.4.1.3.2 Detector H: LSMI Adaptive Filter and Scalar CFAR Detector ("Two-Stage" LAMF)

The expressions for the corresponding diagonally loaded two-stage detector

$$\frac{|\boldsymbol{y}^{\dagger}\hat{\boldsymbol{R}}_{CME}(\beta)^{-1}\boldsymbol{s}|^2}{\boldsymbol{s}^{\dagger}\hat{\boldsymbol{R}}_{CME}(\beta)^{-1}\hat{\boldsymbol{R}}_{CFAR}\hat{\boldsymbol{R}}_{CME}(\beta)^{-1}\boldsymbol{s}} \underset{H_0}{\overset{H_1}{\gtrless}} h > 1 \tag{6.248}$$

$$\hat{\boldsymbol{R}}_{CME}(\beta) = \hat{\boldsymbol{R}}_{CME} + \beta\boldsymbol{I}_M \tag{6.249}$$

are

$$P_{FA} = (1+h)^{-N_{CFAR}} \tag{6.250}$$

and

$$P_d = \left[\frac{1+q^2}{1+q^2+h}\right]^{N_{CFAR}} F_1\left(m-1, N_{CFAR}, -N_{CFAR}, N_{CME}+1; \frac{q^2}{1+q^2+h}, \frac{q^2}{1+q^2}\right). \tag{6.251}$$

We are now in a position to compare numerical results for these eight detectors.

6.4.2 Comparative Detection Performance Analysis

Consider the scenario that was recently analyzed in Reference 1; a 12-sensor uniform linear antenna array that observes a total of only 24 training data samples:

$$M = 12, \quad m = 6, \quad N = 24 = 2M. \tag{6.252}$$

The covariance matrix for $m = 6$ interference sources is

$$\boldsymbol{R} = \sum_{k=1}^{6} \sigma_k^2 \boldsymbol{s}(\theta_k)\boldsymbol{s}^{\dagger}(\theta_k) + \sigma_0^2 \boldsymbol{I}_M \tag{6.253}$$

$$\sigma_0^2 = 1, \quad \sigma_1^2 = \cdots = \sigma_6^2 = 1{,}000 \tag{6.254}$$

$$\sin\boldsymbol{\theta} = [-0.8, -0.4, 0.2, 0.5, 0.7, 0.9] \tag{6.255}$$

(our target is located at $\sin\theta_0 = -0.6$). We see that this scenario meets the "favorable" conditions of (6.6) for the application of the LSMI technique, since $\lambda_6/\lambda_7 = \lambda_6/\lambda_{12} \simeq 40$ dB.

In what follows, most of our figures are split across two subfigures because the number of ROC curves is large; we retain ROCs for detectors A and D as common reference curves in both, and display results for detectors that use conventional CMEs in subfigure (a), with those that use loaded CMEs in subfigure (b).

We begin with the clairvoyant detector A (6.218) whose ROC curve is given analytically by (6.219) (see Figure 6.39, curve labeled "A [ideal + ideal] *"; the asterisk indicates curves/detectors that are independent of partitioning, while the square brackets signify impractical detectors). Note that for the same scenario, in Reference 1 we also calculated this ROC by direct Monte-Carlo simulations for validation reasons. The perfect correspondence seen there between analytic and simulated results proved the sufficient accuracy of the Monte-Carlo setting (i.e. number of trials for false-alarm and detection probability calculations) that is also adopted in this subsection.

Figure 6.39 also reproduces from Reference 1 the ROCs of the conventional AMF detector (E), labeled "simulated E", again calculated by Monte-Carlo simulations; now we augment this with the analytic formula (6.237), labeled "E (CFAR AMF) *." This theoretical curve makes use of the GSL software routine `gsl_sf_hyperg_2F1` [82] to calculate the function ${}_2F_1(\alpha, \beta, \gamma; x)$ in order to compute the false-alarm threshold h for $P_{FA} = 10^{-4}$, followed by the Matlab routine `quad` to numerically evaluate the integral that defines the probability of detection. (For some scenarios, this "E" curve could not be calculated for high SNRs because of numerical computation issues with the routine `quad`.) The accuracy of numerical calculations was verified by the condition $P_d = P_{FA}$ for $q^2 = 0$. Again we observe ideal correspondence

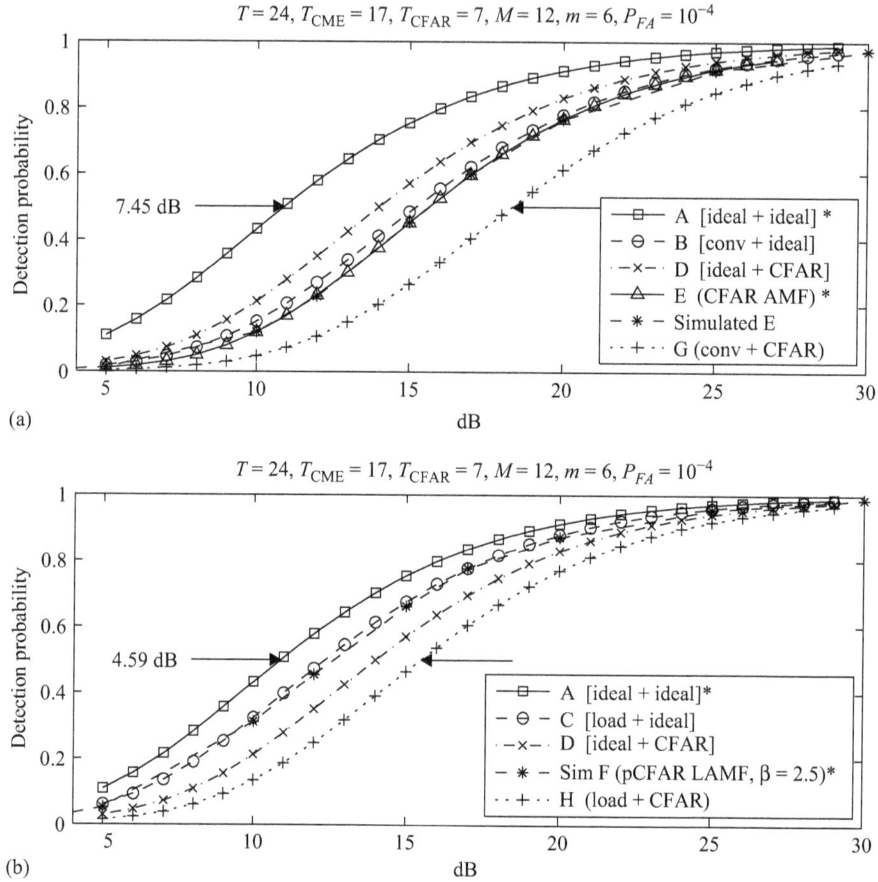

Figure 6.39 ROCs for the optimum partition for conventional CME, 12 sensors, 6 interferers, and 24 snapshots.

between the analytical and "experimental" ROCs for detector E, which demonstrates the accuracy of the computations and the Monte-Carlo simulations. The two curves so far described (A and E) are well known, and so it is with no surprise that we observe a familiar AMF loss factor of 4.75 dB at $P_d = 0.5$ and $P_{FA} = 10^{-4}$ for the conventional ML sample covariance matrix estimate that involves all $N = 24$ snapshots.

The remaining curves in Figure 6.39(a) are the ROCs for the two benchmark detectors B (6.230) and (6.226) (conventional-CME beamformer with ideal thresholding) and D (6.233) (ideal receiver with scalar CFAR), and finally detector G (conventional SMI filter with scalar CFAR).

Figure 6.39(b) repeats benchmark curves A and D and adds ROCs for the loaded-CME detectors C (6.230) and (6.228), F and H (6.251). The ROC for the "practically CFAR" LAMF detector F is reproduced from Reference 1, again calculated by Monte-Carlo simulations, for the loading factor $\beta = 2.5$.

The particular partition $\{N_{CME}, N_{CFAR}\} = \{17, 7\}$ used in Figure 6.39 was found (by exhaustive search) to be optimal for the *conventional*-CME detector G, i.e. resulted in the

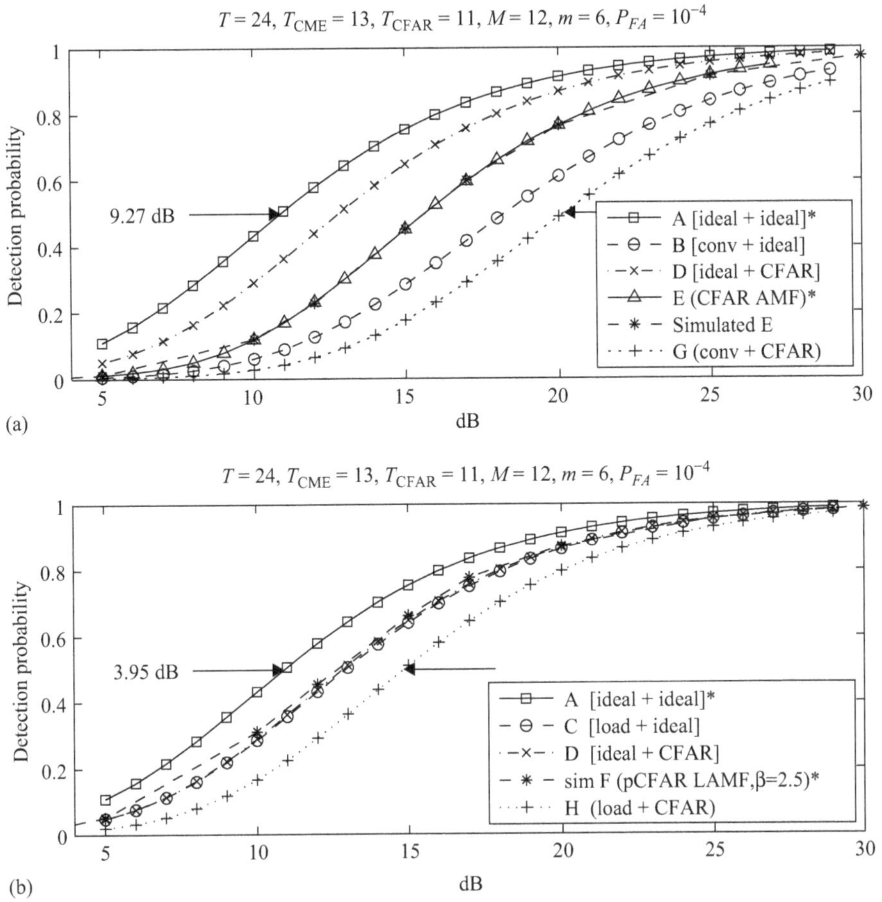

Figure 6.40 ROCs for the optimum partition for loaded CME, 12 sensors, 6 interferers, and 24 snapshots.

minimum SNR loss at $P_d = 0.5$ for detector G compared with the ideal detector A, in this case 7.45 dB. Of course, for a proper comparison, we use 17 snapshots for the adaptive beamformers B and C, and seven snapshots for adaptive thresholding in D. For the reader's convenience, the following list details the various possible sample-support partitions.

Figure 6.39(a)	conventional CME	Figure 6.39(b)	loaded CME
A:	$N_{CME} \to \infty, N_{CFAR} \to \infty$	A:	$N_{CME} \to \infty, N_{CFAR} \to \infty$
B:	$N_{CME} = 17, N_{CFAR} \to \infty$	C:	$N_{CME} = 17, N_{CFAR} \to \infty$
D:	$N_{CME} \to \infty, N_{CFAR} = 7$	D:	$N_{CME} \to \infty, N_{CFAR} = 7$
E:	$N = N_{CME} + N_{CFAR} = 24$	F:	$N = N_{CME} + N_{CFAR} = 24$
G:	$N_{CME} = 17, N_{CFAR} = 7$	H:	$N_{CME} = 17, N_{CFAR} = 7$

Figure 6.40 shows, in the same format, ROCs for the same eight detectors but for the partition optimized for SNR losses in the *loaded*-CME detector H at 50% detection probability, namely $\{N_{CME}, N_{CFAR}\} = \{13, 11\}$ resulting in 3.95 dB.

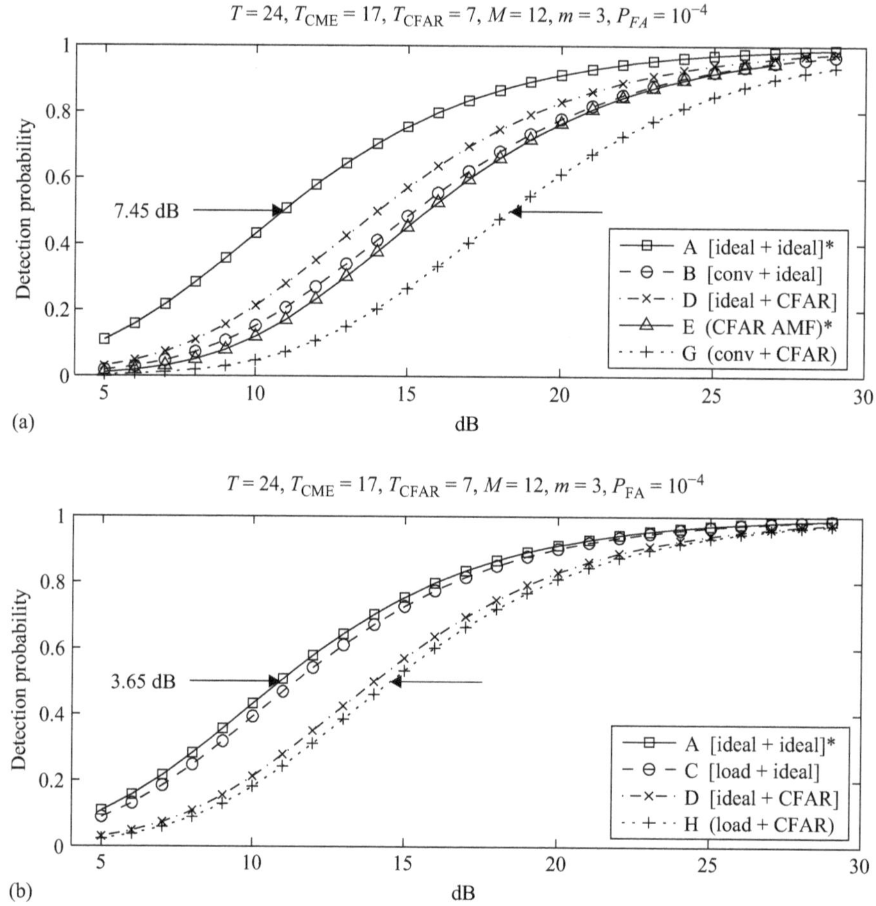

Figure 6.41 ROCs for the optimum partition for conventional CME, 12 sensors, 3 interferers, and 24 snapshots.

Before commenting on these results in detail, we shall introduce the remaining figures. For the same antenna array setting ($M = 12$, $N = 24$), Figures 6.41 and 6.42 show ROCs in a similar format for $m = 3$ dominant interferers (instead of six). Since the RMB [7] loss factor (6.225) and (6.226) for SMI does not depend on scenario, we naturally find the same optimal partition for *conventional* CME as in Figure 6.39. While the ROCs in Figure 6.41(a) therefore repeat those in Figure 6.39(a), we retain this subfigure for convenience of comparison. Apart from curves A and D in Figure 6.41(b), of course, the other two ROCs are different from those (*loaded-CME* detectors) in Figure 6.39(b), despite the same partitioning, due to the reduced number of interference sources.

The search for the optimal partitioning of sample support for (the loaded-CME) detector H here leads to $\{N_{CME}, N_{CFAR}\} = \{10, 14\}$, as shown in Figure 6.42(b). Since $N_{CME} = 10$ is less than the antenna array dimension $M = 12$, it is not possible to utilize conventional-CME detectors. For this reason, we instead introduce in Figure 6.42(a) ROCs for the conventional-CME detectors at their minimum possible partition $\{N_{CME}, N_{CFAR}\} = \{12, 12\}$.

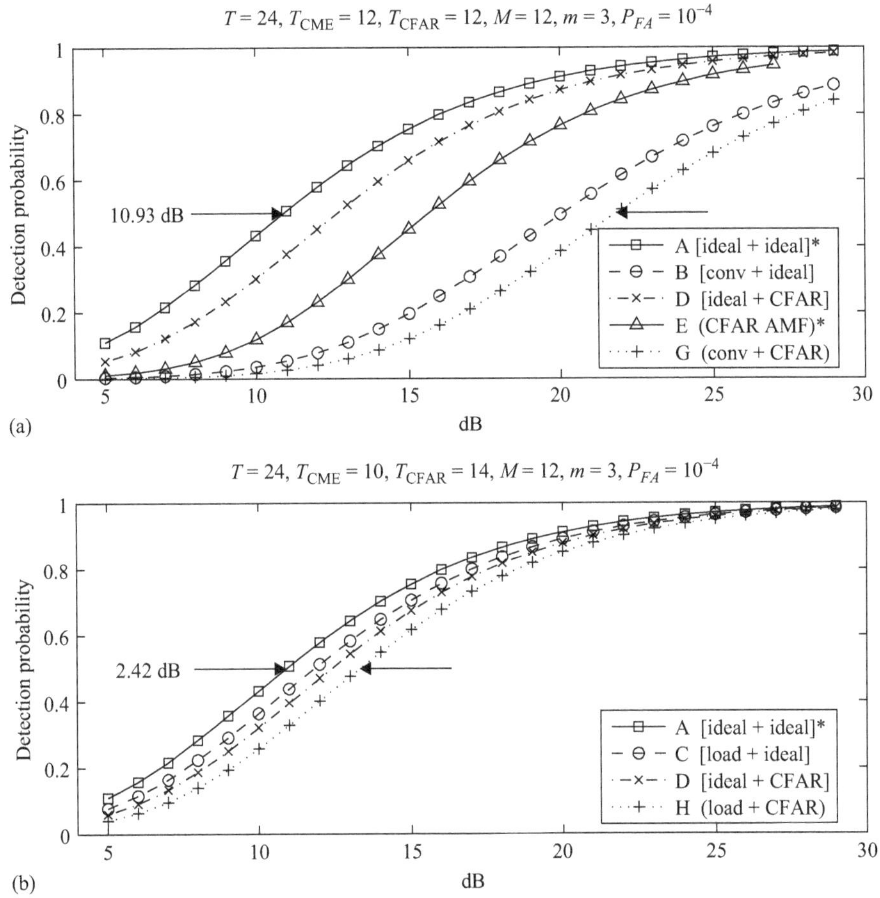

Figure 6.42 ROCs for 12 sensors, 3 interferers, and 24 snapshots for (a) the smallest possible sample support for conventional CME, and (b) the optimum partition for loaded CME.

In the same format, Figures 6.43 and 6.44 show results for an $M = 10$-sensor antenna array, $N = 50$ snapshots, $m = 3$ interferers, and false-alarm probability $P_{FA} = 10^{-6}$, which are the same scenario studied in References 4 and 5.

These results allow us to perform the following comparative analysis:

(1) (one-stage) CFAR AMF detector versus (one-stage) "practically CFAR" LAMF detector, i.e. curve E versus F;

(2) (one-stage) CFAR AMF detector versus two-stage AMF detector with the same conventional CME, for different partitions, i.e. curve E versus G;

(3) (one-stage) "practically CFAR" LAMF detector versus two-stage LAMF detector with the same loaded CME, for different partitions, i.e. curve F versus H;

(4) optimal partitions for two-stage AMF and LAMF detectors, i.e. curves G and H;

(5) most importantly, one-stage CFAR AMF detector versus optimally partitioned two-stage LAMF detector, i.e. curve E versus H.

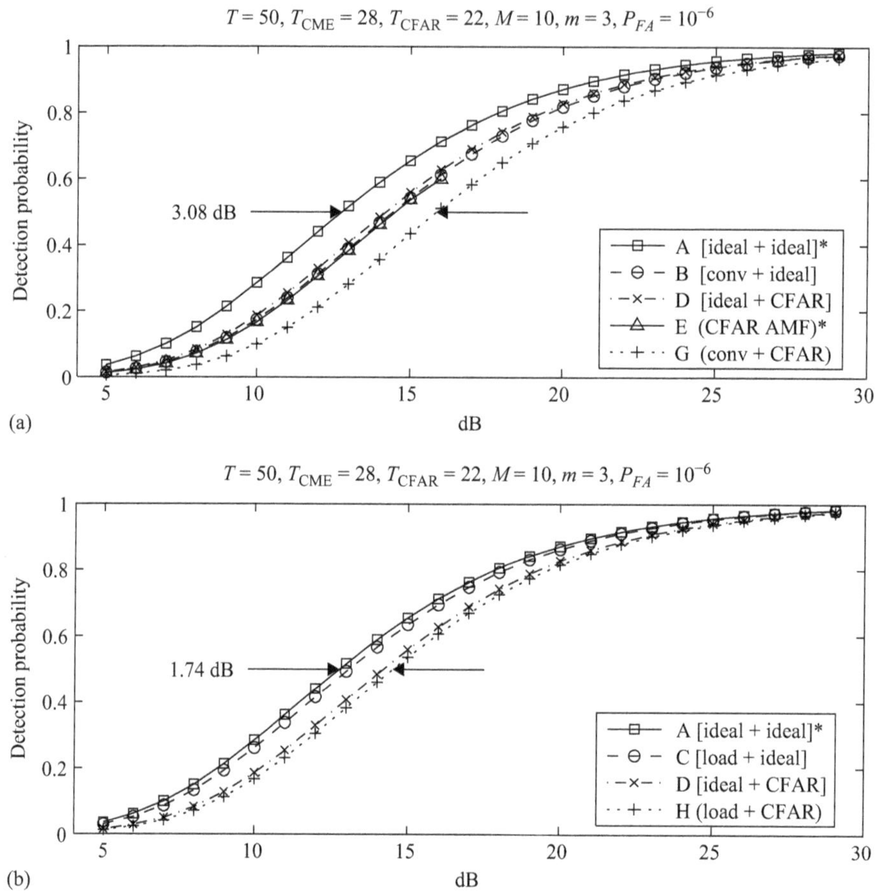

Figure 6.43 ROCs for the optimum partition for conventional CME, 10 sensors, 3 interferers, and 50 snapshots.

First, as already seen, Figure 6.39(a) shows a 4.75-dB SNR degradation in the standard AMF detector (E) compared with the ideal (A); this figure is in agreement with Reference 5. Figure 6.39(b) reveals that the one-stage practically CFAR LAMF detector (F) has a loss of only 1.73 dB. Therefore, when the "practically CFAR" properties of the LAMF detector are sufficiently accurate and properly exploited, we find that it has a significant performance improvement over the traditional AMF detector [1].

Note that the mean SNR improvement due to diagonal loading of the covariance matrix estimate, as described by (6.225) and (6.228), is only 1.55 dB. An additional 1.5-dB SNR improvement seen in the ROCs may be attributed to the significantly different pdfs of the AMF and LAMF detector outputs (the LAMF one "fluctuates less" [1]).

Second, a comparison of the traditional one-stage AMF detector (E) with the two-stage AMF detector (G), which uses the same unloaded CME, demonstrates that even for the *conventional-CME* optimum $\{N_{CME}, N_{CFAR}\} = \{17, 7\}$ in the case of Figure 6.39, the one-stage AMF detector is superior to the two-stage one, by the amount of $7.45 - 4.75 = 2.70$ dB.

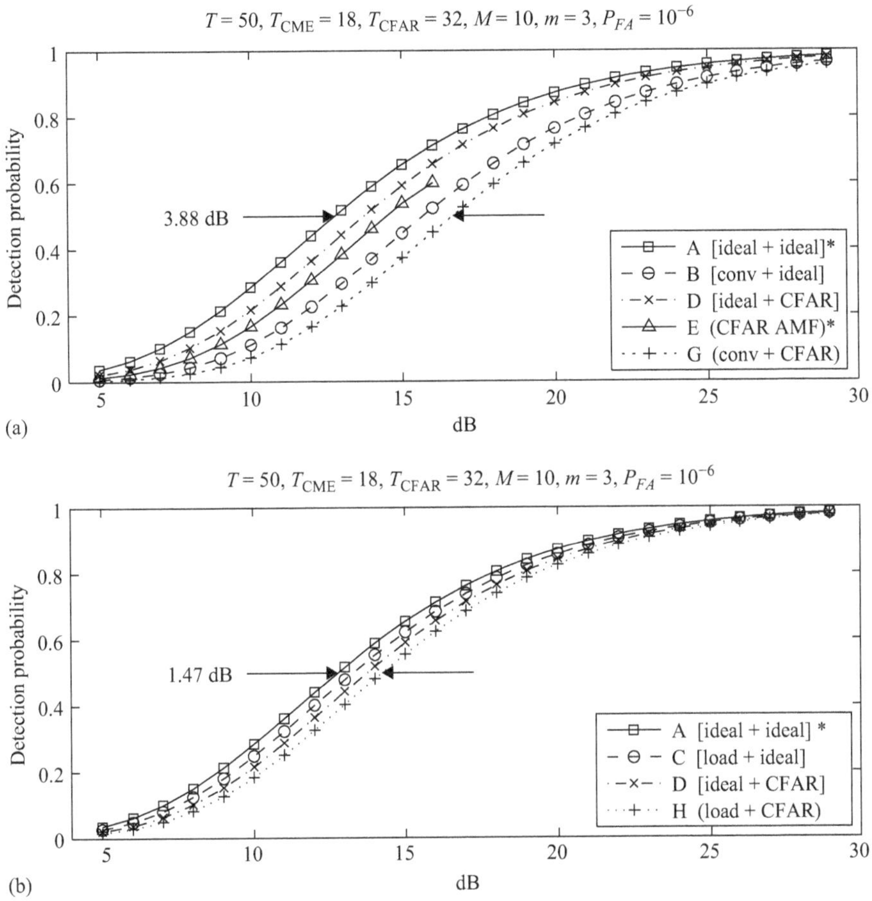

Figure 6.44 ROCs for the optimum partition for loaded CME, 10 sensors, 3 interferers, and 50 snapshots.

According to (6.225) and (6.226), the sample-support reduction from $N = 24$ to $N = 17$ causes a 1.5-dB SNR degradation, hence an additional 1.2-dB degradation must be due to the very limited sample volume $N = 7$ used for adaptive threshold estimation.

Third, we similarly find that the one-stage LAMF detector (F) remains superior to the two-stage LAMF detector (H), even for the *loaded-CME* optimal partition $\{N_{CME}, N_{CFAR}\} = \{13, 11\}$ of Figure 6.40, by the amount of $3.95 - 1.73 = 2.22$ dB. According to (6.225) and (6.228), the sample-support reduction from $N = 24$ to $N = 17$ in the LAMF adaptive filter causes about 1-dB SNR degradation, so that a further 1.2-dB loss is due to the $N = 7$ finite-sample support in the second CFAR stage of the detector. Note that the same losses (1.2 dB) as above, which are associated with the seven snapshots used for output power (adaptive threshold) estimation, are not a surprise, since the conditional pdf describing the output of these two-stage detectors is the same (exponential in this case).

Therefore, all the above-analyzed detectors, which rely on the ideal (AMF) or "practical" (LAMF) CFARness, gain against the two-stage alternatives *that adopt the same CME*. SNR

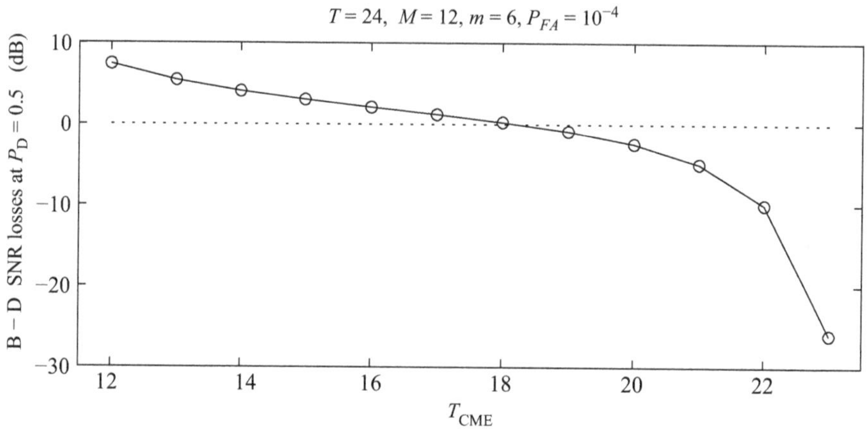

Figure 6.45 Difference between the performance of detectors B and D for the Figure 6.39 scenario as the partitioning varies; the optimal partition is $N_{CME} = 17$.

degradations seen here are due to two reasons: a decreased sample volume (and hence a decreased efficiency of interference mitigation in the adaptive filter/antenna) and a very small sample volume allocated for adaptive threshold evaluation. Our crude analysis, based on the beta-distribution (6.225) with (6.226) and (6.228), shows that the impact of these two causes are about the same for the partitioning $\{N_{CME}, N_{CFAR}\} = \{17, 7\}$.

The introduced results on our impractical "benchmark" detectors provide an additional important insight into the nature of the optimal partitioning, both for conventional and for loaded CME (detectors G and H, respectively). As we already expected, for conventional CME the SNR losses associated with finite sample support in covariance matrix estimation (RMB losses) and the losses associated with finite sample support in adaptive threshold estimation (FJ losses) are both found to be independent of scenario (in particular, the number of interferers m). On the other hand, an examination of the ROCs for the "complementary" benchmark detectors B and D (in the sense that B has ideal thresholding $N_{CFAR} \to \infty$, while D has ideal filtering $N_{CME} \to \infty$) for conventional-CME optimal partitions against non-optimal partitions leads to the following important observation. The difference in SNR losses between detectors B and D at $P_d = 0.5$ for each of our figures is 1.21 dB and 5.43 dB, 1.21 dB and 7.41 dB, 0.23 dB and 1.95 dB, respectively (the first figure in each pair corresponds to the conventional-CME optimal partition, while the second is for a non-optimal partition). Figure 6.45 shows that the optimal partitioning occurs close to where the RMB losses equal the FJ losses. This observation leads us to the realization (of an intuitively expected fact) that the loss factor for detector G is equal to the sum of the losses for the benchmark detectors B and D to a surprisingly high accuracy:

$$(G - A) \simeq (B - A) + (D - A) \tag{6.256}$$

in notational terms.

The analogous result for the loaded case is also true:

$$(H - A) \simeq (C - A) + (D - A). \tag{6.257}$$

These properties suggest a simple approach for partitioning the available N training samples for the two-stage LAMF detector H (if we are prepared to use this detector instead of the "practically CFAR" detector F for the scenario in question). For the "cliff-type" interference eigenspectrum (6.6) considered in this study, the estimate of the number of dominant sources m is accurate whenever $N > m$. Therefore the sample support N may first be used to estimate m, then the beta distribution (6.225) and (6.228) may be used to find a N_{CME} that results in an average SNR loss factor (6.227) that is the same as the FJ loss for N_{CFAR} samples. We have already seen that the optimal sample support N_{CME} in the two-stage LAMF detector (H) is proportional to m, unlike the support required for efficient interference mitigation in the traditional AMF detector (E) that is proportional to M.

Fifth, the most important result of this study is demonstrated by comparing E and H; for the scenario with $M = 12$ and $N = 24$ we got SNR losses (with respect to the ideal detector) for the optimal partitions:

	One-stage AMF (E)	Two-stage LAMF (H):	N_{CME}	N_{CFAR}
$m = 6$	} 4.75 dB	3.95 dB	13	11
$m = 3$		2.42 dB	10	14

It is important to note that, according to References 13 and 14, $N \simeq 2m$ data samples are required to obtain average 3-dB losses when using the LSMI technique, irrespective of M. Hence these first two scenarios with $M/m = 2$ and $M/m = 4$ are not as favorable for revealing LSMI improvement over SMI as they would be for a higher ratio.

Finally, we consider more carefully the scenario studied in References 4 and 5 (Figures 6.43 and 6.44). In this case, the conventional-CME optimal partition is $\{N_{CME}, N_{CFAR}\} = \{28, 22\}$ while the loaded-CME optimum is $\{N_{CME}, N_{CFAR}\} = \{18, 32\}$. According to References 4 and 5, the traditional AMF and GLRT detectors in this case have a loss factor between 1.7 and 1.9 dB, which is in good agreement with our calculated 1.72 dB for detector E. For the loaded-CME optimum, the two-stage LAMF detector (H) has 1.47-dB losses, which is an improvement over the AMF detector, especially as there is a small range in losses for all detectors here. Most importantly, for a fixed number of dominant interferers m, the two-stage LAMF detector (H) will have exactly the same performance for an arbitrarily large antenna array, whereas for $M = 20$ sensors, a GLRT detector would require $N = 100$ snapshots to achieve equal performance as for $M = 10$ and $N = 50$ [4].

6.4.3 Observations

In this section, we considered the problem of adaptive detector performance under very limited training support, when additional *a priori* information on the interference properties allows much better covariance matrix estimation than just the direct sample matrix, but at the expense of losing the CFAR property. To make use of such efficient (but non-CFAR) CMEs for adaptive detectors, some portion of the available training data are taken from covariance matrix estimation and used instead for evaluating the false-alarm threshold. To investigate a trade-off between these two competing systems, we have analyzed a number of "benchmark" and practical "one-stage" (CFAR) and "two-stage" adaptive detectors with very limited training-sample support.

For the Swerling I (fluctuating target) model and a fixed number of dominant interferers in white noise, this section compares the performance of the four "benchmark" (impractical) detectors:

1. clairvoyant,
2. adaptive SMI filter with ideal thresholding,
3. adaptive LSMI filter with ideal thresholding,
4. optimal/Wiener filter with scalar adaptive CFAR,

and the four practical detectors:

1. (one-stage) CFAR AMF [5],
2. (one-stage) "practically CFAR" LAMF [1],
3. "two-stage adaptive SMI + adaptive threshold" (two-stage AMF),
4. "two-stage adaptive LSMI + adaptive threshold" (two-stage LAMF),

under the condition of fixed total training sample size.

Most of the relevant ROCs have been analytically derived, while the demonstrated accurate agreement between the analytically derived and the Monte-Carlo ROCs tends to validate the ROC (for the "practically CFAR" LAMF detector) calculated by Monte-Carlo simulations only.

The first important conclusion is that in all considered cases, when the "strict" or "practical" CFAR property arises from the detector design, such (one-stage) CFAR detectors are found to be better than the corresponding two-stage detection schemes *that use the same covariance matrix estimate*. In other words, the traditional one-stage CFAR AMF detector is always better than the two-stage method comprising adaptive SMI and adaptive thresholding (two-stage AMF), though the relative loss factor can be reasonably small (see the $M = 10$, $N = 50$ case in Figure 6.43(a) where G–E is only 1.36 dB). Similarly for scenarios where we can precalculate the false-alarm threshold with sufficiently high accuracy (by estimating the number of dominant interferers), "practically CFAR" LAMF detectors are better than the two-stage detection schemes that use the same diagonally loaded CME.

At the same time, we found that two-stage LCME detectors outperform conventional one-stage AMF detectors, even for scenarios with relatively many strong interferers ($m \simeq M/2$). We demonstrated that the optimal partition of a given total number of training samples N into covariance matrix estimation (N_{CME}) and scalar CFAR (N_{CFAR}) subsets satisfies an intuitively expected condition: the optimum occurs when the losses associated with adaptive thresholding (only) are about the same as those for CME only (i.e. with ideal thresholding) and that the total loss factor can be accurately estimated as the sum of these two losses. Most importantly, for a given number of interferers, the required sample support does not depend on the antenna array dimension M and therefore the gain that such an adaptive detection scheme provides with respect to CFAR AMF increases with M.

For antenna arrayed radar systems with many more sensors than the number of dominant interferers ($M \gg m$), this saving in sample support is crucial. Moreover, the two-stage detection schemes considered here *do not rely* upon any CFAR properties of the detector, that in most practical applications are not invariant (due to the various mismatches). In addition, two-stage detection schemes pave the way for even more efficient covariance matrix estimates (in terms of sample-support requirements), when additional reliable information on the admissible covariance matrix structure is available (centrosymmetric, Toeplitz, AR(m), ARMA(m), etc.).

In Section 6.3, we demonstrated that for interferences with autoregressive properties and hence "non-cliff-like" eigenspectra, the parametric time-varying autoregressive model $TVAR$(m) gives a significant improvement in the interference mitigation efficiency of adaptive filtering with sample support comparable with the model order m, rather than with the filter (antenna) dimension M ($m \ll M$). We also showed that $TVAR$(m)-based AMF detectors are "practically CFAR" under certain conditions, so that the false-alarm threshold can be precalculated with acceptable accuracy, similar to LAMF detectors. In some cases, however, the required/desired accuracy is not achieved, so that the two-stage adaptive detector analyzed in this section may be considered instead for practical applications.

An even more dramatic sample-support reduction (compared with conventional SMI) was recently demonstrated in References 59 and 60 for STAP applications. There we showed that parametric STAP very efficiently mitigates ground clutter in the KASSPER airborne radar phenomenological dataset, with its 352 degrees of freedom, in some cases requiring only a few training samples. Yet in that practical application, with a parametric covariance matrix model serving as a CME, we cannot expect an invariance of the output statistics that is sufficient for CFARness. In such cases, therefore, the two-stage adaptive detector architecture provides a framework for implementing very efficient but non-CFAR adaptive (antenna, STAP) solutions.

References

[1] Y. Abramovich, N. Spencer, and A. Gorokhov, "Modified GLRT and AMF framework for adaptive detectors," *IEEE Trans. Aero. Elect. Syst.*, vol. 43, no. 3, pp. 1017–1051, Jul. 2007.

[2] Y. I. Abramovich, B. A. Johnson, and N. K. Spencer, "Sample-deficient adaptive detection: Adaptive scalar thresholding versus CFAR detector performance," *IEEE Trans. Aerosp. Electron. Syst.*, vol. 46, no. 1, pp. 32–46, 2010.

[3] Y. I. Abramovich, N. K. Spencer, and B. A. Johnson, "Band-inverse (TVAR) covariance matrix estimation for adaptive detection," *IEEE Trans. Aerosp. Electron. Syst.*, vol. 46, no. 1, pp. 1–22, Jan. 2010.

[4] E. Kelly, "An adaptive detection algorithm," *IEEE Trans. Aerosp. Elect. Syst.*, vol. 22, no. 1, pp. 115–127, Mar. 1986.

[5] F. Robey, D. Fuhrmann, E. Kelly, and R. Nitzberg, "A CFAR adaptive matched filter detector," *IEEE Trans. Aerosp. Elect. Syst.*, vol. 28, no. 1, pp. 208–216, Jan. 1992.

[6] M. Daniels and R. Kass, "Shrinkage estimators for covariance matrices," *Biometrics*, vol. 57, pp. 1173–1184, 2001.

[7] I. Reed, J. Mallett, and L. Brennan, "Rapid convergence rate in adaptive arrays," *IEEE Trans. Aerosp. Elect. Syst.*, vol. 10, no. 6, pp. 853–863, Nov. 1974.

[8] J. Michels, M. Rangaswamy, and B. Himed, "Performance of parametric and covariance based STAP tests in compound-Gaussian clutter," *Digital Signal Processing*, vol. 12, pp. 307–388, 2002.

[9] E. Conte, M. Lops, and G. Ricci, "Asymptotically optimum radar detection in compound-Gaussian clutter," *IEEE Trans. Aerosp. Elect. Syst.*, vol. 31, pp. 611–616, Apr. 1995.

[10] E. Conte, M. Lops, and G. Ricci, "Adaptive matched filter detection in spherically invariant noise," *IEEE Sig. Proc. Lett.*, vol. 3, pp. 248–250, 1996.

[11] H. Finn and R. Johnson, "Adaptive detection mode with threshold control as a function of spatially sampled clutter level estimates," *RCA Rev.*, vol. 29, pp. 414–463, Sep. 1968.

[12] G. Alpargu and G. Styan, "Some remarks and a bibliography on the Kantorovich inequality," in *Proc. Sixth Lukacs Symp.*, 1996, pp. 1–13.

[13] Y. Abramovich, "A controlled method for adaptive optimization of filters using the criterion of maximum SNR," *Radio Eng. Electron. Phys.*, vol. 26, no. 3, pp. 87–95, 1981.

[14] Y. Abramovich and A. Nevrev, "An analysis of effectiveness of adaptive maximization of the signal-to-noise ratio which utilizes the inversion of the estimated correlation matrix," *Radio Eng. Electron. Phys.*, vol. 26, no. 12, pp. 67–74, 1981.

[15] H. Cox, R. Zeskind, and M. Owen, "Robust adaptive beamforming," *IEEE Trans. Acoust. Sp. Sig. Proc.*, vol. 35, pp. 1365–1376, 1987.

[16] B. Carlson, "Covariance matrix estimation errors and diagonal loading in adaptive arrays," *IEEE Trans. Aerosp. Elect. Syst.*, vol. 24, no. 7, pp. 397–401, Jul. 1988.

[17] S. Vorobyov, A. Gershman, and L. Zhi-Quan, "Robust adaptive beamforming using worst-case performance optimization: A solution to the signal mismatch problem," *IEEE Trans. Sig. Proc.*, vol. 51, no. 2, pp. 313–324, Feb. 2003.

[18] J. Li, P. Stoica, and Z. Wang, "On robust Capon beamforming and diagonal loading," *IEEE Trans. Sig. Proc.*, vol. 51, no. 2, pp. 1702–1715, Feb. 2003.

[19] S. Shahbazpanahi, A. Gershman, Z.-Q. Luo, and K. Wong, "Robust adaptive beamforming for general-rank signal models," *IEEE Trans. Sig. Proc.*, vol. 51, no. 9, pp. 2257–2269, Sep. 2003.

[20] S. Kraut and L. Scharf, "The CFAR adaptive subspace detector is a scale-invariant GLRT," *IEEE Trans. Sig. Proc.*, vol. 47, no. 9, pp. 2538–2541, Sep. 1999.

[21] S. Kraut, L. Scharf, and L. McWhorter, "Adaptive subspace detectors," *IEEE Trans. Sig. Proc.*, vol. 49, no. 1, pp. 1–16, Jan. 2001.

[22] O. Besson, L. Scharf, and F. Vincent, "Matched direction detectors and estimators for array processing with subspace steering vector uncertainties," *IEEE Trans. Sig. Proc.*, vol. 53, no. 12, pp. 4453–4463, Dec. 2005.

[23] B. Liu, B. Chen, and J. H. Michels, "A GLRT for multichannel radar detection in the presence of both compound Gaussian clutter and additive white Gaussian noise," *Digital Signal Processing: A Review Journal*, vol. 15, no. 5, pp. 437–454, 2005.

[24] F. Gini, M. Greco, and A. Farina, "Clairvoyant and adaptive signal detection in non-Gaussian clutter: A data-dependent threshold interpretation," *IEEE Trans. Sig. Proc.*, vol. 47, no. 6, pp. 1522–1531, Jun. 1999.

[25] E. Conte, A. De Maio, and C. Galdi, "CFAR detection of multidimensional signals: An invariant approach," *IEEE Trans. Sig. Proc.*, vol. 51, no. 1, pp. 142–151, Jan. 2003.

[26] Y. Jin and B. Friedlander, "A CFAR adaptive subspace detector for second-order Gaussian signals," *IEEE Trans. Sig. Proc.*, vol. 53, no. 3, pp. 871–884, Mar. 2005.

[27] M. Bondarenko and D. Lekhovitsky, "Choice of a learning sample in adaptive detectors of signals against the background of Gaussian interferences," in *Proceedings of CAMSAP-2005*. Puerto Vallarta, Mexico: IEEE, 13 Dec. 2005, pp. 217–220.

[28] P. Chen, W. Melvin, and M. Wicks, "Screening among multivariate normal data," *Journal of Multivariate Analysis*, vol. 69, pp. 10–29, 1999.

[29] W. Melvin, "Space–time adaptive radar performance in heterogeneous clutter," *IEEE Trans. Aerosp. Elect. Syst.*, vol. 36, no. 2, pp. 621–633, Apr. 2000.

[30] M. Rangaswamy, B. Himed, and J. Michels, "Statistical analysis of the nonhomogeneity detector," in *Proc. Asilomar-2000*, vol. 2, Pacific Grove, CA, USA, 2000, pp. 1117–1121.

[31] M. Rangaswamy, B. Himed, and J. Michels, "Performance analysis of the nonhomogeneity detector for STAP applications," in *Proc. IEEE RADAR-2001*, Atlanta, GA, USA, 2001, pp. 193–197.

[32] K. Gerlach, "Outlier resistant adaptive matched filtering," *IEEE Trans. Aerosp. Elect. Syst.*, vol. 38, no. 3, pp. 885–901, Jul. 2002.

[33] Y. Abramovich and N. Spencer, "Expected-likelihood covariance matrix estimation for adaptive detection," in *Proc. RADAR-2005*, Arlington, VA, USA, 2005, pp. 623–628.

[34] Y. Abramovich and N. Spencer, "Two-set expected-likelihood GLRT technique for adaptive detection," in *Proc. CAMSAP-2005*, Puerto Vallarta, Mexico, 2005, pp. 12–15.

[35] B. Porat, *Digital Processing of Random Signals*. New Jersey: Prentice-Hall, 1994, 5th edition.

[36] R. Muirhead, *Aspects of Multivariate Statistical Theory*. New York: Wiley, 1982.

[37] T. Anderson, *An Introduction to Multivariate Statistical Analysis*. New York: Wiley, 1958.

[38] B. Nagarsenker and K. Pillai, "Distribution of the likelihood ratio criterion for testing a hypothesis specifying a covariance matrix," *Biometrica*, vol. 60, no. 2, pp. 359–361, 1973.

[39] L. Blake, "*Prediction of radar range*," in *Radar Handbook*, M. Skolnik, Ed. McGraw-Hill, 1990, 2nd edition.

[40] L. McWhorter, L. Scharf, and L. Griffiths, "Adaptive coherence estimation for radar signal processing," in *Proc. Asilomar-96*, vol. 1, Pacific Grove, CA, USA, 1996, pp. 536–540.

[41] Y. Abramovich, N. Spencer, and A. Gorokhov, "Bounds on maximum likelihood ratio – Part I: Application to antenna array detection-estimation with perfect wavefront coherence," *IEEE Trans. Sig. Proc.*, vol. 52, no. 6, pp. 1524–1536, Jun. 2004.

[42] I. Gradshteyn and I. Ryzhik, *Tables of Integrals, Series, and Products*. New York: Academic Press, 2000, 6th edition.

[43] C. Gierull, "Performance analysis of fast projections of the Hung–Turner type for adaptive beamforming," *Sig. Process. (special issue on Subspace Methods, Part I: Array Signal Processing and Subspace Computations)*, vol. 50, no. 1, pp. 17–28, 1996.

[44] P. Djurić, "A model selection rule for sinusoids in white Gaussian noise," *IEEE Trans. Sig. Proc.*, vol. 44, no. 7, pp. 1744–1757, Jul. 1996.

[45] R. Klemm, *Space-Time Adaptive Processing: Principles and Applications*. UK: IEE, 1998.

[46] O. Cheremisin, "Loading factor selection in the regularised algorithm for adaptive filter optimisation," *Radioteknika i Elektronika*, vol. 30, no. 12, 1985, English translation should be found in *Soviet Journal of Communication Technology and Electronics*.

[47] X. Mestre and M. A. Lagunas, *Robust Adaptive Beamforming*. John Wiley & Sons, 2005, ch. 8 – Diagonal Loading for Finite Sample Size Beamforming: An Asymptotic Approach, pp. 200–257.

[48] L. Scharf, *Statistical Signal Processing*. New York: Addison-Wesley, 1991.

[49] A. Gershman, C. Mecklenbräucker, and J. Böhme, "Matrix fitting approach to direction-of-arrival estimation with imperfect spatial coherence of wavefronts," *IEEE Trans. Sig. Proc.*, vol. 45, no. 7, pp. 1894–1899, Jul. 1997.

[50] A. Paulraj and T. Kailath, "Direction-of-arrival estimation by eigenstructure methods with imperfect spatial coherence of wave fronts," *J. Acoust. Soc. Am.*, vol. 83, no. 3, pp. 1034–1040, Mar. 1988.

[51] Y. Abramovich, N. Spencer, and A. Gorokhov, "Bounds on maximum likelihood ratio – Part II: Application to antenna array detection-estimation with imperfect wavefront coherence," *IEEE Trans. Sig. Proc.*, vol. 53, no. 6, pp. 2046–2058, Jun. 2005.

[52] J. Roman, M. Rangaswamy, D. Davis, Q. Zhang, B. Himed, and J. Michels, "Parametric adaptive matched filter for airborne radar applications," *IEEE Trans. Aerosp. Elect. Syst.*, vol. 36, no. 2, pp. 677–692, Apr. 2000.

[53] P. Parker and A. Swindlehurst, "Space–time autoregressive filtering for matched subspace STAP," *IEEE Trans. Aerosp. Elect. Syst.*, vol. 39, no. 2, pp. 510–520, Apr. 2003.

[54] C. Hawkes and S. Haykin, "Modeling of clutter for coherent pulsed radar," *IEEE Trans. Info. Theory*, vol. 21, no. 6, pp. 703–707, Nov. 1975.

[55] H. Dym and I. Gohberg, "Extensions of band matrices with band inverses," *Linear Algebra Appl.*, vol. 36, pp. 1–24, Mar. 1981.

[56] Y. Abramovich, N. Spencer, and M. Turley, "Time-varying autoregressive (TVAR) models for multiple radar observations," *IEEE Trans. Sig. Proc.*, vol. 55, no. 4, pp. 1298–1311, Apr. 2007.

[57] Y. Abramovich, N. Spencer, and M. Turley, "Order estimation and discrimination between stationary and time-varying autoregressive (TVAR) models," *IEEE Trans. Sig. Proc.*, vol. 55, no. 6, pp. 2861–2876, Jun. 2007.

[58] Y. Abramovich, N. Spencer, and M. Turley, "Time-varying autoregressive (TVAR) adaptive order and spectrum estimation," in *Proc. Asilomar-2005*, Pacific Grove, CA, USA, 2005, pp. 89–93.

[59] Y. Abramovich, M. Rangaswamy, B. Johnson, P. Corbell, and N. Spencer, "Performance of two-dimensional parametric STAP: KASSPER airborne radar data analysis," in *Proc. IRS-2008*, Warsaw, Poland, 2008, CD.

[60] Y. Abramovich, M. Rangaswamy, B. Johnson, P. Corbell, and N. Spencer, "Performance of 2-D mixed autoregressive models for airborne radar STAP: KASSPER-aided analysis," in *Proc. IEEE RadarCon 2008*, Rome, Italy, 2008, pp. 696–700.

[61] Y. Abramovich, M. Rangaswamy, P. Corbell, B. Johnson, and N. Spencer, "KASSPER analysis of 2-D parametric STAP performance: Further results on time-varying autoregressive "relaxations", in *Proc. RADAR-2008*, Adelaide, Australia, 2008, pp. 148–153.

[62] H. Woerdeman, "Matrix and operator extensions," Ph.D. dissertation, Vrije Universiteit, Amsterdam, Netherlands, 1989.

[63] R. Grone, C. Johnson, E. Marques de Sá, and H. Wolkowicz, "Positive definite completions of partial Hermitian matrices," *Linear Algebra Appl.*, vol. 58, pp. 109–124, 1984.

[64] D. Lekhovytskiy, S. Milovanov, I. Rakov, and B. Sverdlov, "Universal adaptive lattice filters: Adaptation for a given root of the estimating correlation matrix," *Radiophys. Quant. Electron.*, vol. 35, nos. 11–12, pp. 621–636, 1992, English translation of *Izvestiya Vysshikh Uchebnykh Zavedenii, Radiofizika*, vol. 35, nos. 11–12, pp. 969–992, Nov–Dec 1992.

[65] Y. Abramovich, M. Rangaswamy, B. Johnson, P. Corbell, and N. Spencer, "Time-varying autoregressive adaptive filtering for airborne radar applications," in *Proc. RADAR-2007*, Boston, MA, USA, 2007, pp. 653–657, invited paper

[66] A. Zoubir and D. Iskander, *Bootstrap Techniques for Signal Processing*. Cambridge, UK: Cambridge University Press, 2004.

[67] Y. Abramovich, A. Gorokhov, and N. Spencer, "Convergence analysis of stochastically-constrained sample matrix inversion algorithms," in *Proc. ISCAS-96*, vol. 2, Atlanta, GA, USA, 1996, pp. 449–452.

[68] Y. Abramovich, N. Spencer, S. Anderson, and A. Gorokhov, "Stochastic-constraints method in nonstationary hot-clutter cancellation – Part I: Fundamentals and supervised training applications," *IEEE Trans. Aerosp. Elect. Syst.*, vol. 34, no. 4, pp. 1271–1292, 1998.

[69] Y. Abramovich, N. Spencer, and S. Anderson, "Stochastic-constraints method in nonstationary hot-clutter cancellation – Part II: Unsupervised training applications," *IEEE Trans. Aerosp. Elect. Syst.*, vol. 36, no. 1, pp. 132–150, 2000.

[70] Y. Abramovich, N. Spencer, and A. Gorokhov, "Sample support analysis of stochastically constrained STAP with loaded sample matrix inversion," in *Proc. IEEE RADAR-2000*, Washington, DC, USA, 2000, pp. 804–808.

[71] Y. Abramovich, N. Spencer, and M. Turley, "Adaptive time-varying processing for stationary target detection in nonstationary interference," in *Proc. ASAP-2005*, MIT Lincoln Laboratory, MA, USA, 2005, CD.

[72] Y. Abramovich and N. Spencer, "Discriminating between stationary and time-varying autoregressive (TVAR) models in array processing," in *Proc. SAM-2006*, Boston, MA, USA, 2006, pp. 132–136.

[73] Y. Abramovich, N. Spencer, and B. Johnson, "Adaptive detection for interferences with a band inverse covariance matrix," in *Proc. DASP-2006*, Fraser Island, Australia, 2006, CD.

[74] J. Bergin and P. Techau, "High-fidelity site-specific radar data set," in *Proc. 2nd DARPA KASSPER Workshop*, Apr. 2002, pp. 1–8.

[75] A. De Maio, G. Foglia, E. Conte, and A. Farina, "CFAR behavior of adaptive detectors: An experimental analysis," *IEEE Trans. Aerosp. Elect. Syst.*, vol. 41, no. 1, pp. 233–251, Jan. 2005.

[76] B. Johnson and Y. Abramovich, "Experimental verification of environmental models for adaptive detection and estimation in HF skywave radar," in *Proc. RADAR-2006*, Verona, NY, USA, 2006, pp. 782–787.

[77] B. Liu, B. Chen, and J. Michels, "A GLRT for radar detection in the presence of compound Gaussian clutter and additive white Gaussian noise," in *Proc. SAM-2002*, Washington, DC, USA, 2002, pp. 87–91.

[78] S. Kalson, "Adaptive array CFAR detection," *IEEE Trans. Aerosp. Elect. Syst.*, vol. 31, no. 2, pp. 534–542, Apr. 1995.

[79] H. Rohling, "Radar CFAR thresholding in clutter and multiple target situations," *IEEE Trans. Aerosp. Elect. Syst.*, vol. 19, no. 4, pp. 608–621, Jul. 1983.

[80] S. Blake, "OS-CFAR theory for multiple targets and nonuniform clutter," *IEEE Trans. Aerosp. Elect. Syst.*, vol. 24, no. 6, pp. 785–790, Nov. 1988.

[81] P. Gandhi and S. Kassam, "Analysis of CFAR processors in nonhomogeneous backgrounds," *IEEE Trans. Aerosp. Elect. Syst.*, vol. 24, no. 4, pp. 427–445, Jul. 1988.

[82] "GNU Scientific Library," `http://www.gnu.org/software/gsl`, 2006, free software under the GNU General Public License.

CHAPTER 7

Compound-Gaussian Models and Target Detection: A Unified View

K. James Sangston[1], *Maria S. Greco*[2], *and Fulvio Gini*[2]

7.1 Introduction

As a radar system operates, it generally receives clutter returns from the environment that must be distinguished from targets of interest. If one assumes that the clutter returns obey complex multivariate Gaussian statistics, then a straightforward application of statistical detection theory leads to an optimal detector in the form of a well-known matched filter (Chap. 2 of this book). The occurrence of Gaussian statistics is often justified on the basis of the central limit theorem (CLT) applied to a phenomenological scattering picture that models the radar return as arising from contributions of a large number of scatterers in the radar resolution cell. In this case, the univariate intensity tail distribution is exponential. For early, low-resolution radars, this model was adequate.

Unfortunately, as resolution capabilities of radars increased, the observed intensity tail distributions were observed to deviate from the exponential model, particularly for sea clutter and low grazing angles. In particular, the intensity was often observed to have tails that exceeded exponential tails, sometimes greatly so, as shown for instance in References 1–10, and references therein. Many researchers proposed several different intensity tail distributions to model the observed radar clutter returns, including the Weibull, log-normal, and K distributions. Generally, such models were chosen because they exhibit larger tails than exponential and thus can provide a better fit to the observed data than the exponential model does. Whereas the exponential model has only one parameter that describes the average value of the intensity, these larger-tailed models have two (or even more) parameters that allow both the average intensity and the shape of the tails to be described. As a result, they give more flexibility in fitting observed data. However, early on, there was little physical motivation for selecting any particular model.

[1]Sensors and Electromagnetic Applications Laboratory, Georgia Tech Research Institute, Georgia Institute of Technology, Atlanta, GA, USA.
[2]Dipartimento di Ingegneria dell'Informazione, Università di Pisa, Pisa, Italy.

In 1972, Trunk at the Naval Research Laboratory [11] proposed that sea clutter returns be modeled by a tail intensity distribution that essentially is a mixture of exponential distributions and gave physical arguments to justify its use. This very general model incorporates not only an extra shape parameter but incorporates an entire probability density function (pdf) and thus is very flexible. The K distribution is known to be of the general form proposed by Trunk, with a particular model, the gamma distribution, for the mixing distribution. In 1982, Jakeman and Pusey [12] pointed out that the K distribution arises from the phenomenological scattering picture if the number of scatterers is allowed to be a particular random variable (r.v.), namely a negative binomial discrete random variable. This gives an alternate physical justification for the appearance of K statistics as well. In 1992, Sangston and Gerlach [13] then pointed out that Jakeman and Pusey's arguments for the K distribution are much more widely applicable and actually can give rise to any member of the general class of models proposed by Trunk.

What emerged from these various investigations was the idea that the general intensity tail distribution of radar clutter returns should be modeled by a mixture of exponential intensity distributions. The number fluctuation idea of Jakeman and Pusey, as extended by Sangston and Gerlach, then leads in a natural way to a multivariate model for the complex clutter returns from multiple pulses. Such models are called compound-Gaussian models and are the focus of this chapter.

In what follows, we start by using the notion of completely monotonic functions to give a unified view of the kinds of intensity tail distributions of interest here. We then show how number fluctuations in the context of a phenomenological picture of the scattering process lead naturally to these intensity tail distributions and their related compound-Gaussian multivariate models. We then discuss the general compound-Gaussian model and show how it may be formulated as various generalizations of the multivariate Gaussian model. These various formulations then lead to various formulations of the general optimal detector for detecting a target against a background of compound-Gaussian clutter. We provide several specific examples of optimal detectors, as well as several examples of suboptimal detectors derived from the various formulations of the compound-Gaussian model. Finally, we discuss a limitation of the compound-Gaussian model and suggest a research area that may lead to the development of more general random process models to describe the radar clutter return process.

7.2 Compound-Exponential Model for Univariate Intensity

7.2.1 Intensity Tail Distribution and Completely Monotonic Functions

Let X be a zero-mean complex random variable with statistics of the univariate intensity $I = |X|^2$ specified by the tail distribution

$$\Pr[I > y] = h_0(y), \quad y \geq 0. \tag{7.1}$$

A straightforward example of such a function is

$$\Pr[I > y] = e^{-y}, \tag{7.2}$$

which is the intensity tail distribution when X is a zero-mean, unit variance complex Gaussian random variable. We are interested in generalizations of this example, and in particular we consider the case where $h_0(y)$ is a *completely monotonic function* of y. A function is completely monotonic on the interval $0 < y < \infty$ if it has derivatives of all orders and satisfies

$$(-1)^n h_0^{(n)}(y) \geq 0, \quad n = 1, 2, \ldots. \tag{7.3}$$

The name arises from the fact that each derivative is alternately non-increasing or non-decreasing and thus monotonic. Furthermore, $h_0(y)$ is said to be completely monotonic on the interval $0 \leq y < \infty$ if it further satisfies $h_0(0^+) < \infty$. In our case, where $h_0(y)$ represents an intensity tail distribution, we have $h_0(0^+) = 1$, and so in what follows we consider only $h_0(y)$s that are completely monotonic on the interval $0 \leq y < \infty$.

To see how the completely monotonic functions generalize the exponential intensity tail distribution and give rise to the class of models initially suggested by Trunk, we consider the Bernstein–Widder theorem [14, 15]. According to the Bernstein–Widder theorem, an intensity tail distribution $h_0(y)$ will be completely monotonic on the interval $0 \leq y < \infty$ if and only if it satisfies

$$h_0(y) = \int_0^\infty e^{-\alpha y} dF_\alpha(\alpha) \tag{7.4}$$

where $F_\alpha(\alpha)$ is the distribution function of a non-negative random variable α [14]. We will call any intensity tail distribution function $h_0(y)$ that satisfies (7.3) or (7.4) (and thus is completely monotonic) a compound-exponential model. This name comes from the representation in (7.4), which shows that $h_0(y)$ is a mixture of exponential tail distributions. It is easy to see that the exponential tail distribution of (7.2) satisfies (7.4) with $F_\alpha(\alpha)$ a distribution function that puts all of its mass at $\alpha = 1$, i.e., $f_\alpha(\alpha) = dF_\alpha(\alpha)/d\alpha = \delta(\alpha - 1)$, where $f_\alpha(\alpha)$ is the pdf of the random variable α and $\delta(\cdot)$ is the Kronecker delta function.

7.2.2 Examples

Completely monotonic functions are much-studied; Miller and Stamko [16] give good summary overviews of many of their properties as well as several examples. Examples that have arisen in the radar noise modeling context include those in Table 7.1.

Table 7.1 Examples of $h_0(y)$ and associated $f_\alpha(\alpha)$.

Distribution	$h_0(y)$	$f_\alpha(\alpha)$
Exponential	$\exp(-y)$	$\delta(\alpha - 1)$
K	$\dfrac{2\left(\sqrt{(\nu)y}\right)^\nu}{\Gamma(\nu)} K_{\nu-1}\left(2\sqrt{(\nu)y}\right), \quad \nu > 0$	$\dfrac{\nu^\nu \exp(-\nu/\alpha)}{\Gamma(\nu)\alpha^{\nu+1}}$
Weibull	$\dfrac{2\left(\sqrt{\nu y}\right)^\nu}{\Gamma(\nu)} K_{\nu-1}\left(2\sqrt{\nu y}\right), \quad \nu > 0$	$\dfrac{2^{\frac{\nu}{2}-1}\nu}{\alpha^{\frac{\nu}{2}+1}}$ times $\sum_{n=0}^{\infty} \dfrac{\left(-2^{\frac{\nu}{2}-1}\alpha^{-\frac{\nu}{2}}\right)^n}{n!\Gamma\left(1 - \frac{\nu}{2}(n+1)\right)}$
Student-t	$\left(1 + \dfrac{y}{\nu}\right)^{-\nu} \quad \nu > 0$	$\dfrac{\nu^\nu \alpha^{\nu-1} \exp(-\nu\alpha)}{\Gamma(\nu)}$

Table 7.2 $\gamma_0(y)$ for selected examples.

Distribution	$\gamma_0(y)$
Exponential	1
K	$\sqrt{\frac{\nu}{y}}\frac{K_{\nu-1}(2\sqrt{\nu y})}{K_\nu(2\sqrt{\nu y})}$
Weibull	$\frac{2^{\frac{\nu}{2}-2}\nu}{y^{1-\frac{\nu}{2}}}$ $\quad \alpha_0 > 0$
Student-t	$\frac{\nu}{\nu+y}$

It is also true that if $h_1(y)$ and $h_2(y)$ are each completely monotonic functions, then both $h_0(y)=a_1h_1(y)+a_2h_2(y)$ and $h_0(y)=h_1(y)h_2(y)$ with $a_1,a_2 \geq 0$ are also completely monotonic functions [16]. Thus, given tail intensity distributions such as those in Table 7.1, we can generate new models as either products or weighted averages $(a_1+a_2=1)$ of such existing models. This gives us a mechanism for developing a completely new class of intensity tail distribution models for comparing to measured data.

It is of interest to study the associated distribution $F_\alpha(\alpha)$, which by (7.4) can be obtained from $h_0(y)$ by inverse Laplace transformation [17]. In particular, it is of interest to examine when $F_\alpha(\alpha)$ is infinitely divisible, since this indicates that α then can be modeled as the sum of independently and identically distributed (IID) random variables (which stems from the definition of infinite divisibility). It is known that $F_\alpha(\alpha)$ is the distribution function of an infinitely divisible random variable if and only if

$$h_0(y) = e^{-w_0(y)}, \tag{7.5}$$

where $w_0(0^+)=0$ and $w_0^{(1)}(y)$ is completely monotonic [17]. In our case, we can always write $w_0(y)=-\ln h_0(y)$, where $w_0(0^+)=0$ since $h_0(0^+)=1$. If we now define

$$\gamma_0(y) = -\frac{d}{dy}\ln h_0(y), \tag{7.6}$$

then $F_\alpha(\alpha)$ is the distribution function of an infinitely divisible random variable if and only if $\gamma_0(y)$ is completely monotonic. Table 7.2 gives $\gamma_0(y)$ for the examples in Table 7.1.

Each $\gamma_0(y)$ in Table 7.2 is completely monotonic (see e.g., [16]). Thus each associated $F_\alpha(\alpha)$ is infinitely divisible. In what follows we will require this condition for any compound-exponential model of interest.

7.3 Role of Number Fluctuations

In this section, we will show that our intensity tail distributions $h_0(y)$ can be obtained from a scattering process picture in which the complex random variable X is obtained as

$$X = \sum_{i=1}^{N} A_i e^{j\varphi_i}, \tag{7.7}$$

where the $A_i > 0$ are IID random variables that are mutually independent of the ϕ_i, which are IID with uniform distribution on $[0, 2\pi]$. This is a standard phenomenological picture of a scattering process giving rise, for example, to a radar return. It is well known that when the number of scatterers N is fixed and large, the CLT leads to the conclusion that X is a complex Gaussian random variable, in which case $h_0(y)$ is exponential. However, here we will let N be a random variable, independent of the A_i and ϕ_i, and "large" in the sense that its mean value $\bar{N}$ is large (i.e., $\bar{N} \gg 1$) This modification to the phenomenological model will lead to the more general completely monotonic $h_0(y)$ as the general intensity tail distribution for X.

7.3.1 Transfer Theorem and the CLT

To gain some insight into the possible effects of including number fluctuations in the formulation of limit theorems, let $X_1, X_2, \ldots$ be a sequence of IID Gaussian random variables with zero mean and unit variance (denoted by $N(0,1)$), let $N_1, N_2, \ldots$ be a sequence of non-negative integer-valued random variables independent of the sequence $X_1, X_2, \ldots$, and define

$$S_n = \frac{1}{\sqrt{n}} \sum_{k=1}^{N_n} X_k, \quad n = 1, 2, \ldots. \tag{7.8}$$

If the value of N_n is fixed, then the conditional random variable S_n is Gaussian with mean 0 and conditional variance given by $var(S_n|N_n) = N_n/n$. The unconditional pdf of S_n may be written as

$$f_{S_n}(s) = \int_0^{\infty} \frac{1}{\sqrt{2\pi t}} e^{-\frac{s^2}{2t}} dF_n(t), \qquad n = 1, 2, \ldots \tag{7.9}$$

where $F_n(\cdot)$ is the distribution function of $t = N_n/n$. If N_n/n converges in distribution to a random variable τ as $n \to \infty$, then (7.9) shows that S_n converges in distribution to a random variable X with pdf given by

$$f_X(x) = \int_0^{\infty} \frac{1}{\sqrt{2\pi\tau}} e^{-\frac{x^2}{2\tau}} dF_\tau(\tau), \quad n = 1, 2, \ldots, \tag{7.10}$$

where $F_\tau(\tau)$ is the distribution function of τ. The intensity tail distribution of X is our completely monotonic function $h_0(y)$, where $\alpha = 1/\tau$. Because the sequence $N_1, N_2, \ldots$ may be chosen so as to make N_n/n converge in distribution to *any* non-negative random variable τ, (7.10) shows that the distribution of X may be significantly non-Gaussian. Moreover, as will be shown, if $F_\tau(\tau)$ is continuous at 0, then N_n converges in probability to ∞ as $n \to \infty$, a result that defines explicitly a sense in which N_n gets large. Thus, a sum of a "large" number of IID Gaussian random variables can give rise to a non-Gaussian limiting random variable! This rather simple example shows that number fluctuations can have profound effects on the results of limit theorems and that an uncritical assessment of the way in which the number of contributions to the sum gets "large" can lead to seriously incorrect results.

We can get a sense of how this result will generalize to sums of IID variables that are not necessarily Gaussian by rewriting (7.8) as

$$S_n = \sqrt{\left(\frac{N_n}{n}\right)}\frac{1}{\sqrt{N_n}}\sum_{k=1}^{N_n} X_k, \qquad n = 1, 2, \ldots. \tag{7.11}$$

If N_n gets large in some sense as $n \to \infty$, we might expect the normalized sum to converge to a Gaussian random variable and the quantity $\sqrt{N_n/n}$ to converge to a random variable $\sqrt{\tau}$. In that case, the pdf of the limiting random variable X is given by (7.10).

The random version of the CLT is a rigorous presentation of the above reasoning for the case when the variables $X_1, X_2, \ldots$ need not be Gaussian but are such that for *non-random* N_n the limiting distribution is Gaussian. This version of the CLT may be formulated for the more general case of random vectors and shows that in general, even if the non-random sum converges to a Gaussian random vector, the sum with a random number of terms may converge to a non-Gaussian random vector. Of course, if $F_\tau(\tau) = U(\tau - \tau_0)$ where $U(\cdot)$ is the Heaviside step function and τ_0 is a fixed constant, then X continues to be a Gaussian random variable. This situation occurs, for instance as will be shown below, if the statistics of N_n are Poisson. More generally, however, it is possible to choose the statistics of N_n in such a way to achieve any possible distribution function $F_\tau(\tau)$ defined for $\tau \geq 0$ and thus a wide range of non-Gaussian limit distributions.

As is well known, if $X_1, X_2, \ldots$ are IID real-valued random variables with zero mean and finite variance σ^2, then the random variable

$$S_N = \frac{1}{\sqrt{N}}\sum_{i=1}^{N} X_i \tag{7.12}$$

converges in distribution as $N \to \infty$ to a Gaussian random variable with zero mean and variance σ^2. The more general problem in which the random variables are not identically distributed, the assumption of finite variance is relaxed, and/or the normalization need not be $1/\sqrt{N}$ is discussed fully in the well-known work of Gnedenko and Kolmogorov [18]. These limit theorems have proven to be quite important in problems of statistical physics. In particular, use of the CLT is almost ubiquitous in physical problems involving stochastic phenomena. A common feature of the classical limit theorems is that N is treated as a deterministic quantity. If N is a random variable, however, these classical results are no longer directly applicable. As N can realistically be treated as a random quantity in many physical problems, limit theorems analogous to the classical theorems but appropriate for random N should prove fruitful and widely applicable.

The first limit theorem for the asymptotic distribution of a sum of a random number of IID real-valued random variables appears to have been given by Robbins [19]. Among other things, he showed that even if the CLT is applicable in the absence of number fluctuations (i.e., in the case of non-random N), inclusion of number fluctuations can lead to non-Gaussian limit distributions. His work gave rise to further investigations, the most relevant of which for the present paper is the theorem, detailed in Appendix 7.A, called *a transfer theorem*, due to Gnedenko and Fahim [20] and applicable for the sum of a random number of IID real-valued random variables.

In essence, the theorem assumes that the sum of N random variables converges in distribution when N is non-random and gives a condition under which the sum converges in distribution when N is random. For instance, let N_n be a random variable with mean n, and let $k_n = n$. Let $\{\boldsymbol{x}_k\}$ be a sequence of IID random vectors, and for each n let

$$\boldsymbol{x}_{nk} = \frac{\boldsymbol{x}_k}{\sqrt{n}}. \tag{7.13}$$

These assumptions give

$$\boldsymbol{s}^n_{k_n} = \frac{1}{\sqrt{n}}\sum_{k=1}^{n}\boldsymbol{x}_k \tag{7.14}$$

and

$$\boldsymbol{s}^n_{N_n} = \frac{1}{\sqrt{n}}\sum_{k=1}^{N_n}\boldsymbol{x}_k. \tag{7.15}$$

Equation (7.14) represents a sum of a deterministic number of random vectors whereas (7.15) is the sum of a random number of random vectors. Assume further that the sum in (7.15) converges to a Gaussian random vector $\boldsymbol{y}$ with zero mean and covariance matrix $\boldsymbol{Q}$. Then the theorem states that when the sum in (7.15) converges, it converges to a random vector $\boldsymbol{y}$ whose characteristic function is given by

$$C_y(\boldsymbol{u}) = \int_0^{\infty} e^{-\frac{\boldsymbol{u}^T\boldsymbol{Q}\boldsymbol{u}}{2}\tau}dF_\tau(\tau). \tag{7.16}$$

Equation (7.16) shows that in general, even if the non-random sum converges to a Gaussian random vector, the sum with a random number of terms may converge to a non-Gaussian random vector.

The most important consequence of this result is that in problems in which the CLT may be invoked to justify the use of the Gaussian distribution, the occurrence of number fluctuations, which is the term we will use for random N_n, may lead to the occurrence of non-Gaussian distributions. In many physical problems, number fluctuations almost certainly occur but are ignored under the belief that "large n" implies Gaussian statistics. A re-examination of such problems with careful consideration of the number fluctuations may lead to new, physically significant results, or it may explain the appearance of non-Gaussian statistics in problems in which the application of the CLT suggests that Gaussian statistics should occur [21].

7.3.2 Models for Number Fluctuations

For each distribution function $F_\tau(\cdot)$ such that $F_\tau(-\infty)=F_\tau(0)=0$ and $F_\tau(\infty)=1$ define a collection of point mass functions (PMFs) parameterized by k_n, $n = 0, 1, 2, \ldots$, by

$$p_c(i;k_n) = \Pr\{N_n = i\} = \int_0^{\infty}\frac{t^i}{i!}e^{-t}dF_\tau(t/k_n) \quad i = 0, 1, 2, \ldots. \tag{7.17}$$

Table 7.3 Examples of number fluctuation models.

Model	Point mass function $p_c(i; k_n)$	Distribution function $F_\tau(\tau)$
Poisson (leads to Rayleigh model)	$\frac{k_n^i}{i!}e^{-k_n}$	$u(\tau-1)$
Negative binomial (leads to K model)	$\frac{\Gamma(i+\nu)}{\Gamma(I+1)\Gamma(\nu)}\frac{r^i}{(r+1)^{i+\nu'}}$ $r=\frac{k_n}{\nu}, \quad \nu>0$	$\int_0^t \frac{\nu^\nu}{\Gamma(\nu)}x^{\nu-1}e^{-\nu x}dx$
Inverse gamma (leads to Student-t model)	$\frac{\left(\frac{\nu}{k_n}\right)^\nu}{\Gamma(\nu)}(\nu k_n)^{\frac{i}{2}}K_{i-\nu}(2\sqrt{\nu k_n})$	$\int_0^t \frac{\nu^\nu}{\Gamma(\nu)}\frac{1}{x^{\nu+1}}e^{-\frac{\nu}{x}}dx$

Table 7.3 gives examples of number fluctuation models that actually lead to a given distribution whose characteristic function is described by (7.16).

In the results above, we assumed that the sequence N_n/k_n converges in distribution to a random variable τ. Discussion on this convergence can be found in the appendix.

7.4 Complex Compound-Gaussian Random Vector

In this section, we present a multivariate model that has the corresponding univariate intensity tail distribution specified by $h_0(y)$. To this end, let $X_1, X_2, \ldots$ be a sequence of identically distributed complex random variables of the type we have been considering where the univariate intensity tail distribution is given by $h_0(y)$ and consider an n-dimensional random vector $\boldsymbol{x} = [X_1, X_2, \ldots, X_n]^T$. We assume for the mean value $\mathbb{E}\{\boldsymbol{x}\} = \boldsymbol{0}$ and for the vector correlation matrix $\mathbb{E}\{\boldsymbol{x}\boldsymbol{x}^\dagger\} = \boldsymbol{M} = \mu\boldsymbol{\Sigma}$, where μ is the power of each vector component and $\boldsymbol{\Sigma}$ is the normalized correlation matrix (all the elements along the main diagonal are equal to 1). Under the stated conditions – i.e., we know $h_0(y)$ and $\boldsymbol{M}$ (or $\boldsymbol{\Sigma}$) – there is not a unique multivariate pdf to describe the behavior of $\boldsymbol{x}$. Thus, we must choose a multivariate model based on some additional criteria. The scattering picture discussed above suggests that, as long as we have a "large" number of elementary scatterers and the actual number of elementary scatterers contributing to the scattered field does not change during the time that the n complex random variables are obtained, then we should choose a multivariate pdf as suggested by the transfer theorem. We do that as follows.

For any integer $n \geq 1$ define

$$h_n(y) = (-1)^n \frac{d^n}{dy^n} h_0(y), \tag{7.18}$$

which we know exists because $h_0(y)$ is completely monotonic. Then, we specify a multivariate pdf by

$$f_x(\boldsymbol{x}) = \frac{h_n(q)}{\pi^n \det(\boldsymbol{\Sigma})} \tag{7.19}$$

with

$$q = \boldsymbol{x}^{\dagger}\boldsymbol{\Sigma}^{-1}\boldsymbol{x}. \tag{7.20}$$

A model with multivariate pdf specified as in (7.19) is called a compound-Gaussian model. With a compound-exponential model for the intensity tail distribution, one may immediately obtain a corresponding multivariate compound-Gaussian model simply by taking derivatives. Of course, it may not be straightforward to compute the required derivatives, but such derivatives always exist and in principle we can obtain the compound-Gaussian model in this fashion. Because computing the derivatives may not always be easy, we also look for other ways to characterize the compound-Gaussian model. To this end, it follows from our earlier considerations that we may also write the multivariate pdf in (7.19) with $h_n(q)$ written in the following forms:

$$h_n(q) = \int_0^{\infty} \alpha^n e^{-\alpha q} dF_{\alpha}(\alpha) = \int_0^{\infty} \frac{1}{\tau^n} e^{-\frac{q}{\tau}} dF_{\tau}(\tau). \tag{7.21}$$

This alternate formulation motivates the physical interpretation of the compound-Gaussian model as one that is locally Gaussian but whose local "power" – i.e., local average intensity specified by τ – is a randomly varying quantity. Thus any particular observed vector $\mathbf{x}$ is Gaussian except that τ (or equivalently α) is both unknown and random, and hence the name compound-Gaussian. If a model for either F_{τ} or F_{α} is available, then the integrations in (7.21) will also lead to the compound-Gaussian model.

Sometimes neither of the above two approaches is straightforward, and even when they are available, they may not give any insight into the operation, for example, of the optimum detector. So we look for other characterizations of $h_n(q)$. Let us now extend our definition in (7.6) to

$$\gamma_k(y) = -\frac{d}{dy} \ln h_k(y), \tag{7.22}$$

for any $k \geq 0$. This definition now leads to

$$h_n(q) = \exp\left[-\int^q \gamma_n(s)ds\right], \tag{7.23}$$

where $\int^q \gamma_n(s)ds$ is the antiderivative of $\gamma_n(q)$. This reformulation of $h_n(q)$ is interesting because from (7.18), (7.22), and (7.23) we find

$$\gamma_n(q) = \frac{h_{n+1}(q)}{h_n(q)} = \frac{\int_0^{\infty} \alpha^{n+1} e^{-\alpha q} dF_{\alpha}(\alpha)}{\int_0^{\infty} \alpha^n e^{-\alpha q} dF_{\alpha}(\alpha)} = \mathbb{E}[\alpha|q], \tag{7.24}$$

i.e., $\gamma_n(q) = \mathbb{E}[\alpha|q]$ is the optimal mean estimate of the random variable α given the observed data vector $\boldsymbol{x}$ (which is completely embodied in the quadratic form q), that is the minimum mean-squared error estimate (MMSE). Thus, since we can write the multivariate pdf in the Gaussian case as

$$f_{\boldsymbol{x}}(\boldsymbol{x}) = \frac{e^{-q/\sigma^2}}{\pi^n \det(\boldsymbol{M})} = \frac{e^{-(\int^q ds)/\sigma^2}}{\pi^n \sigma^{2n} \det(\boldsymbol{\Sigma})}, \tag{7.25}$$

the multivariate compound-Gaussian pdf may be thought of as a multivariate Gaussian pdf with the known γ_0 (in this case $a_0 = 1$) replaced by an optimum estimator $\gamma_n(q)$ evaluated as shown in (7.24).

We may also repeatedly apply (7.18) and (7.23) to rewrite $h_n(q)$ in yet another way:

$$h_n(q) = \left[\prod_{k=0}^{n-1} \gamma_k(q)\right] h_0(q). \tag{7.26}$$

As may be seen, this formulation casts the multivariate pdf into two components – one arising directly from the intensity tail distribution $h_0(y)$ and one arising from a product of estimators $\gamma_k(q)$. As will be discussed further below, these two components play different roles in the detection problem.

Finally, it is straightforward to show

$$\gamma_{n+k}(y) = (D_l + I)^k \gamma_n(y) \tag{7.27}$$

and

$$\prod_{k=0}^{n-1} \gamma_n(y) = (D + \gamma_0(y)I)^{n-1} \gamma_0(y), \tag{7.28}$$

where $D = -\frac{d}{dy}$, $D_l = D\ln$, and I is the identity operator. These results give different ways to obtain the multivariate pdf from the intensity tail distribution.

Practically, independently of any way to write $h_n(q)$, each complex compound-Gaussian vector $\mathbf{x}$ can be written as a product between two independent terms, $\boldsymbol{x} = \sqrt{\tau}\boldsymbol{g}$ where $\boldsymbol{g}$ is a complex Gaussian vector with unit variance, usually called speckle and τ is a positive random variable referred to as texture that represents the local random power. Given a specific value of the texture, $\mathbf{x}$ is a complex, zero-mean Gaussian random vector with conditional covariance matrix given by $\boldsymbol{M}_\tau = \mathbb{E}\{\boldsymbol{x}\boldsymbol{x}^\dagger|\tau\} = \tau\boldsymbol{\Sigma}$, where $\boldsymbol{\Sigma} = \mathbb{E}\{\boldsymbol{g}\boldsymbol{g}^\dagger\}$ [22, 23].

7.5 Optimum Detection of a Signal in Complex Compound-Gaussian Clutter

The detection problem is modeled herein by the following hypothesis test:

$$\begin{cases} H_0: & z = \boldsymbol{x} \\ H_1: & z = \boldsymbol{x} + \beta\boldsymbol{p} \end{cases} \tag{7.29}$$

where $\boldsymbol{x}$ is a n-dimensional complex zero-mean compound-Gaussian clutter vector, $\boldsymbol{p}$ is a n-dimensional complex vector representing a known steering vector for a signal of interest, and β is a complex number representing the generally unknown amplitude and initial phase of the signal of interest. In this problem, H_0 represents the hypothesis that no signal is present – i.e., the observed vector z at the radar comprises only clutter, whereas H_1 represents the hypothesis that a signal is present and so the observed vector z at the radar includes both signal and clutter returns.

The optimal detector, under the Neyman–Pearson criterion, for this problem is given by the likelihood ratio compared with a fixed threshold that is set to yield a desired false alarm probability (P_{FA}):

$$\Lambda(z) = \frac{p_z(z|H_1)}{p_z(z|H_0)} \underset{H_0}{\overset{H_1}{\gtrless}} e^T \tag{7.30}$$

i.e., a ratio of multivariate pdfs under the two different hypotheses H_1 – *signal present* and H_0 – *signal not present.* For the case of detecting a known additive signal **s** against a clutter background described by the compound-Gaussian model we have

$$\Lambda(z) = \frac{h_n(q_1)}{h_n(q_0)} \underset{H_0}{\overset{H_1}{\gtrless}} e^T, \tag{7.31}$$

where $q_0 = z^\dagger \boldsymbol{\Sigma}^{-1} z$ and $q_1 = (z - s)^\dagger \boldsymbol{\Sigma}^{-1}(z - s)$, and $s = \beta \boldsymbol{p}$ is assumed known. That is

$$\Lambda(z) = \frac{\displaystyle\int_0^{+\infty} \frac{1}{\tau^n} \exp\left[-\frac{q_1}{\tau}\right] p_\tau(\tau) d\tau}{\displaystyle\int_0^{+\infty} \frac{1}{\tau^n} \exp\left[-\frac{q_0}{\tau}\right] p_\tau(\tau) d\tau} \underset{H_0}{\overset{H_1}{\gtrless}} e^T \tag{7.32}$$

and after some calculations [24]

$$\Lambda(z) = \int_0^{\infty} \frac{1}{\tau^n} \left[\exp\left(-\frac{q_1}{\tau}\right) - \exp\left(T - \frac{q_0}{\tau}\right)\right] p_\tau(\tau) d\tau \underset{H_0}{\overset{H_1}{\gtrless}} 0. \tag{7.33}$$

In realistic scenarios, the complex signal amplitude β is unknown. If we ignore the probabilistic law of the amplitude fluctuations (i.e., $p_\beta(\beta)$), a possible solution is obtained by means of the generalized likelihood ratio test (GLRT) approach, whereby the unknown parameter β is replaced by its maximum likelihood (ML) estimate $\hat{\beta}_{\mathrm{ML}}$, thereby obtaining the strategy

$$\max_\beta \Lambda(z; \beta) = \Lambda(z; \hat{\beta}_{\mathrm{ML}}) \underset{H_0}{\overset{H_1}{\gtrless}} \exp(T) \tag{7.34}$$

The ML estimate of β has been obtained in References 26 and 27 as $\hat{\beta}_{\mathrm{ML}} = \frac{p^\dagger \boldsymbol{\Sigma}^{-1} z}{p^\dagger \boldsymbol{\Sigma}^{-1} p}$. After some easy algebra, we derive that in this case, the detection strategy is again given by (7.33), provided that now we define $q_1 \triangleq z^\dagger \boldsymbol{\Sigma}^{-1} z - \frac{|p^\dagger \boldsymbol{\Sigma}^{-1} z|^2}{p^\dagger \boldsymbol{\Sigma}^{-1} p}$. Due to the appearance of the integral in (7.32) and (7.33), the structure of the (generalized) likelihood ratio is difficult to implement as it is written and does not provide any hints on the behavior of the detector.

7.5.1 Likelihood Ratio and Data-Dependent Threshold Interpretation

It is well known that the log-likelihood ratio in the case of colored Gaussian noise leads to the whitening-matched filter. Since in our case the disturbance model is a compound-Gaussian process, the matched filter may play a role in the optimum detector even in the case at hand. To explore this possibility, let us consider again (7.31) or equivalently

$$h_n(q_1) \underset{H_0}{\overset{H_1}{\gtrless}} e^T h_n(q_0), \tag{7.35}$$

where $h_n(q)$ is a monotonically decreasing function of q, therefore has an inverse. Equation (7.35) leads to $h_n^{-1}(e^T h_n(q_0)) \underset{H_0}{\overset{H_1}{\gtrless}} q_1$ which may be rewritten as

$$q_0 - q_1 \underset{H_0}{\overset{H_1}{\gtrless}} f_{\text{opt}}(q_0, T), \underset{H_0}{\overset{H_1}{\gtrless}} \lambda, \tag{7.36}$$

where in general

$$f_{\text{opt}}(q_0, T) = q_0 - h_n^{-1}(e^T h_n(q_0)). \tag{7.37}$$

In Gaussian noise, for which $h_n(q) = \exp(-q/\sigma^2)/\sigma^{2n}$, it follows that $f_{\text{opt}}(q_0, T) = \sigma^2 T$ and the log-likelihood ratio leads to

$$q_0 - q_1 \underset{H_0}{\overset{H_1}{\gtrless}} \sigma^2 T. \tag{7.38}$$

One interpretation of these two detectors is that the detection structure, i.e., the left-hand side of (7.38) and (7.36), is the same in both cases. The threshold however changes from Gaussian to compound-Gaussian. Because the left-hand side of (7.38) and (7.36) is essentially the matched filter, this comparison shows that the matched filter is indeed an essential part of the detection structure in the case of compound-Gaussian clutter. In particular, when the target signal is supposed perfectly known, then $q_0 - q_1 = 2\text{Re}(\boldsymbol{s}^\dagger \boldsymbol{\Sigma}^{-1} \boldsymbol{z}) - \boldsymbol{s}^\dagger \boldsymbol{\Sigma}^{-1} \boldsymbol{s}$ and (7.40) leads to

$$\text{Re}(\boldsymbol{s}^\dagger \boldsymbol{\Sigma}^{-1} \boldsymbol{z}) \underset{H_0}{\overset{H_1}{\gtrless}} \frac{f_{\text{opt}}(q_0, T)}{2} - \frac{\boldsymbol{s}^\dagger \boldsymbol{\Sigma}^{-1} \boldsymbol{s}}{2}. \tag{7.39}$$

Because the left-hand side of (7.39) is the whitening-matched filter as applied to complex data, this derivation shows that the optimum detector for a known signal in compound-Gaussian clutter is equivalent to *a matched filter compared with a data-dependent threshold* [24].

In the case in which the complex signal amplitude β is unknown and we replace it in the likelihood ratio with its ML estimate, the structure of the GLRT becomes

$$|\boldsymbol{p}^\dagger \boldsymbol{\Sigma}^{-1} \boldsymbol{z}|^2 \underset{H_0}{\overset{H_1}{\gtrless}} (\boldsymbol{p}^\dagger \boldsymbol{\Sigma}^{-1} \boldsymbol{p}) f_{\text{opt}}(q_0, T). \tag{7.40}$$

It is worth noting that $f_{\text{opt}}(q_0, T)$ is the same as in (7.39).

Table 7.4 Optimal detectors in K and Student-t compound-Gaussian noises.

Model	K	Student-t
Likelihood ratio $h_n(q_1)/h_n(q_0)$	$\left(\sqrt{\frac{q_0}{q_1}}\right)^{n-\nu}\frac{K_{n-\nu}(2\sqrt{\nu q_1})}{K_{n-\nu}(2\sqrt{\nu q_0})}$	$\left(\frac{\nu+q_0}{\nu+q_1}\right)^{n+\nu}$
Estimator-correlator $e^{\int_{q_1}^{q_0}\alpha_n(s)ds}$	$\alpha_n(s)=\sqrt{\frac{\nu}{s}}\frac{K_{n-\nu+1}(2\sqrt{\nu s})}{K_{n-\nu}(2\sqrt{\nu s})}$	$\alpha_n(s)=\frac{\nu+n}{\nu+s}$
Matched filter with data-dependent threshold $q_0-q_1\lessgtr f(q_0,T)$	$f(q_0,T)$ Not generally available in closed form.	$f(q_0,T)=$ $\nu\left(1-e^{-\frac{T}{\nu+n}}\right)\left(1+\frac{1}{\nu}q_0\right)$

7.5.2 Likelihood Ratio and the Estimator-Correlator Interpretation

The estimator-correlator structure has been discussed previously by Sangston et al. [24]. Let us go back to (7.27) and (7.28) where $\alpha=1/\tau$. With these equations the log-likelihood ratio can be written [24] as

$$\Lambda(z)=\ln\frac{h_n(q_1)}{h_n(q_0)}=\int_{q_1}^{q_0}\mathbb{E}[\alpha|s]ds, \tag{7.41}$$

then

$$\Lambda(z)=\exp\left\{\int_{q_1}^{q_0}\mathbb{E}[\alpha|s]ds\right\}. \tag{7.42}$$

In Gaussian case, the likelihood ratio can be written as

$$\Lambda(z)=\exp\left\{\int_{q_1}^{q_0}\frac{1}{\sigma^2}ds\right\}. \tag{7.43}$$

Equations (7.42) and (7.43) reveal that the optimal detector in compound-Gaussian case operates just as the Gaussian detector does except it replaces the known local power level in the Gaussian case with an optimal estimate of the randomly varying inverse local power level. This is the estimator-correlator formulation of the optimum detector.

In Table 7.4, we give the two formulations of the optimal detector for the cases of Student-t-distributed and K-distributed compound-Gaussian models (for the sake of simplicity we have set in both cases the average power equal to 1).

7.6 Suboptimum Detectors in Complex Compound-Gaussian Clutter

As mentioned above, implementation of the optimal detector for the K-distributed compound Gaussian model appears to be quite complicated. A similar observation holds for the Weibull model, where the required derivatives to obtain $h_n(q)$ or $\gamma_n(q) = \mathbb{E}[\alpha|q]$ are tedious to compute and not generally available in closed form. These considerations suggest studying suboptimal implementations of the optimal detector.

7.6.1 Suboptimum Approximations to Likelihood Ratio

From (7.36), the optimum detection structure is given by

$$\Lambda(z) = \frac{\int_0^{+\infty} \frac{1}{\tau^n} \exp\left[-\frac{q_1}{\tau}\right] p_\tau(\tau) d\tau}{\int_0^{+\infty} \frac{1}{\tau^n} \exp\left[-\frac{q_0}{\tau}\right] p_\tau(\tau) d\tau} \underset{H_0}{\overset{H_1}{\gtrless}} e^T. \tag{7.44}$$

From a physical point of view, the difficulty in utilizing this detection structure arises from the fact that the power level τ associated with the conditionally Gaussian clutter is unknown and randomly varying. One approach to overcome this difficulty is to replace the unknown power level with an estimate inside the likelihood ratio. In particular, replace τ in the Gaussian kernel of the likelihood ratio with an estimate $\hat{\tau}_i$, $i = 0,1$. The likelihood ration test then becomes

$$\left(\frac{\hat{\tau}_0}{\hat{\tau}_1}\right)^n \exp\left[\frac{q_0}{\hat{\tau}_0} - \frac{q_1}{\hat{\tau}_1}\right] \underset{H_0}{\overset{H_1}{\gtrless}} e^T. \tag{7.45}$$

This approach is suboptimal and cannot guarantee a good detector. However, intuitively one expects that a good quality estimate should lead to a good detector. The form that the detector takes depends on the type of estimate used for the unknown power level. Candidates are the MMSE, ML, and maximum a posteriori (MAP) estimates [24]. The ML estimate, which is given by $\hat{\tau}_i = q_i/n$, is particularly attractive, as it does not depend on the details of the pdf $p_\tau(\tau)$ and leads therefore to a distribution-free test

$$\left(\frac{q_0}{q_1}\right)^n \underset{H_0}{\overset{H_1}{\gtrless}} e^T. \tag{7.46}$$

This detector has been obtained previously under varying assumptions by Korado [27], Picinbono and Vezzosi [28], Conte et al. [29], Sharf and Lytle [30], and Gini [25] and is called the GLRT. It may be rewritten as

$$\left(\frac{1}{1 - \dfrac{q_0 - q_1}{q_0}}\right)^n \underset{H_0}{\overset{H_1}{\gtrless}} e^T, \tag{7.47}$$

which, using $\hat{\tau}_{0,\mathrm{ML}} = q_0/n$, is equivalent to

$$q_0 - q_1 \underset{H_0}{\overset{H_1}{\gtrless}} \tau_{0,\mathrm{ML}} n(1 - e^{-T/n}). \tag{7.48}$$

Comparison of this latter formulation of GLRT with (7.42) is particularly interesting as it shows that the GLRT is essentially the matched filter with the unknown power level replaced by its ML estimate under the H_0 hypothesis.

7.6.2 Suboptimum Approximations to the Data-Dependent Threshold

Equation (7.39) shows that the optimum detector can be interpreted as a matched filter compared with a data-dependent threshold that is a function of only the quadratic statistic q_0. Quite often it is not possible to write this data-dependent threshold $f_{\text{opt}}(q_0, T)$ in a closed form. In some cases, it is possible, as when the clutter is K-distributed and $\nu - n = 1/2$ and

$$f_{\text{opt}}(q_0, T) = \begin{cases} T\sqrt{\frac{\eta}{2\nu}}\left(\sqrt{q_0} - \frac{T}{2}\sqrt{\frac{\eta}{2\nu}}\right) & \text{for } q_0 > \frac{\eta T^2}{2\nu} \text{ sign } (T) \\ \dfrac{q_0}{2} & \text{elsewhere} \end{cases} \tag{7.49}$$

where $\eta = E\{\tau\}$, the location parameter of the K distribution.

The difficulty of writing the optimum threshold in closed form suggests that some suboptimum detectors can be obtained by replacing the optimum threshold with some suboptimum approximations. The idea is to look for the best (in a mean-square sense) linear, quadratic, or higher order approximation to the threshold and thereby obtain a suboptimum detector that is simpler to implement than the optimum one yet whose performance is close to optimal.

Consider the approximations:

$$f_1(q_0, T) = b_0 + b_1 q_0 \quad \text{and} \quad f_2(q_0, T) = c_0 + c_1 q_0 + c_2 q_0^2. \tag{7.50}$$

The problem to solve may be stated, for example for the linear approximation, as follows:

$$\min_{b_0, b_1}[\mathbb{E}\{[f_{\text{opt}}(q_0, T) - (b_0 + b_1 q_0)]^2\}]. \tag{7.51}$$

To find the coefficients b_0 and b_1, the derivatives of the mean-square error with respect to the unknown coefficients are set equal to zero. This procedure yields a set of linear equations in two unknowns, whose solution in matrix form is given by

$$\begin{bmatrix} b_0 \\ b_1 \end{bmatrix} = \begin{bmatrix} 1 & \mathbb{E}\{q_0\} \\ \mathbb{E}\{q_0\} & \mathbb{E}\{q_0^2\} \end{bmatrix}^{-1} \begin{bmatrix} \mathbb{E}\{f_{\text{opt}}(q_0, T)\} \\ \mathbb{E}\{q_0 f_{\text{opt}}(q_0, T)\} \end{bmatrix}. \tag{7.52}$$

The coefficient for the second approximation is found in a similar manner by solving

$$\min_{c_0, c_1, c_2} [\mathbb{E}\{[f_{\text{opt}}(q_0, T) - (c_0 + c_1 q_0 + c_2 q_0^2)]^2\}] \tag{7.53}$$

yielding

$$\begin{bmatrix} c_0 \\ c_1 \\ c_2 \end{bmatrix} = \begin{bmatrix} 1 & \mathbb{E}\{q_0\} & \mathbb{E}\{q_0^2\} \\ \mathbb{E}\{q_0\} & \mathbb{E}\{q_0^2\} & \mathbb{E}\{q_0^3\} \\ \mathbb{E}\{q_0^2\} & \mathbb{E}\{q_0^3\} & \mathbb{E}\{q_0^4\} \end{bmatrix}^{-1} \begin{bmatrix} \mathbb{E}\{f_{\text{opt}}(q_0, T)\} \\ \mathbb{E}\{q_0 f_{\text{opt}}(q_0, T)\} \\ \mathbb{E}\{q_0^2 f_{\text{opt}}(q_0, T)\} \end{bmatrix}. \tag{7.54}$$

Higher order approximations may be found in a similar manner. However, we believe the second-order approximation represents a good trade-off between performance and complexity.

Note that the solutions of (7.52) and (7.54) require knowledge of the moments and cross-moments of $f_{\text{opt}}(q_0, T)$ and q_0. In practical applications, these moments could be estimated from data and inserted into the detector form. In addition, the threshold could be precomputed and stored in a look-up table to allow detector operation in real time. In Reference 24, we have provided some cases in which the threshold approximations can be written in closed form.

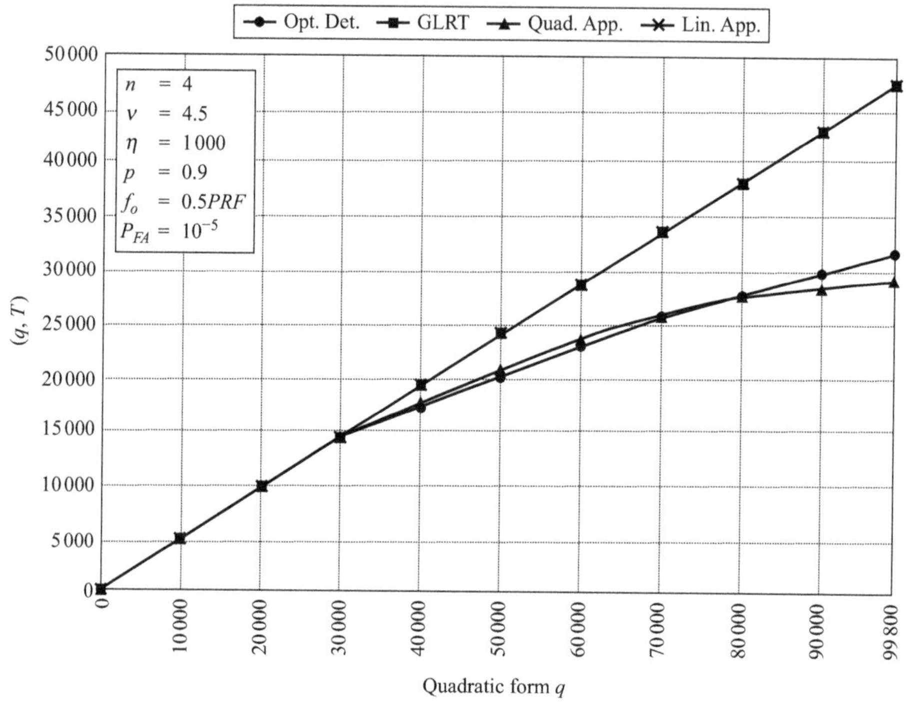

Figure 7.1 Comparison of data-dependent threshold [24].

Figure 7.1 shows the behavior of the various approximations to the data-dependent threshold. The quadratic approximation generally yields the best approximation to the optimum threshold, with the linear approximation being the second best followed by the approximation resulting from the GLRT. In this figure $n = 4$, $\nu = 4.5$, and $\eta = 1\,000$.

7.6.3 Suboptimum Approximations to Estimator-Correlator

In the estimator-correlator structure, the MMSE estimator of α may be difficult to implement in a practical detector. The structure itself suggests that suboptimum detectors may be obtained by replacing the MMSE estimator with a suboptimum estimator. Denote this suboptimum estimator by $\alpha(q_i)$. From (7.42) the estimator-correlator structure becomes

$$\Lambda(z) = \exp\left\{\int_{q_1}^{q_0} \hat{\alpha}(s)ds\right\}. \tag{7.55}$$

As in the case of approximation to the likelihood ratio directly, candidate estimators are the ML and the MAP estimators. It is straightforward to show that the ML estimate of α is given by $\alpha_{i,\mathrm{ML}} = 1\tau_{i,\mathrm{ML}}$. When this estimator is used in (7.55) the resulting detector is given by the GLRT in (7.46).

The MAP estimate on the other hand depends on the details of $p_\alpha(\alpha)$. For K-distributed clutter, the MAP estimator is given by [24]

$$\hat{\alpha}_{i,\mathrm{MAP}} = \frac{(n-\nu-1) + \sqrt{(n-\nu-1)^2 + 4\frac{\nu}{\eta}q_i}}{2q_i} \tag{7.56}$$

and the suboptimum estimator-correlator detector is

$$\Lambda_{\mathrm{MAP}}(z) = \left(\frac{\hat{\alpha}_{1,\mathrm{MAP}}}{\hat{\alpha}_{0,\mathrm{MAP}}}\right)^{(n-\nu-1)} \exp\left[q_0\hat{\alpha}_{0,\mathrm{MAP}} - q_1\hat{\alpha}_{1,\mathrm{MAP}}\right]. \tag{7.57}$$

An indication of how the GLRT of (7.46) and the suboptimum of (7.57) are close to the optimum detector can be obtained by looking at how closely the suboptimum estimates approximate the optimum MMSE estimate, which for K-distributed clutter is given by

$$\hat{\alpha}_{i,\mathrm{MMSE}} = \sqrt{\frac{\nu}{\eta q_i}}\frac{K_{\nu-n-1}\left(\sqrt{\frac{4\nu q_i}{\eta}}\right)}{K_{\nu-n-1}\left(\sqrt{\frac{4\nu q_i}{\eta}}\right)} \qquad i = 0, 1 \tag{7.58}$$

Figure 7.2 shows a comparison of the three estimates as a function of the quadratic form q for the case in which $n = 4$, $\nu = 4.5$, and $\eta = 1\,000$. The figure shows that in this case the MAP estimator is better than the ML and that the detector resulting from using the MAP estimate should be close to optimal. As the number of samples n increases, one may show that

$$\hat{\alpha}_{i,\mathrm{MAP}} \to \frac{n-\nu-1}{q_i} + \frac{\nu}{\eta(n-\nu-1)} \cong \frac{n}{q_i} \tag{7.59}$$

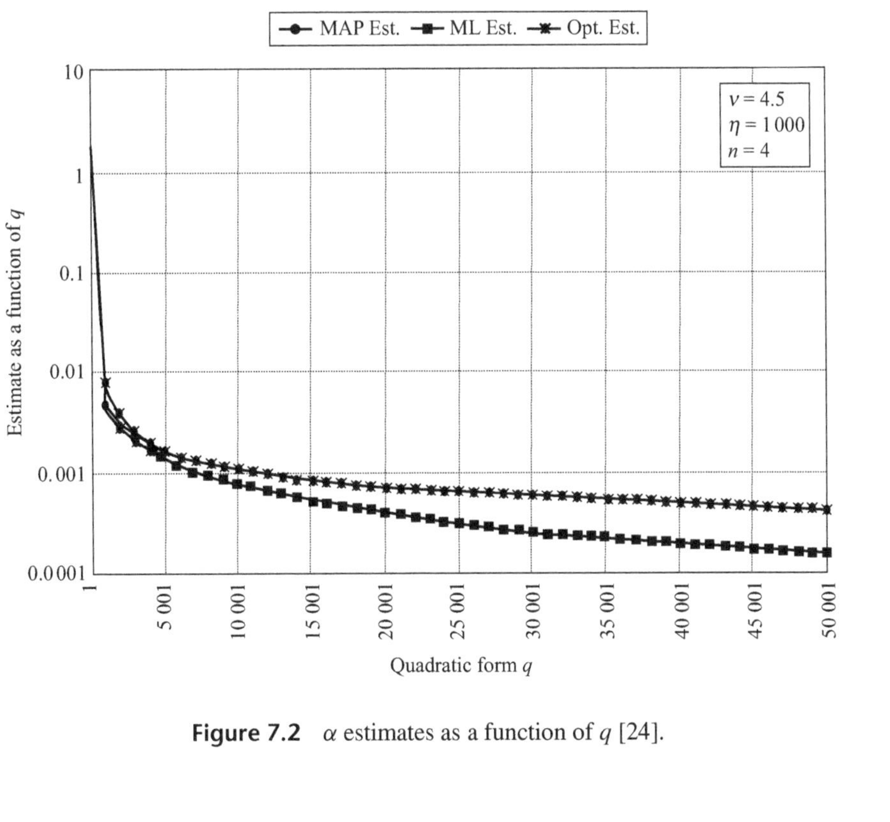

Figure 7.2 α estimates as a function of q [24].

and

$$\hat{\alpha}_{i,\text{MMSE}} \to \frac{n-\nu}{q_i} \cong \frac{n}{q_i} \tag{7.60}$$

which shows that as the number of samples is asymptotically large, the GLRT is equivalent to the optimum detector. This observation is consistent with the results of Picinbono and Vezzosi [28] and Conte et al. [8].

7.6.4 Performance Evaluation of Optimum and Suboptimum Detectors

The performance of the optimum detector defined in (7.31) has been calculated in Reference 21 when the signal fluctuates according to the Swerling I model. The detection performance of the detector (7.31) is given therein by

$$P_{FA} = \int_0^\infty d\tau p_\tau(\tau) \int_0^\infty d\gamma \frac{\gamma^{n-2}}{\tau^{n-1}(n-2)!} \exp\left(-\frac{1}{\tau} h_n^{-1}(e^{-T} h_n(\gamma))\right) \tag{7.61}$$

and

$$P_D = \int_0^\infty d\tau p_\tau(\tau) \int_0^\infty d\gamma \frac{\gamma^{n-2}}{\tau^{n-1}(n-2)!} \exp\left(-\frac{\gamma}{\tau} - \frac{h_n^{-1}(e^{-T} h_n(\gamma)) - \gamma}{\tau + S/C}\right) \tag{7.62}$$

where $S/C = \sigma_s^2 \boldsymbol{p}^\dagger \boldsymbol{\Sigma}^{-1} \boldsymbol{p}$ is the signal-to-clutter ratio. Using (7.37) and replacing it in (7.61) and (7.62) we obtain

$$P_{FA} = \int_0^\infty d\tau p_\tau(\tau) \int_0^\infty d\gamma \frac{\gamma^{n-2}}{\tau^{n-1}(n-2)!} \exp\left(-\frac{1}{\tau}(\gamma - 2f_{\text{opt}}(\gamma, T))\right) \tag{7.63}$$

and

$$P_D = \int_0^\infty d\tau p_\tau(\tau) \int_0^\infty d\gamma \frac{\gamma^{n-2}}{\tau^{n-1}(n-2)!} \exp\left(-\frac{\gamma}{\tau} - \frac{2f_{\text{opt}}(\gamma, T)}{\tau + S/C}\right). \tag{7.64}$$

Detectors that are obtained from the estimator-correlator interpretation may in principle be assessed through the use of (7.61) and (7.62) where $h_n(\gamma)$ is replaced with $\hat{h}_n(\gamma)$ where the optimum estimator of α is replaced by some suboptimum estimator. The detectors obtained from the matched filter interpretation may be assessed through the use of (7.63) and (7.64) where $f_{\text{opt}}(\gamma, T)$ is replaced by some suboptimum $\hat{f}(\gamma, T)$ as in (7.49) and (51).

In the following, we show some performance results for the optimum detector, the GLRT of (7.46) and the two detectors obtained from (7.49) and (51). In all cases, the clutter is assumed to be K-distributed. The clutter covariance matrix is exponential, i.e., $\Sigma_{ij} = \rho^{|i-j|}$ with $\rho = 0.9$. The Doppler frequency of the Swerling I target is $f_D = 0.5PRF$ where PRF is the pulse repetition frequency of the radar. The probability of false alarm has been set to $P_{FA} = 10^{-5}$. In Figure 7.3, the dimension of the processed vector is $n = 4$, and in Figure 7.4 $n = 16$.

In all the examined cases, the quadratic approximation to the data-dependent threshold leads to detection performance almost indistinguishable from the optimum. As this approximation is much simpler than the optimum detector, it represents a good trade-off between performance and ease of implementation. On the other hand, the CFAR property of the GLRT is a highly

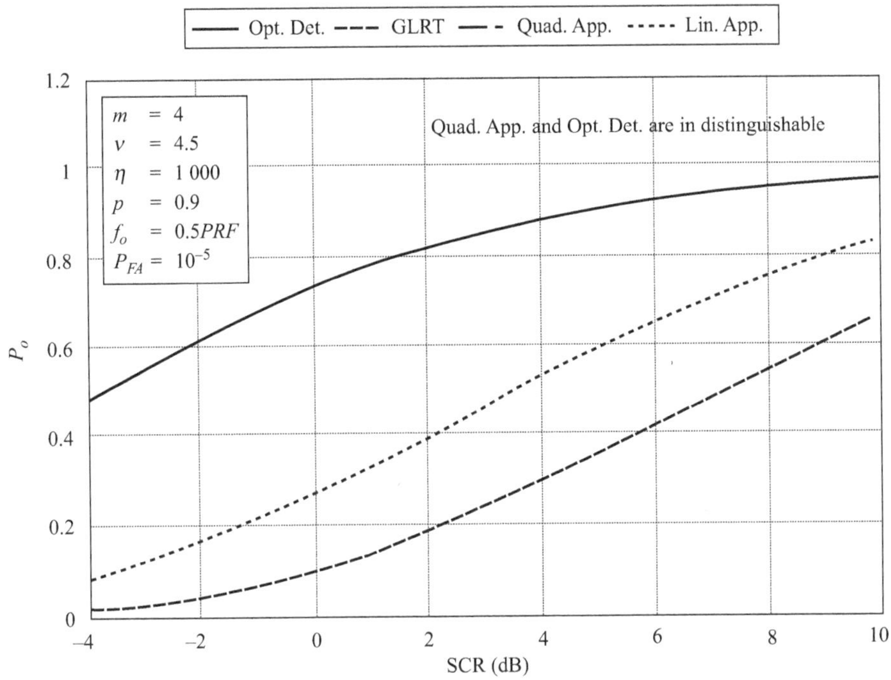

Figure 7.3 Detection performance comparison of optimal and suboptimal detectors.

desirable property for a practical detector. If the number of pulses to be processed for each detection decision is large, then the GLRT has performance within a fraction of a dB of the optimal performance as proved in Figure 7.4. It is easier to implement than the quadratic approximation and hence is preferable when n is large.

7.7 New Interpretation of the Optimum Detector

In this section, we re-examine the optimum detector from a new point-of-view to gain insight into how the detector actually makes a detection decision.

7.7.1 Product of Estimators Formulation

Let $f(x)$ be a univariate amplitude pdf that can be written as a Rayleigh mixture:

$$f(x) = \int_0^\infty \alpha x e^{-\alpha x^2} \, dF_\alpha(\alpha) \quad x \geq 0, \tag{7.65}$$

where $F_\alpha(\alpha)$ is the cumulative distribution function of a positive random variable α, and consider the univariate power distribution of $y = x^2$ given by

$$g(y) = \frac{f(\sqrt{y})}{2\sqrt{y}} = \int_0^\infty \alpha e^{-\alpha y} dF_\alpha(\alpha) \quad y \geq 0. \tag{7.66}$$

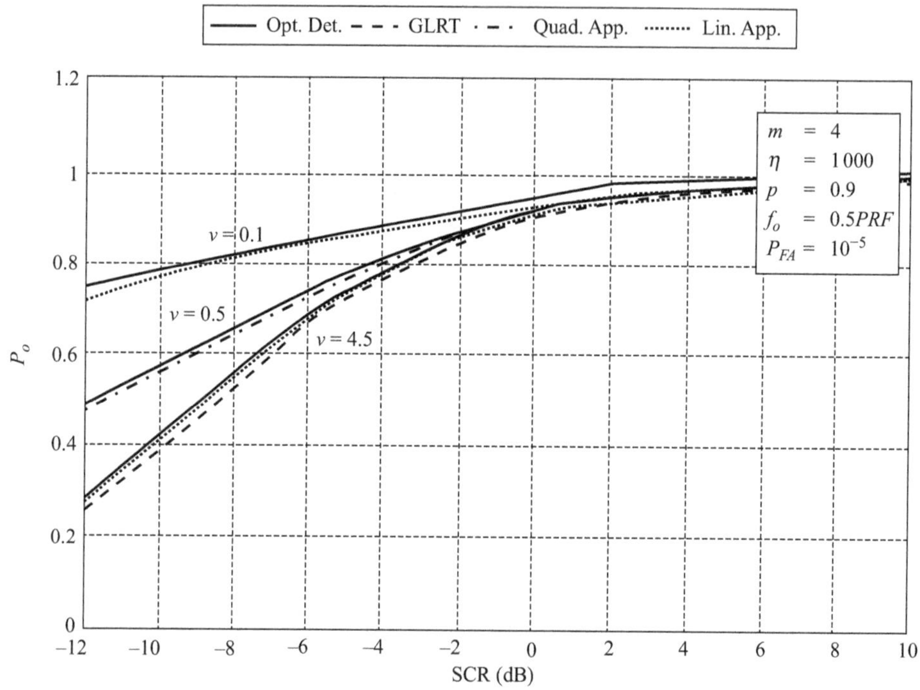

Figure 7.4 Detection performance comparison of optimal and suboptimal detectors.

Define

$$h_0(y) = 1 - \int_0^y g(s)ds = \int_0^\infty e^{-\alpha y} dF_\alpha(\alpha) \quad y \geq 0. \tag{7.67}$$

This is simply the intensity tail distribution. Remembering that $\gamma_0(y) = -\frac{d}{dy} \ln h_0(y)$ and $h_0(y) = \exp\left[-\int \gamma_0(y)dy\right]$, we may rewrite $h_n(y)$ to obtain an alternate formulation of the likelihood ratio. First note that if we define $h_n(y) = Dh_{n-1}(y)$, with $n = 1, 2, \ldots$ then repeated application of this operation to $h_0(y)$ yields

$$h_n(y) = \int_0^\infty \alpha^n e^{-\alpha y} dF_\alpha(\alpha), \quad y \geq 0, \tag{7.68}$$

i.e., we obtain $h_n(y)$, that defines the multivariate compound-Gaussian model. Now examine

$$h_1(y) = Dh_0(y) = \gamma_0(y)h_0(y), \tag{7.69}$$

which follows from (7.68). Continuing, we may write

$$\begin{aligned} h_2(y) &= Dh_1(y) = D\gamma_0(y)h_0(y) = \gamma_0(y)Dh_0(y) + h_0(y)D\gamma_0(y) \\ &= \gamma_0(y)h_1(y) + \frac{h_1(y)}{\gamma_0(y)} D\gamma_0(y) = (\gamma_0(y) + D_l\gamma_0(y))h_1(y), \end{aligned} \tag{7.70}$$

where $D_l = D\ln$. If we now define

$$\gamma_1(y) \triangleq D_l\gamma_0(y) + \gamma_0(y) = (D_l + I)\gamma_0(y), \tag{7.71}$$

then we have $h_2(y) = \gamma_1(y)h_1(y)$. It is apparent that we may repeat this argument and in general write

$$h_n(y) = Dh_{n-1}(y) = \gamma_{n-1}(y)h_{n-1}(y), \tag{7.72}$$

where $\gamma_n(y) = (D_l + I)\gamma_{n-1}(y) = (D_l + I)^n\gamma_0(y)$. From (7.73), it immediately follows that

$$h_n(y) = [\Pi_{k=0}^{n-1}\gamma_k(y)]h_0(y), \tag{7.73}$$

which in turn allows us to rewrite the likelihood ratio as

$$L(q_1, q_0) = \left[\Pi_{k=0}^{n-1}\frac{\gamma_k(q_1)}{\gamma_k(q_0)}\right]\frac{h_0(q_1)}{h_0(q_0)}. \tag{7.74}$$

Equation (7.75) presents a new formulation of the likelihood ratio for the problem of detecting signals against a compound-Gaussian background.

It follows from (7.73) that $\gamma_n(q) = \frac{h_{n+1}(q)}{h_n(q)} = \frac{Dh_n(q)}{h_n(q)} = D_lh_n(q)$, which leads directly to the estimator-correlator formulation of the optimal detector given in Table 7.4. Thus, in the likelihood ratio as given in (7.75), the expression $\gamma_k(q_i)$ represents an estimate of the random variable α under each hypothesis.

7.7.2 General Properties of Product of Estimators

The product of estimators presented here recasts the optimal detector into two different components as shown in (7.74): a product of γ_k terms and the ratio $h_0(q_1)/h_0(q_0)$. From these two components, we see a general feature of the optimal detector in compound-Gaussian clutter. In general, as the statistics of the clutter deviate from Gaussian (i.e., the tails of the amplitude pdf get larger than Rayleigh tails), we expect that the probability that the quadratic form q under either hypothesis will exceed any given value y will grow, and over a broad range of values of q_1 and q_0, the difference in the values of $h_0(q_1)$ and $h_0(q_0)$ will therefore decrease. As a result, as the background becomes more non-Gaussian, we expect that the term $h_0(q_1)/h_0(q_0)$ to contribute less to the likelihood ratio. This behavior is easily seen, for example, for Weibull statistics, where we have $h_0(q) = \exp\left(-\frac{\nu-2}{\eta}q^{\nu/2}\right)$. As $\nu \to 0$, this term becomes a constant and thus does not contribute to the likelihood ratio.

On the other hand, one may easily show for a Gaussian background that the likelihood ratio becomes $L(q_1, q_0) = \frac{h_0(q_1)}{h_0(q_0)} = \exp\left(\frac{q_0 - q_1}{\eta}\right)$, so that the γ_ks do not contribute at all in this case (they become constants and cancel out of the likelihood ratio). Therefore, in a quasi-Gaussian background, the γ_ks are expected to contribute very little to the likelihood ratio.

These observations suggest the optimal detector – as shown by the product of estimators formulation – separates into two components taking importance in different regimes. Namely the ratio $h_0(q_1)/h_0(q_0)$ determines the detection decision in quasi-Gaussian clutter whereas the product of the γ_k terms determines the detection decision in more non-Gaussian clutter. Thus, we have

$$\underbrace{\left[\prod_{k=0}^{n-1} \frac{\gamma_k(q_1)}{\gamma_k(q_0)}\right]}_{\substack{\text{More important} \\ \text{for severely non-} \\ \text{Gaussian clutter}}} \underbrace{\frac{h_0(q_1)}{h_0(q_0)}}_{\substack{\text{More important} \\ \text{for quasi-} \\ \text{Gaussian clutter}}}.$$

As a particular example, let us examine the univariate amplitude statistics be Student-t distributed with the following amplitude pdf, which is obtained by letting the local power level τ have inverse gamma statistics:

$$f_x(x) = \frac{2x}{\eta\left(1 + \dfrac{x^2}{\eta\nu}\right)^{\nu+1}} \qquad 0 < \nu < \infty. \tag{7.75}$$

For the Student-t model, we find

$$h_0(y) = \frac{1}{\left(1 + \frac{y}{\eta\nu}\right)^{\nu}} \quad \text{and} \quad \gamma_k(y) = \frac{\dfrac{1}{\eta} + \dfrac{k}{\eta\nu}}{1 + \dfrac{y}{\eta\nu}}. \tag{7.76}$$

Figures 7.5–7.8 illustrate the behavior of the two components of the likelihood ratio: $h_0(q)$ and $\gamma_k(q) \propto \prod_{k=1}^{K}\left(1 + \frac{q}{\eta\nu}\right)^{-1}$ (we ignore the numerator in each of the estimates since they cancel out the likelihood ratio).

It can be seen that as one of these two component curves gets steeper, the other component curve gets flatter. This demonstrates the changing roles of the two components to the detector as the clutter varies from Gaussian to non-Gaussian. As one or the other of the component curves gets steeper, we expect that component to contribute more to the likelihood ratio. This occurs because a difference between q_0 and q_1 will result in a larger value being contributed to the likelihood ratio from the steeper curve. Conversely, in regions where the curve is flat we do not expect as large a contribution to the likelihood ratio.

To consider a second advantage of the product of estimators formulation, examine the behavior of the γ_ks for Student-t clutter as shown in Figure 7.9.

It is evident from this figure, which describes clutter with very heavy tails, that each $\gamma_k(q)$ – as well as the product of the γ_ks – is steeper for smaller values of q than for larger values of q. Because a small difference between q_1 and q_0 in the steep region where q is small can result in a large value for the ratio $\gamma_k(q_1)/\gamma_k(q_0)$, the γ_k terms, which are expected to dominate the likelihood ratio in severely non-Gaussian clutter, should lead to detections in this steep region. As discussed further below, this behavior appears to be a general feature of any compound-Gaussian model whose tails can become sufficiently large. Note that the quadratic form q is a measure of the local power level of the background. Thus, in the severely non-Gaussian regimes of the compound-Gaussian model, the above considerations suggest when a detection occurs *it is likely to occur when q is small* – i.e., when the local power level is small. This behavior makes sense since a given target is more likely to be detectable when the local

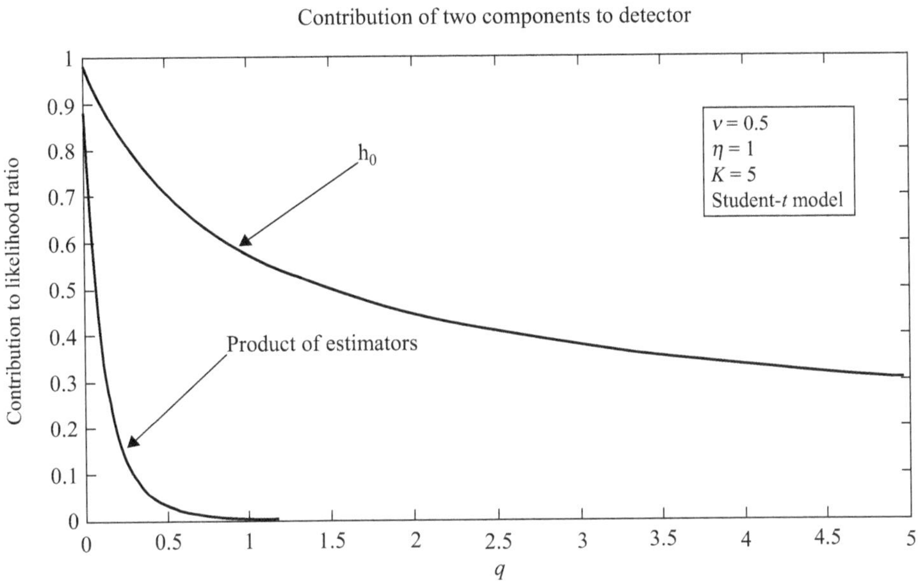

Figure 7.5 Extremely non-Gaussian clutter.

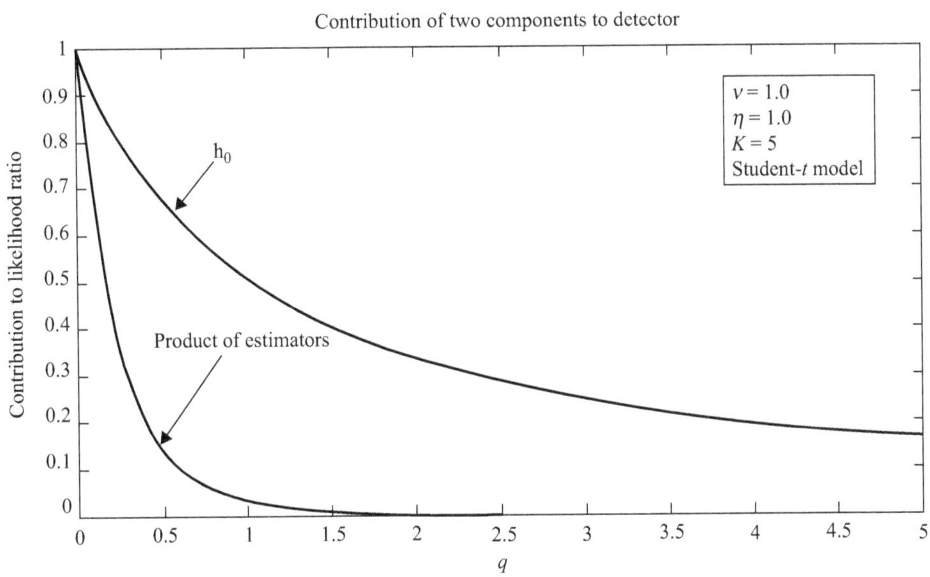

Figure 7.6 Less severely non-Gaussian clutter. The curves have moved closer to each other.

power level is small rather than when it is large. Severely non-Gaussian clutter is sometimes said to be "spiky." The product of estimators formulation suggests that in "spiky" clutter the optimal detector detects targets "between the spikes" – i.e., when the local power level is small.

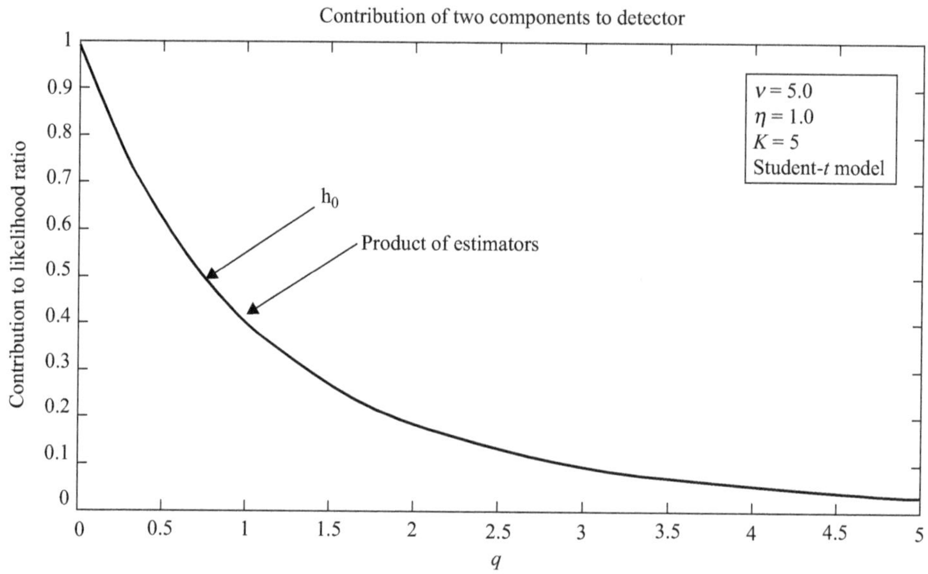

Figure 7.7 Even less severely Gaussian clutter. The curves now coincide.

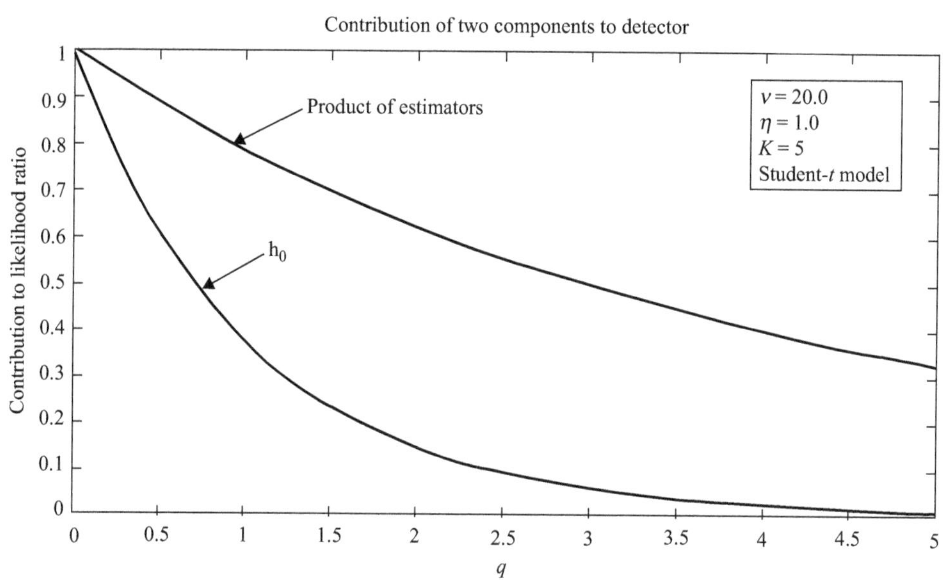

Figure 7.8 Quasi-Gaussian clutter. The curves have changed places.

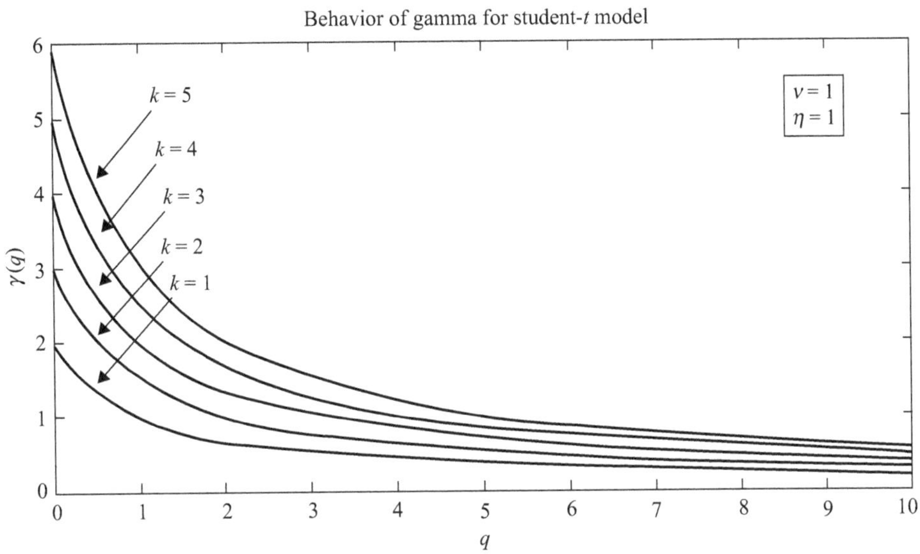

Figure 7.9 Behavior of gamma for Student-t statistics.

7.7.2.1 Examples of Product of Estimators

Consider first the case of Rayleigh univariate amplitude statistics. In this case, we have

$$f_x(x) = \frac{2}{\eta}x\exp\left(-\frac{x^2}{\eta}\right), \tag{7.77}$$

which yields

$$h_0(y) = \exp\left(-\frac{y}{\eta}\right) \quad \text{and} \quad \gamma_0(y) = \frac{1}{\eta} \tag{7.78}$$

It immediately follows that $\gamma_k(y) = \frac{1}{\eta}\ \forall k$ and we find $L(q_1, q_0) = \frac{h_0(q_1)}{h_0(q_0)} = \exp\left(\frac{q_0 - q_1}{\eta}\right)$, as expected.

Consider next the case of K-distributed univariate amplitude statistics. In this case, we have

$$f_x(x) = 4\frac{\left(\sqrt{\nu/\eta}\right)^{\nu+1}}{\Gamma(\nu)}x^{\nu}K_{\nu-1}\left(2x\sqrt{\nu/\eta}\right), \tag{7.79}$$

which leads to

$$h_0(y) = 2\frac{\left(\sqrt{y\nu/\eta}\right)^{\nu}}{\Gamma(\nu)}K_{\nu}\left(2\sqrt{y\nu/\eta}\right) \quad \text{and} \quad \gamma_0(y) = \frac{\left(\sqrt{\nu/\eta}\right)K_{\nu-1}\left(2\sqrt{y\nu/\eta}\right)}{\left(\sqrt{y}\right)K_{\nu}\left(2\sqrt{y\nu/\eta}\right)}. \tag{7.80}$$

It is well known that as $\nu \to \infty$ K-distributed amplitude statistics approach Rayleigh statistics and we recover the results presented above. Let us now examine the case where $\nu \to 0$. In particular, let y be fixed and let ν be small enough such that $\sqrt{y\nu/\eta}$ is small. In that case, we may use the asymptotic expansion of the modified Bessel function: $K_{\nu}(t) \cong \frac{\Gamma(\nu)}{2}\frac{t}{2}^{-\nu}$ for $t \to 0$ to show that $h_0(y) \cong 1$ when $\nu \to 0$. Although this is not a uniformly asymptotic result since ν

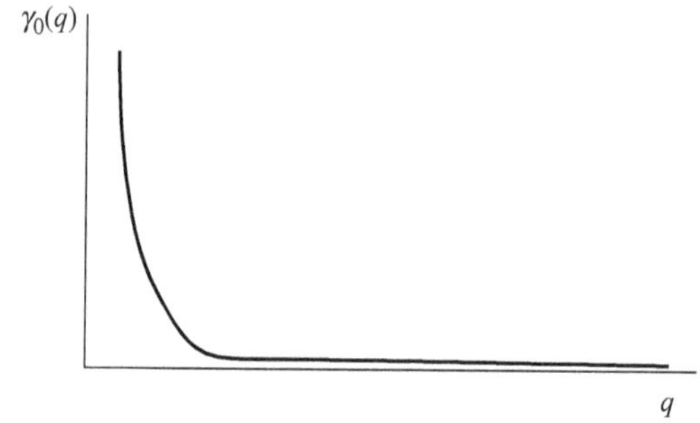

Figure 7.10 General behavior of $\gamma_0(q)$ in severely non-Gaussian clutter.

must decrease as y increases to keep $\sqrt{y\nu/\eta}$ small, it nonetheless shows that the detection decision is based more on the γ_k terms than on the ratio $h_0(q_1)/h_0(q_0)$ as the background becomes more non-Gaussian since $h_0(q_1)/h_0(q_0)$ tends to 1 in this case and thus contributes very little to the likelihood ratio. Note further that as the background becomes more non-Gaussian, i.e., $\nu \to 0$, we find $\gamma_0(y) \cong 1/y$. From this we may find the other γ_ks by differentiation. For example, we find $\gamma_k(y) \cong k/y$.

It is then evident from the structure of $\gamma_0(q)$ that for small values of q, $\gamma_0(q)$ is steeper than for larger values of q (the same observation holds for the γ_k's in general) as the clutter becomes more non-Gaussian. This behavior is shown schematically in Figure 7.10.

As a result, the γ_k terms contribute the most to the likelihood ratio in the steep region where q is small since a small difference between q_1 and q_0 in this region can result in a large value for the ratio $\gamma_k(q_1)/\gamma_k(q_0)$. As $f_x(x) = \frac{2^{\frac{\nu}{2}-1}}{\eta}\nu x^{\nu-1}\exp\left(-\frac{2^{\frac{\nu}{2}-1}}{\eta}x^{\nu}\right)$ we have $\gamma_k(q_1)/\gamma_k(q_0) \cong q_1/q_0$ and the likelihood ratio in this case becomes

$$L(q_1, q_0) = \left(\frac{q_1}{q_0}\right)^n. \tag{7.81}$$

The likelihood ratio in (7.82) defines the well-known GLRT. (This result can also be obtained by letting $\nu \to 0$ in the likelihood ratio.)

Consider next Weibull amplitude statistics:

$$f_x(x) = \frac{2^{\frac{\nu}{2}-1}}{\eta}\nu x^{\nu-1}\exp\left(-\frac{2^{\frac{\nu}{2}-1}}{\eta}x^{\nu}\right). \tag{7.82}$$

This choice leads to

$$\gamma_0(y) = \frac{2^{\frac{\nu}{2}-1}\nu}{\eta y^{1-\frac{\nu}{2}}}, \quad \text{and} \quad h_0(y) = \exp\left(-\frac{2^{\frac{\nu}{2}-1}}{\eta}y^{\frac{\nu}{2}}\right). \tag{7.83}$$

It is immediately evident that this model recovers Rayleigh statistics as $\nu \to 2$. Consider now $\nu \to 0$. In this case, as we have already discussed above, $h_0(q_1)/h_0(q_0)$ tends to 1. We also see from (7.84) that $\gamma_0(y) \cong \frac{2^{\frac{\nu}{2}-1}\nu}{\eta y}$, from which we can obtain $\gamma_k(y) \cong \frac{C_k}{y}$ where C_k is a constant. Thus again as the clutter becomes severely non-Gaussian the optimal detector becomes the GLRT as was discussed above with respect to the K-distributed statistics.

As a final example, consider Student-t amplitude statistics. We gave $h_0(y)$ and $\gamma_k(y)$ in (7.77). The general likelihood ratio for this case becomes

$$L(q_1, q_0) = \left(\frac{1 + \frac{q_0}{\nu\eta}}{1 + \frac{q_1}{\nu\eta}} \right)^n \left(\frac{1 + \frac{q_0}{\nu\eta}}{1 + \frac{q_1}{\nu\eta}} \right)^\nu \tag{7.84}$$

The two terms comprising this likelihood ratio have different behavior as a function of ν, and their relative contributions may be seen to vary as the amplitude statistics vary from Rayleigh to severely non-Rayleigh. In particular, the right-hand term does not contribute as $\nu \to 0$ and the left-hand term does not contribute as $\nu \to \infty$.

It is evident that this model recovers the Rayleigh result as $\nu \to \infty$. Consider now the severely non-Gaussian case. As $\nu \to 0$ we find which again leads to the GLRT. Thus, as was the case for K and Weibull amplitude statistics, the GLRT again arises as the optimal detector for our detection problem with Student-t amplitude statistics as the background becomes increasingly non-Gaussian.

7.7.2.2 Conjectured Behavior for Heavy-Tailed Distributions

Consider a compound-Gaussian model for which as the tails of the intensity tail distribution get heavy. This is the case for three compound-Gaussian models examined here – the K, Weibull, and Student-t distributions – i.e., each of these distributions is parameterized such that the tails can be made extremely large. In this case, we see from the product of estimators formulation that as the tails get heavy, the likelihood ratio behaves as $L(q_1, q_0) = \left(\frac{q_1}{q_0}\right)^n$, which is the GLRT of (7.82). It is known that the GLRT is asymptotically optimal for all compound-Gaussian models as $n \to \infty$ [29]; the considerations above lead to the conjecture that the GLRT is also asymptotically optimal for any value of n for all compound-Gaussian models that can be made to have sufficiently large tails. A heuristic explanation for this result is as follows. The optimal detector attempts to take into account information embodied in the distribution function $F_\tau(\tau)$ (or equivalently in $F_\alpha(\alpha)$) about the fluctuation of the local power level and use this information to make the detection decision. However, unlike in quasi-Gaussian or even moderately non-Gaussian models, where the local power level fluctuates generally within a reasonably small region around a mean value, in extremely non-Gaussian models with heavy tail behavior, the only information embodied in $F_\tau(\tau)$ is that the local power level can essentially take on any value. In this case, the optimal estimate of the inverse of the local power level reduces to an ML estimate, which does not utilize any information from the distribution function $F_\tau(\tau)$, and thus we obtain the GLRT.

7.7.2.3 Observations on Compound-Gaussian Noise as a Random Process

Let us return to our phenomenological scattering picture and explicitly show its dependence on time:

$$E(t) = \sum_{i=1}^{N_s(t)} a_i(t) e^{j\phi_i(t)}. \tag{7.85}$$

In this view, the scattered field at any particular time comprises the contributions of a random number $N_S(t)$ of elementary scatterers encompassed within a "swath," which we indicate by the subscript "S". Note that in general $N_S(t)$ is a random process, i.e., it changes with "time," with also encompasses changes with location as a radar scans. However, if it changes at a rate that is much slower than the rate at which the samples are obtained, then the results obtained herein for the multivariate compound-Gaussian model can be applicable. For example, if a radar samples the environment at a rate faster than the $N_S(t)$ changes, etc., then during a period of time that $N_S(t)$ is equal to a fixed – but random – number, the radar will collect n pulses. That is the situation captured by the compound-Gaussian multivariate model, where all n complex samples are associated with the same – but random – local power level τ. However, in general, n cannot be made arbitrarily large and still have the compound-Gaussian model remain valid because eventually $N_S(t)$ will change while the radar is collecting pulses. This, therefore, puts a limit $n_{\max}$ on the number of pulses n for which the compound-Gaussian model remains applicable. For $n > n_{\max}$, the assumption that $N_S(t)$ – and hence the local power level τ – remains the same for all n pulses breaks down, and new techniques for solving the detection problem are required. Thus it would be beneficial to develop models for the random process $N_S(t)$ and explore the effects of the power level changing during a dwell.

To this end we observe, as discussed above, that each of the intensity tail distributions examined here is associated with an infinitely divisible distribution F_α. For example, the K-distribution is associated with α modeled as an inverse gamma-distributed random variable, whereas the Student-t distribution is associated with α modeled as gamma-distributed random variable. Both of these distributions for α are known to be infinitely divisible, which means they can be obtained from the sum of IID random variables. This suggests that Levy processes, which are processes having stationary and independent increments and are naturally associated with infinitely divisible distributions, be explored as a mechanism to develop random process models of α, and hence of $\tau = 1/\alpha$. (For these two models, Levy processes can be used to develop random process models for τ directly, but in general one would need to develop a random process model for α and then model τ as its inverse.) Such models would then permit further studies of the effects of time-varying local power levels on the kinds of detection problems studied here. An example of this approach is presented in Reference 31.

Appendix 7.A

Transfer Theorem and Its Interpretation

Let $\{N_n\}$ be a sequence of non-negative integer-valued random variables, and let $\{\boldsymbol{x}_{nk}\}$ be a double entry table of random vectors taking values in R^d, $d \geq 1$. Assume for each n that the random vectors $\boldsymbol{x}_{nk}$, $k = 1, 2, \ldots$ are IID with zero mean and that N_n is independent of the sequence $\{\boldsymbol{x}_{nk}\}$. Define

$$s_k^n = \sum_{i=1}^{k} \boldsymbol{x}_{ni}. \tag{7.86}$$

The transfer theorem may be stated as follows.

Theorem 7.1. *Let there exist a sequence $\{k_n\}$ of integers with $k_n \to \infty$ as $n \to \infty$ and distribution functions $F_{\boldsymbol{x}}$ and F_τ such that*

A. *$s_{k_n}^n$ converges in distribution to a random vector $\boldsymbol{x}$ with distribution function $F_{\boldsymbol{x}}$;*

B. *N_n/k_n converges in distribution to a random variable τ with distribution function F_τ. Then*

C. $s^n_{N_n}$ *converges in distribution to a random vector* $\boldsymbol{y}$ *with characteristic function given by*

$$C_y(U) = \int_0^\infty [C_x(U)]^t dF_\tau(t), \tag{7.87}$$

where C_x *is the characteristic function of the distribution* F_x.

One way to understand this theorem is as follows. Under the assumption that condition A holds, the theorem gives condition B the status of a sufficient condition in order for condition C to hold. The natural question that then arises is if condition B is also a necessary condition. In other words, under the assumption that A and C are true, does it follow that B is true? To this end we have [32–34];

Theorem 7.2. *If conditions A and C in the statement of the above theorem hold and* F_τ *does not place all of its mass at 0, then condition B also holds.*

In particular, if $\boldsymbol{x}$ is a Gaussian random vector, then the two theorems may be combined to give:

Theorem 7.3. *Let there exist a sequence* $\{k_n\}$ *of integers with* $k_n \to \infty$ *as* $n \to \infty$ *such that* $s^n_{k_n}$ *converges in distribution to a Gaussian random vector* $\boldsymbol{x}$. *Then* $s^n_{N_n}$ *converges in distribution to a random vector* $\boldsymbol{y}$ *with characteristic function given by (7.2) if and only if* N_n/k_n *converges in distribution to a random variable* τ *with distribution* F_τ.

The results presented above do not quite give a complete result with respect to the relationship between the conditions A–C. In the interpretation given earlier, the results are predicated on the validity of assumption A. In the physical motivation given for examining these problems, A is a natural assumption. It is interesting, however, to ask the following question: if B and C hold, does it follow that A holds? If we restrict our attention to the case of Gaussian $\boldsymbol{x}$ we have the following result due to Szasz and Freyer:

Theorem 7.4. *If conditions B and C in the statement of Theorem 1 hold, with the further restrictions that* $\boldsymbol{x}$ *is a Gaussian random vector and* F_τ *does not place all of its mass at 0, then condition A holds also.*

From these theorems, we see that if the limiting distribution in the absence of number fluctuations is Gaussian, then we have the following relations among the conditions A–C in the presence of number fluctuations (again we assume F_τ does not place all of its mass at 0):

(1) A and B imply C,

(2) A and C imply B,

(3) B and C imply A.

These implications among the conditions A–C give a rather complete account of the influence of number fluctuations on the convergence of partial sums that, in the absence of number fluctuations, converge to a Gaussian random vector.

Convergence of N_n

By analogy with the CLT, a question naturally arises as to how large is N_n. In other words, does N_n in some sense approach ∞, which would support our heuristic reasoning about the convergence in (7.13)? This question is different than the question about the convergence of N_n/k_n. In the CLT, we have $n \to \infty$ so it is important to understand if $N_n \to \infty$ as well.

After all, the appearance of non-Gaussian statistics in the random CLT case could conceivably be the result of N_n (i.e., the number of scatterers being summed) being small in some sense. However, the conditions of the theorem are such that $N_n \to \infty$ in a probabilistic sense under quite general conditions. To see this, first note that because $k_n \to \infty$ as $n \to \infty$, if N_n remains finite as $n \to \infty$ (for instance, N_n might converge to a random variable with finite support), then N_n/k_n will converge to 0. This case was excluded from the earlier theorems. Therefore, at the very least we are interested in cases in which N_n converges in some sense with support that includes ∞. With respect to cases in which $N_n \to \infty$ we have the following easily argued result:

Proposition 7.1. *If N_n/k_n converges in distribution to a random variable τ with distribution function F_τ where $F_\tau(t)$ is continuous at $t = 0$ (hence F_τ has no mass at 0), then N_n converges in probability to ∞ (i.e., for any $K > 0$ $\lim_{n\to\infty} \Pr(N_n \geq K) = 1$).*

Note that if F_τ has some mass at 0, then the resulting distribution function of $\boldsymbol{y}$ will also have mass at 0, which clearly indicates non-Gaussian behavior. Even in the cases where this degeneracy does not occur, however, it is nonetheless possible for the limiting random vector $\boldsymbol{y}$ to be non-Gaussian even though N_n approaches ∞ (in the sense indicated above). This non-Gaussian behavior is directly attributable to the occurrence of number fluctuations.

References

[1] A. Farina, F. Gini, M. Greco, and L. Verrazzani, "High Resolution Sea Clutter Data: A Statistical Analysis of Recorded Live Data," *IEE Proceedings-F*, Vol. 144, No. 3, pp. 121–130, June 1997.

[2] M. Farshchian and F. Posner, "The Pareto Distribution for Low Grazing Angle and High Resolution X-Band Sea Clutter," *2010 IEEE Radar Conference*, Washington, DC, pp. 789–793, May 2010.

[3] T.J. Nohara and S. Haykin, "Canadian East Coast Radar Trails and the K-Distribution," *IEE Proceedings-F*, Vol. 138, No. 2, pp. 80–88, April 1991.

[4] S. Watts, "Radar Detection Prediction in Sea Clutter using the Compound K-Distribution Model," *IEE Proceedings-F*, Vol. 132, No. 7, pp. 613–620, 1985.

[5] K.D. Ward, "Compound Representation of High Resolution Sea Clutter," *Electronics Letters*, Vol. 17, No. 6, pp. 561–563, 1981.

[6] J.B. Billingsley, A. Farina, F. Gini, M. Greco, and L. Verrazzani "Statistical Analyses of Measured Radar Ground Clutter Data," *IEEE Transactions on Aerospace and Electronic Systems*, Vol. 35, No. 2, pp. 579–593, April 1999.

[7] J.B. Billingsley, *Low-Angle Radar Land Clutter – Measurements and Empirical Models*, William Andrew Publishing, Norwich, NY, 2002.

[8] E. Conte, A. De Maio, and C. Galdi, "Statistical Analysis of Real Clutter at Different Range Resolutions," *IEEE Transactions on Aerospace and Electronic Systems*, Vol. 40, No. 3, pp. 903–918, July 2004.

[9] M. Greco and S. Watts, "Radar Clutter Modeling and Analysis," *e-Reference Signal Processing*, Elsevier, October 2013.

[10] K.D. Ward, C.J. Baker, and S. Watts, "Maritime Surveillance Radar Part 1: Radar Scattering from the Ocean Surface," *IEE Proceedings-F*, Vol. 137, No. 2, pp. 51–62, 1990.

[11] G.V. Trunk, "Radar Properties of Non-Rayleigh Sea Clutter," *IEEE Transactions on Aerospace and Electronic Systems*, Vol. 8, No. 2, pp. 196–204, March 1972.

[12] E. Jakeman and P.N. Pusey, "Significance of K Distributions in Scattering Experiments," *Physical Review Letters*, Vol. 4, No. 9, pp. 546–550, 1978.

[13] K.J. Sangston and K.R. Gerlach, "Non-Gaussian Noise Models and Coherent Detection of Radar Targets," *NRL Report 5341-92-9367*, November 1992.

[14] D. Widder, *The Laplace Transform*, Princeton University Press, London, 1941.

[15] L. Bondesson, "A General Result on Infinite Divisibility," *The Annals of Probability*, Vol. 7, No. 6, pp. 965–979, 1979.

[16] K. Miller and S. Samko, "Completely Monotonic Functions" *Integral Transforms and Special Functions*, Vol. 12, No. 4, pp. 389–402, 2001.

[17] W. Feller, *An Introduction to Probability Theory and Its Applications*, Vol. II, John Wiley & Sons, USA, 1971.

[18] B.V. Gnedenko and A.N. Kolmogorov, *Limit Distributions for Sums of Independent Random Variables*, Addison-Wesley, 1954.

[19] H. Robbins, "The Asymptotic Distribution for Sums of a Random Number of Random Variables," *Bulletin of the American Mathematical Society*, Vol. 54, pp. 1151–1161, 1948.

[20] B.V. Gnedenko and H. Fahim, "On a Transfer Theorem," *Doklady Akademii Nauk SSSR*, Vol. 187, pp. 15–17, 1969.

[21] K.J. Sangston and K.R. Gerlach, "Coherent Detection of Radar Targets in a Non-Gaussian Background," *IEEE Transactions on Aerospace and Electronic Systems*, Vol. 30, No. 2, pp. 330–340, April 1994.

[22] E. Conte and M. Longo, "Characterization of Radar Clutter as a Spherically Invariant Random Process," *IEE Proceedings-F*, Vol. 134, No. 2, pp. 191–197, 1987.

[23] M. Rangaswamy, D.D. Weiner, and A. Ozturck, "Non-Gaussian Random Vector Identification Using Spherically Invariant Random Processes," *IEEE Transactions on Aerospace and Electronic Systems*, Vol. 29, No. 1, pp. 111–124, 1993.

[24] K.J. Sangston, F. Gini, M. Greco, and A. Farina, "Structures for Radar Detection in Compound Gaussian Clutter," *IEEE Transactions on Aerospace and Electronic Systems*, Vol. 35, No. 2, pp. 445–458, April 1999.

[25] F. Gini, "Sub-optimum Coherent Radar Detection in a Mixture of K-Distributed and Gaussian Clutter," *IEE Proceedings Radar, Sonar and Navigation*, Vol. 144, No.1, pp. 39–48, February 1997.

[26] F. Gini, M. Greco, and A. Farina, "Clairvoyant and Adaptive Signal Detection in Non-Gaussian Clutter: A Data-Dependent Threshold Interpretation," *IEEE Transactions on Signal Processing*, Vol. 47, No. 6, 1999, pp. 1522–1531.

[27] V.A. Korado, "Optimum Detection of Signals with Random Parameters against the Background of Noise of Unknown Intensity under Conditions of Constant False Alarm Probability," *Radio Engineering and Electronic Physics*, Vol. 13, No. 6, pp. 404–411, 1971.

[28] B. Picinbono and G. Vezzosi, "Detection d'un signal certain dans un bruit non stationnaire et non Gaussien', *Annales des Telecommunications*, Vol. 25, pp. 433–439, 1970.

[29] E. Conte, M. Longo, and G. Ricci, "Asymptotically Optimum Radar Detection in Compound-Gaussian Clutter," *IEEE Transactions on Aerospace and Electronic Systems*, Vol. 31, No. 2, pp. 617–625, 1995.

[30] L. Scharf, and D.W. Lytle, "Signal Detection in Gaussian Noise of Unknown Level: An Invariance Application," *IEEE Transactions on Information Theory*, IT-17, No. 4, pp. 404–411, 1971.

[31] K.J. Sangston, "A Dynamical Model for the Statistical Variation of Clutter," submitted to *IET Radar, Sonar, Navigation* (2014).

[32] D. Szasz and B. Freyer, "A Problem of Summation Theory with Random Indices," *Litovskii Matematicheskii Sbornik*, Vol. 11, pp. 181–187, 1971.

[33] D. Szasz, "On the Limiting Classes of Distributions for Sums of a Random Number of Independent, Identically Distributed Random Variables," *Theory of Probability and Its Applications*, Vol. 17, pp. 401–415; 1972.

[34] D. Szasz, "Stability and Law of Large Numbers for Sums of a Random Number of Random Variables," *Acta Scientiarum Mathematicarum*, Vol. 33, pp. 269–274, 1972.

CHAPTER 8

Covariance Matrix Estimation in SIRV and Elliptical Processes and Their Applications in Radar Detection

Jean-Philippe Ovarlez[1], *Frédéric Pascal*[2], *and Philippe Forster*[3]

8.1 Background and Problem Statement

In adaptive radar detection, the main problem consists in detecting a complex signal $\alpha \boldsymbol{p} \in \mathbb{C}^m$ corrupted by an additive noise $\boldsymbol{c}$ (clutter, thermal noise, etc.). The background parameters are usually estimated using available signal-free secondary data $(\boldsymbol{c}_k)_{k\in\{1,\ldots,K\}}$. In a case of point-like target (see Chapter 9 of this book for extended targets detection scheme), this problem can be stated as the following binary hypothesis test:

$$\begin{cases} H_0 : \boldsymbol{y} = \boldsymbol{c}, & \boldsymbol{y}_k = \boldsymbol{c}_k, \quad \text{for } k = 1, \ldots, K, \\ H_1 : \boldsymbol{y} = \alpha \boldsymbol{p} + \boldsymbol{c}, & \boldsymbol{y}_k = \boldsymbol{c}_k, \quad \text{for } k = 1, \ldots, K, \end{cases} \tag{8.1}$$

where $\boldsymbol{y}$ is the complex m-vector of the received signal, α is an unknown complex target amplitude and $\boldsymbol{p}$ stands for a generally known *steering vector*.

The Probability of False Alarm (PFA), denoted as P_{fa}, is defined as the supremum of the probability of accepting H_1 under H_0 (over the unknown parameters of the distribution of the observables under H_0); if the decision rule has a distribution under H_0 that is independent of the nuisance parameters (covariance matrix for example), the decision rule is said to guarantee to have the Constant False Alarm Rate (CFAR) property. The probability of detection P_d is

[1]CentraleSupélec/SONDRA and French Aerospace Lab, ONERA DEMR/TSI, Gif-sur-Yvette, France
[2]L2S/CentraleSupélec-CNRS-Paris-Sud University, Gif-sur-Yvette, France
[3]SATIE, ENS Cachan, CNRS, Paris-Sud University, Cachan, France

defined as the probability of accepting H_1 when H_1 is true. (The interested reader may refer to Chapter 1 of this book for a detailed discussion on binary hypothesis tests.)

When the parameters of the noise are unknown, the usual procedure to derive the best strategy relies on the Generalized Likelihood Ratio (GLR) statistic, namely the ratio $\Lambda(\boldsymbol{y},\boldsymbol{y}_1,\ldots,\boldsymbol{y}_K)$ between the Probability Density Function (PDF) of the data under H_1 and that under H_0 with the unknown parameters vector $\boldsymbol{\theta}_i$ (i referring to the selected hypothesis H_i) of the PDF (amplitude of the target α, covariance matrix $\boldsymbol{M}$ of the noise, etc.) substituted by their Maximum Likelihood Estimates (MLEs). Then, the GLR Test (GLRT) compares the GLR statistic, $\Lambda(\boldsymbol{y},\boldsymbol{y}_1,\ldots,\boldsymbol{y}_K)$ to a given threshold λ selecting H_1 when the statistic is above the threshold, H_0 otherwise. This test is defined as follows:

$$\Lambda(\boldsymbol{y},\boldsymbol{y}_1,\ldots,\boldsymbol{y}_K) = \frac{\max\limits_{\boldsymbol{\theta}_1} p_c^1(\boldsymbol{y},\boldsymbol{y}_1,\ldots,\boldsymbol{y}_K;\boldsymbol{\theta}_1)}{\max\limits_{\boldsymbol{\theta}_0} p_c^0(\boldsymbol{y},\boldsymbol{y}_1,\ldots,\boldsymbol{y}_K;\boldsymbol{\theta}_0)} \underset{H_0}{\overset{H_1}{\gtrless}} \lambda, \tag{8.2}$$

where p_c^i denote the PDF of the cell under test $\boldsymbol{y}$ under each hypothesis H_i.

Under both hypotheses, it is assumed that $K \geq m$ signal-free data $\boldsymbol{y}_k$ are available for clutter parameters estimation. The $\boldsymbol{y}_k$s are the so-called secondary data where they are assumed independent, but their statistical distribution depends on the noise nature. In this chapter, two cases will be investigated according to the noise statistics: the Gaussian noise and the case of non-Gaussian noise as modeled by Spherically Invariant Random Vector (SIRV).

In the Gaussian case, $\boldsymbol{c}$ and $\boldsymbol{c}_k$ are complex circular zero-mean Gaussian m-vectors sharing the same covariance matrix $\boldsymbol{M}$, with distribution denoted by $\mathcal{CN}(\boldsymbol{0},\boldsymbol{M})$:

$$p_c(\boldsymbol{c}) = \frac{1}{\pi^m \det(\boldsymbol{M})} \exp\left(-\boldsymbol{c}^\dagger \boldsymbol{M}^{-1} \boldsymbol{c}\right). \tag{8.3}$$

This model has been widely used in radar community due to the fact that it is very simple and can generally fit a lot of experimental data.

8.1.1 Background Parameter Estimation in Gaussian Case

When no prior information on the $\boldsymbol{M}$-structure is available, the MLE $\widehat{\boldsymbol{M}}_{SCM}$ of $\boldsymbol{M}$ is the so-called Sample Covariance Matrix (SCM), defined as the solution that maximizes the likelihood function L built with the K signal-free Gaussian secondary data $\boldsymbol{c}_k$:

$$L(\boldsymbol{c}_1,\ldots,\boldsymbol{c}_K) = \prod_{k=1}^{K} p_c(\boldsymbol{c}_k) = \frac{1}{\pi^{mK} \det(\boldsymbol{M})^K} \exp\left(-\sum_{k=1}^{K} \boldsymbol{c}_k^\dagger \boldsymbol{M}^{-1} \boldsymbol{c}_k\right). \tag{8.4}$$

Cancelling the gradient of L with respect to $\boldsymbol{M}$ leads to the well-known solution:

$$\widehat{\boldsymbol{M}}_{SCM} = \frac{1}{K} \sum_{k=1}^{K} \boldsymbol{c}_k \boldsymbol{c}_k^\dagger. \tag{8.5}$$

This estimate has a lot of interesting properties:

- It is a consistent and an unbiased estimate.
- $K\widehat{\boldsymbol{M}}_{SCM}$ follows the complex Wishart distribution $\mathcal{CW}(K,\boldsymbol{M})$ [1].

- The asymptotic distribution of $\widehat{\boldsymbol{M}}_{SCM}$ is given by [2]

$$\sqrt{K}\,\mathrm{vec}(\widehat{\boldsymbol{M}}_{SCM} - \boldsymbol{M}) \xrightarrow{d} \mathcal{GCN}\left(\boldsymbol{0}, \boldsymbol{M}^T \otimes \boldsymbol{M}, \left(\boldsymbol{M}^T \otimes \boldsymbol{M}\right) \boldsymbol{K}_{m^2}\right), \tag{8.6}$$

  where $\boldsymbol{K}_{m^2}$ is a $m^2 \times m^2$-commutation matrix which transforms $\mathrm{vec}(\boldsymbol{A})$ into $\mathrm{vec}(\boldsymbol{A}^T)$, $\mathcal{GCN}(\boldsymbol{\mu}, \boldsymbol{\Sigma}, \boldsymbol{\Omega})$ denotes the Generalized Complex Normal distribution with $\boldsymbol{\Sigma} = \mathbb{E}\left[(\boldsymbol{c} - \boldsymbol{\mu})(\boldsymbol{c} - \boldsymbol{\mu})^\dagger\right]$ the covariance matrix, and $\boldsymbol{\Omega} = \mathbb{E}\left[(\boldsymbol{c} - \boldsymbol{\mu})(\boldsymbol{c} - \boldsymbol{\mu})^T\right]$ the pseudo-covariance matrix.
- The asymptotic distribution of the SCM with its trace equal to m, i.e. $\widehat{\boldsymbol{M}}^m_{SCM} = m\widehat{\boldsymbol{M}}_{SCM}/\mathrm{Tr}(\widehat{\boldsymbol{M}}_{SCM})$ is given by

$$\sqrt{K}\,\mathrm{vec}(\widehat{\boldsymbol{M}}^m_{SCM} - \boldsymbol{M}) \xrightarrow{d} \mathcal{GCN}\left(0, \boldsymbol{\Lambda}, \boldsymbol{\Omega}\right), \tag{8.7}$$

  where $\boldsymbol{\Lambda}$ and $\boldsymbol{\Omega}$ are defined by

$$\begin{cases} \boldsymbol{\Lambda} = \boldsymbol{M}^T \otimes \boldsymbol{M} - \dfrac{1}{m}\mathrm{vec}(\boldsymbol{M})\,\mathrm{vec}(\boldsymbol{M})^\dagger, \\ \boldsymbol{\Omega} = (\boldsymbol{M}^T \otimes \boldsymbol{M})\boldsymbol{K}_{m^2} - \dfrac{1}{m}\mathrm{vec}(\boldsymbol{M})\,\mathrm{vec}(\boldsymbol{M})^T. \end{cases} \tag{8.8}$$

8.1.2 Optimal Detection in Gaussian Case

When $\boldsymbol{M}$ is known, the GLRT for an unknown α is given by

$$\frac{\max\limits_{\alpha} p_c(\boldsymbol{y} - \alpha\boldsymbol{p})}{p_c(\boldsymbol{y})},$$

where p_c is given by (8.3). This leads to the well-known Optimum Gaussian Detector (OGD):

$$\Lambda_{OGD}(\boldsymbol{y}) = \frac{\left|\boldsymbol{p}^\dagger \boldsymbol{M}^{-1}\boldsymbol{y}\right|^2}{\boldsymbol{p}^\dagger \boldsymbol{M}^{-1}\boldsymbol{p}} \underset{H_0}{\overset{H_1}{\gtrless}} \lambda_{OGD}, \tag{8.9}$$

where the detection threshold λ_{OGD} is related to the PFA according to

$$\lambda_{OGD} = -\log(P_{fa}).$$

As for the noise, the covariance matrix $\boldsymbol{M}$ is in general not known and, hence, it has to be estimated from the data. To this end, two adaptive approaches in Gaussian environment can be considered:

- Assuming that the secondary data $(\boldsymbol{c}_k)_{k\in\{1,\ldots,K\}}$ share exactly the same spectral properties of the noise $\boldsymbol{c}$ in the cell under test: this scenario is widely used and is usually referred to as homogeneous environment. In that case, the covariance matrix of the secondary data and the one of the cell under test are the same.
- Assuming that the secondary data $(\boldsymbol{c}_k)_{k\in\{1,\ldots,K\}}$ share, up to an unknown scaling factor, the same spectral properties of the noise $\boldsymbol{c}$. In that case, the covariance matrices of the secondary data and the one of the cell under test differ from an unknown scaling factor σ^2. This approach is referred to as partially homogeneous environment.

8.1.2.1 Homogeneous Gaussian Environment

When the covariance matrix $\boldsymbol{M}$ is unknown, the strategy consists in deriving the exact GLRT procedure [see also Chapter 3 for more details]:

$$\frac{\max_{\alpha,\boldsymbol{M}} p_c(\boldsymbol{y}-\alpha\boldsymbol{p}) \prod_{k=1}^{K} p_c(\boldsymbol{y}_k)}{\max_{\boldsymbol{M}} \prod_{k=1}^{K} p_c(\boldsymbol{y}_k)} \underset{H_0}{\overset{H_1}{\gtrless}} \lambda,$$

leading to the well-known adaptive Kelly's test [3]:

$$\Lambda_{Kelly}(\boldsymbol{y}) = \frac{\left|\boldsymbol{p}^\dagger \widehat{\boldsymbol{M}}_{SCM}^{-1} \boldsymbol{y}\right|^2}{\left(\boldsymbol{p}^\dagger \widehat{\boldsymbol{M}}_{SCM}^{-1} \boldsymbol{p}\right)\left(K + \boldsymbol{y}^\dagger \widehat{\boldsymbol{M}}_{SCM}^{-1} \boldsymbol{y}\right)} \underset{H_0}{\overset{H_1}{\gtrless}} \lambda_{Kelly}. \tag{8.10}$$

The relationship between the PFA P_{fa} and the detection threshold λ_{Kelly} is therefore given by [4]

$$P_{fa} = \left(\frac{1}{\lambda_{Kelly}} - 1\right)^{K+1-m}.$$

On the other hand, the widely used and simpler strategy, called the *Two-Step GLRT*, is to develop first the GLRT for $\boldsymbol{M}$ known, i.e. (8.9), and then to substitute for $\boldsymbol{M}$ the SCM leading to the so-called Adaptive Matched Filter (AMF) test [5]:

$$\Lambda_{AMF}(\boldsymbol{y}) = \frac{\left|\boldsymbol{p}^\dagger \widehat{\boldsymbol{M}}_{SCM}^{-1} \boldsymbol{y}\right|^2}{\boldsymbol{p}^\dagger \widehat{\boldsymbol{M}}_{SCM}^{-1} \boldsymbol{p}} \underset{H_0}{\overset{H_1}{\gtrless}} \lambda_{AMF}. \tag{8.11}$$

The relationship between the PFA P_{fa} and the detection threshold λ_{AMF} is therefore given by [4, 5]

$$P_{fa} = {}_2F_1\left(K-m+1, K-m+2; K+1; -\frac{\lambda_{AMF}}{K}\right), \tag{8.12}$$

where ${}_2F_1(.)$ is the hypergeometric function [6] defined as:

$${}_2F_1(a,b;c;z) = \frac{\Gamma(c)}{\Gamma(b)\,\Gamma(c-b)} \int_0^1 \frac{t^{b-1}(1-t)^{c-b-1}}{(1-t\,z)^a}\, dt.$$

These expressions show that each detection threshold assuring a given PFA is independent of the clutter covariance matrix (CCM) $\boldsymbol{M}$: the Kelly and AMF detectors are said to be CFAR with respect to $\boldsymbol{M}$.

8.1.2.2 Partially Homogeneous Gaussian Environment

In this case, the noise $\boldsymbol{c}$ in the cell under test $\boldsymbol{y}$ is distributed according to $\mathcal{CN}(\boldsymbol{0}, \sigma^2 \boldsymbol{M})$ while the secondary signal-free data $\boldsymbol{c}_k$ are distributed according to $\mathcal{CN}(\boldsymbol{0}, \boldsymbol{M})$ where σ^2 is an unknown scaling factor. The GLRT is therefore given by

$$\frac{\max\limits_{\alpha,\sigma^2,\boldsymbol{M}} \dfrac{1}{\pi^m \det(\boldsymbol{M})\,\sigma^{2m}} \exp\left(-\dfrac{(\boldsymbol{y}-\alpha\boldsymbol{p})^\dagger \boldsymbol{M}^{-1}(\boldsymbol{y}-\alpha\boldsymbol{p})}{\sigma^2}\right) \prod_{k=1}^{K} p_c(\boldsymbol{y}_k)}{\max\limits_{\boldsymbol{M}} \prod_{k=1}^{K} p_c(\boldsymbol{y}_k)} \underset{H_0}{\overset{H_1}{\gtrless}} \lambda. \tag{8.13}$$

This GLRT for partially homogeneous environment yields the Adaptive Cosine (or Coherence) Estimate (ACE) [7], also known as Adaptive Normalized Matched Filter (ANMF) or GLRT-Linear Quadratic (GLRT-LQ) [8]:

$$\Lambda_{ANMF}(\boldsymbol{y}, \widehat{\boldsymbol{M}}_{SCM}) = \frac{\left|\boldsymbol{p}^\dagger \widehat{\boldsymbol{M}}_{SCM}^{-1} \boldsymbol{y}\right|^2}{\left(\boldsymbol{p}^\dagger \widehat{\boldsymbol{M}}_{SCM}^{-1} \boldsymbol{p}\right)\left(\boldsymbol{y}^\dagger \widehat{\boldsymbol{M}}_{SCM}^{-1} \boldsymbol{y}\right)} \underset{H_0}{\overset{H_1}{\gtrless}} \lambda_{ANMF}, \tag{8.14}$$

where the detection threshold λ_{ANMF} is related to the PFA λ_{ANMF} according to:

$$P_{fa} = (1-\lambda_{ANMF})^{K-m+1}\, {}_2F_1\,(K-m+2, K-m+1; K+1; \lambda_{ANMF}). \tag{8.15}$$

When $\boldsymbol{M}$ is known, the corresponding GLRT is known as the Normalized Matched Filter (NMF) and has been proposed by the authors in References 9 and 10: the problem was under a Gaussian environment to determine a detector invariant to the scaling factor σ^2. Previously, other authors [11, 12] have proposed this expression:

$$\Lambda_{NMF}(\boldsymbol{y}) = \frac{\left|\boldsymbol{p}^\dagger \boldsymbol{M}^{-1} \boldsymbol{y}\right|^2}{\left(\boldsymbol{p}^\dagger \boldsymbol{M}^{-1} \boldsymbol{p}\right)\left(\boldsymbol{y}^\dagger \boldsymbol{M}^{-1} \boldsymbol{y}\right)} \underset{H_0}{\overset{H_1}{\gtrless}} \lambda_{NMF}, \tag{8.16}$$

where the detection threshold λ_{NMF} is related to the PFA according to:

$$\lambda_{NMF} = 1 - P_{fa}^{\frac{1}{m-1}}. \tag{8.17}$$

Finally, it can be observed that, under the H_0 hypothesis, the distribution of all the previously presented decision statistics is independent of the unknown parameter $\boldsymbol{M}$. All the detectors OGD, Kelly [3], AMF [5], NMF [9], ANMF [7, 13], possess the CFAR property (i.e. they have a distribution independent of the noise parameters) when the model is valid, i.e. when the statistical data model correspond to the reality. The significance of the CFAR property from a practical point of view is very important: it allows to set the threshold in order to operate at the desired false alarm rate and not at some value of the probability of deciding H_1 when H_0 is valid, lower than the preassigned P_{fa}. It can be shown that OGD, Kelly's detector, and AMF guarantee the CFAR property with respect to $\boldsymbol{M}$ in homogeneous environment and that the NMF and ANMF guarantee the CFAR property with respect to σ^2 and $\boldsymbol{M}$ in the partially homogeneous one. The interested reader may refer to Chapter 7 of this book for optimum detection of a signal in complex compound-Gaussian (CG) clutter.

This hypothesis of a Gaussian noise can be not valid anymore because of the statistical heterogeneity and the non-stationarity of the secondary data used in the estimation process. For example, for high-resolution radars, the resolution cell containing clutter returns is such that the Central Limit Theorem cannot be applied anymore since the number of scatterers become too small and random. Equivalently, it is known that reflected clutter radar returns could be also very impulsive in low grazing angle radar [14–16]. This is why, in the last decades, the radar community has been very interested on problems dealing with non-Gaussian clutter modeling. If the clutter is no more homogeneous (clutter transitions) or become non-Gaussian, these detectors will not guarantee a good false alarm regulation and will degrade their detection performance.

Note that the two detectors (8.14) and (8.16) are homogeneous of degree 0 with respect to $\boldsymbol{p}$, $\boldsymbol{y}$, $\boldsymbol{M}$, or $\widehat{\boldsymbol{M}}_{SCM}$, in the sense that a scale factor multiplying one of those quantity does not modify the resulting detector. They can both be considered, with the appropriate scalar product, as the squared cosine of the angle θ between $\boldsymbol{p}$ and the data vector $\boldsymbol{y}$ collected in the cell under test and lie between 0 and 1 and can be considered as the steering vector angle detectors whereas the OGD and AMF characterize the power detector. These angular detectors will play an important role in the non-Gaussian environment because of this very attractive scale invariance property.

8.2 Non-Gaussian Environment Modeling

In the literature of radar detection and estimation, SIRV modeling and more recently Complex Elliptical Symmetric (CES) distributions have been considered and studied for their good statistical properties and for their good fitting to experimental non-Gaussian radar data.

8.2.1 CES Distribution

In this section, we present the class of CES distributions originally introduced by Kelker [17] and recently used in radar community [18]. They provide a multivariate location-scale family of distributions that primarily serve as long tailed alternatives to the multivariate Gaussian model. They are proven to represent a more accurate characterization of background data (clutter, impulsive noise) than models based on the multivariate Gaussian assumption. A good review on these distributions can be found in References 19 and 20.

In the following, it will be assumed, without loss of generality, that the multivariate process will be zero-mean.

An m-dimensional random complex vector $\boldsymbol{c}$ follows a CES distribution if its characteristic function is of the form:

$$\Phi_c(\boldsymbol{u}) = \phi(\boldsymbol{u}^{\dagger}\, \boldsymbol{\Sigma}\, \boldsymbol{u}),$$

for some function $\phi : \mathcal{R}^{+} \to \mathcal{R}$, called characteristic generator, a positive semi-definite matrix $\boldsymbol{\Sigma}$, called the *scatter* matrix. We shall write $\boldsymbol{c} \sim \mathcal{CES}(\boldsymbol{0}, \boldsymbol{\Sigma}, \phi)$.

Therefore, a random vector $\boldsymbol{c} \sim \mathcal{CES}(\boldsymbol{0}, \boldsymbol{I}_m, \phi)$ is spherically distributed since $\Phi_c(\boldsymbol{u}) = \phi(\boldsymbol{u}^{\dagger}\, \boldsymbol{u})$, and every affine transformation of a spherical random vector has an elliptical distribution. The converse is true, according to the following theorem, when the transformation matrix has full rank.

8.2.1.1 Stochastic Representation Theorem

An m-dimensional random vector $\boldsymbol{c} \sim \mathcal{CES}(\boldsymbol{0}, \boldsymbol{\Sigma}, \phi)$ with $\text{rank}(\boldsymbol{\Sigma}) = k \leq m$ if and only if it admits

$$\boldsymbol{c} \stackrel{d}{=} \mathcal{R} \boldsymbol{A}\, \mathcal{U}^{(k)},$$

where $\mathcal{U}^{(k)}$ is a k-dimensional random vector uniformly distributed on the unit complex k sphere $\mathcal{CS}^k$; $\mathcal{R}$ is a non-negative random variable called *generating variate*, being stochastically independent of $\mathcal{U}^{(k)}$; $\boldsymbol{0} \in \mathbb{C}^m$ and $\boldsymbol{\Sigma} = \boldsymbol{A}\boldsymbol{A}^\dagger$ is a factorization of $\boldsymbol{\Sigma}$ where $\boldsymbol{A} \in \mathbb{C}^{m \times k}$ with rank$(\boldsymbol{A}) = k$.

The generating variate $\mathcal{R}$ determines the distribution's shape, in particular the tail of the distribution. Indeed the generating variate can be connected to the characteristic generator via its cumulative distribution function [20].

The stochastic representation provides a simple manner to simulate elliptically distributed random vectors. The uniform spherical distribution can be easily obtained from a complex normal distributed random vector, $\boldsymbol{y} \sim \mathcal{CN}(\boldsymbol{0}, \boldsymbol{I}_m)$, when dividing it by its length, $\mathcal{U}^{(k)} \stackrel{d}{=} \frac{\boldsymbol{y}}{||\boldsymbol{y}||_2}$. Then, the transformation matrix $\boldsymbol{A}$ on the vector $\mathcal{U}^{(k)}$ produces elliptically contoured density surfaces. Thereupon, some knowledge on the Cumulative Distribution Function (CDF) of $\mathcal{R}$ is required to completely determine the shape of the distribution. Remark that the dispersion on the elliptical distribution is uniquely determined by $\boldsymbol{\Sigma}$, and the particular factorization $\boldsymbol{A}$ adds no information.

From $\boldsymbol{c} \sim \mathcal{CES}(\boldsymbol{0}, \boldsymbol{\Sigma}, \phi)$, it does not follow that $\boldsymbol{c}$ has a PDF. If it exists, it can be related to the density function of the generating variate $\mathcal{R}$, provided $\mathcal{R}$ is absolutely continuous. Then the PDF of $\boldsymbol{c}$ has the form:

$$p_c(\boldsymbol{u}) = \frac{1}{\det(\boldsymbol{\Sigma})} h_m\left(\boldsymbol{u}^\dagger\, \boldsymbol{\Sigma}^{-1}\, \boldsymbol{u}\right), \tag{8.18}$$

where h_m is any function such as (8.18) defines a PDF in $\mathbb{C}^m$. The function h_m is usually called *density generator* and it is assumed to be only approximately known. In this case we shall write $\mathcal{CES}(\boldsymbol{0}, \boldsymbol{\Sigma}, h_m)$ instead of $\mathcal{CES}(\boldsymbol{0}, \boldsymbol{\Sigma}, \phi)$.

The scatter matrix $\boldsymbol{\Sigma}$ describes the shape and orientation of the elliptical equidensity contours. If the second-order moment exists, then $\boldsymbol{\Sigma}$ reflects the structure of the covariance matrix $\boldsymbol{M}$, i.e. the covariance matrix is equal to the scatter matrix up to a scalar constant $\boldsymbol{\Sigma} = \kappa \boldsymbol{M}$. Nevertheless, we can always find an appropriate normalization constraint such that cov$(\boldsymbol{c}) = \boldsymbol{\Sigma}$. This constraint is to take $\mathbb{E}[\mathcal{R}^2] = \text{rank}(\boldsymbol{\Sigma})$. Note that while the scatter matrix is always defined up to a scalar constant, the covariance matrix does not exist for some CES distributions (e.g. Cauchy distribution).

The class of elliptical distributions includes a large number of well-known distributions, as for instance the multivariate Gaussian [1], the K-distribution [21], or the multivariate t-distribution [18].

8.2.2 The Subclass of SIRV

In recent years, there has been an increasing interest for non-Gaussian radar clutter distribution models motivated by experimental radar clutter measurements [14–16] which have shown that the clutter can be perfectly modeled by K-distribution or Weibull distribution. These distributions, called Spherically Invariant Random Processes, characterize the CG distributions [22, 23], i.e. complex CG processes with random power. They represent an important subclass of CES distributions widely used in signal processing applications, e.g. wireless radio propagation problems [24], radar clutter echoes modeling [25–28], hyperspectral background characterization [29–31].

A random vector $\boldsymbol{c}$ is said to have a CG distribution if it can be written as:

$$\boldsymbol{c} \stackrel{d}{=} \sqrt{\tau}\boldsymbol{x}, \tag{8.19}$$

where τ is a positive random variable called *texture* and $\boldsymbol{x}$ an m-dimensional independent zero-mean circular complex Gaussian vector $\mathcal{CN}(\boldsymbol{0},\boldsymbol{M})$ called *speckle*. This vector is called an SIRV and is characterized by its PDF:

$$p_c(\boldsymbol{u}) = \frac{1}{\pi^m \det(\boldsymbol{M})} \int_0^{+\infty} \frac{1}{\tau^m} \exp\left(-\frac{\boldsymbol{u}^\dagger \boldsymbol{M}^{-1} \boldsymbol{u}}{\tau}\right) f_\tau(\tau)\, d\tau, \tag{8.20}$$

where $f_\tau(.)$ is the texture PDF. The covariance matrix $\boldsymbol{M}$ has to be normalized, for example according to $\mathrm{Tr}(\boldsymbol{M}) = m$, for identifiability considerations [32].

This PDF can be put on the more general form:

$$p_c(\boldsymbol{u}) = \frac{1}{\det(\boldsymbol{M})} \tilde{h}_m\left(\boldsymbol{u}^\dagger \boldsymbol{M}^{-1} \boldsymbol{u}\right), \tag{8.21}$$

where $\tilde{h}_m(.)$ is any function such as (8.20) defines a PDF in $\mathbb{C}^m$:

$$\tilde{h}_m(t) = \frac{1}{\pi^m} \int_0^{+\infty} \frac{1}{\tau^m} \exp\left(-\frac{t}{\tau}\right) f_\tau(\tau)\, d\tau. \tag{8.22}$$

It is worth pointing out that, as the SIRV are a subclass of CES distributions, they admit the stochastic representation given by (8.2.1.1). Therefore, $\boldsymbol{c} \stackrel{d}{=} \sqrt{\tau} \boldsymbol{A} \boldsymbol{x}$ where $\boldsymbol{x} \sim \mathcal{CN}(\boldsymbol{0},\boldsymbol{I})$ and $\boldsymbol{M} = \boldsymbol{A}\boldsymbol{A}^\dagger$ any factorization of $\boldsymbol{M}$ with $\mathrm{rank}(\boldsymbol{A}) = \mathrm{rank}(\boldsymbol{M}) = k$.

Conditionally to the cell under test or to the unknown scalar texture τ, the vector $\boldsymbol{c}$ is Gaussian and characterized by the covariance matrix $\boldsymbol{M}$ that identifies the correlation degree existing on the dimension of the vector $\boldsymbol{x}$. From cell to cell, the texture τ models the random power of the observation vector according to the PDF $f_\tau(.)$ and therefore can handle power heterogeneity from cell to cell in the secondary data.

Many other interesting and useful properties related to SIRV subclass can be found in Chapter 7.

8.3 Covariance Matrix Estimation in CES Noise

There has been an intense research activity in robust estimation theory in the statistical community in these last decades [2, 33–36]. Among several solutions, the so-called M-estimators originally introduced by Huber [37] and investigated in the seminal work of Maronna [38] have imposed themselves as an appealing alternative to the classical SCM. They have been introduced within the framework of CES distributions. M-estimators of the covariance matrix are however seldom used in the signal processing community. Only a limited case, the Tyler's estimator [36, 39] also called the Fixed Point Estimator [40] has been widely used as an alternative to the SCM for radar applications. Concerning the M-estimators, notable exceptions are the recent papers by Ollila [19, 41–44] who advocates their use in several applications such as array processing. The M-estimators have also been recently studied in the case of large datasets, where the dimension of the data is of the same order as the dimension of the sample [45].

8.3.1 M-Estimators

When no prior information on the $\boldsymbol{\Sigma}$-structure is available, the MLE of the scatter matrix $\boldsymbol{\Sigma}$ is defined as the matrix which minimizes the negative log-likelihood function:

$$-\sum_{k=1}^{K} \log p_c(\boldsymbol{c}_k) = K \log \det(\boldsymbol{\Sigma}) - \sum_{k=1}^{K} \log h_m\left(\boldsymbol{c}_k^\dagger \, \boldsymbol{\Sigma}^{-1} \, \boldsymbol{c}_k\right). \tag{8.23}$$

Assuming that $h_m(.)$ is continuously differentiable and setting to zero the gradient of of the above expression with respect to $\boldsymbol{\Sigma}$ leads to fixed point equation:

$$\widehat{\boldsymbol{\Sigma}} = \frac{1}{K} \sum_{k=1}^{K} \varphi\left(\boldsymbol{c}_k^\dagger \, \widehat{\boldsymbol{\Sigma}}^{-1} \boldsymbol{c}_k\right) \boldsymbol{c}_k \, \boldsymbol{c}_k^\dagger, \tag{8.24}$$

where $\varphi(t) = -h_m'(t)/h_m(t)$ characterizes a weighting function that depends on the density generator $h_m(.)$ of the underlying CES distribution and where $h_m'(.)$ denotes the real derivative of $h_m(.)$.

Note that (8.24) is an implicit equation, i.e. the solution is on the left-hand side and also on the right-hand side. It characterizes therefore a fixed point equation.

For the complex normal distribution characterized by the density generator $h_m(t) = \pi^{-m} \exp(-t)$, we have $\varphi(t) = 1$ which yields to the SCM defined by (8.5). For the particular case of SIRV distribution, we obtain $\varphi(t) = -\tilde{h}_m'(t)/\tilde{h}_m(t) = \tilde{h}_{m+1}(t)/\tilde{h}_m(t)$. These estimates have a drawback: they depend on the density generator of the background and therefore on the *a priori* knowledge characterizing the noise. The M-estimators provide an alternative since the weighting function does not rely anymore on the knowledge of the distribution. The M-estimators have first been studied in the real case, defined as solution of (8.24) with real samples $\boldsymbol{c}_k$s, and then, results have been extended to the complex case [19, 46].

The Huber estimator is a well-known M-estimator. It is characterized by its corresponding weighting function $\varphi(.)$ defined by

$$\varphi(t) = \frac{1}{\beta} \min(1, \gamma^2/t),$$

where γ^2 and β depend on a single parameter $0 < q < 1$, according to

$$q = F_{2m}(2\gamma^2), \tag{8.25}$$

$$\beta = F_{2m+2}(2\gamma^2) + \gamma^2 \, \frac{1-q}{m}, \tag{8.26}$$

where $F_m(.)$ is the cumulative distribution function of a χ^2 distribution with m degrees of freedom. Thus Huber estimate is the solution of

$$\widehat{\boldsymbol{M}}_{Hub} = \frac{1}{K\beta} \sum_{k=1}^{K} \left[\boldsymbol{c}_k \, \boldsymbol{c}_k^\dagger \, \mathbb{1}_{d_k \le \gamma^2}\right] + \frac{1}{K\beta} \gamma^2 \sum_{k=1}^{K} \left[\frac{\boldsymbol{c}_k \, \boldsymbol{c}_k^\dagger}{d_k} \mathbb{1}_{d_k > \gamma^2}\right], \tag{8.27}$$

where $d_k = \boldsymbol{c}_k^\dagger \, \widehat{\boldsymbol{M}}_{Hub}^{-1} \, \boldsymbol{c}_k$ and $\mathbb{1}(.)$ stands for the indicator function.

The first summation corresponds to unweighted data which are treated as in the SCM; the second one is associated to normalized data treated as outliers. In a complex Gaussian context and when K tends to infinity, it may be shown that the proportion of data treated with the SCM is equal to q. Moreover the choice of γ^2 and β according to (8.25) and (8.25) leads to a consistent M-estimator of the covariance matrix.

8.3.2 Properties of the M-Estimators

Let $(\boldsymbol{c}_1, \ldots, \boldsymbol{c}_K)$ be a K-sample vector independent of the dimension m where $\boldsymbol{c}_k \sim \mathcal{CES}(\boldsymbol{0}, \boldsymbol{\Sigma}, \phi)$ and let us consider the $\boldsymbol{V}_K$ estimate, solution of the following equation:

$$\boldsymbol{V}_K = \frac{1}{K} \sum_{k=1}^{K} \varphi\left(\boldsymbol{c}_k^{\dagger} \boldsymbol{V}_K^{-1} \boldsymbol{c}_k\right) \boldsymbol{c}_k \boldsymbol{c}_k^{\dagger}. \tag{8.28}$$

According to the choice of the function φ in (8.28), the estimate $\boldsymbol{V}_K$ will be different. Existence and uniqueness of the solution of (8.28) has been shown in the real case, provided function φ satisfies a set of general assumptions stated by Maronna [38]. These conditions have been extended to the complex case by Ollila [42]. They are recalled here below:

- φ is non-negative, non-increasing, and continuous on $[0, \infty)$.
- Let $\psi(s) = s\,\varphi(s)$ and $\alpha = \sup_{s \geq 0} \psi(s)$. The function ψ is increasing and strictly increasing on the interval where $\psi(s) < \alpha$ with $m < \alpha$.
- Let $P_K(.)$ denote the empirical distribution of $(\boldsymbol{c}_1, \ldots, \boldsymbol{c}_K)$. There exists $a > 0$ such that for every hyperplane S, dim $(S) \leq m-1$, $P_K(S) \leq 1 - \frac{m}{\alpha} - a$. This assumption can be slightly relaxed as shown in References 47 and 48.

When considering the asymptotic limit of (8.24) which is roughly speaking the limit of (8.28) when K tends to infinity:

$$\boldsymbol{V} = \mathbb{E}\left[\varphi\left(\boldsymbol{c}^{\dagger} \boldsymbol{V}^{-1} \boldsymbol{c}\right) \boldsymbol{c}\,\boldsymbol{c}^{\dagger}\right], \tag{8.29}$$

where $\boldsymbol{c} \sim \mathcal{CES}(\boldsymbol{0}, \boldsymbol{\Sigma}, \phi)$, Maronna [38] and Ollila [42] have shown that:

- the implicit equation (8.29) (resp. (8.28)) admits an unique solution $\boldsymbol{V}$ (resp. $\boldsymbol{V}_K$) and $\boldsymbol{V} = \sigma^{-1}\,\boldsymbol{\Sigma}$ where σ is the solution of the following equation: $\mathbb{E}\left[\psi(\sigma\,||\boldsymbol{t}||^2)\right] = m$ with $\boldsymbol{t} \sim \mathcal{CES}(\boldsymbol{0}, \boldsymbol{I}_m, \phi)$,
- $\boldsymbol{V}_K$ is a consistent estimate of $\boldsymbol{V}$, i.e. $\boldsymbol{V}_K \xrightarrow[n \to \infty]{p.s} \boldsymbol{V}$,
- a simple iterative procedure provides $\boldsymbol{V}_K$.

Some Remarks:

- Although the so-called Fixed Point Estimate (FPE) [49] which is defined by (8.28) for $\varphi(x) = m/x$ and which will be presented in (8.40) has the general form of an M-estimator, this characterizes a limit case that does not verify the conditions given by Maronna: indeed, the function $\varphi(t) = m/t$ is not defined in $t = 0$ and $\psi(t) = m$ is not increasing,
- The SCM given in (8.5) is not an M-estimate according to conditions given by Maronna because the upper limit of function $\psi(.)$ is infinite,

- The FPE and SCM seem to characterize the two limit cases of given M-estimator defined by Maronna: the FPE has the most robust estimate in the CES framework while the SCM is the optimal estimate in the homogeneous case, i.e. Gaussian case.

8.3.3 Asymptotic Distributions of the *M*-Estimators

Let $(\boldsymbol{c}_1, \ldots, \boldsymbol{c}_K)$ be an K-sample of m-dimensional complex independent vectors with $\boldsymbol{c}_k \sim \mathcal{CES}(\boldsymbol{0}, \boldsymbol{\Sigma}, \phi)$, $k = 1, \ldots, K$. We consider the complex M-estimator $\widehat{\boldsymbol{M}}$ which verifies (8.24), and we denote $\boldsymbol{M}$ the solution of (8.28).

The asymptotic distribution of $\widehat{\boldsymbol{M}}$ is given by [46]

$$\sqrt{K}\,\mathrm{vec}(\widehat{\boldsymbol{M}} - \boldsymbol{M}) \xrightarrow{d} \mathcal{GCN}\,(\boldsymbol{0}, \boldsymbol{\Lambda}, \boldsymbol{\Omega}), \tag{8.30}$$

where $\boldsymbol{\Lambda}$ and $\boldsymbol{\Omega}$ represent the asymptotic covariance matrix and the asymptotic pseudo-covariance matrix and are defined by

$$\begin{aligned} \boldsymbol{\Lambda} &= \sigma_1\, \boldsymbol{M}^T \otimes \boldsymbol{M} + \sigma_2\, \mathrm{vec}(\boldsymbol{M})\, \mathrm{vec}(\boldsymbol{M})^{\dagger}, \\ \boldsymbol{\Omega} &= \sigma_1\, (\boldsymbol{M}^T \otimes \boldsymbol{M})\, \boldsymbol{K}_{m^2} + \sigma_2\, \mathrm{vec}(\boldsymbol{M})\, \mathrm{vec}(\boldsymbol{M})^T, \end{aligned} \tag{8.31}$$

with

$$\begin{cases} \sigma_1 = a_1\,(m+1)^2\,(a_2+m)^{-2}, \\ \sigma_2 = a_2^{-2}\left[(a_1-1) - \dfrac{2\,a_1\,(a_2-1)}{(2\,a_2+2\,m)^2}\,[2\,m + (2\,m+4)\,a_2]\right], \end{cases}$$

and

$$\begin{cases} a_1 = [m\,(m+1)]^{-1}\,\mathbb{E}\left[\psi^2(\sigma\,||\boldsymbol{t}||^2)\right], \\ a_2 = m^{-1}\,\mathbb{E}\left[\sigma\,||\boldsymbol{t}||^2\,\psi'(\sigma\,||\boldsymbol{t}||^2)\right], \end{cases}$$

where σ is the solution of $\mathbb{E}[\psi(\sigma\,||\boldsymbol{t}||^2)] = m$, where $\boldsymbol{t} \sim \mathcal{CES}(\boldsymbol{0}, \boldsymbol{I}_m, \phi)$.

This result is also given in Reference 19 with other assumptions but without proof.

The result has to be compared to those obtained for the SCM in (8.6) where $\sigma_1 = 1$ and $\sigma_2 = 0$ and for the SCM normalized according to (8.8) where $\sigma_1 = 1$ and $\sigma_2 = -1/m$. It shows that, asymptotically, the behavior of the M-estimate is practically the same than the one of the SCM, up to properly chosen factors σ_1 and σ_2.

The study of M-estimators shows that the estimated scatter matrix $\widehat{\boldsymbol{M}}$ is reflecting structure-type information but not scale information. This leads to the following important property. Let $\boldsymbol{H}(.)$ be a r-dimensional multivariate function on the set of $m \times m$ positive-definite symmetric matrices with continuous first partial derivatives and such as $\boldsymbol{H}(\boldsymbol{M}) = \boldsymbol{H}(\alpha\,\boldsymbol{M})$ for all $\alpha > 0$. It means that for any homogeneous of degree 0 function $\boldsymbol{H}(.)$:

$$\boldsymbol{H}(\boldsymbol{\Sigma}) = \boldsymbol{H}(\alpha\,\boldsymbol{\Sigma}) = \boldsymbol{H}(\boldsymbol{M}) = \boldsymbol{H}(\alpha\,\boldsymbol{M}). \tag{8.32}$$

Whereas Tyler has derived similar result for real M-estimates [50], Mahot [46] and Ollila [19] have derived the following and very important theorem:

Theorem 8.3.1 *Let $\boldsymbol{M}$ be a fixed complex Hermitian positive-definite matrix and $\widehat{\boldsymbol{M}}$ a sequence of Hermitian positive-definite random matrix estimates of order m satisfying (8.30). Thus, one has*

$$\sqrt{K}\left(\boldsymbol{H}(\widehat{\boldsymbol{M}}) - \boldsymbol{H}(\boldsymbol{M})\right) \xrightarrow{d} \mathcal{GCN}\left(\boldsymbol{0}, \boldsymbol{\Lambda}_H, \boldsymbol{\Omega}_H\right), \tag{8.33}$$

where $\boldsymbol{\Lambda}_H$ and $\boldsymbol{\Omega}_H$ are defined as:

$$\begin{aligned} \boldsymbol{\Lambda}_H &= \sigma_1\, \boldsymbol{H}'(\boldsymbol{M})\,(\boldsymbol{M}^T \otimes \boldsymbol{M})\,\boldsymbol{H}'(\boldsymbol{M})^{\dagger}, \\ \boldsymbol{\Omega}_H &= \sigma_1 \boldsymbol{H}'(\boldsymbol{M})(\boldsymbol{M}^T \otimes \boldsymbol{M})\,\boldsymbol{K}_r\,\boldsymbol{H}'(\boldsymbol{M})^T, \end{aligned} \tag{8.34}$$

and $\boldsymbol{H}'(\boldsymbol{M}) = \dfrac{d\boldsymbol{H}(\boldsymbol{M})}{d\mathrm{vec}(\boldsymbol{M})} = (h'_{ij})$ *with* $h'_{ij} = \dfrac{\partial h_i}{\partial m_j}$ *where* $\mathrm{vec}(\boldsymbol{M}) = (m_i)$.

When the data have a complex Gaussian distribution, the SCM is a complex Wishart matrix. Moreover, the SCM estimator verifies the conditions of the property and its coefficients (σ_1, σ_2) are equal to $(1, 0)$. Complex normalized M-estimators also verify the conditions of the theorem. Thus they have the same asymptotic distribution as the complex normalized Wishart matrix, up to a scale factor σ_1 depending on the considered M-estimator. The same conclusion will hold for the FPE [36, 40] since it verifies the assumptions of Theorem 8.3.1 for a specific standardization.

In practice, $\boldsymbol{H}(.)$ may be a function which associates a parameter of interest to a covariance matrix. This scale-invariant property has also been exploited in Reference 41. The concerned signal and radar processing applications are those in which multiplying the covariance matrix by a positive scalar does not change the result. This is the case for instance for direction of arrival estimation problems when using the MUSIC method. Another example is given by adaptive radar processing in which the parameter is the ANMF test statistic [4, 7]. Here, $\boldsymbol{H}$ is defined by

$$\widehat{\boldsymbol{M}} \to \boldsymbol{H}(\widehat{\boldsymbol{M}}) = \frac{\left|\boldsymbol{p}^{\dagger}\, \widehat{\boldsymbol{M}}^{-1}\, \boldsymbol{y}\right|^2}{\left(\boldsymbol{p}^{\dagger}\, \widehat{\boldsymbol{M}}^{-1}\, \boldsymbol{p}\right)\left(\boldsymbol{y}^{\dagger}\, \widehat{\boldsymbol{M}}^{-1}\, \boldsymbol{y}\right)}.$$

Another example is given by

$$\widehat{\boldsymbol{M}} \to \boldsymbol{H}(\widehat{\boldsymbol{M}}) = m\,\widehat{\boldsymbol{M}} / \,\mathrm{Tr}(\widehat{\boldsymbol{M}}),$$

leading to covariance matrix normalization constraint like $\mathrm{Tr}(\boldsymbol{H}(\widehat{\boldsymbol{M}})) = m$.

Let us give an illustration of Theorem 8.3.1. Let us consider an adaptive radar receiving an m-vector $\boldsymbol{y}$. The estimated covariance matrix of the Gaussian environment is $\widehat{\boldsymbol{M}}$, and the aim is to detect signals of steering vector $\boldsymbol{p}$. This steering vector defines for example to Doppler steering vector. The ANMF $\Lambda_{ANMF}(\boldsymbol{y}, \widehat{\boldsymbol{M}})$ given in (8.14) has been computed for $\widehat{\boldsymbol{M}} = \widehat{\boldsymbol{M}}_{SCM}$ and for the complex Huber's M-estimator $\widehat{\boldsymbol{M}} = \widehat{\boldsymbol{M}}_{Hub}$ defined in (8.27). In Figure 8.1, the vertical scale represents the variance of Λ_{ANMF} obtained with the SCM and the complex Huber's M-estimator. The horizontal scale represents the number of samples used to estimate the covariance matrix. A third curve represents the variance of Λ_{ANMF} for $\sigma_1\, K$ data. As one can see, it overlaps the SCM's curve, illustrating Theorem 8.3.1. The coefficient σ_1 is here equal to 1.067. This results shows, even in Gaussian environment where the optimal MLE

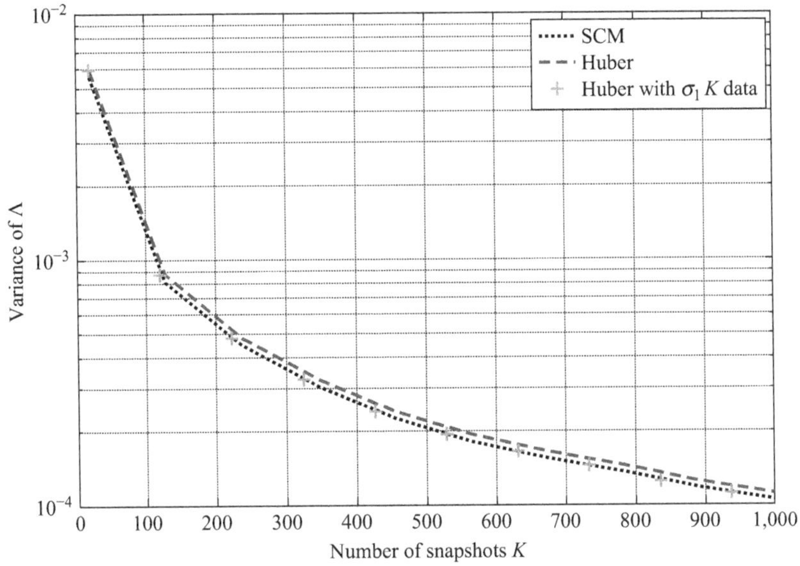

Figure 8.1 Variances on the ANMF detector built with Huber's estimate with $q = 0.75$ and with the SCM estimate, with spatially white Gaussian additive noise.

covariance matrix is the SCM, one can obtain the same performance than the SCM ones with another estimate, here the Huber's estimator, by using a little bit more secondary data. This also means that the loss of performance (comparing to the optimal solution) can be theoretically derived for all M-estimators-based approaches, thanks to Theorem 8.3.1.

Now, let us consider a K-distributed environment, with shape parameter firstly equal to 0.1 and then 0.01 for a more impulsive noise. Figure 8.2 whose scales are the same as in Figure 8.1 shows that the SCM is not robust in a non-Gaussian context contrary to Huber's M-estimator. Indeed, the more the noise differs from a Gaussian noise, the more the detector's variance is deteriorated in that case while it still gives good results with Huber's M-estimator.

8.3.4 Link to *M*-Estimators in the SIRV Framework

In the SIRV background, it becomes clear that the SCM defined by (8.5) is not a good estimate of the speckle covariance matrix $\boldsymbol{M}$. Assuming that K secondary SIRV data $\boldsymbol{c}_k$ are available, the SCM estimate is clearly polluted by the texture information:

$$\widehat{\boldsymbol{M}} = \frac{1}{K}\sum_{k=1}^{K} \boldsymbol{c}_k\, \boldsymbol{c}_k^{\dagger} = \frac{1}{K}\sum_{k=1}^{K} \tau_k\, \boldsymbol{x}_k\, \boldsymbol{x}_k^{\dagger} \neq \frac{1}{K}\sum_{k=1}^{K} \boldsymbol{x}_k\, \boldsymbol{x}_k^{\dagger}. \tag{8.35}$$

A simple solution to build a texture invariant covariance matrix estimate has been proposed by Gini and Conte. It consists in normalizing the data $\boldsymbol{c}_k$ by their norm $\sqrt{\boldsymbol{c}_k^{\dagger}\, \boldsymbol{c}_k}$ before using the SCM estimate. This leads to define the Normalized SCM (NSCM) estimate:

$$\widehat{\boldsymbol{M}}_{NSCM} = \frac{m}{K}\sum_{k=1}^{K} \frac{\boldsymbol{c}_k\, \boldsymbol{c}_k^{\dagger}}{\boldsymbol{c}_k^{\dagger}\, \boldsymbol{c}_k} = \frac{m}{K}\sum_{k=1}^{K} \frac{\boldsymbol{x}_k\, \boldsymbol{x}_k^{\dagger}}{\boldsymbol{x}_k^{\dagger}\, \boldsymbol{x}_k}. \tag{8.36}$$

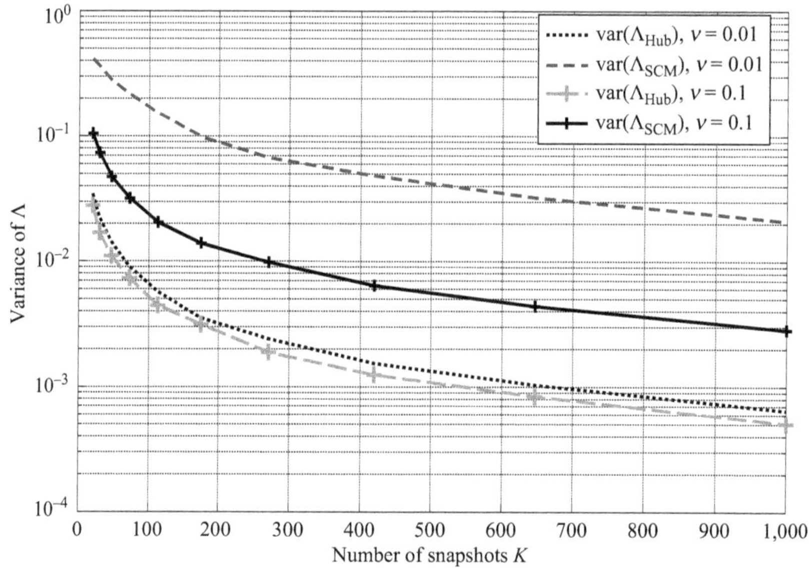

Figure 8.2 Variances on the ANMF detector built with the Huber's estimate and with the SCM, for K-distributed additive noise with various shape parameters $\nu = 0.01$ and $\nu = 0.1$.

This very simple estimate clearly does not depend on the texture τ_k but it is biased and non-consistent (except when $\boldsymbol{M} = \boldsymbol{I}_m$) as proven in Reference 51.

8.3.4.1 Deterministic Texture

In Reference 52, the authors have suggested to set the random parameters $\{\tau_k\}_{k\in\{1,\ldots,K\}}$ as unknown deterministic parameters. In that case, conditionally to the unknown deterministic texture parameter τ, the SIRV PDF is Gaussian with the following PDF:

$$p_c(\boldsymbol{c}_k \,|\, \tau_k, \boldsymbol{M}) = \frac{1}{\pi^m \, \tau_k^m \det(\boldsymbol{M})} \exp\left(\frac{\boldsymbol{c}_k^\dagger \boldsymbol{M}^{-1} \boldsymbol{c}_k}{\tau_k}\right). \tag{8.37}$$

The corresponding likelihood function to maximize with respect to $\boldsymbol{M}$ and τ_ks is given by

$$\prod_{k=1}^{K} p_c(\boldsymbol{c}_k \,|\, \tau_k, \boldsymbol{M}) = \frac{1}{\pi^{m\,K} \det(\boldsymbol{M})^K} \prod_{k=1}^{K} \frac{1}{\tau_k^m} \exp\left(-\frac{\boldsymbol{c}_k^\dagger \boldsymbol{M}^{-1} \boldsymbol{c}_k}{\tau_k}\right). \tag{8.38}$$

Maximization of (8.38) with respect to τ_ks, for a given $\boldsymbol{M}$, leads to

$$\hat{\tau}_k = \frac{\boldsymbol{c}_k^\dagger \boldsymbol{M}^{-1} \boldsymbol{c}_k}{m}, \tag{8.39}$$

and then by replacing the τ_ks in (8.38) by their MLEs $\hat{\tau}_k$s, we obtain the reduced likelihood function:

$$\frac{1}{\pi^{mK}\det(\boldsymbol{M})^K}\prod_{k=1}^{K}\frac{m^m\exp(-m)}{\left(\boldsymbol{c}_k^{\dagger}\boldsymbol{M}^{-1}\boldsymbol{c}_k\right)^m}.$$

Finally, maximizing this latter expression with respect to $\boldsymbol{M}$ leads to true MLE, called the FPE and given by the following equation:

$$\widehat{\boldsymbol{M}}_{FP} = \frac{m}{K}\sum_{k=1}^{K}\frac{\boldsymbol{c}_k\boldsymbol{c}_k^{\dagger}}{\boldsymbol{c}_k^{\dagger}\widehat{\boldsymbol{M}}_{FP}^{-1}\boldsymbol{c}_k}. \quad (8.40)$$

Note that this latter equation can be rewritten from (8.19) as

$$\widehat{\boldsymbol{M}}_{FP} = \frac{m}{N}\sum_{k=1}^{K}\frac{\boldsymbol{x}_k\boldsymbol{x}_k^{\dagger}}{\boldsymbol{x}_k^{\dagger}\widehat{\boldsymbol{M}}_{FP}^{-1}\boldsymbol{x}_k}. \quad (8.41)$$

Equation (8.41) shows that $\widehat{\boldsymbol{M}}_{FP}$ does not depend at all on the texture τ_ks but only on the Gaussian vectors $\boldsymbol{x}_k$s.

8.3.4.2 Random SIRV Texture

In the SIRV framework, the MLE $\widehat{\boldsymbol{M}}$ of the covariance matrix $\boldsymbol{M}$ is obtained following the same procedure given for those of the CES distributions detailed in Section 8.3.1:

$$\widehat{\boldsymbol{M}} = \frac{1}{K}\sum_{k=1}^{K}\frac{\tilde{h}_{m+1}\left(\boldsymbol{c}_k^{\dagger}\widehat{\boldsymbol{M}}^{-1}\boldsymbol{c}_k\right)}{\tilde{h}_m\left(\boldsymbol{c}_k^{\dagger}\widehat{\boldsymbol{M}}^{-1}\boldsymbol{c}_k\right)}\boldsymbol{c}_k\boldsymbol{c}_k^{\dagger}. \quad (8.42)$$

Later, Gini, Conte *et al.* have obtained in References 53 and 54 exactly the same expression. Gini [53] has plotted (this is reported in Figure 8.3), for different values of the K-distribution shape parameter ν, the function $t \to \varphi(t) = \tilde{h}_{m+1}/\tilde{h}_m(t)$ given by

$$\varphi(t) = \frac{\sqrt{\nu}}{\mu\, t}\frac{K_{\nu-m-1}\left(\sqrt{4\,\nu\, t/\mu}\right)}{K_{\nu-m}\left(\sqrt{4\,\nu\, t/\mu}\right)}. \quad (8.43)$$

In the case of Student-t distribution, the weighting function is given by

$$\varphi(t) = \frac{\nu + 2\,m}{\nu + 2\,t}. \quad (8.44)$$

They are plotted in Figure 8.4 for different values of parameter shape ν. Note that $\nu = 0$ leads to FPE while $\nu = +\infty$ leads to SCM.

For the particular choice $\varphi(t) = m/t$, the solution of this implicit equation is the FPE and it is related to the exact MLE given by (8.40) and found when considering texture as deterministic unknown parameter. Gini, Conte *et al.* have also called this estimate the Approximate ML estimate $\widehat{\boldsymbol{M}}_{FP}$. One of the good properties of the FPE is that it remains independent of the *a priori* hypothesis made on the PDF of the noise. Figures 8.3 and 8.4 show that the particular FP function $\varphi(t) = m/t$ has practically the same behavior than many other K-distributions or Student-t φ functions.

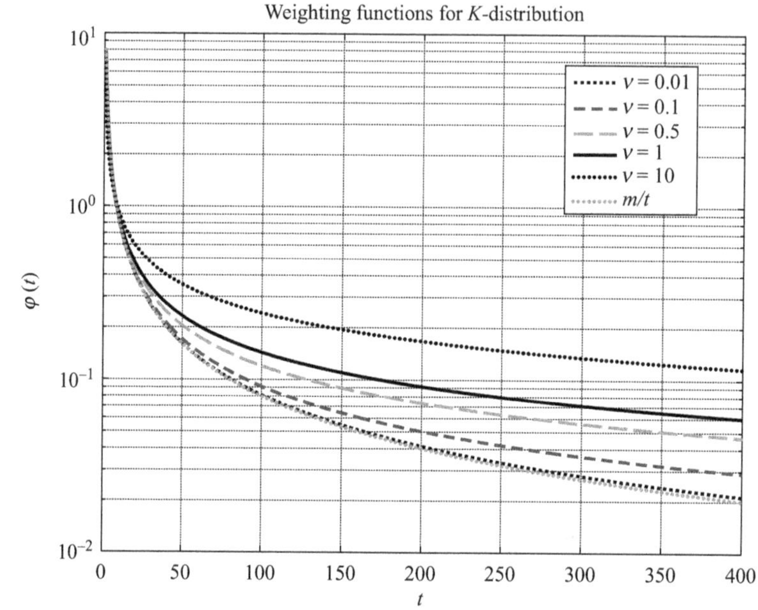

Figure 8.3 Plot of the $\varphi(.)$ functions for various shape parameters ν of the K-distribution ($m = 8$, $\mu = 1$). Comparison with the FP function $\varphi(t) = m/t$.

Existence and uniqueness of the FPE have been proven in Reference 49, while the complete statistical properties of $\widehat{\boldsymbol{M}}_{FP}$ have been derived in Reference 55. The convergence of the following recursive scheme:

$$\boldsymbol{M}_{n+1} = \frac{m}{K}\sum_{k=1}^{K} \frac{\boldsymbol{c}_k\,\boldsymbol{c}_k^\dagger}{\boldsymbol{c}_k^\dagger\,\boldsymbol{M}_n^{-1}\,\boldsymbol{c}_k} n \geq 1, \quad \boldsymbol{M}_0 \in \mathbb{C}^{m\times m},$$

whatever the initialization given by any definite positive matrix $\boldsymbol{M}_0$ has also been proven in Reference 49. Hence, one can choose $\boldsymbol{M}_0 = \widehat{\boldsymbol{M}}_{SCM}$ given in (8.5) but another candidate could be simply $\boldsymbol{M}_0 = \boldsymbol{I}_m$. For this last choice, the first step of the recursive scheme yields to $\boldsymbol{M}_1 = \widehat{\boldsymbol{M}}_{NSCM}$, that is characterizing at the first step of this algorithm a quite good estimate. Note that the solution is always defined up to a scaling factor: if $\boldsymbol{M}$ is solution, $\alpha\,\boldsymbol{M}$, with any $\alpha > 0$, is also solution. The identifiability condition $\mathrm{Tr}(\boldsymbol{M}) = m$ helps in defining a unique solution.

The statistical properties of the FPE have been described in Reference 55. The FPE $\widehat{\boldsymbol{M}}_{FP}$ is unbiased and consistent. Its asymptotic distribution is given by

$$\sqrt{K}\,\mathrm{vec}(\widehat{\boldsymbol{M}}_{FP} - \boldsymbol{M}) \xrightarrow{d} \mathcal{GCN}(\boldsymbol{0}, \boldsymbol{\Lambda}, \boldsymbol{\Omega}),$$

where $\boldsymbol{\Lambda}$ and $\boldsymbol{\Omega}$ are defined by

$$\begin{aligned}\boldsymbol{\Lambda} &= \frac{m+1}{m}\left(\boldsymbol{M}^T \otimes \boldsymbol{M} - \frac{1}{m}\,\mathrm{vec}(\boldsymbol{M})\,\mathrm{vec}(\boldsymbol{M})^\dagger\right),\\ \boldsymbol{\Omega} &= \frac{m+1}{m}\left((\boldsymbol{M}^T \otimes \boldsymbol{M})\boldsymbol{K}_{m^2} - \frac{1}{m}\,\mathrm{vec}(\boldsymbol{M})\,\mathrm{vec}(\boldsymbol{M})^T\right).\end{aligned} \tag{8.45}$$

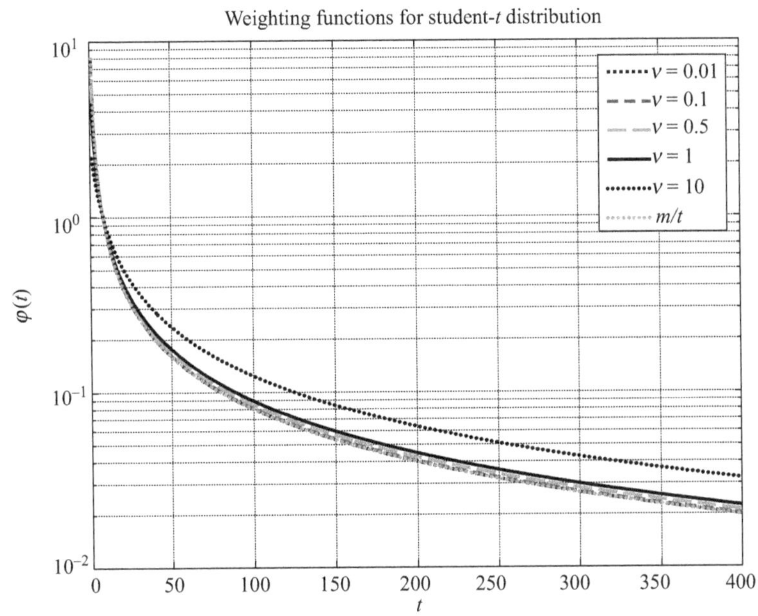

Figure 8.4 Plot of the $\varphi(.)$ functions for various shape parameters ν of the Student-t ($m = 8$). Comparison with the FP function $\varphi(t) = m/t$.

It is important to note the similitude between (8.45) and (8.31) where $\sigma_1 = (m+1)/m$ and $\sigma_2 = -(m+1)/m^2$ and also between (8.45) and (8.8). The FPE and the NSCM have the same asymptotic properties: the same asymptotic distribution with the same asymptotic covariance matrix up to the scalar $(m+1)/m$. It means that $\sqrt{K}\text{vec}(\widehat{\boldsymbol{M}}^m_{SCM} - \boldsymbol{M})$ and $\sqrt{K'}\text{vec}(\widehat{\boldsymbol{M}}_{FP} - \boldsymbol{M})$ converge exactly toward the same distribution with $K' = \frac{m+1}{m}K$. For a higher number of secondary data K, the FPE built with K' data behaves like NSCM with $K = \frac{m}{m+1}K'$ data. The consequence is that all the previous statistical results obtained with the SCM (AMF, ANMF distributions for example) can be extended to those built with the FPE, but with a higher number of secondary data.

Example: it is clear that the ANMF detector defined in (8.14) is invariant to any scale factor on the SCM (for instance, its trace). It means that the two following detectors $\Lambda_{ANMF}(\boldsymbol{y}, \widehat{\boldsymbol{M}}_{SCM})$ and $\Lambda_{ANMF}(\boldsymbol{y}, \widehat{\boldsymbol{M}}^m_{SCM})$ are both equal and have therefore the same detection/threshold relationship in Gaussian noise environment:

$$P_{fa} = (1 - \lambda_{ANMF})^{K-m+1} {}_2F_1\left(K - m + 2, K - m + 1; K + 1; \lambda_{ANMF}\right). \tag{8.46}$$

Consequently, the ANMF $\Lambda_{ANMF-FP}(\boldsymbol{y}, \widehat{\boldsymbol{M}}_{FP})$ built with the FPE:

$$\Lambda_{ANMF-FP}(\boldsymbol{y}) = \frac{\left|\boldsymbol{p}^\dagger \widehat{\boldsymbol{M}}_{FP}^{-1} \boldsymbol{y}\right|^2}{\left(\boldsymbol{p}^\dagger \widehat{\boldsymbol{M}}_{FP}^{-1} \boldsymbol{p}\right)\left(\boldsymbol{y}^\dagger \widehat{\boldsymbol{M}}_{FP}^{-1} \boldsymbol{y}\right)} \underset{H_0}{\overset{H_1}{\gtrless}} \lambda_{ANMF-FP} \tag{8.47}$$

will verify (for K large enough) the following detection/threshold relationship:

$$P_{fa} = (1 - \lambda_{ANMF-FP})^{K'-m+1} \, {}_2F_1\left(K' - m + 2,\ K' - m + 1; K' + 1; \lambda_{ANMF-FP}\right), \quad (8.48)$$

with $K' = \dfrac{m}{m+1} K$.

8.4 Optimal Detection in CES Noise

For many types of SIRV and CES distributions, it can be possible to derive the associated optimal GLRT detector. However, the likelihood ratio test associated to the detection test can rarely be put on a closed-form expression that makes it very difficult to handle for practical ways. Moreover, the parameters defining the density generator have to be estimated each time when the noise characteristics are changing. Several works have allowed to establish from different ways a same form of very interesting and powerful detector:

- The NMF proposed by Scharf [9, 10]: the problem was however, in the Gaussian context, to derive a detector invariant to any scaling factor (the variance). In the SIRV or CES background, this scaling factor is logically the texture. Previously, Korado [11] and Picinbono [12] have already obtained the same expression.
- The GLRT-LQ also called ANMF has been derived by Conte [8] and Gini [32]. It has been obtained when considering asymptotical form of optimal detector for K-distributed noise.
- Sansgton *et al.* in Reference 52 have derived this detector when considering the SIRV texture as deterministic and replacing it by its MLE (GLRT).
- The asymptotic BORD (Bayesian Optimum Radar Detector) defined by Jay [56, 57] has been derived from Bayesian modeling of the texture PDF and in an asymptotical way when number of secondary data K is infinite.

However, all these detectors require a correct estimate of the SIRV covariance matrix.

When $\boldsymbol{M}$ is known and texture τ is unknown, the model has been widely studied through the NMF defined by (8.16).

When $\boldsymbol{M}$ is unknown, one solution is to substitute a given estimator $\widehat{\boldsymbol{M}}$ of $\boldsymbol{M}$ in (8.16) resulting in an adaptive version of the GLRT.

When replacing $\boldsymbol{M}$ by an any estimator $\widehat{\boldsymbol{M}}$, this detector is often called ACE [58] or ANMF.

In this context, SIRV or CES MLE based on secondary data like (8.42) and (8.24) or extended M-estimators could be more judiciously chosen. Some specific M-estimates, like the FPE given in (8.41) are invariant with respect to the texture. As the estimator $\widehat{\boldsymbol{M}}_{FP}$ does not depend on the texture and as the ANMF detector is homogeneous of degree 0, one can obtain a very powerful adaptive detector ANMF which is invariant with the SIRV or CES texture:

$$\Lambda_{ANMF-FP}(\boldsymbol{y}) = \frac{\left|\boldsymbol{p}^\dagger \widehat{\boldsymbol{M}}_{FP}^{-1} \boldsymbol{y}\right|^2}{\left(\boldsymbol{p}^\dagger \widehat{\boldsymbol{M}}_{FP}^{-1} \boldsymbol{p}\right)\left(\boldsymbol{y}^\dagger \widehat{\boldsymbol{M}}_{FP}^{-1} \boldsymbol{y}\right)} \underset{H_0}{\overset{H_1}{\gtrless}} \lambda_{ANMF-FP}. \quad (8.49)$$

Using the asymptotic property (8.45) and Theorem 8.3.1 with $\boldsymbol{H}(.)$ defined by the ANMF detector, the relationship between the PFA P_{fa} and the detection threshold $\lambda_{ANMF-FP}$ can be clearly expressed [59]:

$$P_{fa} = (1 - \lambda_{ANMF-FP})^{a-1}\, {}_2F_1(a, a-1; b-1; \lambda_{ANMF-FP}), \tag{8.50}$$

where $K' = \frac{m}{m+1}K$, $a = K' - m + 2$, and $b = K' + 2$.

For other M-estimates $\widehat{\boldsymbol{M}}$, the relationship between the PFA P_{fa} and the detection threshold λ_{ANMF} of the ANMF built with $\widehat{\boldsymbol{M}}$ can be also expressed:

$$P_{fa} = (1 - \lambda_{ANMF})^{a-1}\, {}_2F_1(a, a-1; b-1; \lambda_{ANMF}), \tag{8.51}$$

where $K' = K/\sigma_1$, $a = K' - m + 2$, and $b = K' + 2$.

8.5 Persymmetric Structured Covariance Matrix Estimation

In order to increase the radar processing performance (for instance, in a detection problem), one solution is to use *a priori* information on the CCM, leading to a more accurate estimate. Toeplitz structure has been addressed by Burg [60] while Fuhrmann [61] used this estimator for radar detection purposes. In radar systems using a symmetrically spaced linear array with constant pulse repetition interval, the CCM exhibits a persymmetric structure. Then, this structure information could be exploited to improve detection performance. In this context, we use a particular linear transformation in order to take into account the persymmetry of the CCM and to study the statistical property of new detectors for both Gaussian and non-Gaussian environments. For Gaussian data, the CCM ML estimator has been derived in Reference 62. The corresponding GLRT has been investigated in Reference 63. For non-Gaussian clutter modeled by SIRV, detection schemes have been proposed in References 64 and 65. In Reference 64, the persymmetry is only exploited to build two sets of independent data in order to derive a SIRV-CFAR detector: the Persymmetric ANMF (P-ANMF). In Reference 65, these sets are used to initialize an iterative algorithm simultaneously proposed in References 53 and 54. This allows to the derivation of the Recursive P-ANMF (RP-ANMF). Our approach, based on the Fixed Point ANMF (FP-ANMF), also called GLRT-FP [53, 54], exploits an original transformation proposed in Reference 66 for Gaussian case and in Reference 67 for non-Gaussian case. This leads to the Persymmetric FP-ANMF (PFP-ANMF), also called GLRT-PFP. The approach is to transform the complex CCM into a real matrix, leading to a simpler problem. Moreover this approach allows the derivation of the statistical analysis of the proposed detection scheme.

It is clear that the estimation accuracy of $\widehat{\boldsymbol{M}}$ has an important impact on the adaptive detection performance in both Gaussian and non-Gaussian clutters. $\widehat{\boldsymbol{M}}_{SCM}$ and $\widehat{\boldsymbol{M}}_{FP}$ defined by (8.5) and (8.40) do not take into account any prior information on the CCM structure. However many applications lead to a CCM which exhibits some particular structure, and considering this structure may lead to an improvement in both estimation and detection performances. Such a situation is frequently met in radar systems using a symmetrically spaced linear array and a symmetrically spaced pulse train for temporal domain processing [60, 63, 64]. In these systems, the CCM $\boldsymbol{M}$ has the persymmetric property, defined as follows:

$$\boldsymbol{M} = \boldsymbol{J}_m \boldsymbol{M}^* \boldsymbol{J}_m,$$

where $\boldsymbol{J}_m$ is the m-dimensional antidiagonal matrix having 1 as non-zero elements. The steering vector of the problem is also persymmetric, i.e. it satisfies

$$\boldsymbol{p} = \boldsymbol{J}_m \boldsymbol{p}^*.$$

Exploiting the persymmetric structure will be done by means of the transformation matrix $\boldsymbol{T}$ introduced in Reference 68 and whose properties are recalled in the following proposition.

Proposition 51 *Let $\boldsymbol{T}$ be the unitary matrix defined as:*

$$\boldsymbol{T} = \begin{cases} \dfrac{1}{\sqrt{2}} \begin{pmatrix} \boldsymbol{I}_{m/2} & \boldsymbol{J}_{m/2} \\ j\,\mathbf{I}_{m/2} & -j\,\mathbf{J}_{m/2} \end{pmatrix} & \textit{for } m \textit{ even}, \\ \dfrac{1}{\sqrt{2}} \begin{pmatrix} \boldsymbol{I}_{(m-1)/2} & 0 & \boldsymbol{J}_{(m-1)/2} \\ 0 & \sqrt{2} & 0 \\ j\,\boldsymbol{I}_{(m-1)/2} & 0 & -j\,\boldsymbol{J}_{(m-1)/2} \end{pmatrix} & \textit{for } m \textit{ odd}. \end{cases} \tag{8.52}$$

Persymmetric vectors and Hermitian matrices are characterized by the following properties:

- $\boldsymbol{p} \in \mathbb{C}^m$ *is a persymmetric vector if and only if $\boldsymbol{T}\boldsymbol{p}$ is a real vector.*
- $\boldsymbol{M}$ *is a persymmetric Hermitian matrix if and only if $\boldsymbol{T}\boldsymbol{M}\boldsymbol{T}^\dagger$ is a real symmetric matrix.*

Using Proposition 51, the original problem (8.1) can be equivalently reformulated. Let us introduce the transformed primary data $\boldsymbol{x}$, the transformed secondary data $\boldsymbol{x}_k$, the transformed clutter vector $\boldsymbol{n}$, and the transformed signal steering vector $\boldsymbol{s}$ defined as: $\boldsymbol{x} = \boldsymbol{T}\boldsymbol{y}$, $\boldsymbol{x}_k = \boldsymbol{T}\boldsymbol{y}_k$, $\boldsymbol{s} = \boldsymbol{T}\boldsymbol{p}$, $\boldsymbol{n} = \boldsymbol{T}\boldsymbol{c}$, $\boldsymbol{n}_k = \boldsymbol{T}\boldsymbol{c}_k$.

It follows that the transformed signal steering vector **s** and the transformed CCM are both real. Then, the original problem (8.1) is equivalent to

$$\begin{cases} H_0 : \boldsymbol{x} = \boldsymbol{n} & \boldsymbol{x}_k = \boldsymbol{n}_k, \text{ for } k = 1, \dots, K, \\ H_1 : \boldsymbol{x} = \alpha\,\boldsymbol{s} + \boldsymbol{n} & \boldsymbol{x}_k = \boldsymbol{n}_k, \text{ for } k = 1, \dots, K, \end{cases} \tag{8.53}$$

where $\boldsymbol{x} \in \mathbb{C}^m$, $\boldsymbol{s}$ is a known real vector.

In the Gaussian case, under hypothesis H_0, $\boldsymbol{n}$ and the K transformed secondary data $\boldsymbol{x}_k$ are i.i.d. and share the same $\mathcal{CN}(\boldsymbol{0}, \boldsymbol{R})$ distribution where $\boldsymbol{R} = \boldsymbol{T}\boldsymbol{M}\boldsymbol{T}^\dagger$ is a real symmetric matrix according to Proposition 51. In the non-Gaussian case, one has

$$\boldsymbol{n} = \sqrt{\tau}\,\boldsymbol{h}, \tag{8.54}$$

$$\boldsymbol{n}_k = \sqrt{\tau_k}\,\boldsymbol{h}_k, \tag{8.55}$$

where $\boldsymbol{h} = \boldsymbol{T}\boldsymbol{g}$ and $\boldsymbol{h}_k = \boldsymbol{T}\boldsymbol{g}_k$ denote the transformed speckle vector with the same real covariance matrix $\boldsymbol{R} = \boldsymbol{T}\boldsymbol{M}\boldsymbol{T}^\dagger$. $\boldsymbol{n}$ and $\boldsymbol{n}_k$ are still SIRVs with the same texture and CCM $\boldsymbol{R} = \boldsymbol{T}\boldsymbol{M}\boldsymbol{T}^\dagger$. From now on, the problem under study is the problem defined by (8.53). The proofs of the results of the next sections can be found in Reference 69.

8.5.1 Detection in Circular Gaussian Noise

Let us first investigate the MLE of the real covariance matrix $\boldsymbol{R}$ from the K secondary data $\boldsymbol{x}_k$. The main motivation for introducing the transformed data is that the resulting distribution of the MLE of $\boldsymbol{R}$ is very simple. This was not the case in Reference 62 when dealing with the original secondary data $\boldsymbol{y}_k$ with persymmetric covariance matrix.

The MLE $\widehat{\boldsymbol{R}}_P$ of the real matrix $\boldsymbol{R}$ is unbiased and is given by

$$\widehat{\boldsymbol{R}}_P = \mathcal{R}e(\widehat{\boldsymbol{R}}_{SCM}), \tag{8.56}$$

where

$$\widehat{\boldsymbol{R}}_{SCM} = \frac{1}{K}\sum_{k=1}^{K} \boldsymbol{x}_k\,\boldsymbol{x}_k^{\dagger} = \boldsymbol{T}\,\widehat{\boldsymbol{M}}_{SCM}\,\boldsymbol{T}^{\dagger}. \tag{8.57}$$

$\widehat{\boldsymbol{R}}_P$ is an unbiased estimator and $K\,\widehat{\boldsymbol{R}}_P$ is real Wishart distributed with parameter matrix $\frac{1}{2}\boldsymbol{R}$ and $2K$ degrees of freedom.

Actually, taking into account the real structure of $\boldsymbol{R}$ (or equivalently the persymmetric structure of $\boldsymbol{M}$) in the ML estimation procedure allows to virtually double the number of secondary data. Let us now consider the AMF for the detection problem (8.53) based on the estimator $\widehat{\boldsymbol{R}}_P$ defined by (8.56). This leads to the following detection test, called the PS-AMF:

$$\Lambda_{PS-AMF} = \frac{\left|\boldsymbol{s}^{\dagger}\,\widehat{\boldsymbol{R}}_P^{-1}\boldsymbol{x}\right|^2}{\boldsymbol{s}^{\dagger}\,\widehat{\boldsymbol{R}}_P^{-1}\,\boldsymbol{s}} \underset{H_0}{\overset{H_1}{\gtrless}} \lambda_{PS-AMF}, \tag{8.58}$$

or equivalently, in terms of the original data:

$$\Lambda_{PS-AMF} = \frac{\left|\boldsymbol{p}^{\dagger}\,\boldsymbol{T}^{\dagger}\,\left[\mathcal{R}e(\boldsymbol{T}\,\widehat{\boldsymbol{M}}_{SCM}\,\boldsymbol{T}^{\dagger})\right]^{-1}\,\boldsymbol{T}\boldsymbol{y}\right|^2}{\boldsymbol{p}^{\dagger}\,\boldsymbol{T}^{\dagger}\,\left[\mathcal{R}e(\boldsymbol{T}\,\widehat{\boldsymbol{M}}_{SCM}\,\boldsymbol{T}^{\dagger})\right]^{-1}\,\boldsymbol{T}\boldsymbol{p}} \underset{H_0}{\overset{H_1}{\gtrless}} \lambda_{PS-AMF}. \tag{8.59}$$

The distribution of (8.58) is well known when $K\,\widehat{\boldsymbol{R}}_P$ is complex Wishart distributed with parameter matrix $K\,\boldsymbol{R}$ and K degrees of freedom: this is the classical AMF distribution [5]. However, in our problem, $K\,\widehat{\boldsymbol{R}}_P$ is real Wishart distributed with parameter matrix $\frac{1}{2}\boldsymbol{R}$ and $2K$ degrees of freedom while $\boldsymbol{x}$ is complex. The relationship between P_{fa} and the detection threshold λ_{PS-AMF} is given by

$$P_{fa} = {}_2F_1\left(\frac{2\,K-m+1}{2}, \frac{2\,K-m+2}{2}; \frac{2\,K+1}{2}; -\frac{\lambda_{PS-AMF}}{K}\right). \tag{8.60}$$

8.5.2 Detection in Non-Gaussian Noise

We address in this section the non-Gaussian case for the detection problem (8.53). The additive SIRV noise $\boldsymbol{n}$ is defined by

$$\boldsymbol{n} = \sqrt{\tau}\,\boldsymbol{h}, \tag{8.61}$$

where τ is a positive random variable, and $\boldsymbol{h}$ is a zero-mean circular complex Gaussian vector with real covariance matrix $\boldsymbol{R}$. The K secondary data $\boldsymbol{n}_k = \sqrt{\tau_k}\,\boldsymbol{h}_k$ are i.i.d. and share the same distribution as $\boldsymbol{n}$.

Since the transformed covariance matrix $\boldsymbol{R}$ is real, its structure may be taken into account in the estimation procedure by retaining only the real part of the FPE. This leads to the proposed covariance estimator called the Persymmetric Fixed-Point (PFP) since it results from the persymmetric structure of the original speckle covariance matrix:

$$\widehat{\boldsymbol{R}}_{PFP} = \mathcal{R}(\widehat{\boldsymbol{R}}_{FP}), \tag{8.62}$$

with

$$\widehat{\boldsymbol{R}}_{FP} = \boldsymbol{T}\,\widehat{\boldsymbol{M}}_{FP}\,\boldsymbol{T}^{\dagger}. \tag{8.63}$$

The statistical properties of the detector $\Lambda_{ANMF-PFP}$ can be investigated under the null hypothesis H_0:

- The distribution of $\widehat{\boldsymbol{R}}_{PFP}$ does not depend on the texture.
- $\widehat{\boldsymbol{R}}_{PFP}$ is a consistent estimator of $\boldsymbol{R}$.
- $\widehat{\boldsymbol{R}}_{PFP}$ is an unbiased estimator of $\boldsymbol{R}$.
- When $\widehat{\boldsymbol{R}}$ is real Wishart distributed with $\frac{m}{m+1}\,2\,K$ degrees of freedom and parameter matrix $\boldsymbol{R}$, $\widehat{\boldsymbol{R}}_{PFP}/\operatorname{Tr}\left(\boldsymbol{R}^{-1}\,\widehat{\boldsymbol{R}}_{PFP}\right)$ and $\widehat{\boldsymbol{R}}/\operatorname{Tr}\left(\boldsymbol{R}^{-1}\,\widehat{\boldsymbol{R}}\right)$ have the same asymptotic distribution.

The adaptive GLRT, for the transformed problem (8.53), based on (8.16) and on the PFP estimator becomes

$$\Lambda_{ANMF-PFP} = \frac{\left|\boldsymbol{s}^{\dagger}\,\widehat{\boldsymbol{R}}_{PFP}^{-1}\,\boldsymbol{x}\right|^2}{\left(\boldsymbol{s}^{\dagger}\,\widehat{\boldsymbol{R}}_{PFP}^{-1}\,\boldsymbol{s}\right)\left(\boldsymbol{x}^{\dagger}\,\widehat{\boldsymbol{R}}_{PFP}^{-1}\,\boldsymbol{x}\right)} \underset{H_0}{\overset{H_1}{\gtrless}} \lambda_{ANMF-PFP}. \tag{8.64}$$

The detector $\Lambda_{ANMF-PFP}$ is SIRV-CFAR. For large K, under hypothesis H_0, $\Lambda_{ANMF-PFP}$ distribution cannot be expressed in a closed form but it has the same distribution as $\Lambda = \frac{\left|\boldsymbol{e}_1^{\dagger}\widehat{\boldsymbol{W}}^{-1}\boldsymbol{w}\right|^2}{\left(\boldsymbol{e}_1^{\dagger}\widehat{\boldsymbol{W}}^{-1}\boldsymbol{e}_1\right)\left(\boldsymbol{w}^{\dagger}\widehat{\boldsymbol{W}}^{-1}\boldsymbol{w}\right)}$ where $\boldsymbol{w} \sim \mathcal{CN}(\boldsymbol{0}, \boldsymbol{I})$, $\boldsymbol{e}_1 = (1, 0, \ldots, 0)^T$ and where $\widehat{\boldsymbol{W}}$ is real Wishart distributed with parameter matrix $\boldsymbol{I}$ and $K' = \frac{m}{m+1}\,2\,K$ degrees of freedom.

In the context of non-Gaussian clutter, Conte and De Maio [64, 65] have proposed two detectors derived respectively from the GLRT with some different estimators: the P-ANMF and the RP-ANMF. In Reference 64, the persymmetry property is only used to separate their original set of secondary data $\boldsymbol{n}_k$ into two new uncorrelated and then independent sets of data $\boldsymbol{r}_{ek}$ and $\boldsymbol{r}_{ok}$, in order to render the detector matrix-CFAR and improve the performance in terms of detection. These new vectors have the same size as the original and share the same texture. Their speckle components are i.i.d. and zero-mean complex Gaussian vectors. These new sets of secondary data allow to introduce their new estimator of the covariance matrix:

$$\widehat{\boldsymbol{\Sigma}} = \frac{1}{K}\sum_{k=1}^{K} \frac{\boldsymbol{r}_{ek}\,\boldsymbol{r}_{ek}^{\dagger}}{\left(\boldsymbol{r}_{ok}\boldsymbol{r}_{ok}^{\dagger}\right)_{i,i}}, \tag{8.65}$$

where $(\boldsymbol{A})_{i,i}$ stands for any (i, i)th element of the matrix $\boldsymbol{A}$.

The previous estimator is then replaced in the classical NMF given by (8.16) and leads to define the P-ANMF detector:

$$\Lambda_{P-ANMF} = \frac{|\boldsymbol{p}^\dagger \, \widehat{\boldsymbol{\Sigma}}^{-1} \boldsymbol{x}|^2}{\left(\boldsymbol{p}^\dagger \, \widehat{\boldsymbol{\Sigma}}^{-1} \boldsymbol{p}\right) \left(\boldsymbol{x}^\dagger \, \widehat{\boldsymbol{\Sigma}}^{-1} \boldsymbol{x}\right)} \underset{H_0}{\overset{H_1}{\gtrless}} \lambda_{P-ANMF}. \tag{8.66}$$

In Reference 65, the same method is used to define two set of secondary data $\boldsymbol{r}_{ek}$ and $\boldsymbol{r}_{ok}$, and the FP matrix estimator $\widehat{\boldsymbol{\Sigma}}_{(\text{inf})}$ is found by using the recursive procedure:

$$\widehat{\boldsymbol{\Sigma}}_{(t+1)} = \frac{N}{K} \sum_{k=1}^{K} \frac{\boldsymbol{r}_{ek} \, \boldsymbol{r}_{ek}^\dagger}{\boldsymbol{r}_{ek}^\dagger \left(\widehat{\boldsymbol{\Sigma}}_{(t)}\right)^{-1} \boldsymbol{r}_{ek}}, \tag{8.67}$$

with starting point:

$$\widehat{\boldsymbol{\Sigma}}_{(0)} = \frac{1}{K} \sum_{k=1}^{K} \frac{\boldsymbol{r}_{ek} \, \boldsymbol{r}_{ek}^\dagger}{\left(\boldsymbol{T} \, \boldsymbol{r}_{ok} \boldsymbol{r}_{ok}^\dagger \, \boldsymbol{T}^\dagger\right)_{i,i}}. \tag{8.68}$$

This estimator is next replaced in the NMF (8.16) to provide the RP-ANMF:

$$\Lambda_{RP-ANMF} = \frac{\left|\boldsymbol{p}^\dagger \, \widehat{\boldsymbol{\Sigma}}_{(\text{inf})}^{-1} \boldsymbol{x}\right|^2}{\left(\boldsymbol{p}^\dagger \, \widehat{\boldsymbol{\Sigma}}_{(\text{inf})}^{-1} \boldsymbol{p}\right) \left(\boldsymbol{x}^\dagger \, \widehat{\boldsymbol{\Sigma}}_{(\text{inf})}^{-1} \boldsymbol{x}\right)} \underset{H_0}{\overset{H_1}{\gtrless}} \lambda_{RP-ANMF}. \tag{8.69}$$

Please note however that, as stated in Reference 49, the solution $\widehat{\boldsymbol{\Sigma}}_{(\text{inf})}$ of the implicit FP matrix equation is unique and does not depend on the starting point.

In order to compare all these detectors, NMF with $\boldsymbol{M}$ known or with the classical SCM, ANMF-FP, ANMF-PFP, P-ANMF, and RP-ANMF, Figure 8.5(a) presents the PFA versus the detection threshold for all these detectors while Figure 8.5(b) shows the PD versus the Signal to Noise Ratio (SNR). The simulated impulsive clutter is settled to be K-distributed.

These figures show the improvement in terms of detection of the RP-ANMF over the conventional ANMF-SCM (which cannot be efficient with non-Gaussian data) but also the improvement of the ANMF-PFP against all the other detectors. Moreover, theoretical results based on the asymptotic Wishart distributions of $\widehat{\boldsymbol{R}}_{FP}$ and $\widehat{\boldsymbol{R}}_{PFP}$ (circle lines) are displayed: it can be noticed that simulated results are in very good agreement with the theory.

8.6 Radar Applications

This section is devoted to the analysis of different radar measurements in which the clutter is strongly impulsive. In a first time, let us give some generalities.

8.6.1 Ground-Based Radar Detection

The ground clutter data presented in this chapter were collected by an operational radar at THALES Air Defence, placed at 13 m above ground and illuminating the ground at low grazing angle. Ground clutter complex echoes were collected in $N = 868$ range bins for 70 different azimuth angles and for $m = 8$ recurrences, which means that vector size is $m = 8$. Near the radar, echoes characterize non-Gaussian heterogeneous ground clutter whereas beyond the

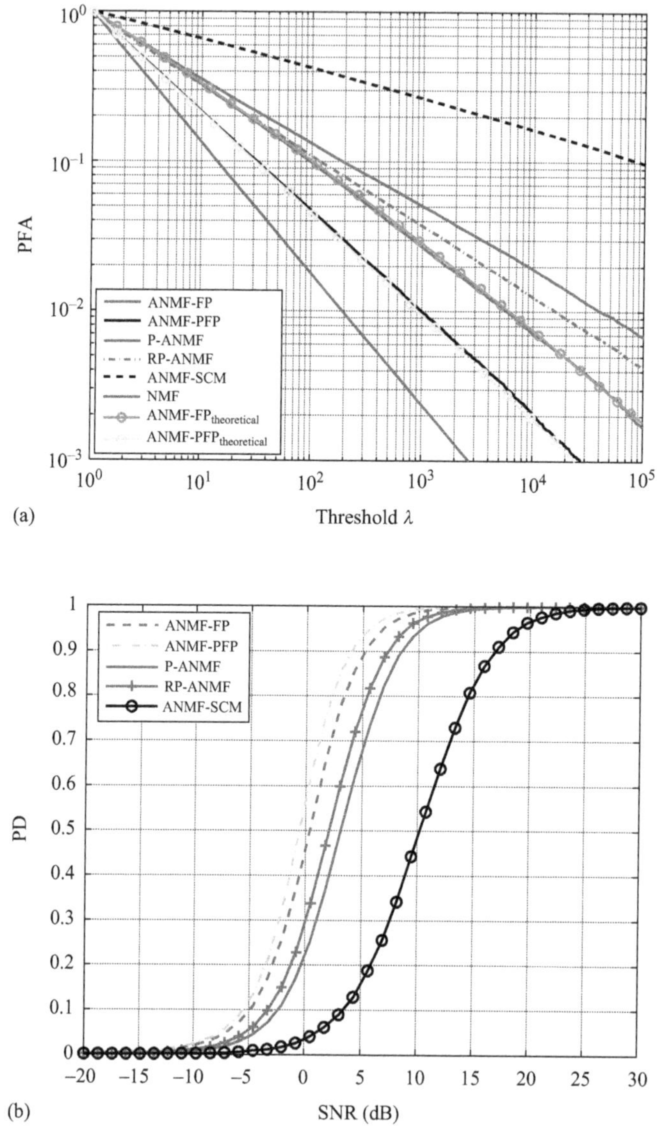

Figure 8.5 Comparison between radar performance for various non-Gaussian detectors in simulated K-distributed clutter characterized by its parameter ν with $P_{fa} = 10^{-3}$, $m = 8, K = 16$, and $\nu = 0.2$. (a) Theoretical and experimental PFA–threshold curves for various non-Gaussian detectors. (b) Probability of detection versus SNR for various non-Gaussian detectors.

radioelectric horizon of the radar (around 15 km) only homogeneous Gaussian thermal noise (the dark part of the map) is present (Figure 8.6). The analysis of these radar data allows to adjust the detection threshold λ for a given PFA. Traditionally, the experimental detection threshold adjustment is determined by counting, by moving a rectangular CFAR-sliding window of size

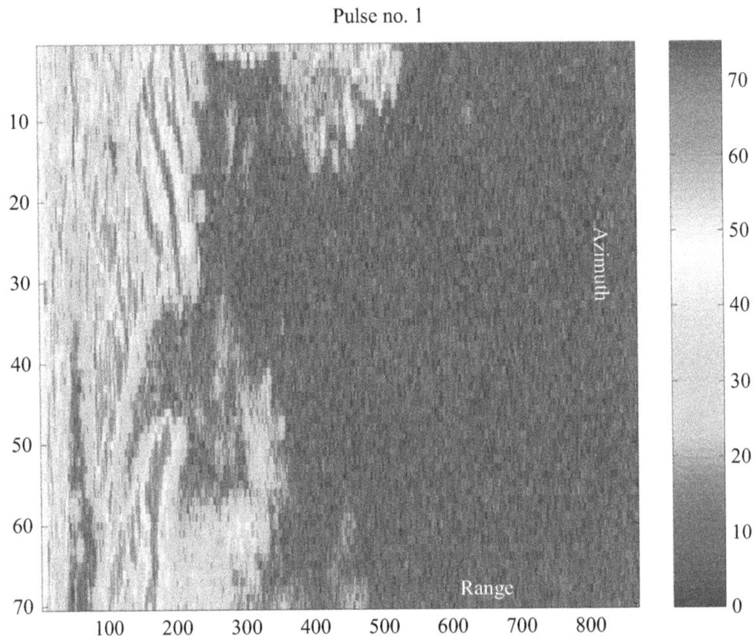

Figure 8.6 Ground clutter data level (in dB) corresponding to the first pulse.

5×5. For all central cells of the mask (i.e. the cell under test) corresponding to the studied observation $\boldsymbol{y}$ (8-vector), a value of $\Lambda(\widehat{\boldsymbol{M}})$ is calculated. The covariance matrix $\widehat{\boldsymbol{M}}$ has so been estimated with the set of $K = 24$ vectors, considered as the secondary data, $\boldsymbol{y}_1, \ldots, \boldsymbol{y}_{24}$, and situated around the tested cell.

To see the CFAR losses, we plotted the PFA–threshold relationship for a perfectly known covariance matrix ($\lambda = 1 - P_{fa}^{\frac{1}{m-1}}$). Notice that this equation has just a theoretical interest because in practice, $\boldsymbol{M}$ is always unknown. It can be used as benchmark for all the adaptive detectors.

On the left of Figure 8.7, the solid curve corresponds to the theoretical relationship "PFA–threshold" if $\boldsymbol{M}$ is known while the dotted curve represents the theoretical relationship "PFA–threshold" when $\boldsymbol{M}$ is assumed unknown and estimated by $\widehat{\boldsymbol{M}}_{FP}$. The curve made of crosses ($\times$) represents the experimental (made with CFAR masks by counting) relationship "PFA–threshold" when $\boldsymbol{M}$ is estimated by $\widehat{\boldsymbol{M}}_{FP}$. It perfectly matches the theoretical relationship. Obtaining this result has been possible only because the detector $\Lambda(\widehat{\boldsymbol{M}}_{FP})$ satisfies the M-CFAR property, essential in an heterogeneous clutter. An essential consequence of this result is that thanks to (8.50), the clutter training is not essential any more for the adjustment of the detection threshold.

8.6.2 Nostradamus Radar Detection

Similar analyses performed on experimental sea clutter data give the same conclusion. Figure 8.8 shows the sea clutter signal (range bins versus pulse repetition interval and range-Doppler) collected by the operational Over The Horizon radar from the French Aerospace Lab (ONERA) illumining the Atlantic ocean, and its associated range-Doppler image. In this

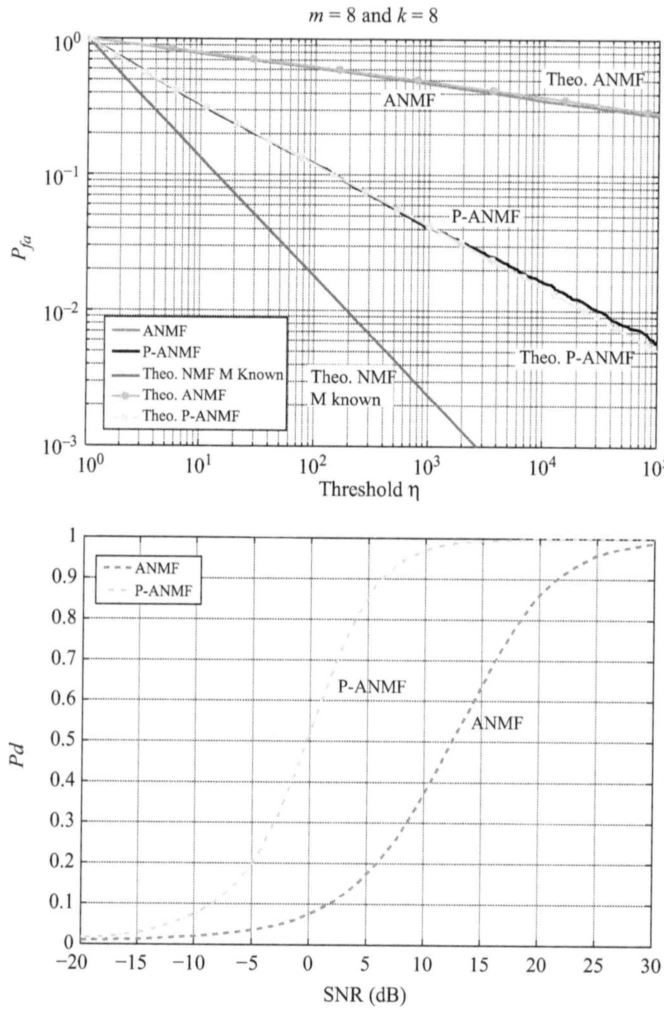

Figure 8.7 Detection performance on THALES data for $\Lambda(\boldsymbol{M}_{FP})$ and $\Lambda(\boldsymbol{R}_{PFP})$ for $P_{fa} = 10^{-2}$, $m = 8$, and $K = 8$. Top: PFA–threshold relationship $\eta = (1 - \lambda)^{-m}$ on THALES data. Bottom: Probability of detection on THALES data.

context, we use a set of $m = 8$ pulses of the signal on the entire range bins group and $K = 16$ reference range bins to estimate the covariance matrix. Figure 8.9 shows the improvement in detection performance on these data and the agreement between theoretical (circle line) and practical (solid line) results.

8.6.3 STAP Detection

Space Time Adaptive Processing (STAP) is a recent technique used in airborne phased array radar to detect moving target embedded in an interference background such as jamming or strong clutter. STAP is a two-dimensional adaptive filtering technique which uses jointly temporal and spatial dimensions to suppress interference and to improve target detection.

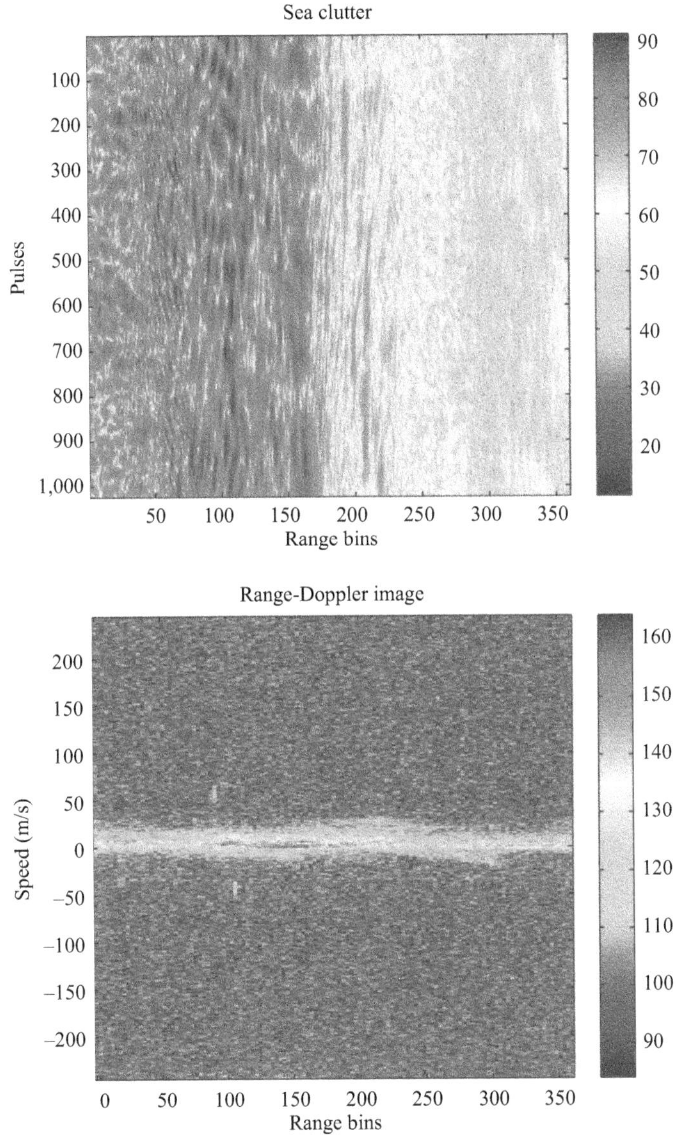

Figure 8.8 Atlantic Ocean sea clutter data collected by ONERA Over The Horizon Radar Nostradamus. Top: Sea clutter data from an over the horizon radar. Bottom: Range-Doppler image of Atlantic Ocean sea clutter.

It can be shown that the STAP algorithm is classically based on the collected data whitening operation when the interference is a Gaussian process (AMF). In that case, the covariance matrix of the noise is often estimated by the classical SCM built from the secondary data collected around the range bin under test. When the noise is non-Gaussian or heterogeneous, detection performance of the AMF-SCM significantly decreases. This section is devoted to the

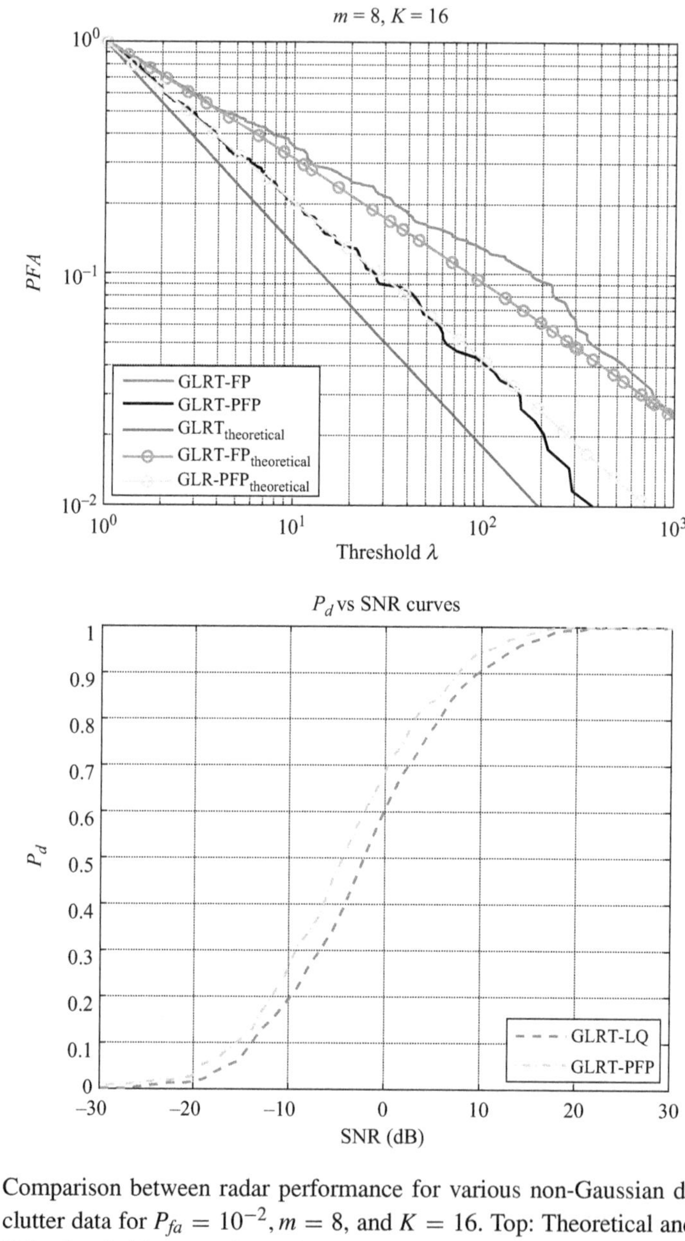

Figure 8.9 Comparison between radar performance for various non-Gaussian detectors in sea clutter data for $P_{fa} = 10^{-2}, m = 8$, and $K = 16$. Top: Theoretical and experimental PFA–threshold curves for various non-Gaussian detectors. Bottom: Probability of detection versus SNR for the ANMF-FP and ANMF-PFP.

improvement of STAP detection performance in a non-Gaussian or non-homogeneous background. Without loss of generality, the problem addressed here below is focused on the GMTI (Ground Moving Target Indicator) case, i.e. when the antenna is looking to the ground.

Let us now consider a source with amplitude A_0, located at azimuth angle θ_0 and in a range gate j with Doppler $f_{d,0}$, the STAP detection problem addressed here consists in deciding between two hypotheses as presented in (8.1). Difference with classical detection is that the so-called *steering vector* contains both temporal and spatial information as it results in a Kronecker product. According to previous discussions, various detectors will be tested on the STAP dataset: AMF with SCM (AMF-SCM), PS-AMF (exploiting the persymmetric structure), ANMF with the FPE (ANMF-FP), and PFP-ANMF.

The STAP data presented here are provided by the DGA/CELAR's simulator that allows to synthetize, in side looking configuration, STAP datacubes from very high-resolution Synthetic Aperture Radar (SAR) collected by ONERA SETHI system. The number of uniform linear array sensors is $L = 4$, and the number of coherent pulses is $M = 64$. The center frequency and the bandwidth are, respectively, equal to $f_0 = 10\,\text{GHz}$ and $B = 5\,\text{MHz}$. The radar velocity is given by $V = 100$ m/s. The inter-element spacing is $d = 0.3$ m and the pulse repetition frequency is $f_r = 1$ kHz. The number of secondary data used to estimate the covariance matrix of size $m = M\,L = 256$ is here $K = 410$ (under the Brennan's rule $K < 2M\,L$) for all the presented results.

The set of data contains ten targets at the range bin 255, with speeds from −4 m/s to 4 m/s. Figures 8.10 and 8.11 refer respectively to AMF-SCM, PS-AMF, ANMF-FP, and PFP-ANMF. Firstly, it may be noted that the AMF-SCM yields very poor results. Secondly, the best result is again provided by the PFP-ANMF which takes into account both the non-Gaussianity of the clutter and the persymmetry structure of the covariance matrix.

The experimental data exploited in this work lead to two main results. Firstly, non-Gaussian detectors based on the ANMF and the FPE outperform conventional Gaussian detectors based on the AMF and the SCM. Secondly, exploiting the persymmetric structure of the covariance matrix yields an additional improvement in terms of detection. This makes the PFP-ANMF an interesting detector for STAP radars.

8.6.4 Robustness of the FPE

To conclude this section, let us present some criteria that allow to quantify the notion of robustness. This is of interest for radar applications with some mismodeling, some missing data, or contaminated data. Standard covariance matrix estimation processes can be very affected by either the presence of outliers in the data or some mismatch on their statistical model. In the SIRV framework, it is possible to analyze the robustness of the SCM, the NSCM, and the FPE in disturbances context by deriving the theoretical bias arising from random disturbances $(\boldsymbol{a}_k)_{k\in\{1,\ldots,P\}}$, with $P < K/2$, in the data $(\boldsymbol{y}_k)_{k\in\{1,\ldots,K\}}$ used to build these estimates. Thus, $\boldsymbol{y}_1, \ldots, \boldsymbol{y}_K$ may be split into two sets:

$$\begin{cases} \boldsymbol{y}_k = \boldsymbol{a}_k & \text{for } 1 \leq k \leq P, \\ \boldsymbol{y}_k = \boldsymbol{c}_k & \text{for } P < k \leq K. \end{cases} \tag{8.70}$$

The superiority of these two estimates is then highlighted when $(\boldsymbol{c}_k)_{k\in\{P+1,\ldots,K\}}$ are Gaussian distributed and corrupted by some P strong random disturbances. This robustness can be of any help for applications such as adaptive radar detection or sources localization methods.

8.6.4.1 SCM Bias Analysis with Disturbances

The difference $\boldsymbol{\Delta}_{SCM}$ between statistical expectation of the contaminated covariance matrix and the bias of the non-contaminated covariance matrix $\mathbb{E}[\widehat{\boldsymbol{M}}_{\text{SCM}}] = \boldsymbol{M}$ can be written as:

$$\boldsymbol{\Delta}_{SCM} = \frac{1}{K} \sum_{k=1}^{P} \mathbb{E}\left[\boldsymbol{a}_k\,\boldsymbol{a}_k^{\dagger}\right] - \frac{P}{K}\boldsymbol{M}.$$

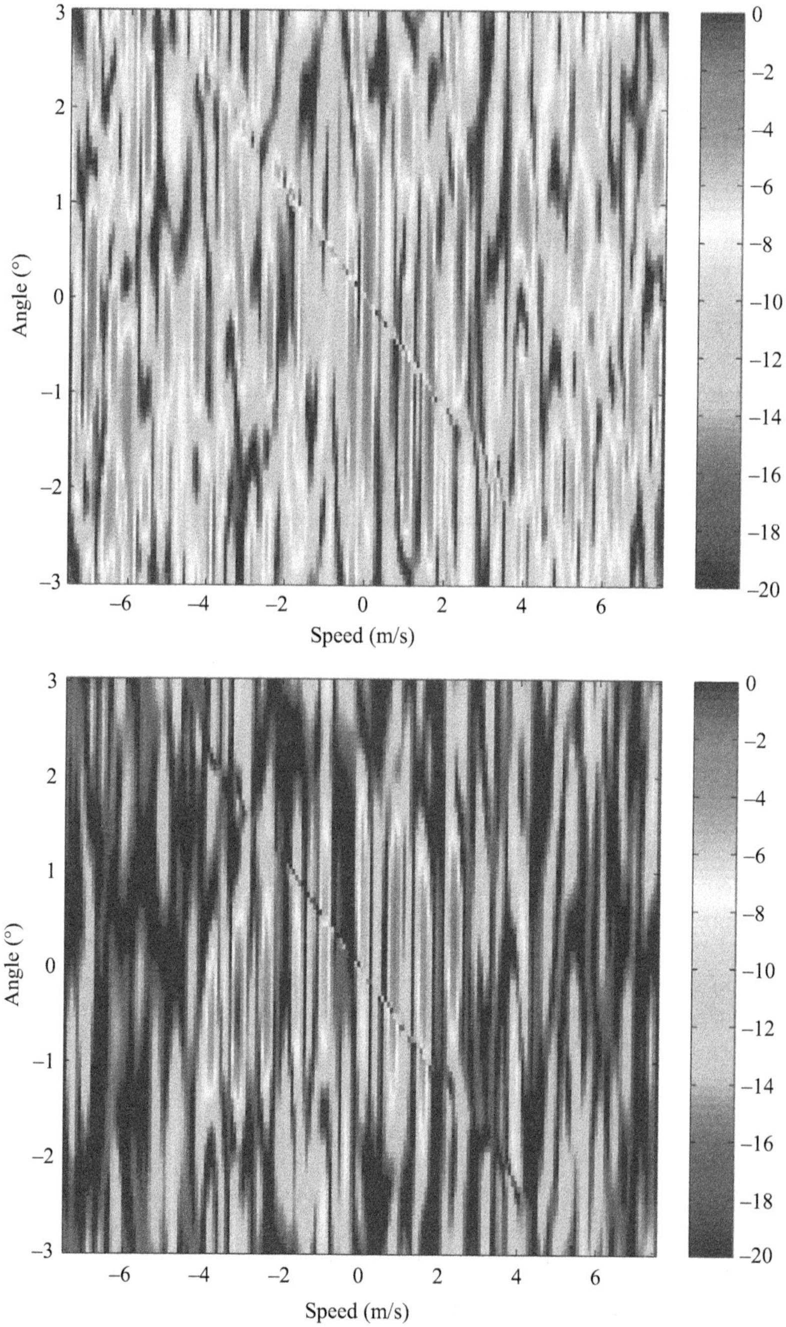

Figure 8.10 AMF Doppler-azimuth detection results in a range bin containing ten targets with different velocities. Top: AMF with SCM. Bottom: PS-AMF.

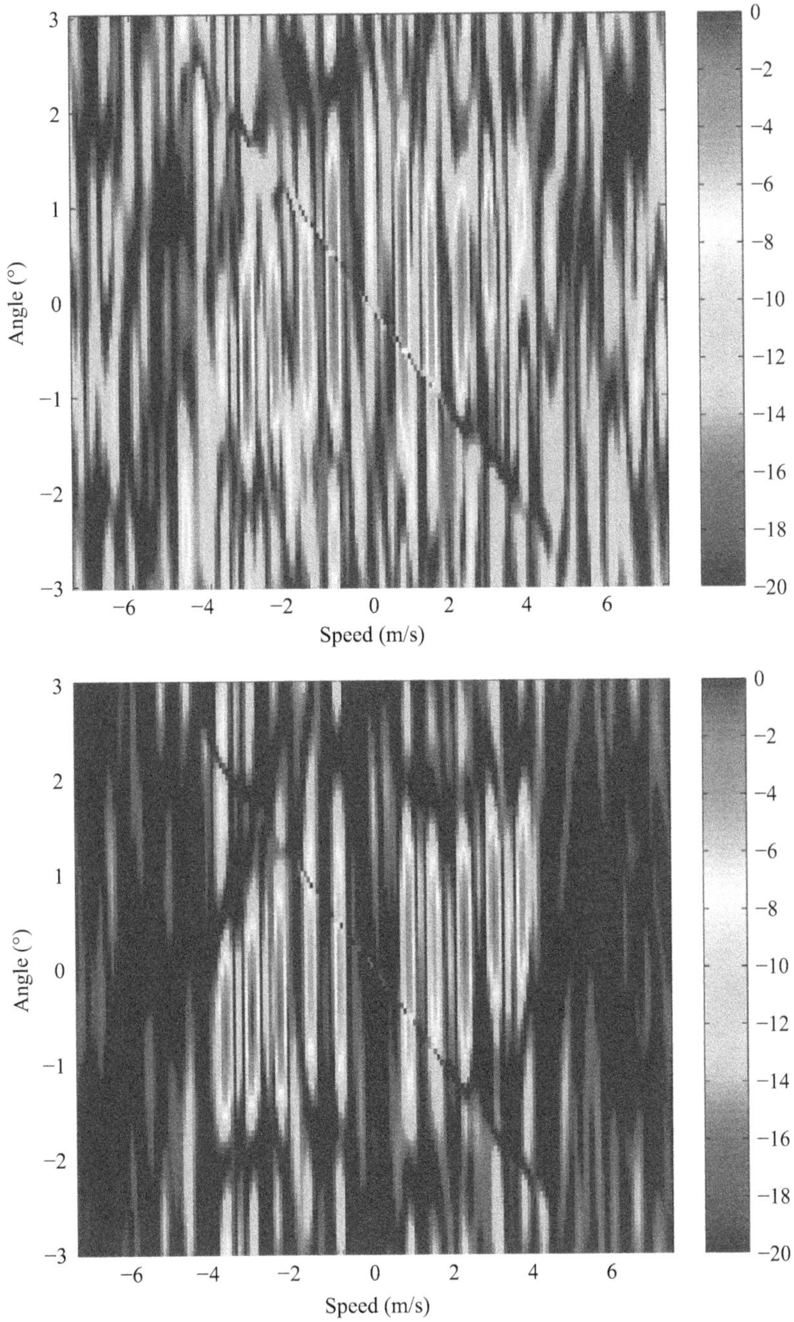

Figure 8.11 ANMF Doppler-azimuth detection results in a range bin containing ten targets with different velocities. Top: ANMF with FPE. Bottom: PFP-ANMF.

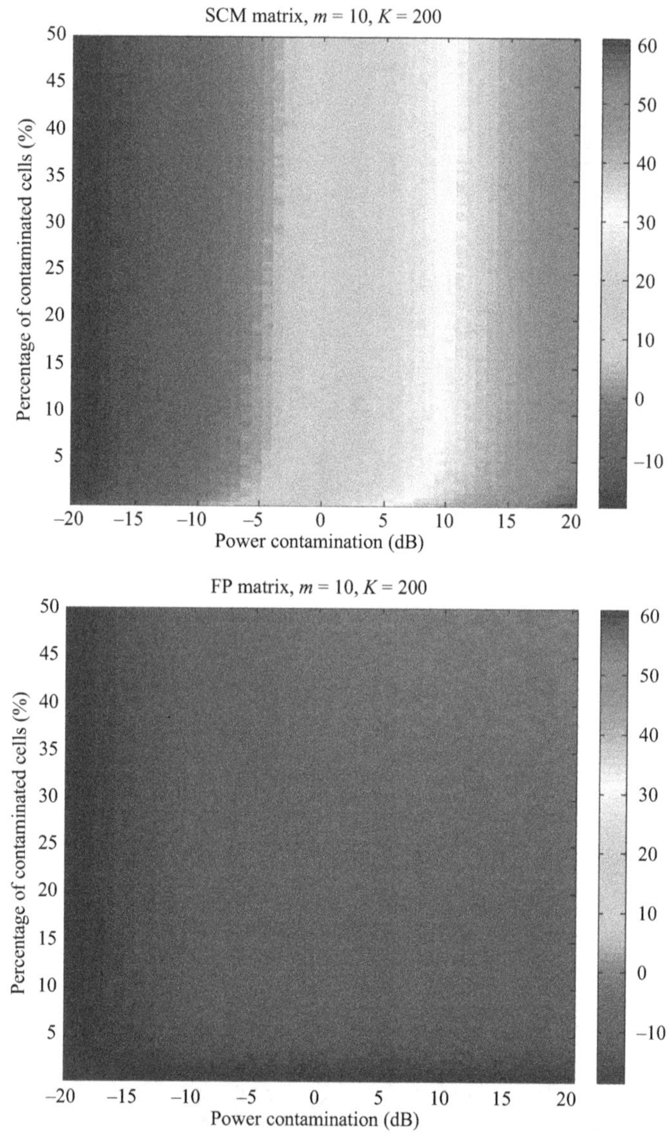

Figure 8.12 SCM (top) and FP (bottom) relative Frobenius norm errors (gray scale) versus disturbances power (x-axis) and percentage P/K of contaminated observations (y-axis) in Gaussian noise.

It can be noted that stronger the disturbances norm $||\boldsymbol{a}_k||$ is, more important the bias is. The SCM covariance matrix is not robust.

8.6.4.2 NSCM Bias Analysis with Disturbances

The difference $\boldsymbol{\Delta}_{NSCM}$ between statistical expectation of the contaminated covariance matrix and the bias of the non-contaminated covariance matrix $\mathbb{E}[\widehat{\boldsymbol{M}}_{\text{NSCM}}]$ ($\neq \boldsymbol{M}$ because NSCM estimate is biased) can be written as:

$$\boldsymbol{\Delta}_{NSCM} = -\frac{P}{K}\boldsymbol{M} + \frac{m}{K}\sum_{k=1}^{P}\mathbb{E}\left[\frac{\boldsymbol{a}_k\,\boldsymbol{a}_k^{\dagger}}{\boldsymbol{a}_k^{\dagger}\,\boldsymbol{a}_k}\right].$$

8.6.4.3 FPE Bias Analysis with Disturbances

In that case, the bias $\boldsymbol{\Delta}_{FP} = \mathbb{E}[\widehat{\boldsymbol{M}}_{\text{FP}}] - \boldsymbol{M}$ of the contaminated FP covariance matrix can be found to be

$$\boldsymbol{\Delta}_{FP} = \frac{m+1}{K}\left(\sum_{k=1}^{P}\mathbb{E}\left[\frac{\boldsymbol{a}_k\,\boldsymbol{a}_k^{\dagger}}{\boldsymbol{a}_k^{\dagger}\,\boldsymbol{M}^{-1}\,\boldsymbol{a}_k}\right] - \frac{P}{m}\boldsymbol{M}\right).$$

In the two last previous cases, it can be noted that the strength $||\boldsymbol{a}_k||$ of the disturbance does not influence the bias results. Every scaled factor on the vector $\{\boldsymbol{a}_k\}$ leaves the bias invariant. The NSCM and FPE are therefore said to be robust to disturbances.

To analyze the robustness of these estimates, some simulations are presented in the contaminated Gaussian context [70]. For all figures, the relative error (Frobenius norm) between contaminated and non-contaminated covariance matrix estimates is represented (in dB) for different percentage numbers P/K (y-axis) and different power level (x-axis) of the disturbances. Note that for a given percentage of contamination, the total power is shared between all the disturbances. Results on Figure 8.12 demonstrate the robustness of the FPE while the SCM estimate is strongly degraded by the disturbances.

8.7 Conclusion

As developed in this chapter, the detection performances are strongly linked to the covariance matrix estimation process. Several estimation methods have been studied through the statistical properties of the estimators. Then, they have been used in various detection problems on simulated data and real datasets. These results have enlightened the interest of using advanced estimation methods. Notice that there was not an exhaustive presentation of the different covariance matrix estimation approaches. Recently, to tackle the problem of few secondary data, as well as to deal with robustness matters, improved regularization techniques have been introduced [71–74]. One can also mentioned the promising framework of the Random Matrix Theory with some recent results in robust covariance matrix estimation for signal processing applications [45, 75, 76].

References

[1] N. R. Goodman, "Statistical analysis based on a certain multivariate complex Gaussian distribution (an introduction)," *The Annals of Mathematical Statistics*, vol. 34, no. 1, pp. 152–177, 1963.

[2] M. Bilodeau and D. Brenner, *Theory of Multivariate Statistics*. Springer Verlag New York, Inc., NY, 1999.

[3] E. J. Kelly, "An adaptive detection algorithm," *Aerospace and Electronic Systems, IEEE Transactions on*, vol. 23, no. 1, pp. 115–127, November 1986.

[4] S. Kraut, L. L. Scharf, and L. T. Mc Whorter, "Adaptive subspace detector," *Signal Processing, IEEE Transactions on*, vol. 49, pp. 1–16, January 2001.

[5] F. C. Robey, D. Fuhrmann, E. J. Kelly, and R. Nitzberg, "A CFAR adaptive matched filter detector," *Aerospace and Electronic Systems, IEEE Transactions on*, vol. 23, pp. 208–216, January 1992.

[6] M. Abramowitz and I. Stegun, *Handbook of Mathematical Functions*. National Bureau of Standard, Applied Mathematics Series – 55, US Government Printing Office, Washington, DC, June 1964.

[7] S. Kraut and L. L. Scharf, "The CFAR adaptive subspace detector is a scale-invariant GLRT," *Signal Processing, IEEE Transactions on*, vol. 47, no. 9, pp. 2538–2541, 1999.

[8] E. Conte, M. Lops, and G. Ricci, "Asymptotically optimum radar detection in compound-Gaussian clutter," *Aerospace and Electronic Systems, IEEE Transactions on*, vol. 31, pp. 617–625, April 1995.

[9] L. L. Scharf and D. W. Lytle, "Signal detection in Gaussian noise of unknown level: An invariance application," *Information Theory, IEEE Transactions on*, vol. 17, pp. 404–411, July 1971.

[10] L. L. Scharf and B. Friedlander, "Matched subspace detectors," *Signal Processing, IEEE Transactions on*, vol. 42, no. 8, pp. 2146–2157, 1994.

[11] V. A. Korado, "Optimum detection of signals with random parameters against the background of noise of unknown intensity under conditions of constant false alarm probability," *Radio Engineering and Electronic Physics*, vol. 13, pp. 969–972, 1968.

[12] B. Picinbono and G. Vezzosi, "Détection d'un signal certain dans un bruit non stationnaire et non gaussien," *Annales des Télécommunications*, vol. 25, pp. 433–439, 1970.

[13] E. Conte, E. Lops, and G. Ricci, "Asymptotically optimum radar detection in compound-Gaussian clutter," *Aerospace and Electronic Systems, IEEE Transactions on*, vol. 31, no. 2, pp. 617–625, 1995.

[14] J. B. Billingsley, *Low-Angle Radar Land Clutter. Measurements and Empirical Models*. William Andrew Publishing, Norwich, NY, 2002.

[15] J. B. Billingsley, "Ground clutter measurement for surface sited radar," Tech. Rep., February 1993.

[16] J. K. Jao, "Amplitude distribution of composite terrain radar clutter and the K-distribution," *Antennas and Propagation, IEEE Transactions on*, vol. 32, no. 10, pp. 1049–1062, October 1984.

[17] D. Kelker, "Distribution theory of spherical distributions and a location-scale parameter generalization," *Sankhyā: The Indian Journal of Statistics, Series A*, vol. 32, no. 4, pp. 419–430, 1970.

[18] P. R. Krishnaiah and J. Lin, "Complex elliptically symmetric distributions," *Communications in Statistics – Theory and Methods*, vol. 15, no. 12, pp. 3693–3718, 1986.

[19] E. Ollila, D. E. Tyler, V. Koivunen, and H. V. Poor, "Complex elliptically symmetric distributions: Survey, new results and applications," *Signal Processing, IEEE Transactions on*, vol. 60, no. 11, pp. 5597–5625, November 2012.

[20] G. Frahm, "Generalized elliptical distributions: Theory and applications," Ph.D. dissertation, Universität zu Köln, 2004.

[21] E. Conte, M. Longo, and M. Lops, "Modelling and simulation of non-Rayleigh radar clutter," *Radar and Signal Processing, IEE Proceedings F*, vol. 138, no. 2. IET, pp. 121–130, 1991.

[22] K. Yao, "A representation theorem and its applications to spherically invariant random processes," *Information Theory, IEEE Transactions on*, vol. 19, pp. 600–608, September 1973.

[23] M. Rangaswamy, D. D. Weiner, and A. Ozturk, "Non-Gaussian random vector identification using spherically invariant random processes," *Aerospace and Electronic Systems, IEEE Transactions on*, vol. 29, no. 1, pp. 111–124, 1993.

[24] K. Yao, M. K. Simon, and E. Bigiieri, "Unified theory on wireless communication fading statistics based on SIRP," in *Signal Processing Advances in Wireless Communications, 2004 IEEE 5th Workshop on*, Lisbon, Portugal, pp. 135–139, 2004.

[25] F. Gini, "Sub-optimum coherent radar detection in a mixture of K-distributed and gaussian clutter," *Radar, Sonar and Navigation, IEE Proceedings*, vol. 144, no. 1, pp. 39–48, February 1997.

[26] S. Watts, "Radar detection prediction in sea clutter using the compound K-distribution model," *Communications, Radar and Signal Processing, IEE Proceedings F*, vol. 132, no. 7, pp. 613–620, December 1985.

[27] E. Conte, M. Longo, M. Lops, and S. L. Ullo, "Radar detection of signals with unknown parameters in K-distributed clutter," *Radar and Signal Processing, IEE Proceedings F*, vol. 138, no. 2, pp. 131–138, April 1991.

[28] F. Gini, M. V. Greco, A. Farina, and P. Lombardo, "Optimum and mismatched detection against K-distributed plus Gaussian clutter," *Aerospace and Electronic Systems, IEEE Transactions on*, vol. 34, no. 3, pp. 860–876, July 1998.

[29] D. Manolakis and D. Marden, "Non Gaussian models for hyperspectral algorithm design and assessment," in *Geoscience and Remote Sensing Symposium (IGARSS), 2002 IEEE International*, vol. 3. IEEE, Toronto, Canada, pp. 1664–1666, 2002.

[30] J.-P. Ovarlez, S. Pang, F. Pascal, V. Achard, and T. Ng, "Robust detection using the SIRV background modelling for hyperspectral imaging," in *Geoscience and Remote Sensing Symposium (IGARSS), 2011 IEEE International*. IEEE, Vancouver, Canada, pp. 4316–4319, 2011.

[31] J. Frontera-Pons, M. Mahot, J.-P. Ovarlez, F. Pascal, S. Pang, and J. Chanussot, "A class of robust estimates for detection in hyperspectral images using elliptical distributions background," in *Geoscience and Remote Sensing Symposium (IGARSS), 2012 IEEE International*. IEEE, Munich, Germany, pp. 4166–4169, 2012.

[32] F. Gini, "Sub-optimum coherent radar detection in a mixture of K-distributed and Gaussian clutter," *Radar, Sonar and Navigation, IEE Proceedings*, vol. 144, pp. 39–48, February 1997.

[33] P. J. Huber and E. M. Ronchetti, *Robust Statistics*. John Wiley & Sons, Inc., Hoboken, NJ, USA, 2009.

[34] F. R. Hampel, E. M. Ronchetti, P. J. Rousseeuw, and W. A. Stahel, *Robust Statistics: The Approach Based on Influence Functions*, Wiley Series in Probability and Statistics, John Wiley & Sons, New York, NY, 1986.

[35] R. A. Maronna, D. R. Martin, and J. V. Yohai, *Robust Statistics: Theory and Methods*, Wiley Series in Probability and Statistics. John Wiley & Sons Ltd., Chichester, England, 2006.

[36] D. E. Tyler, "A distribution-free m-estimator of multivariate scatter," *The Annals of Statistics*, vol. 15, no. 1, pp. 234–251, 1987.

[37] P. J. Huber, "Robust estimation of a location parameter," *The Annals of Mathematical Statistics*, vol. 35, no. 1, pp. 73–101, 1964.

[38] R. A. Maronna, "Robust M-estimators of multivariate location and scatter," *Annals of Statistics*, vol. 4, no. 1, pp. 51–67, January 1976.

[39] D. E. Tyler, "Radial estimates and the test for sphericity," *Biometrika*, vol. 69, no. 2, p. 429, 1982.

[40] F. Pascal, Y. Chitour, J.-P. Ovarlez, P. Forster, and P. Larzabal, "Covariance structure maximum likelihood estimates in compound Gaussian noise: Existence and algorithm analysis," *Signal Processing, IEEE Transactions on*, vol. 56, no. 1, pp. 34–48, January 2008.

[41] E. Ollila and V. Koivunen, "Influence function and asymptotic efficiency of scatter matrix based array processors: Case MVDR beamformer," *Signal Processing, IEEE Transactions on*, vol. 57, no. 1, pp. 247–259, 2009.

[42] E. Ollila and V. Koivunen, "Robust antenna array processing using M-estimators of pseudo-covariance," in *Personal, Indoor and Mobile Radio Communications, 2003. PIMRC 2003. 14th IEEE Proceedings on*, vol. 3. IEEE, Beijing, China, pp. 2659–2663, 2003.

[43] E. Ollila and V. Koivunen, "Influence functions for array covariance matrix estimators," in *Proceedings IEEE Workshop on Statistical Signal Processing (SSP)*. IEEE, St. Louis, MO, USA, pp. 445–448, October 2003.

[44] E. Ollila, L. Quattropani, and V. Koivunen, "Robust space-time scatter matrix estimator for broadband antenna arrays," in *Vehicular Technology Conference, 2003. VTC 2003-Fall. 2003 IEEE 58th*, vol. 1, Orlando, FL, USA, pp. 55–59, October 2003.

[45] R. Couillet, F. Pascal, and J. Silverstein, "Robust m-estimation for array processing: A random matrix approach," *Information Theory, IEEE Transactions on*, vol. 60, no. 11, pp. 7269–7278, 2014.

[46] M. Mahot, F. Pascal, P. Forster, and J.-P. Ovarlez, "Asymptotic properties of robust covariance matrix estimates," *Signal Processing, IEEE Transactions on*, vol. 61, no. 13, pp. 3348–3356, 2013.

[47] D. E. Tyler, "Some results on the existence, uniqueness, and computation of the M-estimates of multivariate location and scatter," *SIAM Journal on Scientific and Statistical Computing*, vol. 9, p. 354, 1988.

[48] J. T. Kent and D. E. Tyler, "Redescending M-estimates of multivariate location and scatter," *Annals of Statistics*, vol. 19, no. 4, pp. 2102–2119, December 1991.

[49] F. Pascal, Y. Chitour, J.-P. Ovarlez, P. Forster, and P. Larzabal, "Covariance structure maximum likelihood estimates in compound Gaussian noise: Existence and algorithm analysis," *Signal Processing, IEEE Transactions on*, vol. 56, pp. 34–38, January 2008.

[50] D. Tyler, "Robustness and efficiency properties of scatter matrices," *Biometrika*, vol. 70, no. 2, p. 411, 1983.

[51] S. Bausson, F. Pascal, P. Forster, J.-P. Ovarlez, and P. Larzabal, "First and second order moments of the normalized sample covariance matrix of spherically invariant random vectors," *IEEE Signal Processing Letters*, vol. 14, no. 6, pp. 425–428, June 2007.

[52] K. J. Sangston, F. Gini, M. V. Greco, and A. Farina, "Structures for radar detection in compound Gaussian clutter," *Aerospace and Electronic Systems, IEEE Transactions on*, vol. 35, no. 2, pp. 445–458, April 1999.

[53] F. Gini and M. V. Greco, "Covariance matrix estimation for CFAR detection in correlated heavy tailed clutter," *Signal Processing, special section on Signal Processing with Heavy Tailed Distributions*, vol. 82, pp. 1847–1859, December 2002.

[54] E. Conte, A. De Maio, and G. Ricci, "Recursive estimation of the covariance matrix of a compound-Gaussian process and its application to adaptive CFAR detection," *Signal Processing, IEEE Transactions on*, vol. 50, pp. 1908–1915, August 2002.

[55] F. Pascal, P. Forster, J.-P. Ovarlez, and P. Larzabal, "Performance analysis of covariance matrix estimates in impulsive noise," *Signal Processing, IEEE Transactions on*, vol. 56, no. 6, pp. 2206–2217, June 2008.

[56] E. Jay, J.-P. Ovarlez, D. Declercq, and P. Duvaut, "BORD: Bayesian optimum radar detector," *Signal Processing*, vol. 83, no. 6, pp. 1151–1162, June 2003.

[57] E. Jay, "Détection en environnement non-gaussien," Ph.D. dissertation, University of Cergy-Pontoise/ONERA, France, June 2002.

[58] L. L. Scharf and L. T. McWhorter, "Adaptive matched subspace detectors and adaptive coherence estimators," in *Proceedings of the 30th Asilomar Conference on Signals, Systems, and Computers*, vol. 2, pp. 1114–1117, November 1996.

[59] F. Pascal, J.-P. Ovarlez, P. Forster, and P. Larzabal, "Constant false alarm rate detection in spherically invariant random processes," in *European Signal Processing Conference, EUSIPCO'04*, Vienna, Austria, pp. 2143–2146, September 2004.

[60] J. P. Burg, D. G. Luenberger, and D. L. Wenger, "Estimation of structured covariance matrices," *Proceedings of the IEEE*, vol. 70, no. 9, pp. 963–974, September 1982.

[61] D. R. Fuhrmann, "Application of Toeplitz covariance estimation to adaptive beamforming and detection," *Signal Processing, IEEE Transactions on*, vol. 39, pp. 2194–2198, October 1991.

[62] R. Nitzberg and J. R. Burke, "Application of maximum likelihood estimation of persymmetric covariance matrices to adaptive detection," *Aerospace and Electronic Systems, IEEE Transactions on*, vol. 25, pp. 124–127, January 1980.

[63] L. Cai and H. Wang, "A persymmetric multiband GLR algorithm," *Aerospace and Electronic Systems, IEEE Transactions on*, vol. 28, no. 3, pp. 806–816, July 1992.

[64] E. Conte and A. De Maio, "Exploiting persymmetry for CFAR detection in compound-Gaussian clutter," *Aerospace and Electronic Systems, IEEE Transactions on*, vol. 39, pp. 719–724, April 2003.

[65] E. Conte and A. De Maio, "Mitigation techniques for non-Gaussian sea clutter," *IEEE Journal of Oceanic Engineering*, vol. 29, pp. 284–302, December 2003.

[66] G. Pailloux, P. Forster, J.-P. Ovarlez, and F. Pascal, "On persymmetric covariance matrices in adaptive detection," in *IEEE International Conference on Acoustics, Speech, and Signal Processing, ICASSP-08*, pp. 2305–2308, April 2008.

[67] G. Pailloux, J.-P. Ovarlez, F. Pascal, and P. Forster, "A SIRV-CFAR adaptive detector exploiting persymmetric clutter covariance structure," in *IEEE Radar Conference'08*, Rome, Italy, pp. 1139–1144, May 2008.

[68] H. L. Van Trees, *Detection, Estimation and Modulation Theory, Part IV: Optimum Array Processing*. John Wiley & Sons, New York, NY, 2002.

[69] G. Pailloux, P. Forster, J.-P. Ovarlez, and F. Pascal, "Persymmetric adaptive radar detectors," *Aerospace and Electronic Systems, IEEE Transactions on*, vol. 47, no. 4, pp. 2376–2390, October 2011.

[70] M. Mahot, P. Forster, J.-P. Ovarlez, and F. Pascal, "Robustness analysis of covariance matrix estimates," in *European Signal Processing Conference, EUSIPCO'10*, Aalborg, Denmark, August 2010.

[71] Y. Chen, A. Wiesel, and A. O. Hero, "Robust shrinkage estimation of high-dimensional covariance matrices," *Signal Processing, IEEE Transactions on*, vol. 59, no. 9, pp. 4097–4107, 2011.

[72] Y. I. Abramovich and O. Besson, "Regularized covariance matrix estimation in complex elliptically symmetric distributions using the expected likelihood approach – Part 1: The over-sampled case," *Signal Processing, IEEE Transactions on*, vol. 61, no. 23, pp. 5807–5818, 2013.

[73] O. Besson and Y. I. Abramovich, "Regularized covariance matrix estimation in complex elliptically symmetric distributions using the expected likelihood approach – Part 2: The under-sampled case," *Signal Processing, IEEE Transactions on*, vol. 61, no. 23, pp. 5819–5829, 2013.

[74] F. Pascal, Y. Chitour, and Y. Quek, "Generalized robust shrinkage estimator and its application to STAP detection problem," *Signal Processing, IEEE Transactions on*, vol. 62, no. 21, pp. 5640–5661, 2014.

[75] R. Couillet, F. Pascal, and J. W. Silverstein, "A joint robust estimation and random matrix framework with application to array processing," in *IEEE International Conference on Acoustics, Speech, and Signal Processing, ICASSP-13*, Vancouver, Canada, vol. 31, pp. 6561–6565, May 2013.

[76] R. Couillet, F. Pascal, and J. W. Silverstein, "The random matrix regime of Maronna's M-estimator with elliptically distributed samples," *Journal of Multivariate Analysis*, vol. 139, no. C, pp. 56–78, 2015.

CHAPTER 9

Detection of Extended Target in Compound-Gaussian Clutter

Augusto Aubry[1], Javier Carretero-Moya[2], Antonio De Maio[1], Antonio Pauciullo[3], Javier Gismero-Menoyo[2], and Alberto Asensio-Lopez[2]

9.1 Introduction

It is a well-established fact that it is possible to achieve a significant performance improvement in maritime surveillance applications by deploying high-resolution radar systems. Recent technological breakthrough in key areas such as real-time digital signal processing hardware, or very high-bandwidth radio frequency (RF) front-ends, is now turning radar systems foreseen two decades ago [1, 2] into a practical reality.

The size reduction of the resolution cell is the straightforward technique to diminish disturbance returns in the very frequent distributed-clutter-limited detection scenario. However, this approach has at least two issues that must be accounted for in the radar design process to benefit from the potential performance improvement. First, if the range resolution goes beyond a certain specific threshold, targets will be over-resolved and their energy will be distributed across several adjacent range bins. Matching the range resolution as close as possible to the physical dimensions of the target to be detected is a possible way to circumvent this problem. However, several works have shown that properly designed detectors for targets distributed across several range cells, or *distributed targets*,[4] outperform lower resolution systems in which the target is

[1]Dipartimento di Ingegneria Elettrica e delle Tecnologie dell'Informazione, Università degli Studi di Napoli "Federico II", Napoli, Italy

[2]Departamento de Senales, Sistemas y Radiocomunicaciones, Escuela Tecnica Superior de Ingenieros de Telecomunicacion, Universidad Politecnica de Madrid, Madrid, Spain.

[3]CNR, IREA, Napoli, Italy.

[4]Also referred to as over-resolved, range-spread, or extended targets.

confined within a single range cell [3–5]. Radar backscatter from range-resolved target scatterers exhibits a less fluctuating behavior that may improve the detection performance [6]. In conventional low-resolution radar systems, the overall target radar cross section (RCS) is the vector sum of randomly phased contributions resulting in large power fluctuations of the signal received by the radar [1]. This has been traditionally characterized by the well-known Swerling target models [7], and it is the source of additional detection losses,[5] which are significantly reduced in high-resolution systems.

Second, high-resolution sea-clutter is characterized by a marked non-Gaussian behavior that makes the development of suitable detection schemes difficult [8]. The distinct statistical properties of high-resolution sea-clutter must be correctly taken into account to guarantee a satisfactory performance. Theoretical considerations [9–11] and the statistical analysis of experimental data [12–16] have shown the validity of the compound-Gaussian framework [17] for high-resolution sea-clutter modeling. Specifically, for short time intervals sea-clutter samples can be considered realizations from spherically invariant random vectors (SIRVs) [18–22]. The interested reader may refer to Chapter 7 of this book for a detailed discussion on the compound-Gaussian models.

In this chapter, experimental data from a Ka-band high-resolution radar system are used to evaluate the performance of several schemes for coherent detection of distributed targets. The importance of this analysis relies on the possibility of exploiting detectors for range-spread targets in modern applications for homeland security. A careful border control is crucial to avoid illegal immigration or drug trade by means of very small boats (such as inflatable or wooden boats). It is thus of great interest to assess the capability of a high-resolution radar system to detect such kind of small targets which, at the resolutions of the data available in this chapter (0.10 m, 0.20 m, and 1 m), appear range-spread.

The chapter is organized as follows. In Section II, a review concerning the state of art in range-distributed radar detection is presented and the detection strategies, whose performances are analyzed in the chapter, are introduced and commented. In Section III, the available clutter data and their main statistical features are described. Therein, some statistical analyses concerning target data are also presented, with emphasis on the capability of common amplitude distributions (usually adopted in radar literature to describe target fluctuations) to represent them. In Section IV, the constant false alarm rate (CFAR) behavior of the considered detectors is studied in correspondence of real clutter data at different sub-meter range resolutions. In Section V, the detection performance is assessed: first the case of a synthetic target injected in real clutter is considered and then the case of a real target (precisely an inflatable boat) embedded in real clutter is tackled. Finally, concluding remarks are provided in Section VI.

9.2 Distributed Target Coherent Detection

9.2.1 Overview

Early theoretical contributions to the topic of range-distributed target detection focused on non-coherent systems and date from well before the widespread development of the technologies required for its practical implementation [2, 6, 23–25]. Experimental evidence already showed

---

[5] With regard to the non-fluctuating case.

that the radar properties of common targets were well modeled as the result of the reflection from a few isolated points. The *few specular reflector* target model and its implications were analyzed by Nitzberg in 1978 [6]. In this reference, it was shown that a better performance is obtained when the radar bandwidth is such that the individual scatterers are just resolved and each becomes a non-fluctuating target. The important concept of *collapsing losses* was defined as the output signal-to-clutter power ratio (SCR) loss associated with the integration of target-/scatterer-free range cells in the detection process. Nitzberg's work was based on a radar target simulation model that consisted of a few equal magnitude and equispaced reflectors. This limitation was soon overcome. A number of different target models were used in Reference 2 to account for non-uniform scatterer configurations in terms of both spatial distribution and amplitude variations. Two high-resolution strategies were compared: an integrated detector, non-coherently adding the power of H_s adjacent range bins prior to thresholding, and a "1 out of H_s" binary detector operating on the thresholded high-resolution bins. Both approaches consistently outperformed the lower resolution counterpart. However, the relative performance of the high-resolution detectors was found to be heavily dependent on the target scattering characteristics, or more precisely, on the range distribution of the target specular or flare points. The integrated detector suffered from collapsing losses for those target models in which the energy of the target was heavily concentrated in range.

A priori knowledge of the spatial scattering density of the target can be incorporated into the detector design to minimize collapsing losses. Gerlach and Steiner [26] derived the single pulse generalized likelihood ratio test (GLRT) detector for a distributed target in Gaussian noise. The use of a specific scattering density function approximately equal to the binomial distribution was found to result in a particularly compact form for the GLRT. The *a priori* knowledge on the spatial scatterer distribution was controlled by a single parameter capable of modeling both extreme situations: a target with returns highly concentrated or evenly distributed in range. For the latter case, it is interesting to note that the GLRT developed becomes the integrated detector previously mentioned [2]. For other non-uniform but still *a priori* known scatterer distributions, the GLRT effectively suppresses the contribution of noise-only range cells in the formation of the detection statistic. As a consequence, the collapsing loss of the conventional integrated detector due to the sparse target structure is minimized. The performance of the GLRT single pulse detector was also found to be better than the binary integrator. However, the results assumed perfect knowledge of the parameters of the target spatial distribution.

A completely different approach was proposed in References 24 and 25 with the distinct feature of considering other scattering centers of the target as an additional source of non-homogeneity in the reference window of the detector. A temporal and spatial clutter map was modified to include a censoring step to eliminate the contamination introduced by other scattering centers of the target.

The interest of the radar community is now mainly focused on the development of distributed target detectors for coherent systems. However, certain non-coherent techniques have been recently proposed. An example is Reference 27 where the fact that sea-clutter spikes are spatially localized is exploited to discriminate between targets and sea-clutter returns through spatial extension.

It is convenient to classify coherent detection strategies for distributed targets in the categories shown in Table 9.1, based on both the target and the clutter models used in their design process, and which result in different detector forms. The detection techniques included in Table 9.1 are briefly described in what follows.

Table 9.1 Classification of distributed target detection schemes.

Target model	Gaussian clutter	Non-Gaussian clutter
Rank-one	One-step GLRT, two-step GLRT [3], MGLRT [29]	NSDD-GLRT, SDD-GLRT [28], adaptive GLRT [5], Rao test and Wald test [30], AMGLRT [31]
Subspace	CFAR tests [32]	DSM-GLRT, GSM-GLRT [4], GLRT, Rao test and Wald test [33]

9.2.2 Rank-One Steering

The problem of adaptive detection of partially known[6] distributed targets in Gaussian clutter with unknown covariance was addressed in Reference 3 relying on two different design procedures: a direct GLRT and a two-step GLRT. Following the latter, two computationally efficient GLRT structures were derived based on the N-dimensional vectors of *primary data* $(\boldsymbol{r}(1), \ldots, \boldsymbol{r}(H_s))$ (returns from H_s adjacent range cells characterized by the potential presence of target scatterers) and assuming that the covariance matrix of the disturbance (two-step GLRT) or its structure (modified two-step GLRT) were known. The adaptive detectors were obtained by substituting the unknown covariance matrix or covariance structure by its estimate based on a *secondary dataset* $(\boldsymbol{r}_1, \ldots, \boldsymbol{r}_K)$ or reference window free of target components $(K \geq N)$. The resultant detectors, under the design assumptions, share the CFAR behavior with respect to the unknown clutter parameters and a performance close (modified two-step GLRT) or even superior (two-step GLRT) than the direct, one-step GLRT when the actual target is matched to the nominal subspace. The results also clearly showed the detection gain associated with higher range resolutions.

Several works have developed suitable detection structures [5, 28, 30] for the non-Gaussian clutter scenario that is very appropriate for the statistical description of echoes backscattered from the sea surface. In Reference 28, Gerlach developed two detectors, the non-scatterer density-dependent GLRT (NSDD-GLRT) and the scatterer density-dependent GLRT (SDD-GLRT). The former, obtained under the assumption that the target scatterers are distributed completely across H_s range cells and whose derivation is reported in Appendix 9.A.1.1 for completeness, has the form:

$$-\sum_{t=1}^{H_s} \ln\left[1 - \frac{\left|\boldsymbol{p}^\dagger \boldsymbol{\Sigma}^{-1}\boldsymbol{r}(t)\right|^2}{(\boldsymbol{p}^\dagger \boldsymbol{\Sigma}^{-1}\boldsymbol{p})(\boldsymbol{r}(t)^\dagger \boldsymbol{\Sigma}^{-1}\boldsymbol{r}(t))}\right] \underset{H_0}{\overset{H_1}{\gtrless}} T_{NSDD} \tag{9.1}$$

where $\boldsymbol{p}$ is the N-dimensional complex vector representing[7] the unit norm temporal steering vector, whose n-th entry is

$$p(n) = \frac{1}{\sqrt{N}} \exp(j2\pi(n-1)f_d) \tag{9.2}$$

---

[6] Known up to a complex constant. A single and common steering vector for all range cells.

[7] Without loss of generality, in this chapter we assume a unit norm steering vector.

with f_d being the normalized Doppler frequency associated with all of the target scatterers and j the imaginary unit. $\boldsymbol{\Sigma}$ is the known structure of the clutter covariance matrix, and T_{NSDD} is a threshold chosen to guarantee a predetermined probability of false alarm (P_{fa}).

It should be noted that in the argument of the $\ln(\cdot)$ function appears the well-known normalized matched filter (NMF) for point-like target detection in compound-Gaussian clutter, which has been derived following different strategies [34–37]. In Reference 34, Conte *et al.* found that the asymptotic evaluation of the probability density function (PDF) of SIRV clutter, for increasingly high number of temporal samples, leads to a GLRT test whose decision statistic is independent of the clutter distribution. The NMF was then obtained by adopting that test even for the case of a finite number of integrated temporal samples. An alternative derivation was proposed by Gini [35], where it was shown that a detector of the same form could be obtained through a GLRT approach considering the texture as an unknown deterministic parameter. Additionally, Scharf [36] found that the NMF is the uniformly most powerful invariant (UMPI) test for detecting a target signal, known up to a multiplicative factor, in Gaussian noise. Finally, in a recent work, De Maio and Conte [37] have proved the UMPI claim to the non-Gaussian SIRV case provided that the texture PDF complies with a mild specific condition, usually satisfied by common practical texture distributions.

As the NSDD-GLRT takes the form of an integrator of identical functions of the data over a pre-specified number of H_s range cells, it suffers from collapsing losses for targets that are not distributed across all of the processed range bins. In the *few specular reflector* target model, only a fraction of the H_s range cells will be occupied by target scatterers, and therefore noise-only contributions to the value of the test statistic will degrade the detection performance. To account for this, and in a similar approach as in Reference 26, *a priori* knowledge about the spatial density of the target scatterers can be incorporated into the detector design resulting in the SDD-GLRT.

The NSDD-GLRT assumed perfect knowledge of the clutter covariance matrix structure $\boldsymbol{\Sigma}$. The development of an adaptive variant was tackled in Reference 5, by resorting to a secondary dataset to estimate the unknown clutter covariance structure. Additionally, the proposed detector also features the possibility of including *a priori* knowledge on the degree of range correlation of the clutter power, ranging from the totally independent case (as for the NSDD-GLRT detector) to an homogeneous scenario. The developed detector, under the design assumptions, ensures the CFAR property with respect to both the structure of the covariance matrix and the power backscattered from each range cell. It is interesting to note that the two-step detector developed ends up coincident with an adaptive version of the NSDD-GLRT for the case that the clutter power is independent from range cell to range cell. In a recent paper [38], the performance of the two-step adaptive GLRT has been compared with that of the optimum Neyman–Pearson detector for K-distributed clutter.

Similar assumptions on the clutter statistics are also made in Reference 30. Therein, two additional detection structures are proposed based on the Rao and Wald tests, as it is shown that no uniformly most powerful test exists for the detection problem under consideration. Both detectors are also CFAR with respect to the unknown statistics. As reported in Appendix 9.A.1.2, assuming again that the clutter power changes from range cell to range cell, the Rao test becomes

$$\sum_{t=1}^{H_s} \frac{\left|\boldsymbol{p}^{\dagger}\boldsymbol{\Sigma}^{-1}\boldsymbol{r}(t)\right|^2}{(\boldsymbol{p}^{\dagger}\boldsymbol{\Sigma}^{-1}\boldsymbol{p})(\boldsymbol{r}(t)^{\dagger}\boldsymbol{\Sigma}^{-1}\boldsymbol{r}(t))} \underset{H_0}{\overset{H_1}{\gtrless}} T_{Rao} \tag{9.3}$$

The equation shows that the Rao test is the arithmetic mean along the range dimension of the NMF test statistic [39]. Similarly, as reported in Appendix 9.A.1.3, the Wald test is

$$\sum_{t=1}^{H_s} \frac{\left|\boldsymbol{p}^{\dagger}\boldsymbol{\Sigma}^{-1}\boldsymbol{r}(t)\right|^2}{(\boldsymbol{p}^{\dagger}\boldsymbol{\Sigma}^{-1}\boldsymbol{p})\left[(\boldsymbol{r}(t)^{\dagger}\boldsymbol{\Sigma}^{-1}\boldsymbol{r}(t)) - \frac{\left|\boldsymbol{p}^{\dagger}\boldsymbol{\Sigma}^{-1}\boldsymbol{r}(t)\right|^2}{(\boldsymbol{p}^{\dagger}\boldsymbol{\Sigma}^{-1}\boldsymbol{p})}\right]} \underset{H_0}{\overset{H_1}{\gtrless}} T_{Wald} \tag{9.4}$$

and its performance, on simulated data, is reported to outperform both the two-step GLRT and the Rao test.

The previously mentioned detectors assume that the target signal is known within a complex scale factor. However, since the target Doppler is usually unknown, in a practical scenario the radar processor usually takes the form of a detector bank, in which each channel is matched to a different hypothetical target Doppler.

9.2.3 Subspace Steering

Robustness to partial uncertainty in the target Doppler response may be incorporated by specifying a multi-dimensional (subspace) target representation [40]. Otherwise stated, the signal to be detected is a deterministic N-dimensional vector that is known to belong to a subspace of dimension r (where $1 \leq r \leq N$). A signal subspace of dimension one represents the case where the signal to be detected is a known N-dimensional vector (precisely, known within a complex multiplicative constant). Similarly, a signal subspace of dimension N (i.e. $r = N$) represents the case where the signal to be detected is an unknown deterministic N-dimensional vector. It is possible to account for partial uncertainty with regard to the signal to be detected by using a value of r between 1 and N.

According to the linear subspace signal model, the target signal at the t-th range cell, $\mathbf{x}(t)$ say, can be modeled as the N-dimensional complex vector:

$$\boldsymbol{x}(t) = \boldsymbol{H}\boldsymbol{\theta}(t) \tag{9.5}$$

where $\boldsymbol{\theta}(t)$ is an $r \times 1$ unknown vector and $\boldsymbol{H}$ is a known $N \times r$ matrix frequently called steering or mode matrix [40–42]. The signal therefore belongs to the linear subspace spanned by the columns of the matrix $\boldsymbol{H}$.

In the context of coherent high-resolution radar detection, it is possible to use the subspace signal model to account for targets distributed both in Doppler and range. The columns of the mode matrix $\boldsymbol{H}$ are temporal steering vectors, complying with the structure (9.2), whose Doppler frequencies can be chosen according to the following criteria:

(1) Discretize the normalized Doppler interval $[-0.5, 0.5]$ into N bins.

(2) For a Doppler bin of interest (DBI), one column of $\boldsymbol{H}$ is tuned to the central Doppler of the DBI, while the other columns are tuned to the central frequencies of $k(h)$ adjacent Doppler bins preceding (following) the DBI. Additionally, $k + h + 1 = r$.

For practical values of N, namely $N \geq 8$, quite often r is chosen equal to 3, namely the column vectors of $\boldsymbol{H}$ are tuned to the central frequencies of the DBI and two adjacent (one preceding and one following) Doppler bins.

Scharf produced key contributions to the problem of detecting signals that comply with a subspace model in Gaussian clutter. In particular, the matched subspace detector (MSD) was developed for detecting an unknown deterministic signal known to lie in a specific subspace [43–46]. The approach was extended by Gini and Farina [40, 42, 47] for CFAR detection of a subspace signal against compound-Gaussian clutter with known covariance structure:

$$\frac{\boldsymbol{r}^{\dagger}\boldsymbol{Q}_2\boldsymbol{r}}{\boldsymbol{r}^{\dagger}\boldsymbol{\Sigma}^{-1}\boldsymbol{r}} \underset{H_0}{\overset{H_1}{\gtrless}} T \tag{9.6}$$

with

$$\boldsymbol{Q}_2 = \boldsymbol{\Sigma}^{-1}\boldsymbol{H}(\boldsymbol{H}^{\dagger}\boldsymbol{\Sigma}^{-1}\boldsymbol{H})^{-1}\boldsymbol{H}^{\dagger}\boldsymbol{\Sigma}^{-1} \tag{9.7}$$

where $\boldsymbol{r}$ denotes the data under test. It can be shown that $\boldsymbol{r}^{\dagger}\boldsymbol{Q}_2\boldsymbol{r}$ is proportional to the energy of the whitened received data vector component which lies within the whitened target signal subspace, whereas the quadratic form $\boldsymbol{r}^{\dagger}\boldsymbol{\Sigma}^{-1}\boldsymbol{r}$ is proportional to the overall energy in the whitened received signal space. Detector (9.6) is the generalized MSD or GMSD. For practical applications, $\boldsymbol{\Sigma}$ should be adaptively estimated [45, 48], in order to get a fully adaptive receiver. The interested reader may refer to Chapter 8 of this book for further details on covariance matrix estimation techniques in SIRV processes.

Subspace detectors have been recently extended to the distributed target scenario [4, 32, 33]. For the Gaussian disturbance case, Jin and Friedlander [32] developed a subspace decomposition-based CFAR structure for the detection of a range-distributed target modeled as a Gaussian vector. In Reference 4, two GLRT detectors are derived for compound-Gaussian clutter: the deterministic scatterer model GLRT (DSM-GLRT) and the Gaussian scatterer model GLRT (GSM-GLRT). In the former, the complex amplitude vector of the target model, i.e. $\boldsymbol{\theta}(t)$, is considered as an unknown deterministic value whereas in the latter $\boldsymbol{\theta}(t)$ is modeled as a complex Gaussian random vector with unknown covariance matrix. Actually, when the steering matrix is reduced to a steering vector, the developed DSM-GLRT reduces to Gerlach's NSDD-GLRT.

In this chapter, the following subspace detectors have been evaluated [33] for $1 \leq r < N$:

$$\sum_{t=1}^{H_s} \frac{\boldsymbol{r}(t)^{\dagger}\boldsymbol{Q}_2\boldsymbol{r}(t)}{\boldsymbol{r}(t)^{\dagger}\boldsymbol{\Sigma}^{-1}\boldsymbol{r}(t)} \underset{H_0}{\overset{H_1}{\gtrless}} T_{RAO} \tag{9.8}$$

$$-\sum_{t=1}^{H_s} \ln\left[1 - \frac{\boldsymbol{r}(t)^{\dagger}\boldsymbol{Q}_2\boldsymbol{r}(t)}{\boldsymbol{r}(t)^{\dagger}\boldsymbol{\Sigma}^{-1}\boldsymbol{r}(t)}\right] \underset{H_0}{\overset{H_1}{\gtrless}} T_{GLRT} \tag{9.9}$$

$$\sum_{t=1}^{H_s} \frac{\boldsymbol{r}(t)^{\dagger}\boldsymbol{Q}_2\boldsymbol{r}(t)}{(\boldsymbol{r}(t)^{\dagger}\boldsymbol{\Sigma}^{-1}\boldsymbol{r}(t)) - (\boldsymbol{r}(t)^{\dagger}\boldsymbol{Q}_2\boldsymbol{r}(t))} \underset{H_0}{\overset{H_1}{\gtrless}} T_{WALD} \tag{9.10}$$

which are, respectively, the natural generalization to a subspace signal model of the Rao test, GLRT, and Wald test considered in the previous subsection. Indeed, if $r = 1$ then $\boldsymbol{H}$ reduces to a column vector and (9.8), (9.9), and (9.10) simplify to, respectively, (9.1), (9.3), and (9.4).

Finally, it should be noted that it is possible to resort to detectors that assume no *a priori* knowledge on the target signal form (i.e. $r = N$). In particular, in Reference 49, some invariant detectors for Gaussian environment are proposed: 1S-GLRT, 2S-GLRT, M2S-GLRT, etc. Unfortunately, this approach cannot be pursued in the presence of heterogeneous non-Gaussian disturbance since modeling the texture and the steering vectors as unknown quantities leads to an ill-posed estimation problem (parameter space for the GLRT computation too large).

9.2.4 Covariance Estimation

The adaptive implementations of any of the detectors discussed requires the on-line estimation of the covariance matrix structure from a secondary dataset. A number of different strategies for the estimation of such a covariance matrix structure have been proposed, and they have a significant effect on the performance of the corresponding detection structure. It is common to exploit the sample covariance matrix (SCM) as the basis of adaptive implementations of coherent detectors for Gaussian noise. However, the SCM is not the maximum likelihood (ML) estimator of the covariance matrix of a SIRV [50–52], but for the case of completely correlated textures among the secondary data [53]. The SCM estimator has the form:

$$\boldsymbol{S} = \frac{1}{K}\sum_{k=1}^{K} \boldsymbol{r}_k \boldsymbol{r}_k^{\dagger} \tag{9.11}$$

in which K is the number of range bins used in the estimation process. Another alternative is the normalized SCM (NSCM) estimator [54,55]:

$$\hat{\boldsymbol{\Sigma}}_{NSCM} = \frac{1}{K}\sum_{k=1}^{K} \frac{\boldsymbol{r}_k \boldsymbol{r}_k^{\dagger}}{\frac{1}{N}\boldsymbol{r}_k^{\dagger}\boldsymbol{r}_k} \tag{9.12}$$

in which the contribution of each range cell within the secondary data is equalized by means of an estimate of its power.

Developing an estimator that guarantees the CFAR property is a complex task. The latest developments are based on recursive algorithms that approximate the ML estimator for the covariance matrix of an SIRV. In this chapter, a recently proposed recursive algorithm is considered: the recursive persymmetric (RP) estimator of Reference 56.

The adaptive NMF (ANMF) using the RP estimator has been shown to be CFAR against SIRV clutter with spatially varying statistics and exploits the persymmetric structure of the covariance matrix [57]. A related covariance estimator, used for example in the recursive ANMF [58] is not included in this chapter as it is suited for symmetric power spectral densities (PSDs), and it has been already reported that it suffers a severe performance degradation when the actual PSD of the clutter does not comply with this assumption. This is the case of sea-clutter, which exhibits a non-zero average Doppler shift related to the average speed of wave crests [15]. For completeness, we report the expression for the RP estimator[8] [56]:

$$\hat{\boldsymbol{\Sigma}}_{RP}(i+1) = \frac{N}{K}\sum_{k=1}^{K} \frac{\boldsymbol{r}_{ek}\boldsymbol{r}_{ek}^{\dagger}}{\boldsymbol{r}_{ek}^{\dagger}\hat{\boldsymbol{\Sigma}}_{RP}^{-1}(i)\,\boldsymbol{r}_{ek}} \tag{9.13}$$

[8]For our purposes, a generic element from the principal diagonal of the matrix in the denominator of (9.14) can be used.

$$\hat{\boldsymbol{\Sigma}}_{RP}(0) = \frac{1}{K}\sum_{k=1}^{K}\frac{\boldsymbol{r}_{ek}\boldsymbol{r}_{ek}^{\dagger}}{\left[\boldsymbol{T}\boldsymbol{r}_{ok}\boldsymbol{r}_{ok}^{\dagger}\boldsymbol{T}^{\dagger}\right]_{(l,l)}} \tag{9.14}$$

with

$$\boldsymbol{r}_{ek} = \frac{1}{\sqrt{2}}(\boldsymbol{r}_k + \boldsymbol{J}\boldsymbol{r}_k^*) \tag{9.15}$$

$$\boldsymbol{r}_{ok} = \frac{1}{\sqrt{2}}(\boldsymbol{r}_k - \boldsymbol{J}\boldsymbol{r}_k^*) \tag{9.16}$$

where $\boldsymbol{J}$ is the permutation matrix

$$\boldsymbol{J} = \begin{bmatrix} 0 & 0 & \cdots & 0 & 1 \\ 0 & 0 & \cdots & 1 & 0 \\ \vdots & \vdots & \vdots & \vdots & \vdots \\ 1 & 0 & \cdots & 0 & 0 \end{bmatrix} \tag{9.17}$$

$\boldsymbol{T} = (\boldsymbol{I}+\boldsymbol{J}) + j(\boldsymbol{I}-\boldsymbol{J})$, and $\boldsymbol{I}$ is the identity matrix. The number of iterations can be set to 3 as it has been shown to provide a good trade-off between computational complexity and performance [56, 59].

The aforementioned detection structures rely on the availability of a target-free *secondary dataset,* sharing with the range cells under test common disturbance statistical properties (covariance matrix or covariance structure). In practical scenarios, this requirement is rarely met. Frequent sources of secondary data heterogeneity include the presence of nearby additional targets, clutter boundaries, and discretes, and even the intrinsic non-stationarity of the clutter background (as is the case of sea-clutter). Several approaches have been proposed to mitigate this problem, such as reducing the sample support required through colored loading [60, 61], fast converging algorithms [62], structured covariance estimators [63–65], knowledge-aided techniques [66–74], data selection techniques [75–82], or parametric approaches [83–85].

In this context, it is worth mentioning that several authors have developed detection techniques capable of operating without the aid of secondary data. This is the case of the detector structure derived with a modified GLRT (MGLRT) technique, proposed in Reference 29 for Gaussian clutter backgrounds. Therein, a procedure to make it upper bounded CFAR is also suggested. A similar approach was followed in Reference 31 to derive the adaptive modified generalized likelihood ratio test (AMGLRT) detector for the detection of range-distributed targets in compound-Gaussian clutter. The AMGLRT, whose derivation is reported for completeness in Appendix 9.A.2, is also part of the detectors evaluated in this chapter and is the following decision rule:

$$\frac{\det(\boldsymbol{Z}\boldsymbol{S}_1^{-1}\boldsymbol{Z}^{\dagger})}{\det_p\left[(\boldsymbol{I}-\boldsymbol{p}\boldsymbol{p}^{\dagger})\boldsymbol{Z}\boldsymbol{S}_1^{-1}\boldsymbol{Z}^{\dagger}(\boldsymbol{I}-\boldsymbol{p}\boldsymbol{p}^{\dagger})\right]} \underset{H_0}{\overset{H_1}{\gtrless}} T_{AMGLRT} \tag{9.18}$$

where

- $Z = [r(1), \ldots, r(H_s)]$, with $H_s \geq N$;
- $S_1 = \frac{1}{N}\text{diag}(r(1)^\dagger(I - pp^\dagger)r(1), \ldots, r(H_s)^\dagger(I - pp^\dagger)r(H_s))$;
- $\det_p(\cdot)$ is the positive determinant, namely the product of the positive eigenvalues of the square matrix argument.

Finally, it should be noted that it is possible to build a target detector resorting to a two-level detection algorithm. This *ad hoc* approach consists of any single range cell detector followed by a "m out of H_s" binary integrator, and its performance has been analyzed in References 29 and 86. This approach has not been considered in this chapter as its performance, evaluated in Reference 29, has been shown to be poorer than the aforementioned techniques that jointly process the complete H_s primary data.

9.3 High-Resolution Experimental Data

An in-house developed high-resolution continuous wave linear frequency modulated (CWLFM) radar was used during 2006 to collect experimental maritime target and sea-clutter data. The main characteristics of the radar system are detailed in Table 9.2 [87]. It is a Ka-band, HH polarized, and fully coherent system, transmitting a CWLFM waveform between 28 and 30 GHz [88]. A bandwidth of up to 2 GHz can be transmitted, and therefore the maximum theoretical range resolution achieved is 7.5 cm. The system has certain distinct features, such as a heterodyne RF architecture that minimizes the required sampling frequency by providing tunable range limits [89]. It is also field-reconfigurable, as the main parameters of the modulating waveform can be easily modified according to the specific needs of each application. This feature has been used to acquire data corresponding to three different range resolutions: 0.10 m, 0.20 m, and 1 m, or equivalently to bandwidths of 2 GHz, 1 GHz, and 200 MHz. Additionally, an optical video camera fitted with 200 mm focal length lens and aligned with the antenna system allows the radar operator to simultaneously perform optical and radar target recordings.

The performance of the system has been tested in different applications, ranging from road traffic monitoring [90] to inverse synthetic aperture radar (ISAR) imaging of vessels [91]. In this chapter, sub-meter range resolution real radar data corresponding to small maritime targets and target-free sea-clutter acquisitions will be used to evaluate the experimental performance of range-distributed target detectors.

Table 9.2 Radar features.

Type	CWLFM
Tx. freq.	28–30 GHz
Tx. bandwidth	2 GHz (Max.)
Modulation freq.	5,000 Hz (Max.)
Beam width	3° (pencil-beam)
Tx. power	30 dBm

Table 9.3 Measurement configurations.

Parameter	Details
Azimuth angle	138° and 180° (Wind 270°)
Bandwidth	200 MHz, 1 GHz, 2 GHz
Range and grazing angle	1,080 m (~2.6°) and 1,755 m (~1.6°)

9.3.1 Sea-Clutter Data

Experimental low-grazing-angle sea-clutter time series were recorded at the south coast of Spain, with the radar system located on top of a 50 m maritime traffic control tower overlooking the Gibraltar Strait. Different measurement configurations were used to assess the potential impact of different parameters in the statistical properties of high-resolution sea-clutter (Table 9.3). The database comprises 95 measurements with different combinations of azimuth angle, bandwidth, and range/grazing angle. The estimated value of the clutter-to-noise ratio in the datafiles is 15 dB. During the measurement field trials, the sea state was inferred from local weather station data to be between 3 and 4 on the Beaufort scale.

The key result of the statistical analysis of the database [15] is the extension of the domain of applicability of the compound-Gaussian clutter model to range resolutions of a few centimeters. The main findings are as follows:

(1) The bi-parametric generalized K-distribution with log-normal (LN) texture usually provides a very good fit to the empirical overall amplitude PDF of the data [12]. However, at the range resolution of 0.10 m, the conventional LN distribution is the most suitable model for the amplitude PDF.

(2) Short bursts of sea-clutter samples can be modeled as SIRVs (except for the case of 0.10 m range resolution).

(3) The statistics of the speckle are Gaussian for a range resolution of a few centimeters. The coherence length of the sea texture has been estimated. The results are physically reasonable and compatible with previous works (100–150 ms). In agreement with the results for the overall PDF, the extracted texture component follows an LN distribution (but for the case of 0.10 m range resolution).

(4) The power-law family of spectral models is a good approximation to the average spectral behavior of the data.

(5) The sea-clutter datafiles exhibit strong fluctuations of both the spectral shape and the clutter amplitude between nearby range cells: there is a severe spectral non-homogeneity as the PSD changes even between adjacent range cells.

Figure 9.1 shows an example of the assessment of the suitability of several different theoretical distributions to model the overall amplitude PDF of the data. Results correspond to a representative upwind 200 MHz bandwidth datafile. The empirical estimated PDF is included in the figure as well as the fits provided by commonly used distributions following a moment matching procedure: Rayleigh, Weibull, LN, K [92], K plus thermal noise (K + N) [93], and generalized K with generalized gamma texture (GK) and LN texture (GK-LNT) [12, 94]. The good fit achieved by the GK-LNT distribution is further supported by the comparison between the theoretical and estimated higher order normalized moments shown in Figure 9.1b.

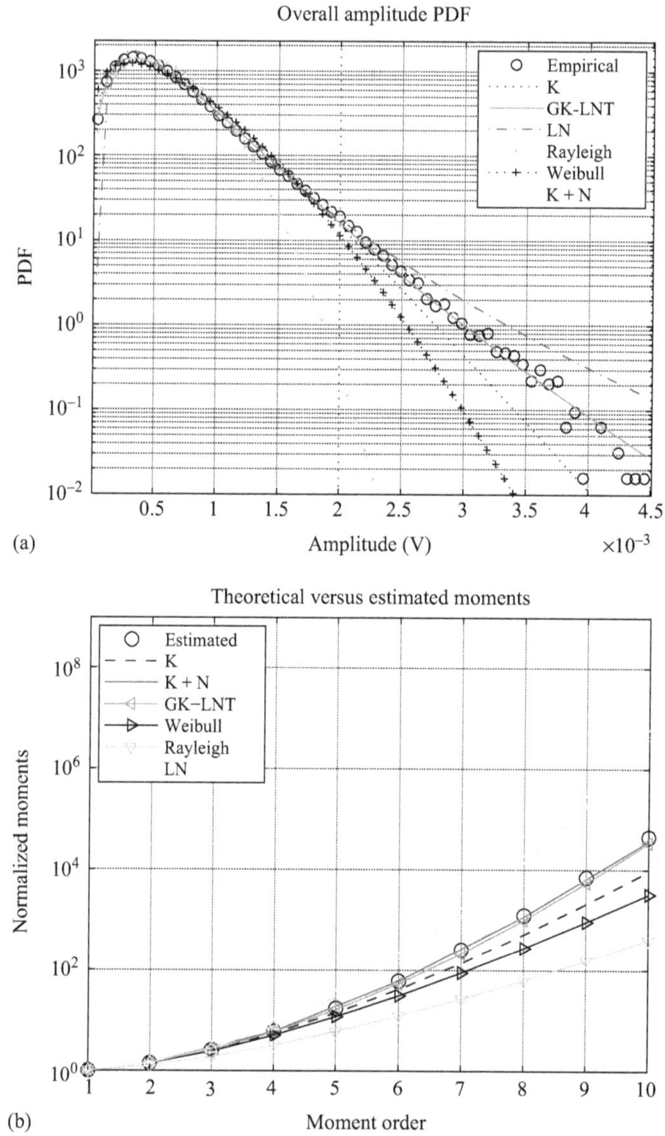

Figure 9.1 Experimental high-resolution sea-clutter data. Overall amplitude PDF. Range resolution 1 m. (a) Overall amplitude PDF analysis and (b) goodness-of-fit test: normalized higher order moments.

A representative example of the spectral heterogeneity of the data is shown in Figure 9.2 as it will be useful for the interpretation and analysis of the detection performance results of the following sections. Sea-clutter amplitude fluctuations are clearly visible in the range cell against time representation of Figure 9.2a. An estimate of the PSD of each range cell is included in Figure 9.2b, by resorting to a Welch modified periodogram with 50% overlap between Hamming weighted samples. The shown periodogram has been normalized to the estimated power in each range cell to focus on the spectral shape as in Reference 95. Strong

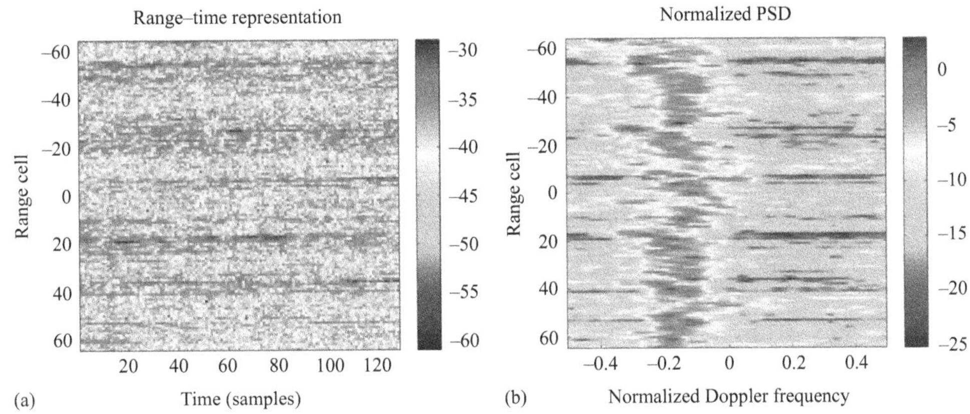

Figure 9.2 Spectral heterogeneity of high-resolution sea-clutter experimental data. Range resolution 0.20 m. (a) Range against time representation and (b) normalized PSD per range cell.

fluctuations of both the spectral shape and the clutter power are evident from the results included in Figure 9.2a and 9.2b. The assessment of the impact of these effects in the performance of coherent distributed target detection schemes is a key aim of this chapter.

9.3.2 Maritime Target Data

Simultaneous video and radar recordings of small non-cooperative maritime targets were performed in several trials. Among the target data within the database, four small vessels have been selected as representative examples of targets of special interest for upcoming maritime surveillance radar systems (in particular, border control and related homeland security applications): a rigid-hulled inflatable boat (Zodiac), a small wooden boat, a patrol boat, and a two mast sailboat. A frame of the video recording of each one is shown in Figure 9.3. An example of the corresponding radar data in range against time representation is shown in Figure 9.4a–9.4d where it is also evident the effect of the target wake.

The few specular reflector model or multiple dominant scatterer (MDS) model was introduced in Section 9.2.1. Figure 9.5 shows the qualitative validity of this model for the experimental target data used in this work. Three range profiles have been extracted from the range–time data matrix and are shown in Figure 9.5b–9.5d. The specific temporal locations are marked in Figure 9.5a. The profiles highlight that the experimental radar returns are sparsely distributed, as supported by other recent high-resolution openly available works such as References 96 and 97. The potential collapsing losses, consequence of the integration of low-SNR range bins in between target flare points, will be discussed in this chapter.

It is also interesting to analyze the target fluctuations. It is well known that the radar backscatter from a target depends on a large number of unknown and time-varying factors, such as geometry, target size, or aspect angle [98, 99]. As a consequence, the target RCS has been traditionally considered a random variable, characterized by different fluctuation models beginning with the early work of Swerling [7], later extended in Reference 100 to account for additional fluctuation laws through the use of the chi-square distribution. Since then, several theoretical distributions for the target amplitude have been proposed as the LN, Weibull, K, and generalized chi [101].

Figure 9.3 Experimental maritime target data: optical video frames. (a) Target A: rigid-hulled inflatable boat; (b) Target B: small wooden boat; (c) Target C: patrol boat; and (d) Target D: two mast sailboat.

In this chapter, an experimental validation of several target fluctuation models has been performed. In what follows, $\alpha = |\alpha| \exp(j\theta)$ is the complex parameter accounting for the target fluctuation. It is usually assumed that θ is uniformly distributed in $(-\pi, \pi]$, whereas the target amplitude $|\alpha|$ is characterized by a specific PDF [101]:

- Swerling chi target fluctuation model

$$p_{|\alpha|}(r) = \frac{2m^m r^{2m-1}}{\Omega^m \Gamma(m)} \exp\left(-\frac{mr^2}{\Omega}\right), \qquad r > 0 \tag{9.19}$$

with the parameter $m > 0$ ruling the depth of the amplitude fluctuation. For $m = 1$, the Swerling chi model coincides with the Rayleigh distribution. The non-random amplitude case is characterized by $m = \infty$.

- LN target fluctuation model

$$p_{|\alpha|}(r) = \frac{1}{r\sigma\sqrt{2\pi}} \exp\left(-\frac{\ln^2 (r/\mu)}{2\sigma^2}\right), \qquad r > 0 \tag{9.20}$$

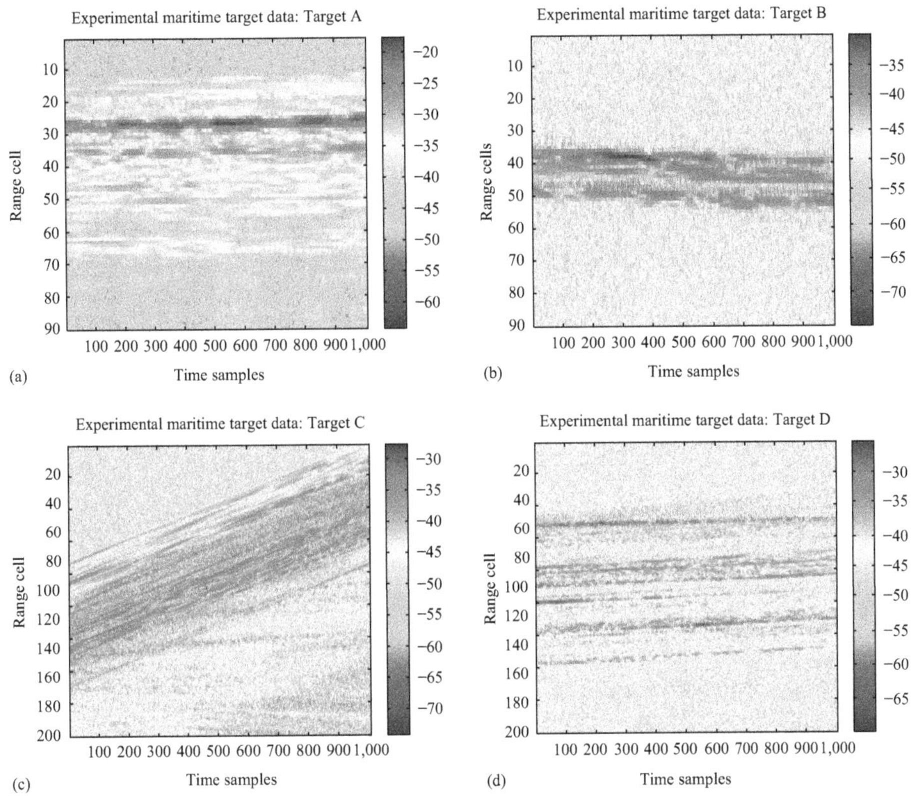

Figure 9.4 Experimental maritime target radar data. (a) Target A: sampling rate 1,000 Hz. Range resolution 0.10 m. (b) Target B: sampling rate 1,000 Hz. Range resolution 0.20 m. (c) Target C: sampling rate 1,000 Hz. Range resolution 0.20 m. (d) Target D: sampling rate 1,000 Hz. Range resolution 0.20 m.

- Weibull fluctuation model

$$p_{|\alpha|}(r) = \frac{c}{b}\left(\frac{r}{b}\right)^{c-1} \exp\left[-\left(\frac{r}{b}\right)^{c}\right], \qquad r > 0 \tag{9.21}$$

- K-distribution fluctuation model

$$p_{|\alpha|}(r) = \sqrt{\frac{2\nu}{a}} \frac{1}{2^{\nu-1}\Gamma(\nu)} \left(\sqrt{\frac{2\nu}{a}}r\right)^{\nu} K_{\nu-1}\left(\sqrt{\frac{2\nu}{a}}r\right), \qquad r > 0 \tag{9.22}$$

- Generalized K-distribution with LN texture (GK-LNT)

$$p_{|\alpha|}(r) = \frac{r}{\sqrt{2\pi\sigma^2}} \int_0^{\infty} \frac{2}{t^2} \exp\left(-\frac{r^2}{t} - \frac{1}{2\sigma^2}[\ln(t/\delta)]^2\right) dt, \qquad r > 0 \tag{9.23}$$

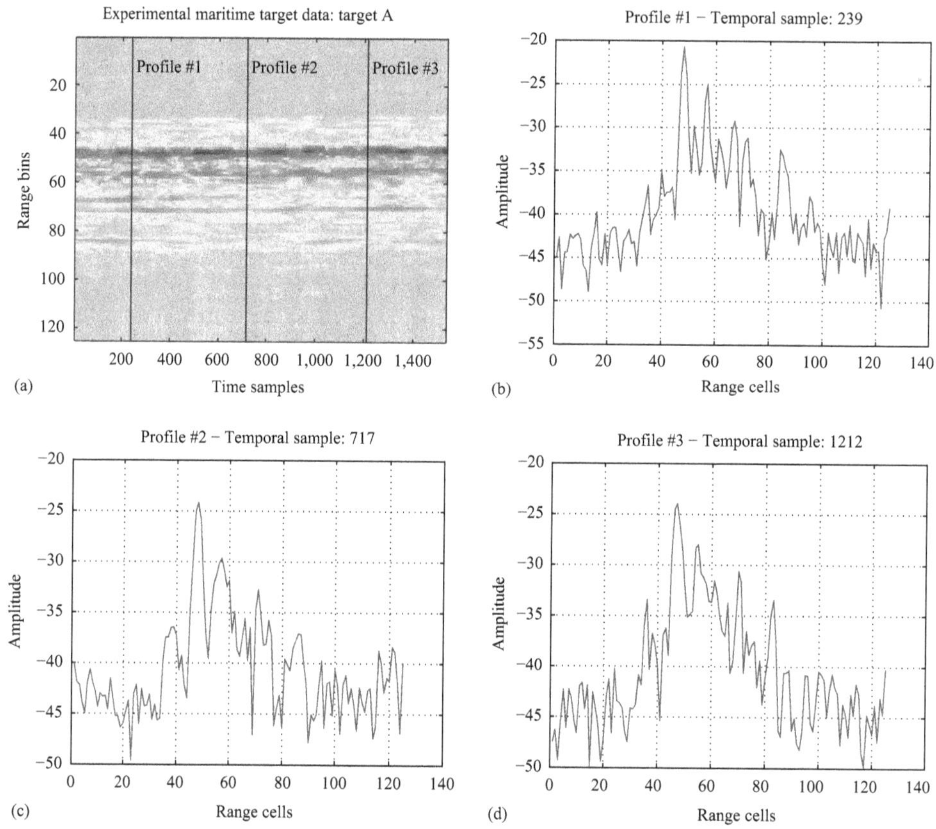

Figure 9.5 MDS model assessment. (a) Extracted range profiles' locations on range–time representation; (b) Range profile 1; (c) Range profile 2; and (d) Range profile 3.

A moment matching procedure has been implemented to fit the empirical distribution with the theoretical models. In the Swerling chi case, the mean square value Ω has been set equal to the one estimated from the real data. The parameter ruling the depth of the amplitude fluctuation m has been chosen so as to minimize the Cramer–Von Mises distance between the empirical and theoretical cumulative distribution functions (CDFs), following the methodology proposed in Reference 101.

To evaluate the suitability of each model, six different techniques have been used:

(1) Evaluation of the root mean square error (RMSE) for each distribution, defined as:

$$\text{RMSE} = \sqrt{\frac{1}{M}\sum_{k=1}^{M}\left|f_{|\alpha|}(x_k \mid \hat{\eta}) - \hat{f}_{|\alpha|}(x_k)\right|^2} \tag{9.24}$$

where $f_{|\alpha|}(x_k \mid \hat{\eta})$ is the theoretical distribution of the data for the estimated parameter vector $\hat{\eta}$ and $\hat{f}_{|\alpha|}(x_k)$ is the estimated PDF from the data, evaluated at the M points x_k's.

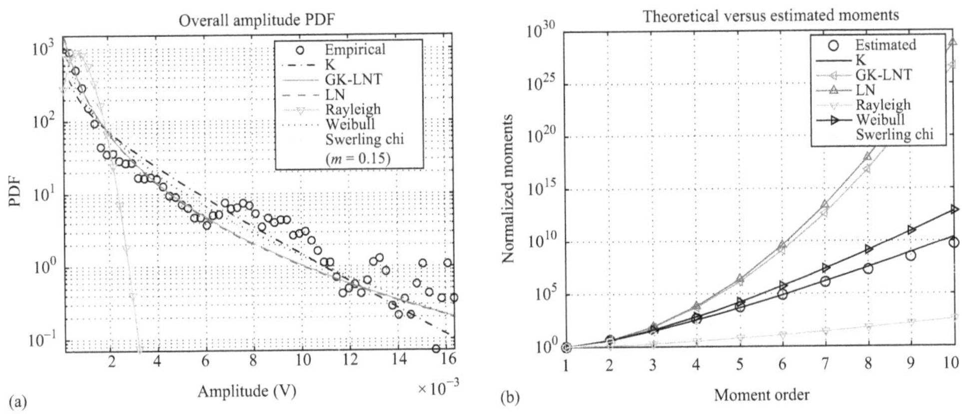

Figure 9.6 Target A: fluctuation model empirical evaluation. (a) Amplitude empirical PDF and theoretical models and (b) normalized higher order moment comparison.

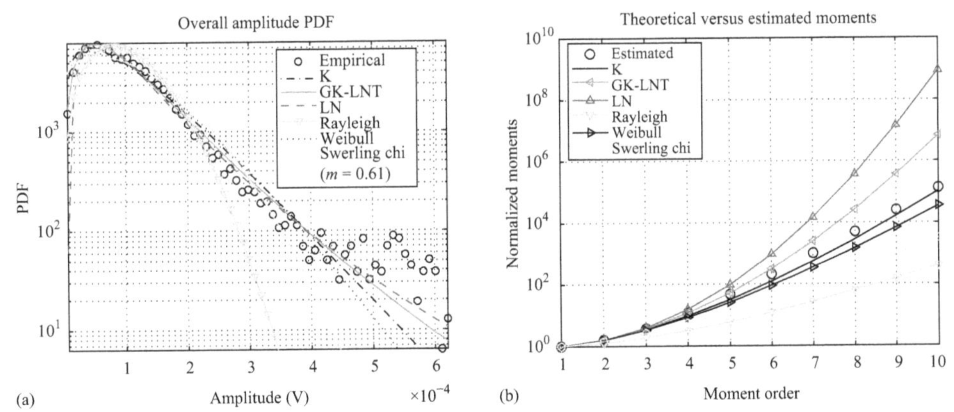

Figure 9.7 Target B: fluctuation model empirical evaluation. (a) Amplitude empirical PDF and theoretical models and (b) normalized higher order moment comparison.

(2) Evaluation of the Cramer–Von Mises distance.

(3) Evaluation of the RMSE at the tail of the distribution, by calculating the RMSE for the rightmost $k = M/2 \ldots M$ bins of the PDF.

(4) Graphical comparison between estimated and theoretical higher order normalized moments:

$$m_X(n) = \frac{E\{X^n\}}{E^n\{X\}} \tag{9.25}$$

(5) Graphical comparison of the goodness of fit of each theoretical PDF in a semilogarithmic plot.

(6) Graphical comparison of the goodness of fit of each theoretical CDF.

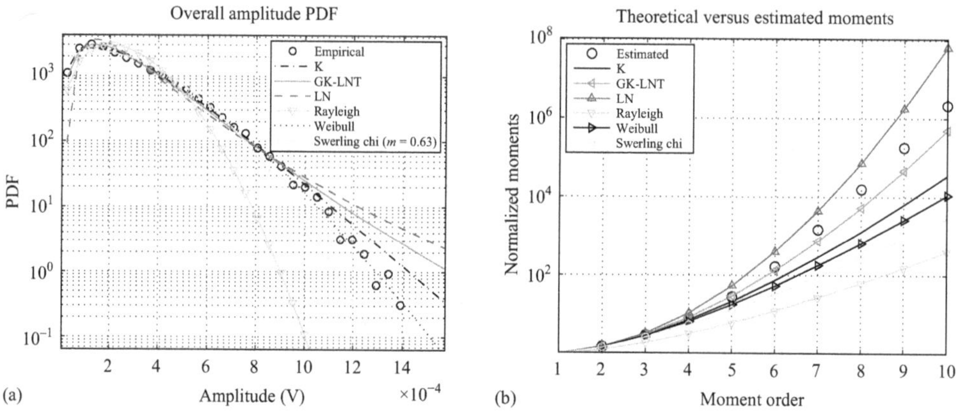

Figure 9.8 Target C: fluctuation model empirical evaluation. (a) Amplitude empirical PDF and theoretical models and (b) normalized higher order moment comparison.

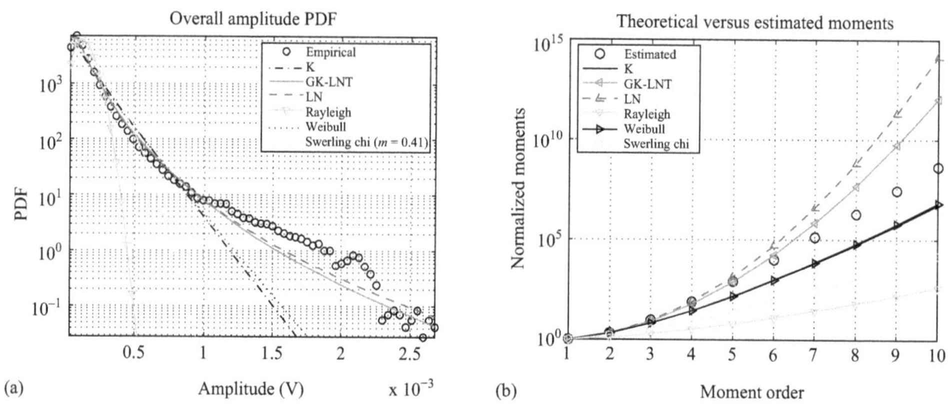

Figure 9.9 Target D: fluctuation model empirical evaluation. (a) Amplitude empirical PDF and theoretical models and (b) normalized higher order moment comparison.

The evaluation procedure has been applied to target datafiles A–D with different results in each case. The best fit is not achieved by the same theoretical model in all cases, although heavy-tailed distributions such as the generalized K-distribution with LN texture, the K-distribution and the LN distribution match the real data better than the traditional Swerling chi model. Figures 9.6–9.9 show the results of the fitting process applied to target datafiles A–D, respectively. It should be noted that the datafiles used for this analysis are characterized by a low clutter level (low sea state). As a consequence, the effect of clutter in the results has been neglected.

9.4 Experimental CFAR Behavior

The experimental CFAR behavior of a subset of the range-distributed coherent detectors detailed in Section 9.2 has been evaluated by means of a direct Monte Carlo approach.

Table 9.4 Experimental datafiles.

	Waveform modulation rate (Hz)	Bandwidth (MHz)	Range resolution (m)
Datafile 1	5,000	200	1
Datafile 2	1,000	1,000	0.20
Datafile 3	1,000	2,000	0.10

Table 9.4 details the parameters associated with each of the three upwind sea-clutter datafiles that have been used. For each parameter set, care has been taken to ensure that the file selected is statistically representative of the behavior of its class within the complete sea-clutter database. Each datafile consists of 120 range cells and 100,000 temporal samples (200 MHz bandwidth files), or 655 range cells and 20,000 time samples (1,000 MHz and 2,000 MHz case). The total number of samples is therefore 12×10^6 (200 MHz bandwidth files), or 13.1×10^6 samples (1,000 MHz and 2,000 MHz case). Depending on the configuration of each particular analysis, the number of available trials ranges between $10/P_{fa}$ and $100/P_{fa}$, for $P_{fa} = 10^{-3}$ [102].

A sliding window of N temporal samples, K reference range cells, and G guard cells at each side of the group of H_s cells under test has been used. The processing window is moved over each datafile with 50% temporal overlap and on a range cell by range cell basis along the range axis. For each position, the following detection statistics are evaluated: Rao test, GLRT, Wald test, subspace versions of the aforementioned detectors, and the AMGLRT.

With the exception of AMGLRT, the detectors considered require an estimate of the disturbance covariance matrix. In this chapter, the three covariance estimation techniques described in Section 9.2.4 have been used, resulting in three detection schemes per detector structure. In each case, a Doppler detector bank has been implemented for 16 steering vectors corresponding to the discrete normalized central Doppler frequencies, uniformly spaced between -0.5 and 0.5. Similarly, the steering matrix for subspace-based tests has been built with three steering vectors: one corresponding to a central Doppler frequency and the two adjacent ones. The P_{fa} per detector scheme and central Doppler channel has been directly estimated normalizing the false alarm count to the number of Monte Carlo trials.

The number of temporal samples has been set to $N = 16$, with $K = 32$ reference cells and $G = 1$ guard range bin per side of the group of cells under test. The value $N = 16$ has been chosen as a compromise between computational cost and Doppler selectivity. Following the Reed–Mallet–Brennan rule [52, 103], which states that the convergence measure of effectiveness (MOE) for the SCM algorithm is $K = 2N$, $K = 32$ has been used[9]. Two values for H_s have been considered, $H_s = 16$ and $H_s = 50$. For the simulation configurations characterized by these two parameter sets, probability of false alarm against threshold curves are obtained per Doppler channel and for each combination of detector and covariance estimation technique. A plot of the empirical P_{fa}, for a specific nominal threshold chosen to guarantee a pre-defined P_{fa} against white Gaussian noise [104], is a convenient graphical representation of the extensive

[9]The MOE is defined as the number of independent and identically distributed secondary data per input sample so that the adaptive processor performance is close (nominally -3 dB) to the optimum.

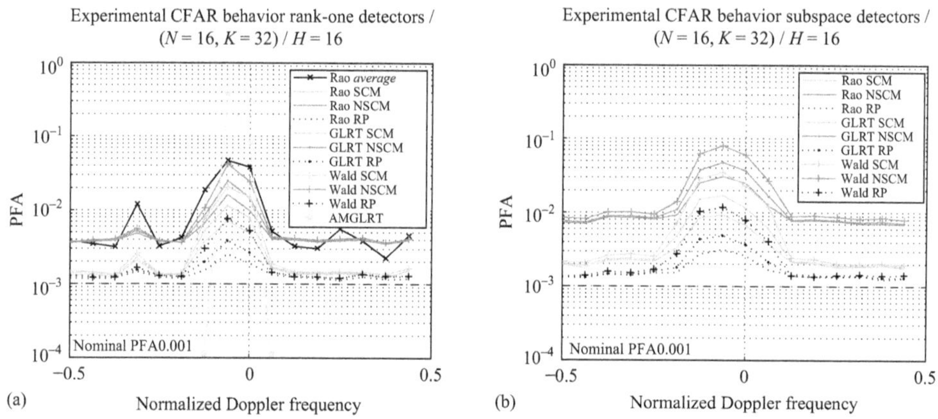

Figure 9.10 Datafile 1 CFAR behavior. Range resolution 1 m. Best fit: GK-LNT distribution. (a) Rank-one detectors and (b) subspace detectors.

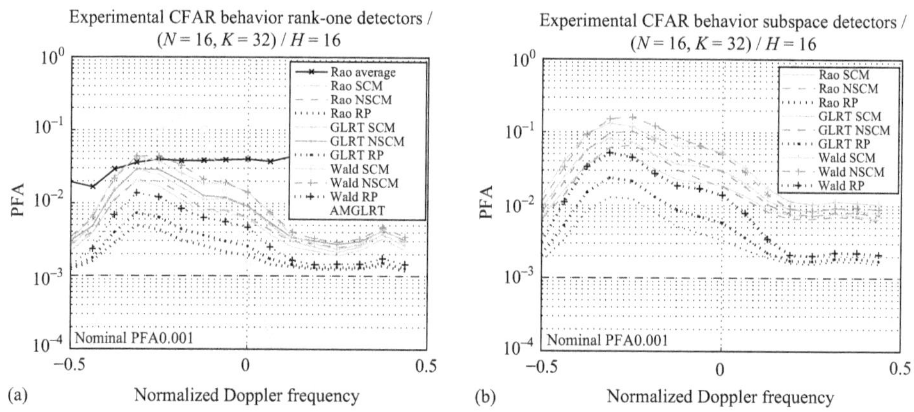

Figure 9.11 Datafile 2 CFAR behavior. Range resolution 0.20 m. Best fit: GK-LNT distribution. (a) Rank-one detectors and (b) subspace detectors.

numerical results obtained. For the detectors for which no theoretical expression relating P_{fa} and threshold is available,[10] simulations against white Gaussian noise have been used to set the nominal thresholds.

The results corresponding to $H_s = 16$ are shown in Figures 9.10–9.12, for datafiles 1–3, respectively, with reference to both rank-one (subplot a) and subspace signal (subplot b) detectors. Thresholds have been set to guarantee for each detection scheme a nominal P_{fa} of 10^{-3}.

[10]The definition of nominal threshold can be different. For example, another alternative was followed in Reference 95.

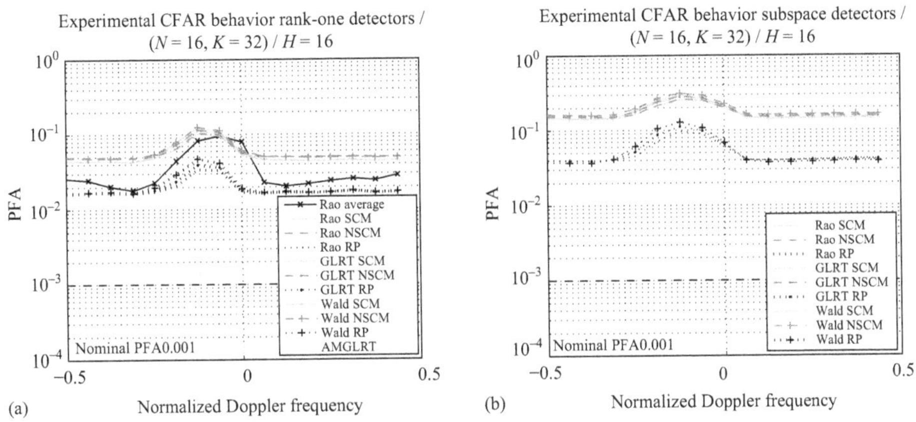

Figure 9.12 Datafile 3 CFAR behavior. Range resolution 0.10 m. Best fit: LN distribution. (a) Rank-one detectors and (b) subspace detectors.

The detector labeled as "Rao average" consists of the Rao test form using the whole datafile for SCM covariance estimation. The performance associated to this detector is generally poorer than the remaining alternatives. This implies that local covariance estimation is beneficial in terms of P_{fa} regulation regardless of the intrinsic heterogeneity of sea-clutter.

In this representation a perfect CFAR behavior is denoted by a constant P_{fa} against Doppler experimental curve. Results clearly show that the CFAR property is not practically achieved by any of the detectors considered, with degradations of up to two orders of magnitude in some specific decision rules. An increased degradation of P_{fa} in the regions of high-clutter PSD is also a common trend. This has been reported before for point-like target detectors [104] and is likely to be related to the higher non-stationarity of the clutter at that spectral zone [95]. The Welch estimated average PSD, for the experimental clutter data used here, supports this conclusion [15].

It is interesting to notice that the CFAR degradation experienced in correspondence of datafile 3 is much more severe than the one in datafiles 1 and 2. This could be explained observing that the distribution better modeling the disturbance in datafiles 1 and 2 is a GK-LNT, which is compatible with an SIRV representation (a basic assumption for all the considered receivers). On the contrary, the clutter in datafile 3 is much more spikier than that in datafiles 1 and 2 and is statistically described by an LN model which is not amenable to a compound representation.

Nevertheless, the best performance is consistently provided by the detector variants resorting to RP covariance estimation technique, even in dataset 3 where there is a statistical model mismatch. Specifically, the Rao test detector with RP covariance estimation features the most robust CFAR behavior across all the simulations performed, with a maximum actual P_{fa} of 0.5×10^{-2}. Subspace detectors have wider regions of degraded CFAR performance that is related to their increased effective Doppler *bandwidth.* Finally, results for $H_s = 16$ and $H_s = 50$ are nearly identical suggesting a relative independence between the size of the processed range window and the CFAR degradation to be expected.

It is interesting to compare the results obtained in correspondence of datafile 3 with those for simulated data with the same spikiness but compatible with a compound-Gaussian

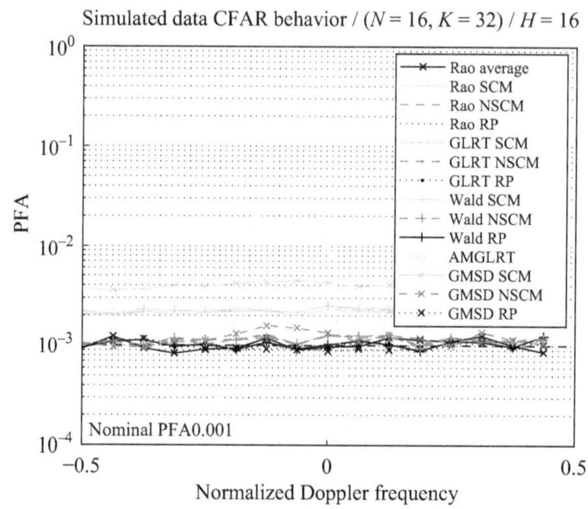

Figure 9.13 Simulated data following datafile C statistics. CFAR behavior.

representation. To this end, a convenient simulation technique has been implemented to generate a dataset with the following properties [105, 106]:

(1) The overall amplitude PDF follows a GK-LNT distribution. The simulated datafile matches, in each range cell, the shape and scale parameters profile estimated from the real data.

(2) The average covariance structure of the empirical data is used as the covariance matrix of the simulated datafile.

The P_{fa} of the detectors against simulated data based on the statistics of datafile 3 is detailed in Figure 9.13 and should be compared with Figure 9.12a. The degradation in CFAR behavior between both figures can be attributed both to the spectral heterogeneity within the reference window and to the statistical model mismatch.

Additionally, a comparison with a practical *ad hoc* detection scheme has been performed, allowing the trade-off between increased computational complexity and improvement in performance over conventional approaches to be assessed. The conventional receiver has been implemented as follows:

(1) Fast Fourier transform (FFT) of the coherent processing interval (CPI) followed by square-law processing in each Doppler channel normalized using a range-acting standard Cell-averaging CFAR.

(2) Normalization using data from the same Doppler channel of nearby range cells.

(3) The overall range-distributed detection statistic is obtained by arithmetic integration of the normalized value of each cell within the considered group of H_s range cells.

The experimental performance of this detector (FFTCFAR) against datafile 1 is compared in Figure 9.14 with that of Rao-SCM, Rao-NSCM, Rao-RP, and GMSD-SCM. The conventional approach presents the poorest CFAR behavior among the coherent detectors evaluated in this chapter.

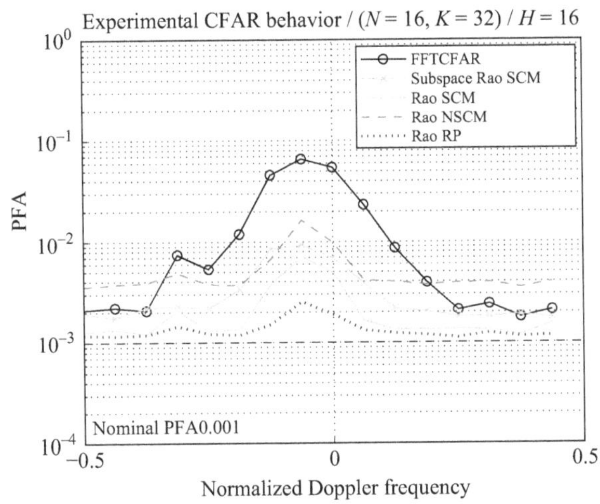

Figure 9.14 Comparison with baseline conventional Doppler receiver.

9.5 Detection Performance

The CFAR property across all Doppler channels is not practically achieved by any of the detectors under evaluation. Therefore, nominal thresholds cannot be used for the evaluation of the detection performance as the actual operating P_{fa} point would not be constant. However, taking advantage of the empirical P_{fa} against threshold and Doppler curves, it is possible to use a different *clairvoyant* threshold for each Doppler channel. With this approach, a constant P_{fa} of 10^{-3} across the Doppler spectrum has been set, and the performance comparison between the different detection strategies has been performed.

Two different approaches have been used for the evaluation of the detection performance: detection probability (P_d) of a synthetic target embedded in real clutter and experimental high-resolution detection maps for the case of real target (inflatable boat) and clutter data.

9.5.1 Detection Probability: Simulated Target and Real Clutter

In this section, the detection performance of rank-one detectors is analyzed injecting a synthetic target in a clutter-only datafile with range-resolution 0.20 m. Precisely, the synthetic target is simulated as follows $\boldsymbol{x}(t) = \alpha_t \boldsymbol{p}$, $t = 1, \ldots, H_s$, with $\boldsymbol{p}$ the temporal steering vector ($f_d = -0.125$, namely a target signal very close to the peak of the clutter PSD), and $H_s = 16$. As to the α_t's, they comply with the following three MDS models: Model 1 (target uniformly spread within the H_s primary cells), Model 2 (target uniformly spread within the first four primary cells), and Model 3 (point-like target concentrated in the first range bin of the primary cells). Additionally, the phases of the target scatterers are modeled as independent random variables uniformly distributed in $(0, 2\pi]$.

The P_{fa} is set to 10^{-3} and the SCR is defined as:

$$\text{SCR} = \sum_{t=1}^{H_s} |\alpha_t|^2 \frac{(\boldsymbol{p}^\dagger \hat{\boldsymbol{M}}_{avg}^{-1} \boldsymbol{p})}{N} \tag{9.26}$$

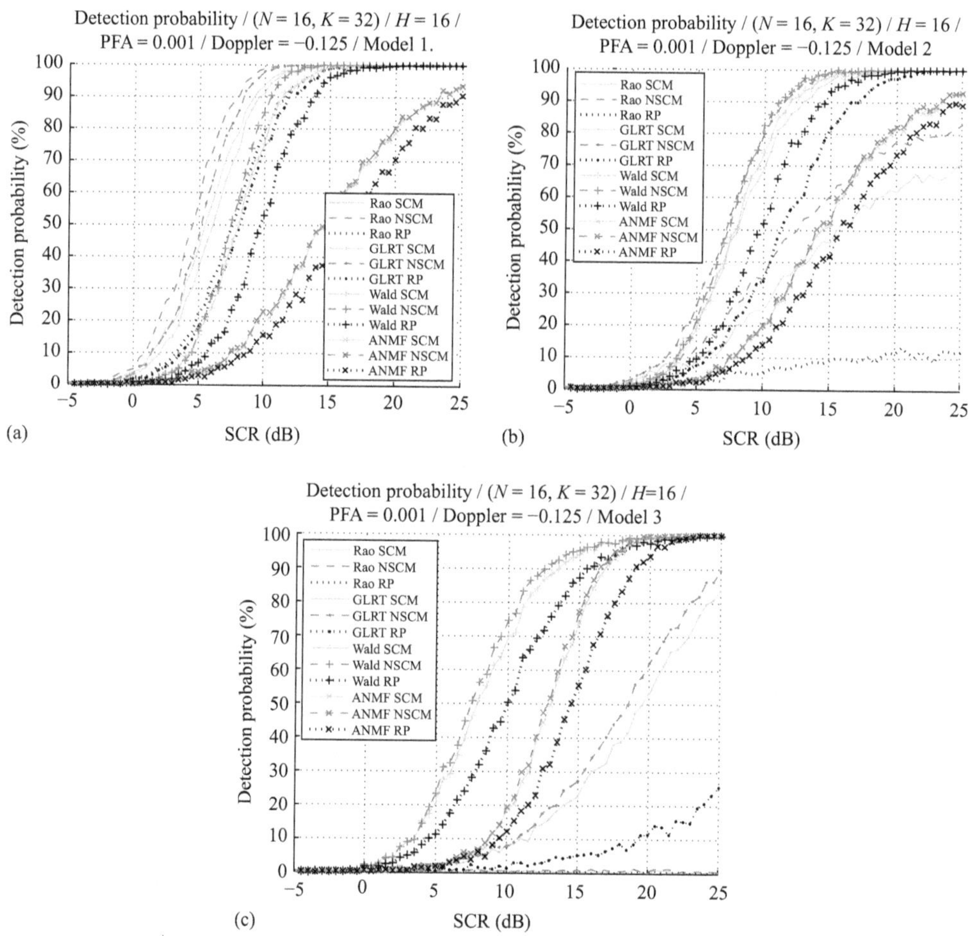

Figure 9.15 Detection probability. Simulated target and real clutter. $N=16$, $H_s=16$, $K=32$. $P_{fa}=10^{-3}$. (a) Target Model 1: uniformly spread target within the complete primary data. (b) Target Model 2: uniformly spread target within the first four primary range cells. (c) Target Model 3: point-like target concentrated in the first primary range cell.

where $\hat{M}_{avg}^{-1}$ is the SCM estimated from the entire clutter dataset. Finally, the number of integrated pulses is $N=16$, and the number of secondary data used for covariance estimation is $K=32$.

In Figure 9.15a–9.15c, P_d is plotted versus SCR with reference to target Models 1–3, respectively. Adaptive versions of the Rao test, Wald test, and GLRT are considered based on the three covariance estimation techniques already discussed in previous sections.

For comparison purposes, the performance of a low-resolution system ($H_s \times 0.2 \simeq 3$ m, which is matched to the typical size of some maritime small targets) using adaptive versions of the NMF is reported too in all the figures. The low-resolution primary data (under both hypotheses) has been obtained adding together the returns from the 16 range cells contained

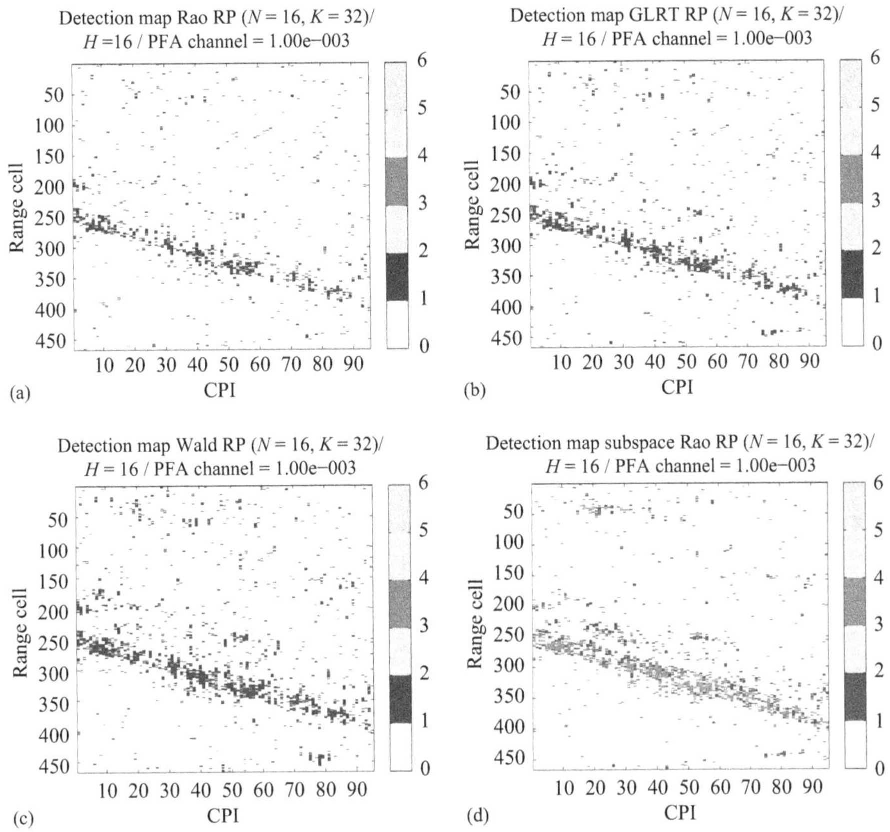

Figure 9.16 High-resolution detection maps. RP covariance estimation. Target A. (a) Rao test; (b) GLRT; (c) Wald test; and (d) Subspace Rao test.

in the primary dataset of the original high-resolution datafile. A similar procedure has been implemented to get a low-resolution secondary dataset from H_sK synthetic target-free high-resolution range cells.

Results clearly show the effect of each specific covariance estimation technique. NSCM consistently provides the best detection performance, closely followed by SCM. The additional loss associated with RP covariance estimation technique is approximately 3 dB with respect to the same detector based on the NSCM.

Given a specific covariance estimator, the relative performance between detectors is strongly dependent on the adopted MDS model. For evenly distributed target scatterers (Model 1), Rao detector features the best performance, with the GLRT approach within 1 dB below it. Wald test presents a P_d approximately 3 dB worse than Rao test. However, the joint analysis of Figure 9.15a–9.15c shows that the three approaches are differently affected by collapsing losses. The comparison clearly shows that Rao test exhibits a strong performance loss when the target is concentrated into a small fraction of the cells under test. In particular, for target Model 3, the P_d associated with Rao detectors is nearly zero (Figure 9.15c). However, in this case Wald test still guarantees a satisfactory detection performance. This clearly supports the inherent robustness of Wald test against collapsing losses. The behavior of GLRT variants is between that of Rao test and Wald test.

The results also highlight that over-resolving the target improves the detection performance. This can be shown by comparing the performance associated with ANMF variants to that of range-distributed detectors. ANMF variants represent the conventional approach of matching the range resolution of the radar system to the physical dimension of the target to be detected. Figure 9.15a shows that the low-resolution approach is outperformed by range-distributed detectors applied to the over-resolved target returns. Even for the sparse target pattern of target Model 2, in which only 4 out of the 16 processed range cells contain useful target energy (Figure 9.15b), the performance of both GLRT and Wald test is still better than the low-resolution counterpart. In the extreme situation given by Model 3, Wald test still outperforms conventional low-resolution ANMF detectors.

9.5.2 Detection Maps: Real Target and Clutter Data

The test statistic of each detection structure has been calculated for each position of the sliding window over a particular target datafile and per Doppler channel. The resultant range against CPI matrix can be thresholded to generate a high-resolution detection map. The thresholding process is performed using clairvoyant thresholds obtained from target-free areas of each particular target datafile producing a high-resolution detection map per detector and Doppler channel. It is convenient to combine the detection maps associated with each Doppler channel. In this representation, the number of Doppler channels exceeding the threshold is indicated per range cell and CPI number.

Example results for a radar acquisition of Target A (with significative radial velocity), for RP covariance estimation technique, are shown in Figure 9.16. The P_{fa} per Doppler channel is set to $P_{fa} = 10^{-3}$, and the range interval shown corresponds to the surroundings of the target. The figures highlight that all the techniques are able to detect the target even if the robust CFAR behavior of RP variants is at the expense of additional detection losses. However, as P_{fa} regulation is a key requirement in any practical system, RP covariance estimation variants can be considered a viable candidate for real operation.

It is interesting to note that the target generates detections in several adjacent Doppler channels. Figure 9.17a highlights an example: at CPI 48 and range index 1710, the threshold is exceeded for three Doppler channels and three detections are declared. For the specific configuration of the example, a group of 16 range cells contribute to the value of the test statistic. The short-time Fourier transform (STFT) of the returns from two range cells, among the $H_s = 16$ under test, is shown in Figure 9.17b and 9.17c. Additionally, the corresponding Rao test statistic is shown in the time-Doppler plane of Figure 9.17d. There is significant energy at several adjacent Doppler channels that exceed the associated clairvoyant threshold.

For the case of subspace detectors and due to the higher actual bandwidth associated with the three steering vectors in the implemented mode matrix, this behavior is more pronounced with detections at up to five Doppler channels.

Summarizing, the obtained detection maps show that for the specific SCR of the considered real datasets, all of the receiving structures achieve a satisfactory detection of the target. This *nominal* SCR has been estimated to be between 10 and 15 dB depending on the particular target datafile.

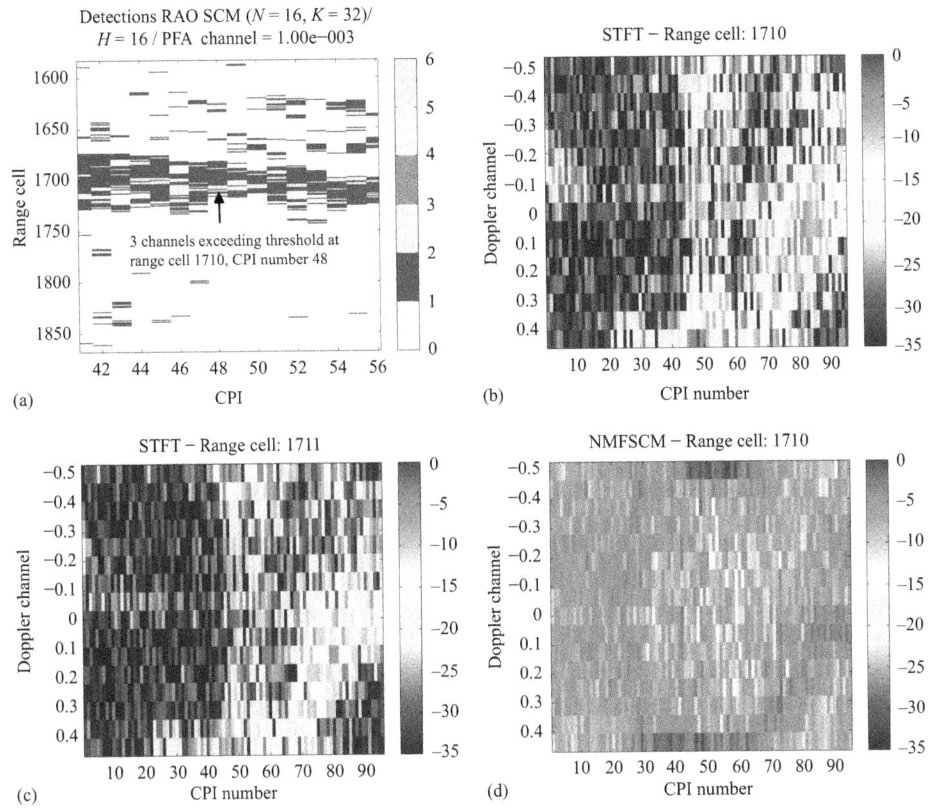

Figure 9.17 Doppler against time analysis for range cell 1710. (a) Rao SCM detector. Multiple Doppler detections at range index 1710, CPI number 48; (b) STFT; (c) STFT; and (d) Test statistic.

9.6 Conclusions

This chapter has considered the performance analysis of range-distributed target detectors, in the presence of high-resolution sea-clutter. To this end, the state of the art concerning the problem of interest has been reviewed. Due to their theoretical CFAR behavior with respect to the clutter power in the range cells under test, rank-one and subspace detectors based on Rao test, Wald test, and GLRT techniques have been chosen as suitable schemes for the analysis. Additionally, a reference data-free range-spread target detector, the AMGLRT, is also considered.

The statistical description of the clutter datafiles as well as the target datasets, corresponding to small maritime vessels of special interest in homeland security applications such as an inflatable or a small wooden boat, has been presented.

Then, the experimental CFAR behavior of the adaptive versions of the selected detectors has been assessed exploiting three covariance matrix estimation approaches and three real clutter datafiles, at different sub-meter range resolutions. Interestingly, one of the datafiles also exhibits statistical model mismatches with respect to the SIRV design hypothesis common to all the detectors considered. As a result of the analysis, it can be claimed that (on the considered datafiles), none among the considered detectors achieves rigorously the CFAR property. However, the Rao detector with RP covariance estimation is the most robust exhibiting an acceptable degradation in the CFAR behavior for the datasets that comply with the SIRV model.

Two approaches have been followed to assess the detection capabilities of the processors. First, the detection performance of each scheme has been evaluated by injecting a synthetic range-distributed target in real high-resolution clutter. Results show the different impact of collapsing losses in each of the detection schemes under evaluation. In particular, the Wald test is robust, achieving a satisfactory performance even operating under a severe mismatch between the actual and hypothesized target range extension. Additionally, the performance improvement associated with over-resolved targets has been quantified through a direct comparison to a low-resolution NMF-based detection approach. Finally, the range–time detection maps of some considered receivers have been presented showing that detection of small maritime targets, such as an inflatable boat, embedded in spiky sea-clutter is possible using high-resolution sub-meter radar systems equipped with suitable detection strategies.

The lesson learned from this experimental analysis can be summarized as follows:

- Even if all the considered receivers do not exhibit the CFAR behavior on real data, the detectors based on RP covariance estimation are the most robust. Indeed, the discrepancy between the theoretical and the actual P_{fa} is acceptable provided that the range resolution is greater than 0.20 m. As a consequence, this is the recommended covariance estimation strategy also for clutter at very high resolution ($\geq$0.20 m). Nevertheless, the problem of achieving a robust CFAR behavior at range resolutions smaller than 0.20 m, where a strong model mismatch is present, represents still a relevant practical open issue.
- The analyzed receivers for range-spread targets outperform the classic FFTCFAR that is the conventional system used in many modern radar systems.
- The analysis of the detection capabilities highlights that suitably designed detectors for range-distributed targets can provide performance improvements over receivers synthesized for point-like target scenarios.
- The experimental results show that certain detectors are extremely robust against collapsing losses (Wald and GLRT detectors). This is a key factor that makes these detectors ideally suited for practical implementation, provided that they are combined with a suitable covariance estimation technique. Therefore, it can be concluded that Wald or GLRT detectors with RP covariance estimation seem to be the most suitable detection schemes.

Appendix 9.A

9.A.1 Derivation of GLRT, Rao Test, and Wald Test

In this section, we derive the GLRT, the Rao test, and the Wald test for the detection problem:

$$\begin{cases} H_0 : \begin{cases} \boldsymbol{r}(t) = \boldsymbol{n}(t), & t = 1, \ldots, H_s \\ \boldsymbol{r}_k = \boldsymbol{n}_k, & k = 1, \ldots, K \end{cases} \\ \\ H_1 : \begin{cases} \boldsymbol{r}(t) = \alpha_t \boldsymbol{p} + \boldsymbol{n}(t), & t = 1, \ldots, H_s \\ \boldsymbol{r}_k = \boldsymbol{n}_k & k = 1, \ldots, K \end{cases} \end{cases} \tag{9.27}$$

where $\boldsymbol{p}$ denotes the (known) unitary norm steering vector, the α_t's, $t=1,\ldots,H_s$, are (unknown) deterministic parameters accounting for both the target and the channel effects, and the random vectors $\boldsymbol{n}(t)$'s, $t=1,\ldots,H_s$, and $\boldsymbol{n}_k$'s, $k=1,\ldots,K$, are independent zero-mean complex circular Gaussian vectors sharing the same positive definite structure of the covariance matrix; precisely, $\boldsymbol{n}(t) \sim \mathcal{N}(0, s(t)^2\boldsymbol{\Sigma})$, $t=1,\ldots,H_s$ and $\boldsymbol{n}_k \sim \mathcal{N}(0, s_k^2\boldsymbol{\Sigma})$, $k=1,\ldots,K$. At the design stage, we assume that the structure $\boldsymbol{\Sigma}$ of the covariance matrix is known. Besides, we define the following two vectors:

- $\boldsymbol{\theta}_r = [\alpha_{R,1}, \alpha_{I,1}, \ldots, \alpha_{R,H_s}, \alpha_{I,H_s}]^T$ a $2H_s$-dimensional vector, with $\alpha_{R,t}$ and $\alpha_{I,t}$ the real and the imaginary part of α_t, $t=1,\ldots,H_s$;
- $\boldsymbol{\theta}_s = [s(1)^2, \ldots, s(H_s)^2]^T$ a H_s-dimensional vector.

9.A.1.1 GLRT

Assuming that the structure $\boldsymbol{\Sigma}$ of the covariance matrix is known, the detection algorithm implementing the GLRT is tantamount to substituting the unknown parameters α_t's, $t=1,\ldots,H_s$, and $s(t)^2$, $t=1,\ldots,H_s$, appearing in the likelihood ratio test with their ML estimates under each hypothesis [107]. Otherwise stated, the GLRT is the following decision rule:

$$\frac{\max\limits_{\boldsymbol{\theta}_r,\boldsymbol{\theta}_s} p(\boldsymbol{r}(1),\ldots,\boldsymbol{r}(H_s)|H_1,\boldsymbol{\theta}_r,\boldsymbol{\theta}_s,\boldsymbol{\Sigma})}{\max\limits_{\boldsymbol{\theta}_s} p(\boldsymbol{r}(1),\ldots,\boldsymbol{r}(H_s)|H_0,\boldsymbol{\theta}_s,\boldsymbol{\Sigma})} \underset{H_0}{\overset{H_1}{\gtrless}} \gamma_{GLRT} \tag{9.28}$$

with γ_{GLRT} the detection threshold to be set according to the desired value of the false alarm probability P_{fa}.

Subsequent developments require specifying the PDFs

$$p(\boldsymbol{r}(1),\ldots,\boldsymbol{r}(H_s)|H_0,\boldsymbol{\theta}_s,\boldsymbol{\Sigma})$$

and

$$p(\boldsymbol{r}(1),\ldots,\boldsymbol{r}(H_s)|H_1,\boldsymbol{\theta}_r,\boldsymbol{\theta}_s,\boldsymbol{\Sigma})$$

Previous assumptions imply that the aforementioned PDFs may be written as:

$$p(\boldsymbol{r}(1),\ldots,\boldsymbol{r}(H_s)|H_0,\boldsymbol{\theta}_s,\boldsymbol{\Sigma}) = \prod_{t=1}^{H_s} \frac{\pi^{-N}}{s(t)^{2N}\det(\boldsymbol{\Sigma})} \exp\left(-\boldsymbol{r}(t)^\dagger \frac{\boldsymbol{\Sigma}^{-1}}{s(t)^2}\boldsymbol{r}(t)\right) \tag{9.29}$$

under H_0 and

$$p(\boldsymbol{r}(1),\ldots,\boldsymbol{r}(H_s)|H_1,\boldsymbol{\theta}_r,\boldsymbol{\theta}_s,\boldsymbol{\Sigma}) = \prod_{t=1}^{H_s} \frac{\pi^{-N}}{s(t)^{2N}\det(\boldsymbol{\Sigma})} \exp\left(-(\boldsymbol{r}(t)-\alpha_t\boldsymbol{p})^\dagger \frac{\boldsymbol{\Sigma}^{-1}}{s(t)^2}(\boldsymbol{r}(t)-\alpha_t\boldsymbol{p})\right) \tag{9.30}$$

under H_1. Maximizing $p(\boldsymbol{r}(1),\ldots,\boldsymbol{r}(H_s)|H_1,\boldsymbol{\theta}_r,\boldsymbol{\theta}_s,\boldsymbol{\Sigma})$ over $\boldsymbol{\theta}_r$ yields

$$\widehat{\boldsymbol{\theta}}_{r,1} = \left[\mathrm{Re}\left(\frac{\boldsymbol{p}^\dagger\boldsymbol{\Sigma}^{-1}\boldsymbol{r}(1)}{\boldsymbol{p}^\dagger\boldsymbol{\Sigma}^{-1}\boldsymbol{p}}\right), \mathrm{Im}\left(\frac{\boldsymbol{p}^\dagger\boldsymbol{\Sigma}^{-1}\boldsymbol{r}(1)}{\boldsymbol{p}^\dagger\boldsymbol{\Sigma}^{-1}\boldsymbol{p}}\right), \ldots, \mathrm{Re}\left(\frac{\boldsymbol{p}^\dagger\boldsymbol{\Sigma}^{-1}\boldsymbol{r}(\mathrm{H_s})}{\boldsymbol{p}^\dagger\boldsymbol{\Sigma}^{-1}\boldsymbol{p}}\right), \mathrm{Im}\left(\frac{\mathbf{p}^\dagger\boldsymbol{\Sigma}^{-1}\mathbf{r}(\mathrm{H_s})}{\boldsymbol{p}^\dagger\boldsymbol{\Sigma}^{-1}\boldsymbol{p}}\right)\right]^T \tag{9.31}$$

and

$$\widehat{\boldsymbol{\theta}}_{s,1} = \operatorname{argmax}_{\boldsymbol{\theta}_s}\left\{\prod_{t=1}^{H_s}\frac{1}{s(t)^{2N}\det(\boldsymbol{\Sigma})}\exp\left[-\frac{1}{s(t)^2}\left(\boldsymbol{r}(t)^\dagger\boldsymbol{\Sigma}^{-1}\boldsymbol{r}(t)-\frac{|\boldsymbol{p}^\dagger\boldsymbol{\Sigma}^{-1}\boldsymbol{r}(t)|^2}{\boldsymbol{p}^\dagger\boldsymbol{\Sigma}^{-1}\boldsymbol{p}}\right)\right]\right\} \tag{9.32}$$

Moreover the maximizer of (9.32) is

$$\widehat{\boldsymbol{\theta}}_{s,1} = \frac{1}{N}\left[\left(\boldsymbol{r}(1)^\dagger\boldsymbol{\Sigma}^{-1}\boldsymbol{r}(1)-\frac{|\boldsymbol{p}^\dagger\boldsymbol{\Sigma}^{-1}\boldsymbol{r}(1)|^2}{\boldsymbol{p}^\dagger\boldsymbol{\Sigma}^{-1}\boldsymbol{p}}\right),\ldots,\left(\boldsymbol{r}(H_s)^\dagger\boldsymbol{\Sigma}^{-1}\boldsymbol{r}(H_s)-\frac{|\boldsymbol{p}^\dagger\boldsymbol{\Sigma}^{-1}\mathbf{r}(H_s)|^2}{\boldsymbol{p}^\dagger\boldsymbol{\Sigma}^{-1}\boldsymbol{p}}\right)\right]^T \tag{9.33}$$

Analogously, maximizing $p(\boldsymbol{r}(1),\ldots,\boldsymbol{r}(H_s)|H_0,\boldsymbol{\theta}_s,\boldsymbol{\Sigma})$ over $\boldsymbol{\theta}_s$ yields

$$\widehat{\boldsymbol{\theta}}_{s,0} = \frac{1}{N}\left[\boldsymbol{r}(1)^\dagger\boldsymbol{\Sigma}^{-1}\boldsymbol{r}(1),\ldots,\boldsymbol{r}(H_s)^\dagger\boldsymbol{\Sigma}^{-1}\boldsymbol{r}(H_s)\right]^T \tag{9.34}$$

Hence, substituting the above ML estimates (9.31), (9.33), and (9.34) in (9.28), the GLRT decision rule can be recast as:

$$\prod_{t=1}^{H_s}\left[\frac{1}{1-\dfrac{|\boldsymbol{p}^\dagger\boldsymbol{\Sigma}^{-1}\boldsymbol{r}(t)|^2}{\boldsymbol{r}(t)^\dagger\boldsymbol{\Sigma}^{-1}\boldsymbol{r}(t)\boldsymbol{p}^\dagger\boldsymbol{\Sigma}^{-1}\boldsymbol{p}}}\right]^N \underset{H_0}{\overset{H_1}{\gtrless}} \gamma_{GLRT} \tag{9.35}$$

which is statistically equivalent to (9.1).

9.A.1.2 Rao Test

Starting from the primary data (i.e. $\boldsymbol{r}(t)$, $t = 1,\ldots H_s$) and assuming that $\boldsymbol{\Sigma}$ is known, the detection algorithm implementing the Rao test can be obtained expanding the likelihood ratio in the neighborhood of the ML estimates of the parameters. Specifically, the Rao test for the problem of interest is the following decision rule:

$$\left\|[\boldsymbol{J}^{-1}(\widehat{\boldsymbol{\theta}}_{r,0},\widehat{\boldsymbol{\theta}}_{s,0})]^{\frac{1}{2}}_{\boldsymbol{\theta}_r,\boldsymbol{\theta}_r}\left.\frac{\partial\ln p(\boldsymbol{r}(1),\ldots,\boldsymbol{r}(H_s)|H_1,\boldsymbol{\theta}_r,\boldsymbol{\theta}_s,\boldsymbol{\Sigma})}{\partial\boldsymbol{\theta}_r}\right|_{\substack{\boldsymbol{\theta}_r=\widehat{\boldsymbol{\theta}}_{r,0}\\ \boldsymbol{\theta}_s=\widehat{\boldsymbol{\theta}}_{s,0}}}\right\|^2 \underset{H_0}{\overset{H_1}{\gtrless}} \gamma_{Rao} \tag{9.36}$$

where

- γ_{Rao} is the detection threshold to be set according to the desired value of the false alarm probability P_{fa};
- $\widehat{\boldsymbol{\theta}}_{r,0}$ and $\widehat{\boldsymbol{\theta}}_{s,0}$ are the ML estimates of $\boldsymbol{\theta}_r$ and $\boldsymbol{\theta}_s$ under H_0;
- $\boldsymbol{J}(\boldsymbol{\theta}_r,\boldsymbol{\theta}_s)$ is the Fisher information matrix [36] that can be partitioned as:

$$\boldsymbol{J}(\boldsymbol{\theta}_r,\boldsymbol{\theta}_s) = \begin{bmatrix}\boldsymbol{J}_{\boldsymbol{\theta}_r,\boldsymbol{\theta}_r}(\boldsymbol{\theta}_r,\boldsymbol{\theta}_s) & \boldsymbol{J}_{\boldsymbol{\theta}_r,\boldsymbol{\theta}_s}(\boldsymbol{\theta}_r,\boldsymbol{\theta}_s)\\ \boldsymbol{J}_{\boldsymbol{\theta}_s,\boldsymbol{\theta}_r}(\boldsymbol{\theta}_r,\boldsymbol{\theta}_s) & \boldsymbol{J}_{\boldsymbol{\theta}_s,\boldsymbol{\theta}_s}(\boldsymbol{\theta}_r,\boldsymbol{\theta}_s)\end{bmatrix}$$

- $[\boldsymbol{J}^{-1}(\boldsymbol{\theta}_r,\boldsymbol{\theta}_s)]_{\boldsymbol{\theta}_r,\boldsymbol{\theta}_r} = [\boldsymbol{J}_{\boldsymbol{\theta}_r,\boldsymbol{\theta}_r}(\boldsymbol{\theta}_r,\boldsymbol{\theta}_s)-\boldsymbol{J}_{\boldsymbol{\theta}_r,\boldsymbol{\theta}_s}(\boldsymbol{\theta}_r,\boldsymbol{\theta}_s)\boldsymbol{J}^{-1}_{\boldsymbol{\theta}_s,\boldsymbol{\theta}_s}(\boldsymbol{\theta}_r,\boldsymbol{\theta}_s)\boldsymbol{J}_{\boldsymbol{\theta}_s,\boldsymbol{\theta}_r}(\boldsymbol{\theta}_r,\boldsymbol{\theta}_s)]^{-1}$.

Subsequent developments require the evaluation of the gradient of $p(\boldsymbol{r}(1),\ldots,\boldsymbol{r}(H_s)|H_1,\boldsymbol{\theta}_r,\boldsymbol{\theta}_s,\boldsymbol{\Sigma})$, given in (9.30), with respect to $\boldsymbol{\theta}_r$. Specifically, it can be shown that

$$\frac{\partial \ln p(\boldsymbol{r}(1),\ldots,\boldsymbol{r}(H_s)|H_1,\boldsymbol{\theta}_r,\boldsymbol{\theta}_s,\boldsymbol{\Sigma})}{\partial \alpha_{R,t}} = 2\mathrm{Re}\left(\boldsymbol{p}^\dagger \frac{\boldsymbol{\Sigma}^{-1}}{\mathrm{s(t)}^2}(\boldsymbol{r}(\mathrm{t}) - \alpha_\mathrm{t}\boldsymbol{p})\right)$$
$$\frac{\partial \ln p(\boldsymbol{r}(1),\ldots,\boldsymbol{r}(H_s)|H_1,\boldsymbol{\theta}_r,\boldsymbol{\theta}_s,\boldsymbol{\Sigma})}{\partial \alpha_{I,t}} = 2\mathrm{Im}\left(\boldsymbol{p}^\dagger \frac{\boldsymbol{\Sigma}^{-1}}{\mathrm{s(t)}^2}(\boldsymbol{r}(\mathrm{t}) - \alpha_\mathrm{t}\boldsymbol{p})\right) \tag{9.37}$$

As to the blocks of the Fisher information matrix, as shown in Appendix 9.A.1.4, they are given by

$$\boldsymbol{J}_{\boldsymbol{\theta}_r,\boldsymbol{\theta}_r}(\boldsymbol{\theta}_r,\boldsymbol{\theta}_s) = 2\boldsymbol{p}^\dagger\boldsymbol{\Sigma}^{-1}\boldsymbol{p}\ \mathrm{diag}\left(\frac{1}{s(1)^2},\frac{1}{s(1)^2},\ldots,\frac{1}{s(H_s)^2},\frac{1}{s(H_s)^2}\right)$$

$$\boldsymbol{J}_{\boldsymbol{\theta}_r,\boldsymbol{\theta}_s}(\boldsymbol{\theta}_r,\boldsymbol{\theta}_s) = \boldsymbol{0}_{2H_s,H_s}$$

with $\boldsymbol{0}_{n,m}$ an $n\times m$ matrix of zeros. It follows that

$$[\boldsymbol{J}^{-1}(\boldsymbol{\theta}_r,\boldsymbol{\theta}_s)]_{\boldsymbol{\theta}_r,\boldsymbol{\theta}_r} = (\boldsymbol{J}(\boldsymbol{\theta}_r,\boldsymbol{\theta}_s)_{\boldsymbol{\theta}_r,\boldsymbol{\theta}_r})^{-1} = \frac{1}{2\boldsymbol{p}^\dagger\boldsymbol{\Sigma}^{-1}\boldsymbol{p}}\ \mathrm{diag}\left(s(1)^2,s(1)^2,\ldots,s(H_s)^2,s(H_s)^2\right) \tag{9.38}$$

Finally, the ML estimate of $\boldsymbol{\theta}_r$ under H_0 is $\widehat{\boldsymbol{\theta}}_{r,0} = \boldsymbol{0}_{1,2H_s}$, while $\widehat{\boldsymbol{\theta}}_{s,0}$, the ML estimate of $\boldsymbol{\theta}_s$ under H_0, is the H_s-dimensional column vector

$$\begin{aligned}\widehat{\boldsymbol{\theta}}_{s,0} &= \mathrm{argmax}_{\boldsymbol{\theta}_s}\left(\prod_{t=1}^{H_s}\frac{1}{s(t)^{2N}\det(\boldsymbol{\Sigma})}\exp\left[-\boldsymbol{r}(t)^\dagger\frac{\boldsymbol{\Sigma}^{-1}}{s(t)^2}\boldsymbol{r}(t)\right]\right)\\ &= \frac{1}{N}\left[\boldsymbol{r}(1)^\dagger\boldsymbol{\Sigma}^{-1}\boldsymbol{r}(1),\ldots,\boldsymbol{r}(H_s)^\dagger\boldsymbol{\Sigma}^{-1}\boldsymbol{r}(H_s)\right]^T\end{aligned}$$

Substituting (9.37), (9.38), $\widehat{\boldsymbol{\theta}}_{r,0}$, and $\widehat{\boldsymbol{\theta}}_{s,0}$ into (9.36), after some algebraic manipulations, we come up with the following decision rule:

$$\sum_{t=1}^{H_s}\frac{|\boldsymbol{p}^\dagger\boldsymbol{\Sigma}^{-1}\boldsymbol{r}(t)|^2}{\boldsymbol{p}^\dagger\boldsymbol{\Sigma}^{-1}\boldsymbol{p}\,\boldsymbol{r}(t)^\dagger\boldsymbol{\Sigma}^{-1}\boldsymbol{r}(t)} \underset{H_0}{\overset{H_1}{\gtrless}} T_{Rao} \tag{9.39}$$

where T_{Rao} is the appropriate modification of the original threshold in (9.36).

9.A.1.3 Wald Test

The Wald test, based on the primary data, can be obtained exploiting the asymptotic efficiency of the ML estimate. Precisely, it is the following decision rule:

$$\widehat{\boldsymbol{\theta}}_{r,1}^T\left([\boldsymbol{J}^{-1}(\widehat{\boldsymbol{\theta}}_{r,1},\widehat{\boldsymbol{\theta}}_{s,1})]_{\boldsymbol{\theta}_r,\boldsymbol{\theta}_r}\right)^{-1}\widehat{\boldsymbol{\theta}}_{r,1} \underset{H_0}{\overset{H_1}{\gtrless}} \gamma_{Wald} \tag{9.40}$$

where γ_{Wald} is the detection threshold to be set according to the desired value of the false alarm probability P_{fa}, and $\widehat{\boldsymbol{\theta}}_{r,1}$ and $\widehat{\boldsymbol{\theta}}_{s,1}$ are the ML estimates of $\boldsymbol{\theta}_r$ and $\boldsymbol{\theta}_s$ under H_1, given in (9.31) and (9.33), respectively. Thus substituting (9.31), (9.33), and (9.38) into (9.40), after some algebraic manipulations, we obtain

$$\sum_{t=1}^{H_s} \frac{|\boldsymbol{p}^\dagger \boldsymbol{\Sigma}^{-1}\boldsymbol{r}(t)|^2}{\boldsymbol{p}^\dagger \boldsymbol{\Sigma}^{-1}\boldsymbol{p}\left(\boldsymbol{r}(t)^\dagger \boldsymbol{\Sigma}^{-1}\boldsymbol{r}(t) - \frac{|\boldsymbol{p}^\dagger \boldsymbol{\Sigma}^{-1}\boldsymbol{r}(t)|^2}{\boldsymbol{p}^\dagger \boldsymbol{\Sigma}^{-1}\boldsymbol{p}}\right)} \underset{H_0}{\overset{H_1}{\gtrless}} T_{Wald} \tag{9.41}$$

where T_{Wald} is the appropriate modification of the original threshold in (9.40).

9.A.1.4 Fisher Information Matrix Evaluation

In this subsection, we evaluate the blocks of the Fisher information matrix $\boldsymbol{J}(\boldsymbol{\theta}_r, \boldsymbol{\theta}_s)$. We begin with $\boldsymbol{J}_{\boldsymbol{\theta}_r,\boldsymbol{\theta}_r}(\boldsymbol{\theta}_r, \boldsymbol{\theta}_s)$ that is the $2H_s \times 2H_s$ matrix whose (i,j)-th element is

$$\boldsymbol{J}_{\boldsymbol{\theta}_r,\boldsymbol{\theta}_r}(\boldsymbol{\theta}_r, \boldsymbol{\theta}_s)(i,j) = -E\left[\frac{\partial \ln p(\boldsymbol{r}(1),\ldots,\boldsymbol{r}(H_s)|H_1,\boldsymbol{\theta}_r,\boldsymbol{\theta}_s,\boldsymbol{\Sigma})}{\partial\theta_r(i)\partial\theta_r(j)}\right] \qquad i,j \in (1,\ldots,2H_s)$$

where $\theta_r(l)$ denotes the l-th element of the vector $\boldsymbol{\theta}_r$.

It can be shown that the following equations hold true $\forall\, t,h \in (1,\ldots,H_s)$.

$$\frac{\partial \ln p(\boldsymbol{r}(1),\ldots,\boldsymbol{r}(H_s)|H_1,\boldsymbol{\theta}_r,\boldsymbol{\theta}_s,\boldsymbol{\Sigma})}{\partial\alpha_{R,t}\partial\alpha_{R,h}} = -\frac{2}{s(t)^2}\boldsymbol{p}^\dagger \boldsymbol{\Sigma}^{-1}\boldsymbol{p}\,\delta[t-h]$$

$$\frac{\partial \ln p(\boldsymbol{r}(1),\ldots,\boldsymbol{r}(H_s)|H_1,\boldsymbol{\theta}_r,\boldsymbol{\theta}_s,\boldsymbol{\Sigma})}{\partial\alpha_{R,t}\partial\alpha_{I,h}} = 0$$

$$\frac{\partial \ln p(\boldsymbol{r}(1),\ldots,\boldsymbol{r}(H_s)|H_1,\boldsymbol{\theta}_r,\boldsymbol{\theta}_s,\boldsymbol{\Sigma})}{\partial\alpha_{I,t}\partial\alpha_{I,h}} = -\frac{2}{s(t)^2}\boldsymbol{p}^\dagger \boldsymbol{\Sigma}^{-1}\boldsymbol{p}\,\delta[t-h]$$

$$\frac{\partial \ln p(\boldsymbol{r}(1),\ldots,\boldsymbol{r}(H_s)|H_1,\boldsymbol{\theta}_r,\boldsymbol{\theta}_s,\boldsymbol{\Sigma})}{\partial\alpha_{I,t}\partial\alpha_{R,h}} = 0$$

where $\delta[\cdot]$ is the Kroneker function. It follows that $\boldsymbol{J}_{\boldsymbol{\theta}_r,\boldsymbol{\theta}_r}(\boldsymbol{\theta}_r, \boldsymbol{\theta}_s)$ can be expressed as:

$$\boldsymbol{J}_{\boldsymbol{\theta}_r,\boldsymbol{\theta}_r}(\boldsymbol{\theta}_r, \boldsymbol{\theta}_s) = 2\boldsymbol{p}^\dagger \boldsymbol{\Sigma}^{-1}\boldsymbol{p}\ \mathrm{diag}\left(\frac{1}{s(1)^2}, \frac{1}{s(1)^2}, \ldots, \frac{1}{s(H_s)^2}, \frac{1}{s(H_s)^2}\right)$$

As to $\boldsymbol{J}_{\boldsymbol{\theta}_r,\boldsymbol{\theta}_s}(\boldsymbol{\theta}_r, \boldsymbol{\theta}_s)$, it is a $2H_s \times H_s$ matrix whose (i,j)-th entry is

$$\boldsymbol{J}_{\boldsymbol{\theta}_r,\boldsymbol{\theta}_s}(\boldsymbol{\theta}_r, \boldsymbol{\theta}_s)(i,j) = -E\left[\frac{\partial \ln p(\boldsymbol{r}(1),\ldots,\boldsymbol{r}(H_s)|H_1,\boldsymbol{\theta}_r,\boldsymbol{\theta}_s,\boldsymbol{\Sigma})}{\partial\theta_r(i)\partial\theta_s(j)}\right] \quad i\in(1,\ldots,2H_s)\, j\in(1,\ldots,H_s)$$

where $\theta_s(l)$ denotes the l-th element of the vector $\boldsymbol{\theta}_s$. Moreover, it can be shown that

$$\frac{\partial \ln p(\boldsymbol{r}(1),\ldots,\boldsymbol{r}(H_s)|H_1,\boldsymbol{\theta}_r,\boldsymbol{\theta}_s,\boldsymbol{\Sigma})}{\partial\alpha_{R,t}\partial\theta_s(h)} = -\frac{2}{s(t)^4}\mathrm{Re}\left(\boldsymbol{p}^\dagger \boldsymbol{\Sigma}^{-1}(\boldsymbol{r}(\mathrm{t}) - \alpha_{\mathrm{t}}\boldsymbol{p})\right)\delta\,[\mathrm{h}-\mathrm{t}]$$

$$\frac{\partial \ln p(\boldsymbol{r}(1),\ldots,\boldsymbol{r}(H_s)|H_1,\boldsymbol{\theta}_r,\boldsymbol{\theta}_s,\boldsymbol{\Sigma})}{\partial\alpha_{I,t}\partial\theta_s(h)} = -\frac{2}{s(t)^4}\mathrm{Im}\left(\boldsymbol{p}^\dagger \boldsymbol{\Sigma}^{-1}(\boldsymbol{r}(\mathrm{t}) - \alpha_{\mathrm{t}}\boldsymbol{p})\right)\delta\,[\mathrm{h}-\mathrm{t}]$$

Thus, observing that

$$E[(\boldsymbol{r}(t) - \alpha_t \boldsymbol{p})] = \boldsymbol{0}_{\{N,1\}}$$

we can conclude that $\boldsymbol{J}_{\boldsymbol{\theta}_r,\boldsymbol{\theta}_s}(\boldsymbol{\theta}) = \boldsymbol{0}_{2H_s,H_s}$.

9.A.2 Derivation of AMGLRT

In this section, we face with the detection problem (9.27), assuming that $\boldsymbol{\Sigma}$ is not known, $K = 0$, and $H_s \geq N$. The procedure followed to derive the AMGLRT is based on the subsequent steps:

- first, assuming the power levels $s(1)^2, \ldots, s(H_s)^2$ to be known, the MGLRT [29] is derived, resorting to the *method of sieves*, namely restricting the parameter space $(\boldsymbol{\Sigma}, \boldsymbol{\alpha})$ to a subspace such that the ML estimate of the parameters exists and is unique;
- second, substituting suitable estimates of the unknown parameters $s(1)^2, \ldots, s(H_s)^2$ in place of their exact values, the fully adaptive test statistic is synthesized.

In order to proceed further, it is necessary to specify the PDFs of the data under both hypotheses. To this end, resorting to Reference 108, and denoting by $\boldsymbol{a} = [\alpha_1^*, \ldots, \alpha_{H_s}^*]^T$, we can write

$$p(\boldsymbol{r}(1), \ldots, \boldsymbol{r}(H_s)|H_0, \boldsymbol{\theta}_s, \boldsymbol{\Sigma}) = \frac{c\, \exp\left[-\mathrm{Tr}\left(\boldsymbol{\Sigma}^+ \boldsymbol{Z} \bar{S}^{-1} \boldsymbol{Z}^\dagger\right)\right]}{\prod_{t=1}^{H_s} \det_p\left(s(t)^2 \boldsymbol{\Sigma}\right)} \delta(\boldsymbol{\Phi}_2^\dagger \boldsymbol{Z}) = m_0(\boldsymbol{Z}|H_0, \boldsymbol{\theta}_s, \boldsymbol{\Sigma}) \delta(\boldsymbol{\Phi}_2^\dagger \boldsymbol{Z}) \tag{9.42}$$

under H_0 and

$$\begin{aligned} p(\boldsymbol{r}(1), \ldots, \boldsymbol{r}(H_s)|H_1, \boldsymbol{\theta}_r, \boldsymbol{\theta}_s, \boldsymbol{\Sigma}) &= \frac{c\, \exp\left[-\mathrm{Tr}\left(\boldsymbol{\Sigma}^+ (\boldsymbol{Z} - \boldsymbol{p}\boldsymbol{a}^\dagger) \bar{S}^{-1} (\boldsymbol{Z} - \boldsymbol{p}\boldsymbol{a}^\dagger)^\dagger\right)\right]}{\prod_{t=1}^{H_s} \det_p\left(s(t)^2 \boldsymbol{\Sigma}\right)} \delta\left(\boldsymbol{\Phi}_2^\dagger (\boldsymbol{Z} - \boldsymbol{p}\boldsymbol{a}^\dagger)\right) \\ &= m_1(\boldsymbol{Z}|H_1, \boldsymbol{\theta}_r, \boldsymbol{\theta}_s, \boldsymbol{\Sigma}) \delta\left(\boldsymbol{\Phi}_2^\dagger (\boldsymbol{Z} - \boldsymbol{p}\boldsymbol{a}^\dagger)\right) \end{aligned} \tag{9.43}$$

under H_1, where

- c is a normalization constant;
- $\bar{S} = \mathrm{diag}(s(1)^2, \ldots, s(H_s)^2)$;
- $\delta(\,\cdot\,)$ is the product of the Dirac delta functions of the matrix elements of the argument;
- $\boldsymbol{\Sigma}^+$ is the Moore–Penrose inverse of $\boldsymbol{\Sigma}$;
- $\boldsymbol{\Phi}_2$ is a $N \times (N - \mathrm{Rank}(\boldsymbol{\Sigma}))$ matrix whose columns form an orthonormal basis for the null space of the columns of $\boldsymbol{\Sigma}$;
- $\boldsymbol{\theta}_r = [\alpha_{R,1}, \alpha_{I,1}, \ldots, \alpha_{R,H_s}, \alpha_{I,H_s}]^T$ is $2H_s$-dimensional vector, with $\alpha_{R,t}$ and $\alpha_{I,t}$ the real and the imaginary parts of α_t, $t = 1, \ldots, H_s$;
- $\boldsymbol{\theta}_s = [s(1)^2, \ldots, s(H_s)^2]^T$ is a H_s-dimensional vector.

Thus, we derive the MGLRT for known $\boldsymbol{\theta}_s$ applying the method of sieves. Specifically, we restrict, under the H_1 hypothesis, the parameter space to the set

$$\mathbf{H}_1 = \left\{ (\boldsymbol{\theta}_r, \boldsymbol{\Sigma}) : \rho = \text{Rank}(\widehat{\boldsymbol{\Sigma}}) \leq N-1, \quad \boldsymbol{\Sigma} = \widehat{\boldsymbol{\Psi}}^{\dagger} \begin{bmatrix} \boldsymbol{B} & \boldsymbol{0}_{\rho,N-\rho} \\ \boldsymbol{0}_{\rho,\rho} & \boldsymbol{0}_{N-\rho,N-\rho} \end{bmatrix} \widehat{\boldsymbol{\Psi}} \right\}$$

where $\widehat{\boldsymbol{\Sigma}} = (\boldsymbol{Z} - \boldsymbol{p}\boldsymbol{a}^{\dagger})\bar{\boldsymbol{S}}^{-1}(\boldsymbol{Z} - \boldsymbol{p}\boldsymbol{a}^{\dagger})^{\dagger}$, $\boldsymbol{B} \succeq \boldsymbol{0}_{\{\rho,\rho\}}$ is a $\rho \times \rho$ matrix and $\widehat{\boldsymbol{\Psi}}$ is the matrix of the normalized eigenvectors of $\widehat{\boldsymbol{\Sigma}}$. Hence, we consider the following decision rule:

$$\frac{\sup\limits_{(\boldsymbol{\theta}_r, \boldsymbol{\Sigma}) \in \mathbf{H}_1} m_1(\boldsymbol{Z}|H_1, \boldsymbol{\theta}_r, \boldsymbol{\theta}_s, \boldsymbol{\Sigma})}{\sup\limits_{\boldsymbol{\Sigma} \succeq \boldsymbol{0}} m_0(\boldsymbol{Z}|H_0, \boldsymbol{\theta}_s, \boldsymbol{\Sigma})} \underset{H_0}{\overset{H_1}{\gtrless}} \gamma_{MGLRT} \tag{9.44}$$

where γ_{MGLRT} is the detection threshold to be set according to the desired value of the false alarm probability P_{fa}.

Assuming $H_s \geq N$ and exploiting Lemma 1 of Reference 29, expression (9.44) can be recast as

$$\sup_{\boldsymbol{\theta}_r : \rho \leq N-1} \frac{\det(\boldsymbol{Z}\bar{\boldsymbol{S}}^{-1}\boldsymbol{Z}^{\dagger})}{\det_{\rho}\left((\boldsymbol{Z} - \boldsymbol{p}\boldsymbol{a}^{\dagger})\bar{\boldsymbol{S}}^{-1}(\boldsymbol{Z} - \boldsymbol{p}\boldsymbol{a}^{\dagger})^{\dagger}\right)} \underset{H_0}{\overset{H_1}{\gtrless}} \gamma_{MGLRT}$$

Moreover, the application of Theorem 1 of Reference 29 leads to

$$\frac{\det(\boldsymbol{Z}\bar{\boldsymbol{S}}^{-1}\boldsymbol{Z}^{\dagger})}{\det_{\rho}\left((\boldsymbol{I}_N - \boldsymbol{p}\boldsymbol{p}^{\dagger})\boldsymbol{Z}\bar{\boldsymbol{S}}^{-1}\boldsymbol{Z}^{\dagger}(\boldsymbol{I}_N - \boldsymbol{p}\boldsymbol{p}^{\dagger})\right)} \underset{H_0}{\overset{H_1}{\gtrless}} \gamma_{MGLRT} \tag{9.45}$$

In order to make the derived detector fully adaptive, we plug into the left-hand side of (9.45) suitable estimates of the unknown power levels, $\hat{s}(t)^2$'s say, in place of their exact values $s(t)^2$. Precisely, we assume that

$$\hat{s}(t)^2 = \frac{1}{N}\boldsymbol{r}(t)^{\dagger}\left(\boldsymbol{I}_N - \boldsymbol{p}\boldsymbol{p}^{\dagger}\right)\boldsymbol{r}(t) \qquad t = 1, \ldots, H_s \tag{9.46}$$

Note that the data used for estimation purposes are previously projected onto the null space of $\boldsymbol{p}$ in order to get target-free observations [109]. In fact, the vectors $\boldsymbol{r}(t)$'s are pre-multiplied for the projection matrix $\boldsymbol{I}_N - \boldsymbol{p}\boldsymbol{p}^{\dagger}$ that spans the null space of $\boldsymbol{p}$. Interestingly, it can be shown that the proposed estimates are unbiased, but for a multiplicative factor, and their variance tends to zero as N diverges (see Appendix 9.A.2.1 for the proof). Notice also that the quoted bias is irrelevant for detection purposes as it factors out between the numerator and the denominator of the resulting test statistic under both hypotheses. Finally, substituting (9.46) in (9.45) we come up with AMGLRT (9.18)

$$\frac{\det\left(\boldsymbol{Z}\boldsymbol{S}_1^{-1}\boldsymbol{Z}^{\dagger}\right)}{\det_{\rho}\left((\boldsymbol{I}_N - \boldsymbol{p}\boldsymbol{p}^{\dagger})\boldsymbol{Z}\boldsymbol{S}_1^{-1}\boldsymbol{Z}^{\dagger}(\boldsymbol{I}_N - \boldsymbol{p}\boldsymbol{p}^{\dagger})\right)} \underset{H_0}{\overset{H_1}{\gtrless}} T_{AMGLRT} \tag{9.47}$$

9.A.2.1 Properties of the Estimates of the Textures

In this subsection, we prove that the estimates (9.46) are unbiased (but for a multiplicative factor), and their variance tends to zero as N diverges. To this end, we assume that $\boldsymbol{\Sigma}$ is bounded in spectral norm [110], namely

$$\forall N \ \lambda_{MAX}(\boldsymbol{\Sigma}) < \eta \in \mathcal{R} \tag{9.48}$$

with $\lambda_{MAX}(\cdot)$ the maximum eigenvalue of the argument.

We first show that the estimates are unbiased but for a multiplicative factor. In fact, for $t = 1, \ldots, H_s$

$$E\left[\hat{s}(t)^2\right] = \frac{s(t)^2}{N}\mathrm{Tr}\left[\left(\boldsymbol{I}_{\mathrm{N}} - \boldsymbol{p}\boldsymbol{p}^{\dagger}\right)\boldsymbol{\Sigma}\right]$$

As to the variances of $\widehat{s}(t)^2$, it can be shown that for $t = 1, \ldots, H_s$

$$Var\left[\hat{s}(t)^2\right] = \frac{s(t)^4}{N^2}\sum_{l=1}^{N}\lambda_l^2\left[\left(\boldsymbol{I}_N - \boldsymbol{p}\boldsymbol{p}^{\dagger}\right)\boldsymbol{\Sigma}\right]$$

with $\lambda_l\left[\left(\boldsymbol{I}_N - \boldsymbol{p}\boldsymbol{p}^{\dagger}\right)\boldsymbol{\Sigma}\right], l = 1, \ldots, N$, the eigenvalues of the matrix $\left(\boldsymbol{I}_N - \boldsymbol{p}\boldsymbol{p}^{\dagger}\right)\boldsymbol{\Sigma}$. In order to prove that the variance of the estimates tends to zero as N diverges we first highlight that

$$\begin{aligned}\lambda_{MAX}\left[\left(\boldsymbol{I}_N - \boldsymbol{p}\boldsymbol{p}^{\dagger}\right)\boldsymbol{\Sigma}\right] &= \lambda_{MAX}\left[\begin{pmatrix}\boldsymbol{I}_{N-1} & \boldsymbol{0}_{\{N-1,1\}} \\ \boldsymbol{0}_{\{1,N-1\}} & 0\end{pmatrix}\boldsymbol{U}\boldsymbol{\Sigma}\boldsymbol{U}^{\dagger}\begin{pmatrix}\boldsymbol{I}_{N-1} & \boldsymbol{0}_{\{N-1,1\}} \\ \boldsymbol{0}_{\{1,N-1\}} & 0\end{pmatrix}\right] \\ &\leq \lambda_{MAX}\left(\boldsymbol{U}\boldsymbol{\Sigma}\boldsymbol{U}^{\dagger}\right) = \lambda_{MAX}(\boldsymbol{\Sigma})\end{aligned} \tag{9.49}$$

where $\lambda_{MAX}[\boldsymbol{A}]$ denotes the maximum eigenvalue of $\boldsymbol{A} \in \mathbb{H}^N$, $\boldsymbol{U}$ is a unitary matrix aimed at rotating the vector $\boldsymbol{p}$ into the N-dimensional vector $\boldsymbol{e}_N = [0, \ldots, 0, 1]^{\dagger}$, and the last inequality follows from the *inclusion principle* [110, p. 189, 4.3.15].

Thus, exploiting (9.48) in conjunction with (9.49) and the following inequality

$$\sum_{l=1}^{N}\lambda_l^2\left[\left(\boldsymbol{I}_N - \boldsymbol{p}\boldsymbol{p}^{\dagger}\right)\boldsymbol{\Sigma}\right] \leq N\lambda_{MAX}^2\left[\left(\boldsymbol{I}_N - \boldsymbol{p}\boldsymbol{p}^{\dagger}\right)\boldsymbol{\Sigma}\right]$$

we can conclude that

$$\lim_{N\to\infty} Var\left[\hat{s}(t)^2\right] = 0 \qquad t = 1, \ldots, H_s$$

References

[1] A. Farina and F. A. Studer, "Detection with high resolution radar: Great promise, big challenge," *Microwave Journal*, vol. 34, pp. 263–273, May 1991.

[2] P. Hughes, "A high-resolution radar detection strategy," *IEEE Transactions on Aerospace and Electronic Systems*, vol. 19, no. 5, pp. 663–667, September 1983.

[3] E. Conte, A. De Maio, and G. Ricci, "GLRT-based adaptive detection algorithms for range-spread targets," *IEEE Transactions on Signal Processing*, vol. 49, no. 7, pp. 1336–1348, July 2001.

[4] N. Bon, A. Khenchaf, and R. Garello, "GLRT subspace detection for range and Doppler distributed targets," *IEEE Transactions on Aerospace and Electronic Systems*, vol. 44, no. 2, pp. 678–696, April 2008.

[5] E. Conte, A. De Maio, and G. Ricci, "CFAR detection of distributed targets in non-Gaussian disturbance," *IEEE Transactions on Aerospace and Electronic Systems*, vol. 38, no. 2, pp. 612–621, April 2002.

[6] R. Nitzberg, "Effect of a few dominant specular reflectors target model upon target detection," *IEEE Transactions on Aerospace and Electronic Systems*, vol. 14, no. 4, pp. 670–673, July 1978.

[7] P. Swerling, "Probability of detection for fluctuating targets," *IRE Transactions on Information Theory*, vol. 6, no. 2, pp. 269–308, April 1960.

[8] K. Ward, C. Baker, and S. Watts, "Maritime surveillance radar. Part I: Radar scattering from the ocean surface," *Radar and Signal Processing, IEE Proceedings F*, vol. 137, no. 2, pp. 51–62, April 1990.

[9] S. Haykin, R. Bakker, and B. Currie, "Uncovering nonlinear dynamics – The case study of sea clutter," *Proceedings of the IEEE*, vol. 90, no. 5, pp. 860–881, May 2002.

[10] M. Di Bisceglie and C. Galdi, "Random walk based characterisation of radar backscatter from the sea surface," *Radar, Sonar and Navigation, IEE Proceedings*, vol. 145, no. 4, pp. 216–225, August 1998.

[11] K. Sangston and K. Gerlach, "Coherent detection of radar targets in a non-Gaussian background," *IEEE Transactions on Aerospace and Electronic Systems*, vol. 30, no. 2, pp. 330–340, April 1994.

[12] M. Greco, F. Bordoni, and F. Gini, "X-band sea-clutter nonstationarity: Influence of long waves," *IEEE Journal of Oceanic Engineering*, vol. 29, no. 2, pp. 269–283, April 2004.

[13] E. Conte, A. De Maio, and C. Galdi, "Statistical analysis of real clutter at different range resolutions," *IEEE Transactions on Aerospace and Electronic Systems*, vol. 40, no. 3, pp. 903–918, July 2004.

[14] A. Farina, F. Gini, M. Greco, and L. Verrazzani, "High resolution sea clutter data: Statistical analysis of recorded live data," *Radar, Sonar and Navigation, IEE Proceedings*, vol. 144, no. 3, pp. 121–130, June 1997.

[15] J. Carretero-Moya, J. Gismero-Menoyo, A. Blanco-del Campo, and A. Asensio-Lopez, "Statistical analysis of a high-resolution sea-clutter database," *IEEE Transactions on Geoscience and Remote Sensing*, vol. 48, no. 4, pp. 2024–2037, April 2010.

[16] P. L. Herselman, C. J. Baker, and H. J. de Wind, "An analysis of x-band calibrated sea clutter and small boat reflectivity at medium-to-low grazing angles," *International Journal of Navigation and Observation*, vol. 2008, pp. 1–14, 2008.

[17] K. Ward, "Compound representation of high resolution sea clutter," *Electronics Letters*, vol. 17, no. 16, pp. 561–563, August 1981.

[18] K. Yao, "A representation theorem and its applications to spherically-invariant random processes," *IEEE Transactions on Information Theory*, vol. 19, no. 5, pp. 600–608, September 1973.

[19] E. Conte and M. Longo, "Characterisation of radar clutter as a spherically invariant random process," *Communications, Radar and Signal Processing, IEE Proceedings F*, vol. 134, no. 2, pp. 191–197, April 1987.

[20] M. Rangaswamy, D. Weiner, and A. Ozturk, "Non-Gaussian random vector identification using spherically invariant random processes," *IEEE Transactions on Aerospace and Electronic Systems*, vol. 29, no. 1, pp. 111–124, January 1993.

[21] B. Picinbono, "Spherically invariant and compound Gaussian stochastic processes (corresp.)," *IEEE Transactions on Information Theory*, vol. 16, no. 1, pp. 77–79, January 1970.

[22] G. Wise and N. Gallagher, "On spherically invariant random processes (corresp.)," *IEEE Transactions on Information Theory*, vol. 24, no. 1, pp. 118–120, January 1978.

[23] J. Tang and Z. Zhu, "Analysis of extended target detectors," in *Proceedings of the IEEE 1996 National Aerospace and Electronics Conference, 1996 (NAECON 1996)*, vol. 1, pp. 364–368, May 1996.

[24] E. Conte and M. Lops, "Clutter-map CFAR detection for range-spread targets in non-Gaussian clutter. Part I: System design," *IEEE Transactions on Aerospace and Electronic Systems*, vol. 33, no. 2, pp. 432–443, April 1997.

[25] E. Conte, M. Di Bisceglie, and M. Lops, "Clutter-map CFAR detection for range-spread targets in non-Gaussian clutter. Part II: Performance assessment," *IEEE Transactions on Aerospace and Electronic Systems*, vol. 33, no. 2, pp. 444–455, April 1997.

[26] K. Gerlach, M. Steiner, and F. Lin, "Detection of a spatially distributed target in white noise," *Signal Processing Letters, IEEE*, vol. 4, no. 7, pp. 198–200, July 1997.

[27] S. Blunt, K. Gerlach, and J. Heyer, "HRR detector for slow-moving targets in sea clutter," *IEEE Transactions on Aerospace and Electronic Systems*, vol. 43, no. 3, pp. 965–974, July 2007.

[28] K. Gerlach, "Spatially distributed target detection in non-Gaussian clutter," *IEEE Transactions on Aerospace and Electronic Systems*, vol. 35, no. 3, pp. 926–934, July 1999.

[29] K. Gerlach and M. Steiner, "Adaptive detection of range distributed targets," *IEEE Transactions on Signal Processing*, vol. 47, no. 7, pp. 1844–1851, July 1999.

[30] E. Conte and A. De Maio, "Distributed target detection in compound-Gaussian noise with Rao and Wald tests," *IEEE Transactions on Aerospace and Electronic Systems*, vol. 39, no. 2, pp. 568–582, April 2003.

[31] A. De Maio, "Blind adaptive detection of distributed targets in compound-Gaussian clutter," in *Radar Conference, 2003. Proceedings of the 2003 IEEE*, May 2003, pp. 291–297.

[32] Y. Jin and B. Friedlander, "A CFAR adaptive subspace detector for second-order Gaussian signals," *IEEE Transactions on Signal Processing*, vol. 53, no. 3, pp. 871–884, March 2005.

[33] J. Guan and X. Zhang, "Subspace detection for range and Doppler distributed targets with Rao and Wald tests," *Signal Processing*, in press. vol. 91, no. 1, pp. 51–60, January 2011.

[34] E. Conte, M. Lops, and G. Ricci, "Asymptotically optimum radar detection in compound-Gaussian clutter," *IEEE Transactions on Aerospace and Electronic Systems*, vol. 31, no. 2, pp. 617–625, April 1995.

[35] F. Gini, "Sub-optimum coherent radar detection in a mixture of K-distributed and Gaussian clutter," *Radar, Sonar and Navigation, IEE Proceedings*, vol. 144, no. 1, pp. 39–48, February 1997.

[36] L. L. Scharf, *Statistical Signal Processing: Detection Estimation, and Time Series Analysis*. New Jersey: Prentice Hall, 1991.

[37] A. De Maio and E. Conte, "Uniformly most powerful invariant detection in spherically invariant random vector distributed clutter," *Radar, Sonar & Navigation, IET*, vol. 4, no. 4, pp. 560–563, August 2010.

[38] X. Shuai, L. Kong, and J. Yang, "Performance analysis of GLRT-based adaptive detector for distributed targets in compound-Gaussian clutter," *Signal Processing*, vol. 90, no. 1, pp. 16–23, 2010. [Online]. Available: http://www.sciencedirect.com/science/article/B6V18-4WBK76V-2/2/19a1e770eb577e84210beef39754bd60

[39] E. Conte, A. De Maio, and G. Ricci, "Covariance matrix estimation for adaptive CFAR detection in compound-Gaussian clutter," *IEEE Transactions on Aerospace and Electronic Systems*, vol. 38, no. 2, pp. 415–426, April 2002.

[40] F. Gini and A. Farina, "Matched subspace CFAR detection of hovering helicopters," *IEEE Transactions on Aerospace and Electronic Systems*, vol. 35, no. 4, pp. 1293–1305, October 1999.

[41] F. Gini, M. Greco, and A. Farina, "Radar detection and preclassification based on multiple hypothesis," *IEEE Transactions on Aerospace and Electronic Systems*, vol. 40, no. 3, pp. 1046–1059, July 2004.

[42] F. Gini and A. Farina, "Vector subspace detection in compound-Gaussian clutter. Part I: Survey and new results," *IEEE Transactions on Aerospace and Electronic Systems*, vol. 38, no. 4, pp. 1295–1311, October 2002.

[43] L. Scharf and B. Friedlander, "Matched subspace detectors," *IEEE Transactions on Signal Processing*, vol. 42, no. 8, pp. 2146–2157, August 1994.

[44] L. Scharf, S. Kraut, and M. McCloud, "A review of matched and adaptive subspace detectors," in *Adaptive Systems for Signal Processing, Communications, and Control Symposium 2000. AS-SPCC. The IEEE 2000*, pp. 82–86, 2000.

[45] S. Kraut, L. Scharf, and L. McWhorter, "Adaptive subspace detectors," *IEEE Transactions on Signal Processing*, vol. 49, no. 1, pp. 1–16, January 2001.

[46] S. Kraut and L. Scharf, "The CFAR adaptive subspace detector is a scale-invariant GLRT," in *Proceedings of the Ninth IEEE SP Workshop on Statistical Signal and Array Processing, 1998*, pp. 57–60, September 1998.

[47] F. Gini, A. Farina, and M. Montanari, "Vector subspace detection in compound-Gaussian clutter. Part II: Performance analysis," *IEEE Transactions on Aerospace and Electronic Systems*, vol. 38, no. 4, pp. 1312–1323, October 2002.

[48] A. Younsi, M. Greco, F. Gini, and A. Zoubir, "Performance of the adaptive generalised matched subspace constant false alarm rate detector in non-Gaussian noise: An experimental analysis," *Radar, Sonar & Navigation, IET*, vol. 3, no. 3, pp. 195–202, June 2009.

[49] E. Conte, A. De Maio, and C. Galdi, "CFAR detection of multidimensional signals: An invariant approach," *IEEE Transactions on Signal Processing*, vol. 51, no. 1, pp. 142–151, January 2003.

[50] E. Kelly, "An adaptive detection algorithm," *IEEE Transactions on Aerospace and Electronic Systems*, vol. 22, no. 2, pp. 115–127, March 1986.

[51] F. Gini and J. Michels, "Performance analysis of two covariance matrix estimators in compound-Gaussian clutter," *Radar, Sonar and Navigation, IEE Proceedings*, vol. 146, no. 3, pp. 133–140, June 1999.

[52] I. Reed, J. Mallett, and L. Brennan, "Rapid convergence rate in adaptive arrays," *IEEE Transactions on Aerospace and Electronic Systems*, vol. 10, no. 6, pp. 853–863, November 1974.

[53] C. Richmond, "A note on non-Gaussian adaptive array detection and signal parameter estimation," *Signal Processing Letters, IEEE*, vol. 3, no. 8, pp. 251–252, August 1996.

[54] E. Conte, M. Lops, and G. Ricci, "Adaptive radar detection in compound-Gaussian clutter," in *Proceedings of Eusipco '94*, pp. 526–529, September 1994.

[55] F. Gini, M. Greco, and L. Verrazzani, "Detection problem in mixed clutter environment as a Gaussian problem by adaptive preprocessing," *Electronics Letters*, vol. 31, no. 14, pp. 1189–1190, July 1995.

[56] E. Conte and A. De Maio, "Mitigation techniques for non-Gaussian sea clutter," *IEEE Journal of Oceanic Engineering*, vol. 29, no. 2, pp. 284–302, April 2004.

[57] R. Nitzberg, "Application of maximum likelihood estimation of persymmetric covariance matrices to adaptive processing," *IEEE Transactions on Aerospace and Electronic Systems*, vol. 16, no. 1, pp. 124–127, January 1980.

[58] E. Conte, A. De Maio, and G. Ricci, "Recursive estimation of the covariance matrix of a compound-Gaussian process and its application to adaptive CFAR detection," *IEEE Transactions on Signal Processing*, vol. 50, no. 8, pp. 1908–1915, August 2002.

[59] F. Gini and M. Greco, "Covariance matrix estimation for CFAR detection in correlated heavy tailed clutter," *Signal Processing*, vol. 82, no. 12, pp. 1847–1859, 2002. [Online]. Available: http://www.sciencedirect.com/science/article/B6V18-46FPV88-3/2/6031e18cdd4a35a79e0e75fc329df27d

[60] J. Hiemstra, "Colored diagonal loading," in *Proceedings of the IEEE 2002 Radar Conference*, pp. 386–390, 2002.

[61] B. Carlson, "Covariance matrix estimation errors and diagonal loading in adaptive arrays," *IEEE Transactions on Aerospace and Electronic Systems*, vol. 24, no. 4, pp. 397–401, July 1988.

[62] M. Steiner and K. Gerlach, "Fast converging adaptive processor or a structured covariance matrix," *IEEE Transactions on Aerospace and Electronic Systems*, vol. 36, no. 4, pp. 1115–1126, October 2000.

[63] A. Aubry, A. De Maio, L. Pallotta, and A. Farina, "Maximum likelihood estimation of a structured covariance matrix with a condition number constraint," *IEEE Transactions on Signal Processing*, vol. 60, no. 6, pp. 3004–3021, June 2012.

[64] A. Aubry, A. De Maio, L. Pallotta, and A. Farina, "Radar detection of distributed targets in homogeneous interference whose inverse covariance structure is defined via unitary invariant functions," *IEEE Transactions on Signal Processing*, vol. 61, no. 2, pp. 4949–4961, October 2013.

[65] B. Kang, V. Monga, and M. Rannfaswamy, "Rank-constrained maximum likelihood estimation of structured covariance matrices," *IEEE Transactions on Aerospace and Electronic Systems*, vol. 50, no. 1, pp. 501–515, January 2014.

[66] E. Conte, A. De Maio, A. Farina, and G. Foglia, "Design and analysis of a knowledge-aided radar detector for Doppler processing," *IEEE Transactions on Aerospace and Electronic Systems*, vol. 42, no. 3, pp. 1058–1079, July 2006.

[67] A. De Maio, G. Foglia, A. Farina, and M. Piezzo, "Estimation of the covariance matrix based on multiple a-priori models," in *2010 IEEE Radar Conference*, pp. 1025–1029, May 2010.

[68] A. De Maio, A. Farina, and G. Foglia, "Knowledge-aided Bayesian radar detectors and their application to live data," *IEEE Transactions on Aerospace and Electronic Systems*, vol. 46, no. 1, pp. 170–183, January 2010.

[69] A. De Maio, S. D. Nicola, L. Landi, and A. Farina, "Knowledge-aided covariance matrix estimation: A MAXDET approach," *Radar, Sonar & Navigation, IET*, vol. 3, no. 4, pp. 341–356, 2009.

[70] W. Melvin and G. Showman, "An approach to knowledge-aided covariance estimation," *IEEE Transactions on Aerospace and Electronic Systems*, vol. 42, no. 3, pp. 1021–1042, July 2006.

[71] W. Melvin and J. Guerci, "Knowledge-aided signal processing: A new paradigm for radar and other advanced sensors," *IEEE Transactions on Aerospace and Electronic Systems*, vol. 42, no. 3, pp. 983–996, July 2006.

[72] S. Blunt, K. Gerlach, and M. Rangaswamy, "STAP using knowledge-aided covariance estimation and the FRACTA algorithm," *IEEE Transactions on Aerospace and Electronic Systems*, vol. 42, no. 3, pp. 1043–1057, July 2006.

[73] J. Bergin, C. Teixeira, P. Techau, and J. Guerci, "Improved clutter mitigation performance using knowledge-aided space–time adaptive processing," *IEEE Transactions on Aerospace and Electronic Systems*, vol. 42, no. 3, pp. 997–1009, July 2006.

[74] A. Aubry, V. Carotenuto, A. De Maio, and G. Foglia, "Exploiting multiple a priori spectral models for adaptive radar detection," *Radar, Sonar & Navigation, IET*, vol. 8, no. 7, pp. 695–707, October 2014.

[75] M. Rangaswamy, F. Lin, and K. Gerlach, "Robust adaptive signal processing methods for heterogeneous radar clutter scenarios," in *Proceedings of the 2003 IEEE Radar Conference*, pp. 265–272, May 2003.

[76] M. Rangaswamy, J. H. Michels, and B. Himed, "Statistical analysis of the nonhomogeneity detector for STAP applications," *Digital Signal Processing*, vol. 14, no. 3, pp. 253–267, 2004. [Online]. Available: http://www.sciencedirect.com/science/article/B6WDJ-48SP09H-1/2/837051deaa6e5d1d0dfb01d952a95a37

[77] M. Rangaswamy, F. C. Lin, and K. R. Gerlach, "Robust adaptive signal processing methods for heterogeneous radar clutter scenarios," *Signal Processing*, vol. 84, no. 9, pp. 1653–1665, 2004, special Section on New Trends and Findings in Antenna Array Processing for Radar. [Online]. Available: http://www.sciencedirect.com/science/article/B6V18-4CHRK2X-D/2/aedb3bfea93c2fb53c3edc8e7bb3ccd8

[78] M. Rangaswamy, "Statistical analysis of the nonhomogeneity detector for non-Gaussian interference backgrounds," *IEEE Transactions on Signal Processing*, vol. 53, no. 6, pp. 2101–2111, June 2005.

[79] K. Gerlach, "Outlier resistant adaptive matched filtering," *IEEE Transactions on Aerospace and Electronic Systems*, vol. 38, no. 3, pp. 885–901, July 2002.

[80] D. Rabideau and A. Steinhardt, "Improved adaptive clutter cancellation through data-adaptive training," *IEEE Transactions on Aerospace and Electronic Systems*, vol. 35, no. 3, pp. 879–891, July 1999.

[81] E. Aboutanios and B. Mulgrew, "Hybrid detection approach for STAP in heterogeneous clutter," *IEEE Transactions on Aerospace and Electronic Systems*, vol. 46, no. 3, pp. 1021 –1033, July 2010.

[82] A. Aubry, A. De Maio, L. Pallotta, and A. Farina, "Covariance matrix estimation via geometric barycenters and its application to radar training data selection," *Radar, Sonar & Navigation, IET*, vol. 7, no. 6, pp. 600–614, July 2013.

[83] M. Rangaswamy and J. Michels, "A parametric multichannel detection algorithm for correlated non-Gaussian random processes," in *1997 IEEE National Radar Conference*, pp. 349–354, May 1997.

[84] M. Rangaswamy, J. Michels, and D. Weiner, "Multichannel detection for correlated non-Gaussian random processes based on innovations," *IEEE Transactions on Signal Processing*, vol. 43, no. 8, pp. 1915–1922, August 1995.

[85] J. Roman, M. Rangaswamy, D. Davis, Q. Zhang, B. Himed, and J. Michels, "Parametric adaptive matched filter for airborne radar applications," *IEEE Transactions on Aerospace and Electronic Systems*, vol. 36, no. 2, pp. 677–692, April 2000.

[86] K. Gerlach and F. Lin, "Convergence performance of binary adaptive detectors," *IEEE Transactions on Aerospace and Electronic Systems*, vol. 31, no. 1, pp. 329–340, January 1995.

[87] A. Blanco-del Campo, A. Asensio-Lopez, J. Gismero-Menoyo, B. Dorta-Naranjo, and J. Carretero-Moya, "Instrumental CWLFM high-range resolution radar in millimeter waveband for ISAR imaging," *Sensors Journal, IEEE*, vol. 11, no. 2, pp. 418–429, 2011.

[88] A. del Campo, A. Lopez, B. Naranjo, J. Menoyo, D. Moran, and C. Duarte, "CWLFM millimeter-wave radar for ISAR imaging with range coverage," in *2005 IEEE International Radar Conference*, pp. 933–938, May 2005.

[89] A. Asensio-Lopez, A. Blanco-del Campo, J. Gismero-Menoyo, D. Ramirez-Moran, G. Torregrosa-Penalva, B. Dorta-Naranjo, and C. Carmona-Duarte, "High range-resolution radar scheme for imaging with tunable distance limits," *Electronics Letters*, vol. 40, no. 17, pp. 1085–1086, August 2004.

[90] J. Munoz-Ferreras, F. Perez-Martinez, J. Calvo-Gallego, A. Asensio-Lopez, B. Dorta-Naranjo, and A. Blanco-del Campo, "Traffic surveillance system based on a high-resolution radar," *IEEE Transactions on Geoscience and Remote Sensing*, vol. 46, no. 6, pp. 1624–1633, June 2008.

[91] J. Munoz-Ferreras and F. Perez-Martinez, "Pitch estimation for non-cooperative maritime targets in ISAR scenarios," *Radar, Sonar & Navigation, IET*, vol. 3, no. 5, pp. 521–529, October 2009.

[92] E. Jakeman and P. Pusey, "A model for non-Rayleigh sea echo," *IEEE Transactions on Antennas and Propagation*, vol. 24, no. 6, pp. 806–814, November 1976.

[93] S. Watts, "Radar detection prediction in K-distributed sea clutter and thermal noise," *IEEE Transactions on Aerospace and Electronic Systems*, vol. 23, no. 1, pp. 40–45, January 1987.

[94] M. Greco, F. Gini, and M. Rangaswamy, "Statistical analysis of measured polarimetric clutter data at different range resolutions," *Radar, Sonar and Navigation, IEE Proceedings*, vol. 153, no. 6, pp. 473–481, December 2006.

[95] M. Greco, F. Gini, A. Younsi, M. Rangaswamy, and A. Zoubir, "Non-stationary sea clutter: Impact on disturbance covariance matrix estimate and detector CFAR," in *2008 International Conference on Radar*, pp. 558–562, September 2008.

[96] T. Moon and P. Bawden, "High resolution RCS measurements of boats," *Radar and Signal Processing, IEE Proceedings F*, vol. 138, no. 3, pp. 218–222, June 1991.

[97] G. Brooker, C. Lobsey, and R. Hennessy, "Low cost measurement of small boat RCS at 94 GHz," in *9th International Conference on Control, Automation, Robotics and Vision, 2006 (ICARCV '06)*, pp. 1–8, December 2006.

[98] M. A. Richards, J. A. Scheer, and W. A. Holm, *Principles of Modern Radar: Basic Principles*. Raleigh, NC: Scitech Publishing, 2010.

[99] M. Skolnik, *Radar Handbook*, 3rd ed. New York: McGraw-Hill, 2008.

[100] P. Swerling, "Radar probability of detection for some additional fluctuating target cases," *IEEE Transactions on Aerospace and Electronic Systems*, vol. 33, no. 2, pp. 698–709, April 1997.

[101] A. De Maio, A. Farina, and G. Foglia, "Target fluctuation models and their application to radar performance prediction," *Radar, Sonar and Navigation, IEE Proceedings*, vol. 151, no. 5, pp. 261–269, October 2004.

[102] J. Echard, "Estimation of radar detection and false alarm probability," *IEEE Transactions on Aerospace and Electronic Systems*, vol. 27, no. 2, pp. 255–260, March 1991.

[103] R. Nitzberg, "Detection loss of the sample matrix inversion technique," *IEEE Transactions on Aerospace and Electronic Systems*, vol. 20, no. 6, pp. 824–827, November 1984.

[104] A. De Maio, G. Foglia, E. Conte, and A. Farina, "CFAR behavior of adaptive detectors: An experimental analysis," *IEEE Transactions on Aerospace and Electronic Systems*, vol. 41, no. 1, pp. 233–251, January 2005.

[105] M. Rangaswamy, D. Weiner, and A. Ozturk, "Computer generation of correlated non-Gaussian radar clutter," *IEEE Transactions on Aerospace and Electronic Systems*, vol. 31, no. 1, pp. 106–116, January 1995.

[106] E. Conte, M. Longo, and M. Lops, "Modelling and simulation of non-Rayleigh radar clutter," *Radar and Signal Processing, IEE Proceedings F*, vol. 138, no. 2, pp. 121–130, April 1991.

[107] H. L. V. Trees, *Detection, Estimation, and Modulation Theory, Pt. 1*. John Wiley & Sons, New York, 1968.

[108] K. Mille, *Multidimensional Gaussian Distributions*. John Wiley & Sons, New York, 1964.

[109] M. Greco, F. Gini, and M. Diani, "Robust CFAR detection of random signals of random signals in compound-Gaussian clutter plus thermal noise," *Radar, Sonar and Navigation, IEE Proceedings*, vol. 148, no. 4, pp. 227–232, August 2001.

[110] R. A. Horn and C. R. Johnson, *Matrix Analysis*. United Kingdom: Cambridge University Press, 1985.

Index

Adaptive Beamformer Orthogonal Rejection Test (ABORT) 104
  AMF-ABORT 105–7
  Whitened ABORT 89, 104, 111, 117
Adaptive Coherence Estimator (ACE) test 44, 136, 298
  subspace version of 72–4
  probability of detection (P_D or P_d) and probability of false alarm (P_{FA} or P_{fa}) 56
Adaptive Matched Filter (AMF) 5, 44, 89, 136, 166, 298
  AMF-ABORT 105–7
  AMF-KWA 113–17
  AMF-Rao 108–10
  AMF-WA 111–13
  Bayesian AMF (B-AMF) 138
  versus generalized likelihood ratio test 171–3
  knowledge-aided parametric AMF (KA-PAMF) 152, 154, 156–8
  PS-AMF 315
  subspace version of 72
    probability of detection (P_D or P_d) and probability of false alarm (P_{FA} or P_{fa}) 55–6
Adaptive Normalized Matched Filter (ANMF) 312, 340
  Fixed Point ANMF (FP-ANMF) 313, 323
  Persymmetric ANMF (P-ANMF) 313, 316
  Recursive P-ANMF (RP-ANMF) 313, 316–17
Adaptive Radar, Subspace Detection for 43
  maximum invariants 53–4
  P_D and P_{FA}, analytical expressions for 54
    for subspace ACE test 56
    for subspace AMF test 55–6
    for subspace GLRT 54–5
  performance results 57–69
  subspace detectors for adaptive radar 43
Adaptive Radar Detection
  using covariance matrix structure 212–14
    TVAR(m) approximation of Hermitian covariance matrix 215–18
    TVAR(m)-based adaptive detectors, simulation results of 223–37
    TVAR(m)-based adaptive filters, performance analysis of 218–23
    TVAR(m) or AR(m) interferences, adaptive detectors for 218–23
  using data partitioning 239–42
    analysis performance of "one-stage" adaptive CFAR detectors versus "two-stage" adaptive processing 205, 242–7
    Comparative Detection Performance Analysis 247–55
  using EL-selected loading 168
    adaptive thresholding using primary data 168–70
    different adaptive process per test cell 170–209
    single adaptive filter formed with secondary data 168–70
  for sample-starved Gaussian training conditions 165
Adaptive Sidelobe Blanker (ASB) 93–8
  KWAS-ASB 119–23
  modifications towards
    selectivity 104
    selectivity and robustness 117
  SRao-ASB 123–5
  Subspace-based ASB (S-ASB) 98–104
  WAS-ASB 117–19

adaptive SMI/LSMI filter with ideal thresholding 243–4
adaptive thresholding 168–70
AMGLRT, derivation of 365–6
AMGLRT detector 341, 359
Amplitude empirical PDF and theoretical models 349, 350–1
"AR(m)-like" model 213
AR(m) models 213, 234
Autoregressive Model (AR) 167, 213, 226, 234

Band-inverse Hermitian covariance matrices 237
baseband equivalent of received signal 22
Bayesian AMF (B-AMF) 138–43, 150, 156
Bayesian GLRT (B-GLRT) 138–43
Bayesian radar detection in interference 133
  KA-STAP models 133, 136
    Bayesian AMF (B-AMF) 138
    Bayesian GLRT (B-GLRT) 138
    hyperparameter selection 140–3
    knowledge-aided homogeneous model 137
    partially homogeneous and compound-Gaussian models 144–6
  KA-STAP techniques 133
  knowledge-aided parametric STAP model 151–9
  knowledge-aided two-layered STAP model 147–51
Bayesian STAP detectors 147, 150
  versus conventional STAP detectors 140
binary hypothesis test 7, 21–2, 70, 105, 111, 205
BORD (Bayesian Optimum Radar Detector) 312

Central Limit Theorem (CLT) 263, 299
  transfer theorem and 267–9
clairvoyant filter 212, 238
clairvoyant optimum filter 169, 206
clairvoyant Wiener filter 218, 224, 243
"cliff-like" scenarios 234, 237
Clutter Covariance Matrix (CCM) 213, 280, 313, 337
Coherent Processing Intervals (CPIs) 4, 57
Complex Elliptical Symmetric (CES)
  CES noise, optimal detection in 312–13
  CES distribution 5, 300
  stochastic representation theorem 300–1
compound-exponential model, for univariate intensity 264–6
  examples 265–6
  intensity tail distribution and completely monotonic functions 264–5
Compound-Gaussian (CG) models 16, 145
  CG distributions 301
  knowledge-aided 147, 160–2
  new interpretation of optimum detector 281
    general properties of product of estimators 283–90
    product of estimators formulation 281–3
  number fluctuations 266
    models for 269–70
    transfer theorem and CLT 267–9
  optimum detection of signal in complex compound-Gaussian clutter 272
    likelihood ratio and data-dependent threshold interpretation 273–4
    likelihood ratio and estimator-correlator interpretation 275
  suboptimum detectors in 275
    performance evaluation of optimum detectors and 280–1
    suboptimum approximations to data-dependent threshold 277–8
    suboptimum approximations to estimator-correlator 278–80
    suboptimum approximations to likelihood ratio 276
    and target detection 263
  univariate intensity, compound-exponential model for 264
    examples 265–6
    intensity tail distribution and completely monotonic functions 264–5
compound-Gaussian clutter 333
  detection performance
    real target and clutter data 358–9
    simulated target and real clutter 355–8
  distributed target coherent detection 334
    covariance estimation 340–2
    rank-one steering 336–8
    subspace steering 338–40
  experimental CFAR behavior 349–55
  high-resolution experimental data 342
    maritime target data 345–9
    sea-clutter data 343–5
compound-Gaussian STAP model 135
  knowledge-aided 147

conic acceptance/rejection idea 87
Constant False Alarm Rate (CFAR)
CFAR property 5, 45, 70, 166, 167–8, 188, 214, 240, 295, 299
and invariance in detection theory 13–14
experimental CFAR behavior 349–55, 359
Continuous Wave Linear Frequency Modulated (CWLFM) radar 342
conventional Doppler receiver 355
conventional SMI adaptive filter and scalar CFAR detector 246
conventional STAP detectors
versus Bayesian STAP detectors 140
covariance estimation 340–2, 353, 357, 360
covariance matrix estimation in SIRV 295
background and problem statement 295–9
background parameter estimation in Gaussian case 296–7
heterogeneous Gaussian environment 298–9
homogeneous Gaussian environment 297–8
covariance matrix estimation in CES noise 302
asymptotic distributions of M-estimators 305–7
link to M-estimators in SIRV framework 308
M-estimators 302–3
properties of M-estimators 304
non-Gaussian environment modeling 300
CES distribution 300–1
subclass of SIRV 301–2
optimal detection in CES noise 312–13
persymmetric structured covariance matrix estimation 313
circular Gaussian noise, detection in 314–15
non-Gaussian noise, detection in 315–17
radar applications 317
Fixed Point Estimate (FPE), robustness of 323
ground-based radar detection 317–19
Nostradamus radar detection 319
Space–Time Adaptive Processing (STAP) detection 319–23
Covariance Matrix Taper (CMT) 50–1, 69
Cumulative Distribution Functions (CDFs) 80, 93, 300, 348
data-dependent threshold
likelihood ratio and 273–4
suboptimum approximations to 277–8
data partitioning, adaptive detection improvement using 239–42
analysis performance of "one-stage" adaptive CFAR detectors versus "two-stage" adaptive processing 242–7
comparative detection performance analysis 247–55
observations 255–7
detection probability (P_D or P_d) 22, 355–8
detection theory 6
basic concepts 9–10
CFAR property and invariance in detection theory 13–14
detector design criteria 11–13
signal and interference models 7–9
Deterministic Scatterer Model GLRT (DSM-GLRT) 339
diagonal loading 167, 170, 173
diagonally loaded and FML adaptive detectors 185
homogeneous interference training conditions 186–7
non-homogeneous interference training conditions 187–91
Doppler filtering 169
Doppler tolerant pulse 24
duplexer 2
dwell time 4, 22
Dym–Gohberg band-matrix extension technique 213, 215

EC-1 2
EL-AMF detector 178, 194, 207
EL-selected loading, adaptive detection improvement using 168
adaptive thresholding using primary data 168–70
different adaptive process per test cell 170
2S-GLRT, AMF, and LAMF detectors, detection performance analysis of 191–209
2S GLRT and AMF detectors for Gaussian models 178–85
AMF versus GLRT detectors 171–3
diagonally loaded and FML adaptive detectors 185–91
EL framework 173–8
observations 209–12

single adaptive filter formed with secondary data 168–70
estimates of textures, properties of 367
estimator-correlator interpretation 275
suboptimum approximations to 278–80

false alarm probability (P_{FA} or P_{fa}) 22, 32, 231, 295
false-alarm threshold calculation 188, 234
Fast Fourier Transform (FFT) 354
finite-subspace (FML) adaptive detectors 185
homogeneous interference training conditions 186–7
non-homogeneous interference training conditions 187–91
finite-subspace (FML) covariance 173
Fisher information matrix evaluation 364–5
Fisher–Neyman factorization theorem 24
Fixed Point Adaptive Normalized Matched Filter (FP-ANMF) 313, 323
Fixed Point Estimate (FPE) 309–11
robustness of 323
FPE bias analysis with disturbances 327
Normalized SCM (NSCM) bias analysis with disturbances 323–6
Sample Covariance Matrix (SCM) bias analysis with disturbances 323
FREYA radars 3

Gaussian case 296
background parameter estimation in 296–7
optimal detection in 297
heterogeneous Gaussian environment 298–9
homogeneous Gaussian environment 297–8
Gaussian models 5
2S GLRT and AMF detectors for 178–85
Gaussian noise 299
circular 314–15
colored: *see* colored Gaussian noise
non-Gaussian noise, detection in 315–17
Gaussian Scatterer Model GLRT (GSM-GLRT) 339
Generalized Inner Product (GIP) test 190
Generalized K-distribution with LN Texture (GK-LNT) 343, 347
Generalized Likelihood Ratio (GLR) statistic 295–6
Generalized Likelihood Ratio Test (GLRT) 5, 12, 14, 44, 105, 136, 166–7, 273, 276, 289, 296, 313, 335, 361–2
Adaptive Matched Filter (AMF) versus 171–3
CFAR property of 280–1
design 21, 28–32, 34, 38
GLRT-FP: *see* Fixed Point Adaptive Normalized Matched Filter (FP-ANMF)
GLRT-Linear Quadratic (GLRT-LQ) 298, 312
Knowledge-Aided Parametric GLRT (KA-PGLRT) 154–8
subspace version of 71–2
probability of detection (P_D or P_d) and probability of false alarm (P_{FA} or P_{fa}) 54–5
generating variate 300
GMTI (Ground Moving Target Indicator) 320
GNU Scientific Library (GSL) 193
grazing angle 8, 343
ground-based radar detection 317–19
ground clutter 4, 257, 317, 319

Hadamard product 50
heavy-tailed distributions, conjectured behavior for 289
Hermitian covariance matrix 213
band-inverse 237
TVAR(m) approximation of 215–18
Hermitian positive definite matrix 51
Hertz, Heinrich 1
heterogeneous environment 12
Himmelbett Operations Room 3
historical background and terminology 1–5
homogeneous environment 12, 91
homogeneous STAP model 135
Huber estimator 303, 307
Huber's M-estimator 306
hyperparameter selection 140–3
hypothesis test 51–3
hypothesized steering vector 50
Hülsmeyer, Christian 1, 2

intensity tail distribution 263
and completely monotonic functions 264–5
interference-plus-noise covariance matrix 48, 54
interference-plus-noise in test vector 49
Interference-to-Noise Ratio (INR) 169
Italian Radar System 2

KA-ACE, derivation of 159–60
Kantorovich inequality 170
KASSPER dataset 228
K distribution 264, 290, 301, 309, 349
  fluctuation model 347
Kelly's GLRT detector 166, 210
knowledge-aided compound-Gaussian model 145
  Bayesian detectors in 160–2
knowledge-aided homogeneous model 137, 144
Knowledge-Aided Parametric AMF (KA-PAMF) 152, 154, 156–8
Knowledge-Aided Parametric GLRT (KA-PGLRT) 154–8
knowledge-aided parametric STAP model 151–9
Knowledge-Aided Space–Time Adaptive Processing (KA-STAP) 133
knowledge-aided two-layered STAP model 147–51
Kronecker delta 78
KWAS-ASB 119–23

Levy processes 290
Likelihood Ratio (LR) 177, 284
  and data-dependent threshold interpretation 273–4
  and estimator-correlator interpretation 275
  suboptimum approximations to 276
Likelihood Ratio Test (LRT) 24, 26, 46, 48
LN target fluctuation model 346
Loaded AMF (LAMF) adaptive detectors 171–2, 175, 206–7
Loaded Sample Matrix Inversion (LSMI) adaptive filter 171
  and scalar CFAR detector 246–7
  LSMI-based LAMF detector 185
Loaded Sample Matrix Inversion (LSMI) algorithm 170
loss factor 92
LSMI/FML technique 211

MANDREL 3
Marconi, Guglielmo 1
Marcum Q-function 48
maritime target data 345–9
Matched Subspace Detector (MSD) 73
maximal invariant statistic 44, 45, 75
  distribution of 77–9
Maximum A Posteriori (MAP) estimate 138, 146, 150, 161, 278–9
Maximum Likelihood
  ML-based RMB SMI technique 170
  ML covariance matrix estimate 210
Maximum Likelihood Estimate (MLE) 12, 73, 138, 166, 170, 276, 296, 310
Maxwell, James Clerk 1
Measure Of Effectiveness (MOE) 351
mesa plots 97
M-estimators 302–3
  asymptotic distributions of 305–7
  link in Spherically Invariant Random Vector (SIRV) framework 308
    deterministic texture 308–9
    random SIRV texture 309–11
  properties of 304
Minimum Mean-Squared Error estimate (MMSE) 138, 142, 150, 271, 278
ML-GLRT ROCs 197–8
ML model identification and order estimation 215–18
MLTVAR(μ)CME 238–9
Modified GLRT (MGLRT) technique 341
Monte-Carlo ROCs 256
Monte-Carlo simulations 174, 180, 200, 211, 214, 247
Most Powerful (MP) test 8
Moving Target Indicator (MTI) 169
Multiple Dominant Scatterer (MDS) model 345, 348
MUSIC method 306

Naval Research Laboratory (NRL) 2, 264
Neyman–Pearson (NP) criterion 21, 45–6
Neyman–Pearson (NP) detector 26–8, 39–40, 44, 46, 166
Neyman–Pearson (NP) lemma 11
non-Gaussian environment modeling 300
  Complex Elliptical Symmetric (CES) distribution 300
    stochastic representation theorem 300–1
  Spherically Invariant Random Vector (SIRV), subclass of 296, 301–2
non-Gaussian noise, detection in 315–17
Non-Scatterer Density-Dependent GLRT (NSDD-GLRT) 336–7
Normalized Matched Filter (NMF) 94, 299, 312, 317, 337

Normalized Sample Covariance Matrix (NSCM) 308, 311
bias analysis with disturbances 323–6
estimator 340
Nostradamus radar detection 319
nuisance parameters, defined 48
number fluctuations 264, 266
models for 269–70
transfer theorem and CLT 267–9

"one-stage" adaptive CFAR detectors versus "two-stage" adaptive processing 242–7
"benchmark" detectors 242–4
"one-stage" adaptive detectors 245–6
"two-stage" adaptive detectors 246–7
optimum detectors 26
new interpretation of 281
general properties of product of estimators 283–90
product of estimators formulation 281–3
performance evaluation of 280–1
optimum NP detector and existence of UMP test 26
coherent case 26
non-coherent case 27–8
Over The Horizon radar from the French Aerospace Lab (ONERA) 319, 321

parametric receivers 87
partially homogeneous and compound-Gaussian models 144–6
partially homogeneous environment 12
partially homogeneous STAP model 135
performance analysis, and comparisons 91
performance results of adaptive subspace detectors 57–69
Persymmetric Adaptive Normalized Matched Filter (P-ANMF) 313, 316
Persymmetric Fixed-Point (PFP) 315
persymmetric structured covariance matrix estimation 313
circular Gaussian noise, detection in 314–15
non-Gaussian noise, detection in 315–17
phase error vector 50
power of a test 10
Power Spectral Density (PSD) 21, 340, 353
"practically CFAR" detector 167, 214, 219, 257
"practically CFAR" diagonally loaded (LAMF) detector 209, 245–6
probability of detection (P_D or P_d) 10, 46–8
and probability of false alarm (P_{FA} or P_{fa}) 54
for subspace ACE test 56
for subspace AMF test 55–6
for subspace GLRT 54–5
probability of false alarm (P_{FA} or P_{fa}) 10, 15, 44, 47, 48, 57
Pulse Repetition Frequency (PRF) 8
Pulse Repetition Interval (PRI) 134

Radioecometro 2
Radio Research Station 2
range-spread targets 5, 7
rank-one steering 336–8
Rao test 12, 14, 338, 362–3
Rayleigh statistics 287, 288
real target and clutter data 358–9
Receiver Operating Characteristic (ROC) 32, 35, 39, 169, 195, 241
Recursive P-ANMF (RP-ANMF) 313, 316–17
Recursive Persymmetric (RP) estimator 340
Reed, Mallet, and Brennan (RMB) test 5
Reed–Mallett–Brennan (RMB) detector 136
Regio Instituto Electrotecnico e delle Comunicazioni (RIEC) 2
Ricean distribution 48
Robey, Fuhrmann,Kelly, and Nitzberg (RFKN) 166, 171–2
robust receivers 85
robust two-stage decision schemes 89
Root Mean Square Error (RMSE) 348

Sample Covariance Matrix (SCM) 296, 340
bias analysis with disturbances 323
Scatterer Density-Dependent GLRT (SDD-GLRT) 336–7
scatter matrix 301
sea-clutter data 343–5
secondary data vectors 49
selective receivers 85, 89
selective two-stage detectors 125–8
selective two-stage receivers 89
Short-Time Fourier Transform (STFT) 358
SideLobe Blanker (SLB) 87
SideLobe Canceler (SLC) 88
signal and interference models 7–9
Signal Corps Laboratories 2
signal detection in interference and noise 45
in colored Gaussian noise 46–7
with unknown phase in zero-mean colored Gaussian noise 47–8
Signal Of Interest (SOI) 43, 272
Signal-to-Clutter power Ratio (SCR) 335

Signal-to-Interference-plus-Noise Ratio (SINR) 45, 139
Signal-to-Noise Ratio (SNR) 31, 33, 86, 169, 317
  equivalent SNR per pulse 37
  loss factor 238
Signal-to-White-Noise Ratio (SWNR) 191
simulated target and real clutter 355–8
single adaptive filter formed with secondary data 168–70
Singular Value Decomposition (SVD) 52, 76
"small-rank interference" models 213
small-rank techniques 212
Space–Time Adaptive Processing (STAP) 5, 147, 192, 257
  detection 319–23
  general STAP signal model 134–6
  knowledge-aided STAP (KA-STAP) 133
speckle 301, 343
spherically invariant random processes 301
Spherically Invariant Random Vector (SIRV) 334
  covariance matrix estimation in: *see* covariance matrix estimation in SIRV
  subclass of 301–2
square-law non-coherent integrator 31
SRao-ASB 123–5
steering vector 7, 238, 295, 306, 321
Stellung 3
stochastic representation theorem 300–1
Student-*t* distribution 290, 310
suboptimum detectors in complex compound-Gaussian clutter 275
  performance evaluation of optimum detectors and 280–1
  suboptimum approximations to data-dependent threshold 277–8
  suboptimum approximations to estimator-correlator 278–80
  suboptimum approximations to likelihood ratio 276
Subspace-based ASB (S-ASB) 98–104
subspace detection, for adaptive radar: *see* adaptive radar, subspace detection for
Subspace Detection Paradigm (SDP) 7
Subspace Detector (SD) 50, 89
subspace signal detection, maximum invariants for 53–4
subspace signal model 43, 44, 49, 338
  rationale for 49–51
subspace steering 338–40
Swerling chi target fluctuation model 346
Synthetic Aperture Radar (SAR) 4

taper matrix 50
target reflectivity 8, 21
telemetro radiofonico del rivelatore (RDT) 2
THALES data 320
Theory of Invariance, in hypothesis testing 14
Tiberio, Lieutenant Ugo 2
Toeplitz covariance matrices 167, 173, 211, 213
training data vectors 49
transfer theorem and CLT 267–9
tunable receivers 5
TVAR(m)-based adaptive detectors 213–14
  simulation results of 223–37
TVAR(m) model 167, 213
TVAR(m) or AR(m) interferences, adaptive detectors for 218–23
TVAR(μ)-based adaptive detectors 214, 239
TVAR(μ)-based adaptive filter 218, 239
TVAR(μ) model 228
two-set GLRT (2S-GLRT) framework 176
two-set -GLRT, AMF, and LAMF detectors, detection performance analysis of 191–209
  "favorable" interference scenario 191–3
  homogeneous interference training conditions 193–200
  non-homogeneous interference training conditions 200–5
  "unfavorable" interference scenario 205–9
two-set GLRT and AMF detectors for Gaussian models 178
  arbitrary scaling factors for interference matrices 181–5
  homogeneous interference training conditions
    fluctuating target with known power 178–80
    fluctuating target with unknown power 180–1
two-stage detectors
  architecture description 91–3
    Adaptive Sidelobe Blanker (ASB) 93–8
    ASB modifications towards selectivity 104
    ASB modifications towards selectivity and robustness 117
    selective two-stage detectors 125–8
    Subspace-based ASB (S-ASB) 98–104
  in Gaussian interference with unknown spectral properties 85
  for point-like targets 87

"two-stage" adaptive detectors 246–7
two-step GLRT 298

Uniformly Most Powerful (UMP) test 8, 21
  optimum NP detector and existence of 26
    coherent case 26
    non-coherent case 27–8
Uniformly Most Powerful Invariant (UMPI) test 337
univariate intensity, compound-exponential model for 264
  examples 265–6
  intensity tail distribution and completely monotonic functions 264–5

VLSI circuitry 239

Wald test 12, 14, 363–4
WAS-ASB 117–19
Watson-Watt, Robert Alexander 2
Weibull distribution 301
Weibull fluctuation model 347
white Gaussian noise, radar detection in 21
  Generalized Likelihood Ratio Test (GLRT) design 28–32
  optimum NP detector and existence of UMP test 26
    coherent case 26
    non-coherent case 27–8
  performance analysis 32
    coherent case 32–6
    non-coherent case 36–40
  problem formulation 22–4
  reduction by sufficiency 24–5
Whitened Adaptive Beamformer Orthogonal Rejection Test (W-ABORT) 89, 104, 111, 117
white-noise equivalent model 214, 222, 230–1
"white-noise optimal" filter 212
"white-noise" loss factor 238
"white-noise" theoretical approximations 194–5
Wishart normalization factor 138
WÜRZBURG 3
WÜRZBURG-RIESE 3

www.ingramcontent.com/pod-product-compliance
Lightning Source LLC
LaVergne TN
LVHW080644210926
848485LV00003B/322